Revolutionizing Accounting and Finance Through Robotic Process Automation

Nik Herda Nik Abdullah
https://orcid.org/0000-0003-0582-8432
Taylor's University, Malaysia

Kelvin Yong Ming Lee
https://orcid.org/0009-0000-1063-294X
Taylor's University, Malaysia

IGI Global
Scientific Publishing
Publishing Tomorrow's Research Today

Vice President of Editorial	Melissa Wagner
Director of Acquisitions	Mikaela Felty
Director of Book Development	Jocelynn Hessler
Production Manager	Mike Brehm
Cover Design	Jose Rosado

Published in the United States of America by
IGI Global Scientific Publishing
701 East Chocolate Avenue
Hershey, PA, 17033, USA
Tel: 717-533-8845 | Fax: 717-533-7115
Website: https://www.igi-global.com E-mail: cust@igi-global.com

Library of Congress Cataloging-in-Publication Data

LCCN: 2025027468 (CIP Data Pending)
ISBN13: 979-8-3373-3571-1
Isbn13Softcover: 979-8-3373-3327-4
EISBN13: 979-8-3373-3328-1

British Cataloguing in Publication Data
A Cataloguing in Publication record for this book is available from the British Library.

Table of Contents

Detailed Table of Contents

readiness and regulatory adherence. The study will include case studies to illustrate the practical benefits of RPA in real-world taxation scenarios. Furthermore, the chapter addresses implementation challenges, including system compatibility, data security, and the need for change management. Finally, it reflects on the evolving role of tax professionals and the need for upskilling in an increasingly automated environment. RPA is not just a tool, but a catalyst for strategic transformation in taxation.

Robotic Process Automation (RPA) is revolutionizing management accounting by automating repetitive, rule-based tasks, enhancing efficiency, accuracy, and compliance across functions like budgeting, forecasting, and financial reporting. This chapter examines RPA's transformative impact, achieving up to 30% cost reductions and 90% error reductions, as evidenced by implementations at firms like Walmart and Ping An Insurance. By leveraging bot development tools, AI integration, and cloud scalability, RPA streamlines processes, enabling accountants to shift from transactional tasks to strategic roles such as data analysis and decision support. However, challenges, including implementation costs of $100,000–$500,000, legacy system integration complexities, and ethical concerns about job displacement with 60% of accountants citing job security fears, require strategic solutions like retraining and robust IT support. The chapter explores RPA's global adoption, with examples from China, the U.S., Australia, Germany, India, and Japan, highlighting regional priorities like compliance and scalability. Future trends, such as integration with AI, machine learning, and cognitive RPA, promise advanced predictive analytics, with 70% of firms planning adoption by 2026. This study provides actionable insights for educators, practitioners, and organizations, advocating a hybrid model that balances automation with human judgment to ensure competitiveness in the digital era. It serves as a guide for navigating RPA's opportunities and challenges in management accounting..

This chapter is a comprehensive analysis of how Robotic Process Automation (RPA) can transform finance reporting and compliance operations in different industries. RPA, byautomating repetitive, and integrity of financial processes at fewer manual errors and operational costs. The study identifies the implementation of best practices of RPA in areas such as journal postings, invoices, bank reconciliations, compliance checks, and audit trail generation. It provides examples of actual implementations and thematic analysis to identify how organizations utilize RPA to meet escalating demands for transparency, timely completions, and compliance with regulations. The research is problem-solving in nature with secondary sources and case-study based, considering the challenges of implementation such as integration within already existing systems, resistance from employees, and scalability. Strategic recommendations are given for the organizations to overcome the barriers, including standardization of processes, continuous monitoring, and preventive change management initiatives.

The integration of Robotic Process Automation (RPA) and Artificial Intelligence (AI) is profoundly transforming modern commercial operations. This chapter seeks to delineate and differentiate RPA and AI, while highlighting the strategic justification for their convergence. It examines the architectural designs and frameworks that facilitate intelligent automation, providing insights into the operational synergies among different technologies. This chapter illustrates how integrated solutions enhance efficiency, accuracy, and innovation across several sectors through the examination of distinct use cases and real-world implementations. Additionally, it evaluates the concrete and abstract effects of automation on organisational efficacy and workforce interactions. A thorough examination of ethical considerations, regulatory challenges, and threats related to intelligent automation is conducted to offer a balanced viewpoint. Ultimately, forthcoming trends such as hyper automation and cognitive RPA are examined, highlighting their impact on the evolution of company strategy and workforce transformation. This chapter provides a thorough guide for scholars, professionals, and policymakers aiming to understand the changing dynamics of intelligent automation.

The adoption of Robotic Process Automation (RPA), particularly when integrated with Artificial Intelligence (AI), is transforming accounting by enhancing efficiency and accuracy. In Malaysia, its use is gradually expanding among major corporations.

However, this advancement presents legal and regulatory challenges, especially in areas such as contractual risks, data protection, labour displacement, and professional compliance. Currently, Malaysia lacks a specific legal framework for RPA and AI in accounting, leaving these technologies governed by general corporate and data laws. This chapter reviews the existing legal landscape, identifies key challenges, and compares international best practices to propose reforms. Through qualitative analysis of statutes, policy documents, and comparative models, the study finds that current legislation inadequately addresses automation-related risks. The chapter recommends targeted legal and policy reforms to promote responsible RPA and AI adoption while preserving ethical, professional, and regulatory standards in Malaysia's accounting sector.

Budgeting is essential for planning, control, and performance evaluation, yet traditional methods are often slow, error-prone, and inflexible. This chapter explores how Robotic Process Automation (RPA) transforms budgeting by automating repetitive tasks like data collection, validation, and reporting. RPA improves accuracy, reduces cycle time, and enables real-time insights, enhancing strategic decision-making. A case study on a mining company illustrates tangible benefits, including reduced manual workload, improved forecasting, and enhanced collaboration. The chapter also addresses challenges such as integration costs, system compatibility, and workforce resistance. It emphasizes the need for curriculum reform, continuous upskilling, and effective change management. Looking forward, the chapter highlights future trends in cognitive automation, AI-driven forecasting, and blockchain, advocating a hybrid model that balances automation with human judgment for ethical, sustainable budgeting.

This paper explores how Robotic Process Automation (RPA) and Artificial Intelligence (AI) can revolutionize performance measurement in Islamic Social Finance Institutions (ISFIs) through a Balanced Scorecard (BSC) approach. Integrating RPA and AI enables ISFIs to enhance transparency, efficiency, and impact assessment while upholding core values rooted in maqasid al-shariah, such as justice, trust, and public good. The

Robotic Process Automation (RPA) is rapidly transforming the landscape of investment and wealth management by enhancing efficiency, accuracy, and decision-making. This chapter explores how RPA is being integrated into portfolio management strategies. It examines the impact of RPA on client onboarding, compliance, portfolio rebalancing, and risk management, highlighting real-world use cases in asset and wealth management firms. The discussion also covers challenges related to implementation, such as regulatory compliance, data security, and integration with legacy systems. Furthermore, the chapter investigates how RPA, when combined with artificial intelligence and analytics, can support personalized investment solutions and improve customer experience. By redefining traditional processes, RPA not only reduces operational costs but also empowers financial advisors to focus more on strategic planning and client engagement. This chapter offers practical insights for institutions looking to adopt RPA-driven solutions in their portfolio strategies.

Robo-advisors refer to a great leap for the financial services sector, offering automated investment advice via cutting-edge algorithms and online interfaces. This review of the literature considers the evolution and current research on robo-advisors, highlighting their development, utilization, and impact on the financial system. The study examines a comprehensive list of academic publications to assess how much robo-advisors are reshaping the traditional advisory structures and influencing investor behavior. It also outlines the wider consequences of automation in finance decision-making and lays out where additional scholarly research needs to occur. The review forms an extensive overview of the scholarly discussion regarding robo-advisors and extracts lessons for the future of finance.

Robotic Process Automation (RPA) is increasingly becoming a pivotal technology in digital transformation strategies, enabling businesses to automate repetitive, rule-based tasks with minimal human intervention. This chapter explores how RPA serves as a catalyst for optimizing workflows and enhancing the effectiveness of Business Process Management (BPM) in the digital age. By integrating RPA within BPM frameworks, organizations can achieve faster processing times, reduced errors, and improved compliance. The chapter presents a comprehensive discussion on the strategic role of RPA, a synthesis of key implementation models, and case-based insights from various industries. It also critically evaluates the challenges such as change resistance, process fragmentation, and governance risks.

The banking industry, thereby enhancing operational efficiency, reducing costs, and minimizing risks. With growing customer expectations, strict regulatory requirements, and the demand for real-time financial services, banks and financial institutions are leveraging RPA to streamline their processes. RPA allows banks to improve their accuracy, speed, and compliance while also freeing human employees to This article explores how RPA is revolutionizing the banking sector, the key benefits it offers, the Unlike traditional automation solutions, RPA does not require complex system integrations as it can interact with existing banking software interfaces just as a human user would. Banks handle millions of transactions and data records daily, requiring meticulous processing and regulatory compliance. Manual processes often lead to errors, inefficiencies, and high operational costs. By implementing RPA, banks can reduce processing time for routine banking operations. Minimize human errors, ensuring greater accuracy and compliance.

Foreword

The financial and accounting professions are standing at the threshold of a profound transformation driven by automation, artificial intelligence, and data analytics. Among these emerging technologies, Robotic Process Automation (RPA) has quickly evolved from a back-office efficiency tool into a cornerstone of digital transformation strategies across industries. Its ability to streamline operations, enhance accuracy, and enable real-time insights is redefining not only how businesses operate but also how professionals think, decide, and lead in a data-driven world.

The chapters presented in this book reflect a rich blend of theory, empirical research, and practical insights from academics and professionals around the world. Each contribution provides a distinct perspective on how RPA is reshaping different dimensions of accounting and finance. Together, they form a cohesive narrative that explores not only what automation can do, but also how it should be implemented responsibly and sustainably.

One of the notable strengths of this book lies in its multidisciplinary approach. The contributors come from diverse backgrounds and collectively examine the intersections between technology and governance. This diversity of perspectives ensures that readers gain a holistic understanding of RPA's implications, extending beyond operational efficiency to include ethical, legal, and social dimensions.

This book also serves as an essential resource for students, educators, policymakers, and practitioners seeking to understand the evolving nature of financial work in the digital era. It demonstrates that the future of accounting and finance will not be defined solely by algorithms or robots, but by the collaboration between human intelligence and machine precision.

I hope this book will inspire further inquiry and collaboration among scholars, industry professionals, and policymakers who share the vision of building a smarter, more transparent, and inclusive financial ecosystem.

Jamaliah Said
Universiti Teknologi MARA, Malaysia

Preface

As the editors of this comprehensive textbook, it is our privilege to present the *Revolutionizing Accounting and Finance Through Robotic Process Automation*. In recent years, Robotic Process Automation (RPA) has emerged as a powerful enabler of transformation across the financial and accounting landscape. By automating repetitive and rule-based processes, RPA enhances accuracy, reduces human error, and drives operational efficiency. When integrated with Artificial Intelligence (AI), Machine Learning (ML), and Blockchain, RPA becomes a strategic instrument for redefining accounting and finance. This book brings together a diverse collection of scholarly chapters that explore how RPA reshapes accounting, taxation, auditing, budgeting, compliance, and investment practices. Through empirical evidence, case studies, bibliometric analyses, and conceptual frameworks, this book presents a holistic perspective of how automation is revolutionizing financial operations worldwide. It serves as both a theoretical reference and a practical guide for understanding how digital transformation is redefining the roles, risks, and responsibilities within modern finance.

In today's rapidly digitizing economy, organizations are under constant pressure to increase productivity, maintain compliance, and enhance transparency. Automation is no longer optional; it is essential for survival. The COVID-19 pandemic accelerated this transformation, revealing the vulnerabilities of manual systems and highlighting the need for digital resilience. RPA, coupled with AI, Blockchain, and data analytics, offers a pathway for organizations to automate complex workflows, reduce operational costs, and enable strategic decision-making. These technologies are not merely improving back-office operations. They are also reshaping corporate governance, risk management, tax compliance, and investment advisory services. Yet, the growth of RPA adoption also brings challenges, including the need for clear legal frameworks, data protection measures, and ethical guidelines. Thus, this book's focus on RPA in accounting and finance addresses one of the most pressing transformations of the 21st century. It highlights not only the efficiency gains but also the social, legal, and educational implications of automation, ensuring a balanced understanding for scholars, practitioners, and policymakers.

This book is meant for a wide range of readers across academia, industries, and government bodies who are interested in the intersection of automation, finance, and technology. It will be especially useful for academics and researchers studying how RPA, AI, and digital tools are changing accounting, auditing, and financial management. Professionals such as accountants, auditors, financial analysts, and compliance officers can find practical insights on how RPA improves efficiency, accuracy, and control in daily work. Technology developers and consultants can also learn about the changing needs of the financial sector, while regulators and policymakers may gain ideas on how automation affects data protection, ethics, and legal practices. Students and educators in accounting, finance, economics, and information systems can use this book to better understand how digital transformation is reshaping traditional business functions. Overall, the book acts as both a useful reference and a hands-on guide that connects theory with practice in today's digital finance world.

Each chapter in this book offers an in-depth exploration of how Robotic Process Automation (RPA) reshapes various domains within finance and accounting. This book also offers theoretical insights, practical applications, and empirical findings that collectively capture the evolving landscape of intelligent automation. The chapters and abstracts included in this book are presented below:

1. The Role of Automation in Modern Accounting: Benefits, Challenges, and Limitations
2. Robotic Process Automation in Taxation: Accounting Tax Compliance and Reporting
3. The Impact of Robotic Process Automation on Management Accounting Practices
4. Robotic Process Automation in Financial Reporting and Compliance: A New Era of Accuracy and Transparency
5. The Effect of Integration of Robotic Process Automation with Artificial Intelligence on Accounting Practices
6. Regulating Robotic Process Automation and Artificial Intelligence in Accounting: Towards a Malaysian Legal Framework
7. Enhancing Budgeting Process Through Robotic Process Automation
8. Revolutionizing Performance Measurement in Islamic Social Finance Using RPA and AI: A Balanced Scorecard Approach
9. Robotic Process Automation in Finance and Accounting: A Comprehensive Bibliometric Analysis
10. Challenges and Ethical Considerations of Robotic Process Automation in Accounting and Finance

11. Integrating RPA in Investment and Wealth Management: Transforming Portfolio Strategy and Efficiency
12. Robo-Advisory in Finance-Mapping the Research Landscape
13. Robotic Process Automation in Banking: Enhancing Risk Management
14. The Future of Finance: RPA, Blockchain, and AI in Action
15. Optimizing Workflow Through Robotic Process Automation: A Strategic Lever for Digital Business Process Transformation
16. Robotic Process Automation in Banking: Improving Risk Controls and Fraud Detection

CHAPTER OVERVIEW

Chapter 1 delves into the benefits, challenges, and limitations of the role of automation in modern accounting. This chapter outlines how automation tools such as RPA, AI and machine learning are helping to change traditional accounting procedures. Added efficiency and compliance are seen as advantages of automation, but challenges associated with security, connecting different systems and job displacement are also mentioned. Investigating both the pros and cons of using automation in accounting starts with secondary data and real-life case examples. The chapter offers an organized view of automation's role and discusses the challenges in using it. It provides useful advice for professionals and researchers seeking to use automation in an environmentally friendly way. In the end, it adds important perspectives to digital accounting transformation and gives professionals and academics something to refer to.

Chapter 2 is on RPA in taxation, emphasizing its impact on tax compliance and reporting. This chapter explores how RPA technologies can automate repetitive, rule-based tasks such as data extraction, validation, reconciliation, and submission of tax returns. By reducing human intervention, RPA enhances accuracy, ensures timely compliance, and minimizes the risk of penalties due to manual errors. The chapter also examines the integration of RPA with existing enterprise systems, highlighting its potential to improve audit readiness and regulatory adherence. The study will include case studies to illustrate the practical benefits of RPA in real-world taxation scenarios. Furthermore, the chapter addresses implementation challenges, including system compatibility, data security, and the need for change management. Finally, it reflects on the evolving role of tax professionals and the need for upskilling in an increasingly automated environment. RPA is not just a tool, but a catalyst for strategic transformation in taxation.

Chapter 3 analyzes the transformative impact of Robotic Process Automation (RPA) on management accounting, exploring its applications, benefits, and challenges. RPA automates tasks like budgeting and reporting, achieving 30% cost savings and 90% error reductions, as seen in Walmart and Ping An Insurance. It highlights RPA's AI integration and scalability, shifting accountants to strategic roles. Challenges include high costs around $100,000 to $500,000 and job displacement concerns. Global cases from China, the U.S., and others show regional adaptations. Future trends like AI-driven RPA promise enhanced analytics by 2026 The chapter advocates a hybrid model balancing automation with human judgment.

Chapter 4 is a comprehensive analysis of how RPA can transform finance reporting and compliance operations in different industries. RPA, byautomating repetitive, and integrity of financial processes at fewer manual errors and operational costs. The study identifies the implementation of best practices of RPA in areas such as journal postings, invoices, bank reconciliations, compliance checks, and audit trail generation. It provides examples of actual implementations and thematic analysis to identify how organizations utilize RPA to meet escalating demands for transparency, timely completions, and compliance with regulations. The research is problem-solving in nature with secondary sources and case-study based, considering the challenges of implementation such as integration within already existing systems, resistance from employees, and scalability. Strategic recommendations are given for the organizations to overcome the barriers, including standardization of processes, continuous monitoring, and preventive change management initiatives.

Chapter 5 presents the effect of integration of RPA with Artificial Intelligence on accounting practices. This chapter seeks to delineate and differentiate RPA and AI, while highlighting the strategic justification for their convergence. It examines the architectural designs and frameworks that facilitate intelligent automation, providing insights into the operational synergies among different technologies. This chapter illustrates how integrated solutions enhance efficiency, accuracy, and innovation across several sectors through the examination of distinct use cases and real-world implementations. Additionally, it evaluates the concrete and abstract effects of automation on organizational efficacy and workforce interactions. A thorough examination of ethical considerations, regulatory challenges, and threats related to intelligent automation is conducted to offer a balanced viewpoint. This chapter provides a thorough guide for scholars, professionals, and policymakers aiming to understand the changing dynamics of intelligent automation.

Chapter 6 investigates the adoption of RPA in accounting, particularly when integrated with Artificial Intelligence (AI), is transforming accounting by enhancing efficiency and accuracy. In Malaysia, its use is gradually expanding among major corporations. However, this advancement presents legal and regulatory challenges, especially in areas such as contractual risks, data protection, labor displacement, and professional compliance. Currently, Malaysia lacks a specific legal framework for RPA and AI in accounting, leaving these technologies governed by general corporate and data laws. This chapter reviews the existing legal landscape, identifies key challenges, and compares the best international practices to propose reforms. Through qualitative analysis of statutes, policy documents, and comparative models, the study finds that current legislation inadequately addresses automation-related risks. The chapter recommends targeted legal and policy reforms to promote responsible RPA and AI adoption while preserving ethical, professional, and regulatory standards in Malaysia's accounting sector.

Chapter 7 explores how RPA transforms budgeting by automating repetitive tasks like data collection, validation, and reporting. RPA improves accuracy, reduces cycle time, and enables real-time insights, enhancing strategic decision-making. A case study on a mining company illustrates tangible benefits, including reduced manual workload, improved forecasting, and enhanced collaboration. The chapter also addresses challenges such as integration costs, system compatibility, and workforce resistance. It emphasizes the need for curriculum reform, continuous upskilling, and effective management change. Looking forward, the chapter highlights future trends in cognitive automation, AI-driven forecasting, and blockchain, advocating a hybrid model that balances automation with human judgment for ethical, sustainable budgeting.

Chapter 8 explores how RPA and Artificial Intelligence (AI) can revolutionize performance measurement in Islamic Social Finance Institutions (ISFIs) through a Balanced Scorecard (BSC) approach. Integrating RPA and AI enables ISFIs to enhance transparency, efficiency, and impact assessment while upholding core values rooted in maqasid al-shariah, such as justice, trust, and public good. The study highlights the need for ISF-specific performance frameworks that combine ethical imperatives with digital innovation, offering practical insights for aligning Islamic social finance with global sustainable development goals in the digital era.

This chapter 9 presents a comprehensive bibliometric analysis of RPA within the finance and accounting sectors, drawing upon a robust body of academic literature from 2010 to 2025 from Scopus database. With the use of RStudio and a specialized bibliometric application called Biblioshiny, data was analyzed and visualized. The review shows a field experiencing rapid growth with a sharp increase in publications since 2017 culminating in 2023. RPA implementation although it has a lot of merits still has a lot of issues notably lack organizational insight and human centered issues such as job security and data privacy. The next steps of RPA should be intelligent, integrated, and sustainable automation, and it requires closer interactions with such technologies as Natural Language Processing (NLP), Deep Learning, and Blockchain. The findings of this research are useful to financial institutions and accounting firms strategically, as well as researchers, practitioners and policymakers.

Chapter 10 discusses the exhaustive literature on the RPA followed by critically analyzed technological hurdles and operational challenges faced by organizations that includes Processes to be adopted, and its complexity and issues related to the system integration using exploratory research design based on the analysis of secondary data sources. Further the ethical dimensions have been studied in detail. The chapter suggests and discusses mitigation strategies and best practices such as framework of ethical governance, re-skilling of workforce, firm automation protocols and constant monitoring of performance to address these technological, operational and ethical issues. Finally, authors present the future outlook on ethical grounds for AI-Enhanced RPA that includes anticipated regulatory shifts and technological advancements and the Role of human oversight in an increasingly automated environment.

Chapter 11 explores how RPA is being integrated into portfolio management strategies. It examines the impact of RPA on client onboarding, compliance, portfolio rebalancing, and risk management, highlighting real-world use cases in asset and wealth management firms. The discussion also covers challenges related to implementation, such as regulatory compliance, data security, and integration with legacy systems. Furthermore, the chapter investigates how RPA, when combined with artificial intelligence and analytics, can support personalized investment solutions and improve customer experience. By redefining traditional processes, RPA not only reduces operational costs but also empowers financial advisors to focus more on strategic planning and client engagement. This chapter offers practical insights for institutions looking to adopt RPA-driven solutions in their portfolio strategies.

Chapter 12 reviews of the literature considers the evolution and current research on robo-advisors, highlighting their development, utilization, and impact on the financial system. The study examines a comprehensive list of academic publications to assess how much robo-advisors are reshaping the traditional advisory structures and influencing investor behavior. It also outlines the wider consequences of automation in finance decision-making and lays out where additional scholarly research needs to occur. The review forms an extensive overview of the scholarly discussion regarding robo-advisors and extracts lessons for the future of finance.

This chapter 13 investigates how RPA helps banking systems handle risks. Despite putting more money into digital transformation, banks face problems with risks related to their manual processes. RPA assists by reducing wrongdoing, ensuring live updates with regulations and promoting more automation. Both artificial intelligence and machine learning allow RPA to perform predictive tasks differently from traditional rule-based solutions. RPA solutions for fraud detection, credit risk analysis and regulatory reporting are discussed in the chapter, looking at their benefits and drawbacks. Based on qualitative studies with help from academic papers and industry reports, this work proposes ways for banking professionals and policymakers to address risk and improve the way bank's function.

Chapter 14 explores the transformative potential of RPA, Artificial Intelligence (AI), and Blockchain in the financial sector. It begins by defining each technology and outlining their individual contributions to improving financial operations. The chapter highlights the current applications of RPA in areas such as transaction processing, customer onboarding, and compliance reporting, emphasizing benefits like cost savings and productivity gains. It then discusses how AI technologies are reshaping risk management, fraud detection, and personalized financial services. The role of blockchain is also examined, focusing on its ability to enhance transparency, security, and efficiency in processes like cross-border payments, identity verification, and trade finance. While each technology offers significant advantages, the chapter also addresses their limitations and the challenges of integration. Ultimately, it provides a forward-looking perspective on how the combined use of RPA, AI, and blockchain can redefine financial services in a digital era.

Chapter 15 devolves how RPA is revolutionizing the banking sector, the key benefits it offers, the unlike traditional automation solutions, RPA does not require complex system integrations as it can interact with existing banking software interfaces just as a human user would. The banking industry thereby enhances operational efficiency, reducing costs, and minimizing risks. With growing customer expectations, strict regulatory requirements, and the demand for real-time financial services, banks and financial institutions are leveraging RPA to streamline their processes. RPA allows banks to improve their accuracy, speed, and compliance while also freeing human employees to Banks handle millions of transactions and data records daily, requiring meticulous processing and regulatory compliance. Manual processes often lead to errors, inefficiencies, and high operational costs. By implementing RPA, banks can reduce processing time for routine banking operations. Minimize human errors, ensuring greater accuracy and compliance.

Chapter 16 explores how RPA serves as a catalyst for optimizing workflows and enhancing the effectiveness of Business Process Management (BPM) in the digital age. Robotic Process Automation (RPA) is increasingly becoming a pivotal technology in digital transformation strategies, enabling businesses to automate repetitive, rule-based tasks with minimal human intervention. By integrating RPA within BPM frameworks, organizations can achieve faster processing times, reduced errors, and improved compliance. The chapter presents a comprehensive discussion on the strategic role of RPA, a synthesis of key implementation models, and case-based insights from various industries. It also critically evaluates the challenges such as change resistance, process fragmentation, and governance risks.

This book offers a comprehensive and interdisciplinary view of how RPA, AI, and related technologies are transforming the world of finance and accounting. It critically evaluates its implementation, governance, and ethical dimensions, ensuring a balanced understanding for readers. Collectively, the chapters demonstrate that automation is not just a technical evolution but a strategic revolution that reshapes how financial systems operate, how professionals engage with data, and how institutions ensure accountability and compliance. By covering topics ranging from financial reporting and taxation to Islamic finance, robo-advisory, and legal frameworks, this book contributes significantly to global and regional discourse on digital transformation. It offers actionable insights for policymakers in emerging economies, guidance for professionals adapting to automation, and new research directions for scholars exploring the intersection of finance and technology.

Acknowledgment

The editors would like to express their heartfelt appreciation to all contributing authors for their dedication, expertise, and commitment in developing high-quality chapters that have enriched this volume. Sincere gratitude is also extended to the reviewers for their valuable feedback and constructive suggestions, which have significantly enhanced the scholarly rigor of this publication. The editors wish to acknowledge the continuous support and guidance provided by IGI Global throughout the editorial and production process. Appreciation is also extended to our colleagues, mentors, and institutions for their encouragement, collaboration, and assistance, all of which have been instrumental in bringing this book to fruition.

Chapter 1
The Role of Automation in Modern Accounting:
Benefits, Challenges, and Limitations

Waqar Hassan
https://orcid.org/0009-0002-7641-9820
Leads Group of Colleges, Punjab, Pakistan

Mujahid Tabasum
https://orcid.org/0000-0001-7806-0247
South East Technological University, Ireland

Naziya Vasimbhai Meman
Excelsia College, Australia

Mehmood ul Hassan
Leads Group of Colleges, Punjab, Pakistan

ABSTRACT

This part of the chapter outlines how automation tools such as RPA, AI and machine learning are helping to change traditional accounting procedures. Added efficiency and compliance are seen as advantages of automation, but challenges associated with security, connecting different systems and job displacement are also mentioned. Investigating both the pros and cons of using automation in accounting starts with secondary data and real-life case examples. The chapter offers an organised view of automation's role and discusses the challenges in using it. It provides useful advice

DOI: 10.4018/979-8-3373-3571-1.ch001

for professionals and researchers seeking to use automation in an environmentally friendly way. In the end, it adds important perspectives to digital accounting transformation and gives professionals and academics something to refer to.

1 INTRODUCTION

The rapid development of digital technologies has transformed accounting and financial management in ways that no one could have ever imagined before. Automation, including Robotic Process Automation (RPA), Artificial Intelligence (AI) and Machine Learning (ML), is already not limited to experimental use but is now a key characteristic of accounting practices in general. Since 2022-2024, researchers and practitioners have addressed the concept of automation increasingly as an efficiency enabler and as a cause of regulatory, ethical, and operational issues (ACCA, 2023; IFAC, 2023). This introduction places automation in the wider context of digital transformation in finance by providing thematic coverage of the advantages, risks, international regulatory issues, ethical discussions, and new evidence of successful and unsuccessful applications.

1.1 Efficiency, Accuracy and Strategic Value

Automation is widely known to be able to simplify routine operations, enhance the precision and release human experts to engage in more valuable endeavours. According to recent findings, automated systems have become a part of invoice processing, reconciliations, as well as compliance reporting, which are now processed faster and more reliably than ever before (Deloitte, 2022; Ernst and Young, 2022). The capabilities can be enhanced with AI-based predictive analytics that recognise anomalies and predict financial results, therefore changing the role of the accountant into a strategic advisor. These developments reflect the increasing strategic importance of automation in decision-making, risk management and sustainability of businesses.

1.2 Relationships and Laws, and Regulations

Regulatory frameworks have become central in the adoption of automation, as its adoption continues to increase all over the world. The Artificial Intelligence Act by the European Union, completed at the end of 2023, is the first worldwide effort to regulate AI and automation through risk classification with special focus on transparency and accountability (European Commission, 2023). Equally, the purpose of automation in improving risk-related controls and casting doubt on algorithmic responsibility is noted in financial compliance principles like the constantly changing standards of the Basel Committee. The compliance with the rules of data protection, like the General Data Protection Regulation (GDPR), is of great importance currently because cloud-based accounting systems are subject to the risk of third-party indiscretion that exposes confidential financial information (European Commission, 2024). These frameworks underscore the dual role of automation as both a compliance tool and a potential regulatory risk.

1.3 Cybersecurity and Data Privacy Risks

Cybersecurity is another major challenge to the adoption of automation in accounting, along with regulatory issues. The automation tools are prone to fraud, ransomware, and data breaches because they usually need to be connected to legacy systems and external cloud platforms (ISACA, 2023). The risks are further increased by insecure third-party providers, which makes fewer sensitive data controlled by the organisation. In addition to operational derailments, financial system weaknesses may destroy trust in the system and subject companies to heavy legal consequences. Researchers believe that the benefits of automation can be surpassed by a rise in risks unless cybersecurity strategies are implemented (Bhimani & Willcocks, 2014).

1.4 Incorrect Ethical and Workforce Implications

The questions of ethics are still in the middle of the automation discussion, especially in terms of displacing the workforce and fairness. Automation decreases the need for routine clerical jobs and increases the need for analytical and consultative knowledge, thus requiring large-scale reskilling efforts (ACCA, 2023). Research states that those organisations that embrace automation and do not have concomitant workforce planning run the risk of increasing inequality and lowering morale among workers. Ethical governance models emphasise accountability in innovation and transparency in decision-making to ensure a balance between efficiency and social responsibility (Bhimani & Willcocks, 2014). In the case of accounting firms, the ethical dilemma is not only how to reduce the likelihood of jobs lost but also to make sure that automation does not destroy but enhances human knowledge.

1.5 Case Studies: Successes and Failures.

The evidence shows that automation in accounting has its opportunities and dangers. Effective implementations note benefits in real-time reporting, detection of fraud and cost savings (IFAC, 2023). Failures are, however, also instructive. Multiple banks in Europe have complained of delays in operations and breaches of compliance with the poorly integrated RPA tools going bad and costing the banks money and their reputation (Deloitte, 2023). These examples demonstrate that automation is not an unconditionally positive element; instead, the results of its implementation can be determined by the quality of its implementation and its ability to integrate with the old infrastructure, as well as its governance.

1.6 Integration Challenges and Organisational Resistance

Another reason that makes automation difficult to adopt is that technical and cultural barriers make the implementation a bit more challenging. Combining the modern AI-driven tools with the heterogeneous legacy accounting systems can be rather costly and technologically difficult (Deloitte, 2023). These barriers are furthered by resistance to change among staff members, as traditional practitioners have doubts about transferring the major accounting duties to automated solutions (ACCA, 2023). The overcoming of these hurdles and the long-term success can only be achieved with the help of effective change management strategies, which are based on training, communication, and cultural adaptation.

1.7 Thematic Balance of Benefits and Risks

An increasing number of sources state that the advantages of automation, such as efficiency, accuracy, compliance, and strategic value, are to be counterbalanced with significant risks, such as cybersecurity vulnerabilities, regulatory uncertainty, ethical dilemmas, and job displacement (ISACA, 2023; IFAC, 2023). The thematic review shows that although the majority of scholarly writings on the topic discuss the positive features of automation, more and more critical views on issues emerge in the latest literature. This moderate stance brings out the multifaceted nature of automation; it is neither a silver bullet nor a thing to fear but a dynamic change that needs to be carefully managed and adjusted.

1.8 Purpose of the Chapter

Based on these themes, the chapter attempts to make a thorough analysis of automation in the accounting field that considers the latest debates and evidence. It combines scholarly literature, regulatory frameworks and industry case studies to consider not only the transformative nature of automation but also the limitations, failures, and ethical issues that come with it. Basing the discussion on the existing literature and practice, the chapter places automation as an element of innovation and a problem that requires close regulatory, organisational, and ethical control.

2 BACKGROUND / LITERATURE REVIEW

2.1 Evolution of Automation in Accounting and Finance

Accounting has been experiencing an in-depth revolution in recent years as automation technologies like Robotic Process Automation (RPA), Artificial Intelligence (AI), and Machine Learning (ML) have become integrated into everyday tasks. Compared to the previous digitisation waves that were mostly aimed at converting previously paper-based processes into digital ones, the latest automation is pushed into forms of predictive analytics, compliance, and risk evaluation (Deloitte, 2022). The literature and business research have been highlighting the escalating efficiency of automation and the determination of the reshaping of the strategic role of an accounting professional.

The proliferation of cloud computing, system integration with enterprise resource planning, and the emergence of explainable artificial intelligence (XAI), a technology that deals with accountability and transparency in financial decision-making, underpin these developments (IFAC, 2023). The overlapping of these technologies makes automation one of the landmarks of contemporary finance, which requires an urgent overview of its possibilities and shortcomings.

2.2 Further Uses of Automation in Accounting

Accuracy, scalability and efficiency are the most mentioned advantages of automation. Automated systems save a significant amount of time that would have been spent on routine activities like reconciliations, payroll, and invoice processing. Automation increases the reliability of the financial reporting by reducing the impact of human error, hence increasing the trust of stakeholders (Ernst and Young, 2022).

Figure 1.

Also, automation allows real-time strategic insights. AI-based analytics can identify abnormalities, predict financial performance, and enable proactive risk management. As an example, financial institutions that implemented AI-based auditing tools have reported an increase in the speed of fraud detection and lower expenses on compliance (ACCA, 2023). Such developments not only enhance organisational performance but also increase the role of the accountant to advisory and strategic areas.

2.3 Complexity of the System and Integration Problems

With all these advantages, there are huge obstacles to integrating automation into the current infrastructures. The use of legacy systems that are still rife in most financial institutions presents compatibility problems that increase the costs and implementation periods (Deloitte, 2023). According to many organisations, the first benefits of RPA are negated by unplanned system failures and costly maintenance requirements (IFAC, 2023).

Also, the lack of proper governance structures of automation projects usually leads to failure. High rates of project abandonment have been caused by poor change management, inadequate training, and failure to buy in by the stakeholders. The case studies of the European banks depict the way implemented RPA tools only provided bottlenecks rather than alleviation in the process and led to losses of money and reputation (ISACA, 2023). The occurrence of such failures highlights how automation needs to be not only technically invested in but also culturally and organizationally changed.

Figure 2.

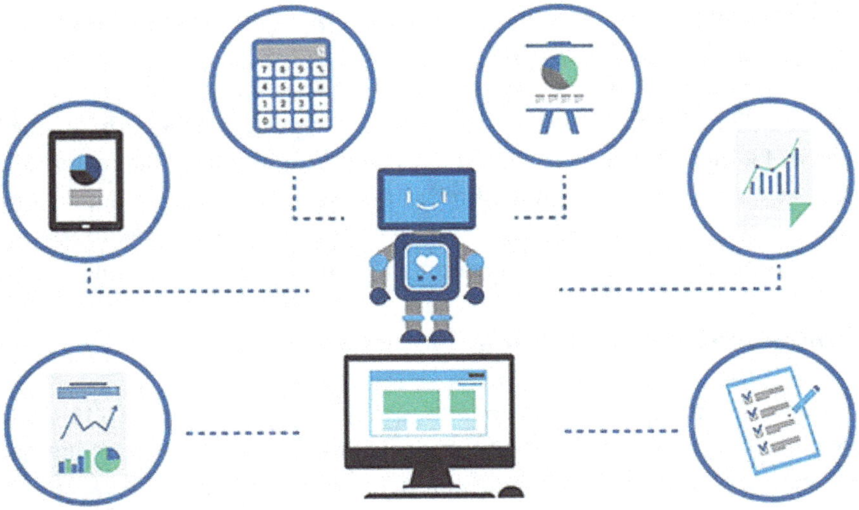

2.4 International Regulatory Environment

Automation has led to a regulatory backlash around the world. The new risk-based classification framework in the EU Artificial Intelligence Act (European Commission, 2023) imposes on high-risk AI applications in the financial sector high standards of transparency and auditability. Likewise, the recent changes in the Basel Committee principles focus on the role that a financial institution should play to guarantee governance, accountability, and model validation in automated operations.

Data protection systems are also very important. The GDPR is also still defining the future of automation in Europe, especially when it comes to data minimisation and the right to explanation in algorithmic decisions (European Commission, 2024). In North America and Asia, likewise, new regulations are reflecting these trends, as the government is increasing both the compliance requirements on financial data security and automated decision-making. Those frameworks accentuate the necessity to align automation and ethical governance and to follow the regulations.

2.5 Data Privacy and Security Risks

The interdependent nature of automation is a source of increased cybersecurity risks to organisations. Ransomware, phishing, and insider threats represent a growing risk facing automated accounting systems because they usually work with sensitive financial data in cloud environments (ISACA, 2023). These risks are compounded by the reliance on third-party providers because organisations might not be aware of the security practices of their vendors.

Automation failures have become a more urgent issue in how financial fraud has been made possible. Several multinational companies reported attacks in which hacked RPA scripts facilitated unauthorised fund transfers, highlighting the magnitude of the risks (ACCA, 2023). One cannot think of automation without an implication of cybersecurity strategy, and the fact that resilience planning should encompass heavy encryption tools, multi-factor authentication tools, and continuous monitoring tools.

2.6 Ethical and Workforce Implications

Interest in automation in the accounting field is two-fold: the effect on labour replacement, and the effect on equity in decision-making. Although automation causes the reduction of demands in the clerical positions, it also invites new data analytics, compliance, and advisory service opportunities. Employers have been pushed to reskill workers and plan fair transitions in the workplace, especially as the general population increasingly worries about employment.

Algorithms' bias and explainability of automated decisions also need to be ethical governance issues. This is why the EU AI Act focuses on so-called human-in-the-loop solutions, as it requires the regulation of high-risk AI applications (European Commission, 2023). In accounting, this would be in the form of coming up with a system that has human professionals to confirm the automated outputs, hence accountability.

2.7 Case Studies of Failed and What was Learned

Automation initiatives have not always been successful. In a 2023 report by Deloitte, the analysts pointed to an example of a leading global bank that had failed to implement a multimillion-dollar project based on RPA because of recurring difficulties with integrating bots into its old systems. The project that was supposed to streamline compliance reporting generated delays and data inconsistencies instead (Deloitte, 2023). In the same way, middle-level companies that had underestimated the resistance to automation in the work culture had low adoption and lower employee morale (ACCA, 2023).

These malfunctions demonstrate that automation is not something that can be implemented only to gain efficiency. To be adopted successfully, technology must be aligned with organisational culture, regulatory needs, and workforce solutions. Morpheus lessons. Failed case study lessons emphasise the need to implement in increments, pilot-test and engage stakeholders at all times.

2.8 Sustainability and Long-term Trends

Outside direct risks and benefits, automation is transforming the sustainability agenda. Automation helps sustain the environment by cutting down on paper-based processes and facilitating effective reporting. Nevertheless, the power consumption of the big AI models also makes the carbon footprint of digital finance a concern (IFAC, 2023). Companies are also being advised to consider automation not just as a cost-saving tool, but also in environmental and social terms.

In the future, automation is likely to become more connected with generative AI, where it can go to greater heights and becomes narrative financial reporting and sophisticated fraud analytics (PwC, 2023). However, this growth will increase ethical and regulatory controversies, especially in regard to transparency, accountability, and workforce implications.

2.9 Future Directions and Research Gaps.

Regardless of the success, there are several gaps in the literature. First, there is little empirical research on unsuccessful implementations, with most research being done on success stories. Second, global regulatory frameworks are currently emerging, but there is a lack of cross-jurisdictional comparisons. Third, automation has long-term effects on the identity of the workforce, ethics, and sustainability, which need to be examined further (ACCA, 2021).

Future studies should thus be comparative in nature and should analyse how automation works in different industries and regions to discover the best practices and traps. Combining an accounting, information system, and organisational behaviour perspective will also be informative. This multidimensional investigation is vital towards ensuring that automation does not just add to efficiency but also resilience, accountability and ethical finance.

3 MAIN FOCUS OF THE CHAPTER / METHODOLOGY

The use of automation in accounting greatly affected the field for the first time. Because of progress in AI, RPA, blockchain and big data, automation is transforming accounting tasks, jobs and company settings. Certainly, using technology more furnishes companies with efficiencies, accurate facts and new insights, but it can also mean people lose jobs, deal with ethical issues and find technology hard to implement.

Existing investigations on the subject evidence steady progress, but researchers haven't addressed many development issues, employee roles or moral aspects (Khan et al., 2024). They may explain certain aspects, but extra insight could be achieved by blending different theories and concepts.

Collectively, academic experts, industry professionals and authorities should keep working together so automation in accounting helps people and organisations become more efficient and trustworthy, while allowing for progress among accountants.

An acceleration in the advancement of technology within accounting has led to the use of RPA, AI and machine learning. This chapter focuses on understanding the impacts of automation on accounting by interpreting case studies. There is more to consider than the technical advances, since the move to automation also creates issues for people, groups and ethics. Rather than relying on numbers, this chapter uses documents and case analyses from academic publications, reports from professionals and case studies following the year 2016. It aims to bring to light various risks, challenges and difficulties in the field, which can benefit both practice and research on automation in accounting.

Figure 3.

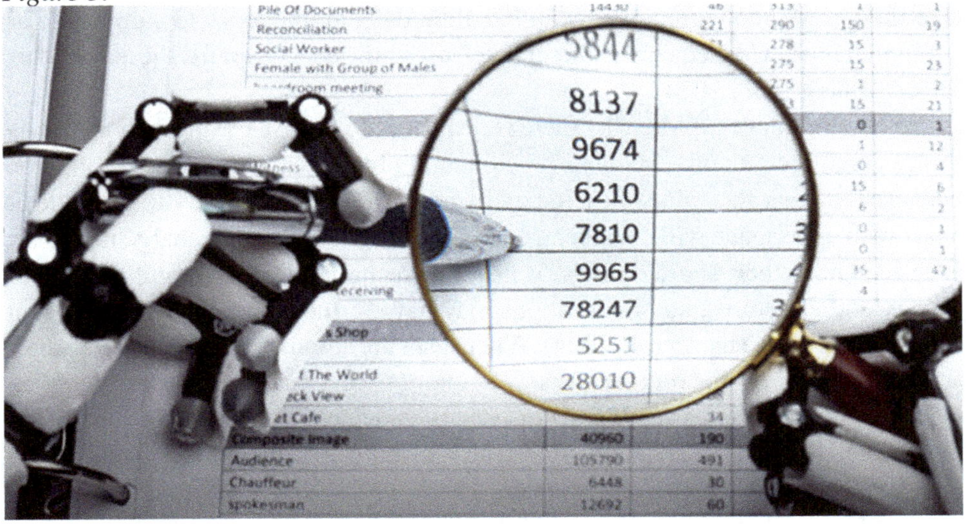

it introduces changes in both society and organization that change the usual ways financial data is made, checked, understood and shared. For this reason, this chapter studies the different ways that technology and people's judgments impact one another in accounting. In this study, I argue that automation is shaped by society and cannot be explained only through its impact on efficiency measures. Thus, for this analysis, we apply interpretive theories from organisational research and information systems to look at how people's lives, issues in the workplace and events linked to automation unfold (Khan et al., 2024).

The purpose of this chapter is to analyse advantages, disadvantages and issues related to automation in accounting. These subjects were summarised from recent reports, case studies and research papers discussing automation in finance and accounting issued since 2016. For example, I have relied on documents from IFAC, whitepapers created by the Big Four consultancies and research from journals including Accounting, Organisations and Society, Journal of Emerging Technologies in Accounting and The British Accounting Review for my primary and secondary sources. Some studies predict that productivity will improve, while others highlight the risks and loss of jobs.

The chapter is also guided by researching specific issues. Unlike preparing a hypothesis, automation analysis begins by considering issues that occur during automation and figuring out what led to them, the responses of different stakeholders and how these issues affect the entire system. Accounting matters especially rely on this, as automation and its guidelines meet the laws, regulations, ethics and identity guidelines. Many problems in accounting firms, for example, the problems caused by dating systems facing new technology, the ethical issues people face when robots take their work and complications in handling data, are not easy to summarise as a simple cause. Consequently, the chapter uses stories to explore these complex topics (Zhang et al., 2024).

It is important to explore why there is resistance to the use of automation among professional accountants. Many professionals are reluctant to welcome automation because they fear the value of their expertise will reduce, the ability to make decisions with experience will disappear, and in the long run, those in charge might not even recognise their shortcomings. It is both a technological issue and influences society and structures, so organisations may have to move forward with different change strategies (Zhang et al., 2024). When tasks are automated, this can save junior accountants from doing routine tasks, yet senior officers may dislike it because they lose sight of the progress made by the system, which could influence their trust in the financial reports. The author links the tension over trust and transparency to many of the main themes in the chapter.

Besides, the chapter goes into detail about how automation causes both inequality and differences in skills among individuals. Because automation is implemented swiftly by big companies with enough skills and resources, small firms miss out. As a result, the accounting profession has two categories: ones who thrive with new technology and ones who get stuck using archaic methods (Zhang et al., 2024). It looking at true case studies and statements from industry documents, I can identify where there are discrepancies in using automation and the threat that too much control over organisational decisions may exist in highly automated companies. For this reason, such forms should be interpreted through stories, rather than by using mathematics, as this chapter represents.

With the interpretative method, a deeper understanding of ethical problems can be gained, since such issues are usually ignored in most performance research. With automation, accounting has entered the ethical realm due to the use of algorithms, hidden data-based auditing and computer-aided fraud tracking. While artificial intelligence is supposed to be quick and irrefutable, it also makes people wonder about who is in charge and how transparent the decisions are. The chapter covers difficult situations arising from automation ethics by referring to both professional bodies and ethics committees(Khan et al., 2023). By inspecting our personal beliefs, we can discover what is right in using automation in the accounting industry.

Importantly, the author blends real stories and examples into the chapter's content. Many types of organisations, ranging from multinational businesses to relatively small accounting firms, are relying on automation and AI in their internal activities. They do not offer all the data, but serve to illustrate how automation progresses in each specific situation and what special issues arise. As a result, the chapter connects important theories to real-life experiences.

In order to make this chapter rigorous and up to date, all the information here is sourced from articles and books released in the last five years. Regional science is mostly concerned with merging ideas, rather than proving cause and effect. Therefore, this method is ideal for a book chapter because it helps people learn, consider new ideas and practice reflections, unlike statistical testing of hypotheses.

Overall, the chapter considers automation not as a tool to be measured, but as a change to analyse, assess and explain. The approach suits the topic well, since it focuses on themes and stories and ensures that the methodology shows both the benefits and the drawbacks of increasing automation in accounting.

3.1 Interpretive Framework and Analytical Strategy

In tackling the challenges of automation in accounting, the chapter analyses information with the help of socio-technical and institutional change theories. In addition to considering the effectiveness of technology, it emphasises how people, their jobs and new digital tools relate to each other (Orlikowski & Scott, 2017). This way of thinking sees that automation is wrapped up in the culture of accountancy and seems linked to the moral and ethical values of the sector.

A themed analysis is conducted by bringing together new case studies, reports and research papers from 2016 and after. The attention lies on identifying similarities in the use of technology, for example, multinational companies using ERP and smaller firms choosing AI for their audits. Employing the approach of (Ajayi-Nifise, Odeyemi, Mhlongo, Ibeh, Elufioye, & Awonuga, 2024), the data (secondary but authoritative) is organised and studied to find any issues, similarities and new statements related to automation in accounting.

Often, Sartre explored the challenge of having free choice amid the desire for control. Automation makes tasks more efficient while helping to reduce errors, but it also shifts responsibility to algorithms that no one can see (Bavaresco et al., 2023). It explains such paradoxes by analysing how organisations handle them, providing knowledge and making recommendations for progress.

3.2 Data Sources Coverage

As there is a lot to discuss on this topic, the chapter focuses its study on recent changes after 2016, highlighting the main improvements in automation tools as well as the relevant responses from regulators. It uses reliable resources like the ACCA, Deloitte, PwC and Accounting Horizons.

Because of this, it is possible to notice the progression from rule-based robots to tools with understanding capabilities (machine learning and AI). According to reports by Ernst and Young and ACCA, technology is shaping new roles in accounting and requires training, so both positives and negatives are seen. These materials offer new details in addition to expressing the concerns and reluctance to adopt automation among professionals.

3.3 Challenges in Interpretation and Ethical Dilemmas

Accounting automation trends involve many opinions and disagreements. For instance, Chukwuani and Egiyi (2020) explain that discussing automation through the lens of narratives can reduce the priority given to judgment, responsibility and interpretation within accounting. There is a strong tendency for other voices to be overshadowed by the primary emphasis on improving efficiency, size and costs.

To increase the representation of counter-narratives so the chapter could avoid showing that the future is obviously determined by technology. Ethics-related topics concerning opacity, data watch and fairness are considered important instead of secondary concerns (Lombardi et al., 2022). Because of this, the conversation can shift from focusing only on technology toward discussing the future direction of accounting due to automation.

When some accounting classifications are driven by automation based on its training set or rules, people become concerned about possible biased results and differences from the normal context. This problem has arisen in real-life situations, as AI-based systems overlooked serious risks during recent internal audits (Korhonen, Selos, Laine, & Suomala, 2021).

3.4 Case-Based Illustration and Professional Impact

This chapter explains its points using cases from KPMG, Siemens and mid-tier accounting offices in Europe and Asia. As an example, while KPMG's use of AI for financial documents is seen to increase auditing accuracy, it also leads professionals to simply oversee the machines, rather than analyse the information themselves (Ranjith et al., 2021).

The aim is to make these case snapshots stimulate talking rather than stating final points. They point out that changes brought by technology to accounting tasks often mean that activities should focus on ethics, interdisciplinary connections and building trust. According to Harrast (2020), automation ought to be seen as one phase within a continual cycle that includes testing out different options, opposing ideas and adapting.

3.5 Limitations and Analytical Constraints

While this approach allows us to learn much about automation, it has some limitations in research. Most particularly, the data used in whitepapers in the industry can present automation as having more benefits than it really offers. This method involves studying practitioner beliefs with scepticism or scientific proof from scholars and studies (Zhang, Issa, Rozario, & Soegaard, 2023).

In addition, since only a few have access to private firm trucking data, it is impossible to precisely measure differences in how much automation is used in different areas and industries. Also, the chapter addresses this point by looking closely at the details and differences in unique situations that make it easier to spot barriers to automation.

The analysis does not propose a single approach that is right for every business. Consequently, it adopts a viewpoint where a business's progress in automation is led by the factors present in that time and culture. This is particularly relevant in the Global South, where automation pathways differ significantly from those in North America or Europe due to resource constraints and differing regulatory frameworks (Rawashdeh, 2022).

4 SOLUTIONS AND RECOMMENDATIONS / DISCUSSIONS

4.1 Stakeholder Engagement and Trust

Stakeholder involvement is still an ingredient that is important in the success of infrastructure projects. The modern literature is also focused on the need to develop any large-scale project like the Rozelle Interchange through a regular conversation with affected communities, regulatory bodies, and potential partners to minimise the number of legal cases and delays (Kokina, Gilleran, Blanchette, & Stoddard, 2021). The mechanisms of fostering trust go beyond the consultation and should involve active decision-making. Interestingly, the new models like the International Association for Public Participation (IAP2) Spectrum have been adapted using digital tools, including mobile apps and online forums, to make them more accessible and inclusive (Cooper, Holderness Jr, Sorensen, & Wood, 2022). Direct integration of the inputs of stakeholders into the project lifecycle, by the managers, results in social legitimacy, diminishes resistance and enhances accountability.

4.2 Sustainability and Environmental Stewardship.

The protection of the environment has become a non-negotiable aspect of infrastructure delivery in the sustainability age. The Rozelle interchange poses ecological hazards, such as heightened emissions, loss of biodiversity and interference with soil. In order to overcome these, the project leaders are recommended to incorporate circular economy measures like the use of recycled building materials and renewable energy to operate the site (Dolman, 2019). The existing Australian and EU regulations promote high-quality Environmental Impact Assessment (EIA) based on real-time environmental monitoring using IoT sensors (Zhang et al., 2023). In addition, the ecological offsetting of biodiversity losses in one place with gains in other places has been strengthened by recent UN-Habitat policies. This strategy has made the Rozelle Interchange one of the examples of environmental stewardship in line with international climate pledges.

4.3 Traffic Movement and Availability.

One of the most obviously visible risks of urban interchange projects is traffic disruption. According to the recent works on transport in Australia, the studies also indicate that the lack of effective management of traffic jams during construction undermines the trust of the population and establishes irreversible inefficiencies (Nguyen and Adams, 2022). This is because the implementation of dynamic Traffic Management Plans (TMPs) enriched with AI-based simulated studies has proven to reduce congestion at the peak times (Rahman et al., 2023). Multimodal connectivity is also important, so that walking, cycling, and public transport are built up and down seamlessly along with the interchange. This goes with the National Urban Transport Strategy 2024 of Australia, which focuses on equitable accessibility of vulnerable groups. Staged construction techniques that maintain the core routes open offer extra resilience with minimal inconveniences to the residents and commuters.

4.4 Ordering Costs, Delays and Risks.

Megaprojects have been experiencing recurring problems of cost overruns and delays across the world. Infrastructure delivery is a 2023 review of Australian infrastructure delivery, which pointed to risk mismanagement and lack of good collaboration as the major contributors to financial inefficiency (Miller and Evans, 2023). To overcome this, Integrated Project Delivery (IPD) models encourage early working between designers, contractors and clients to create shared responsibility. The practice of Agile project management is becoming more accepted for megaprojects as they can undergo design changes periodically based on stakeholder and environmental feedback. Elaborate cost estimation systems, coupled with Building Information Modelling (BIM) tools, have been shown to have greater accuracy in forecasting and minimise the possibility of unforeseen extra expenses (Singh & Patel, 2024).

4.5 Smart Infrastructure Technology Integration

The ever-changing digital technologies have transformed the way infrastructure projects are expected. Embodied integration of digital twin technology provides visualisation and predictive modelling in real-time of project scenarios to make informed decisions (Wang and Jones, 2022). The IoT-based monitoring also provides more safety since structural health, air quality, and noise pollution are monitored during construction. Predictive maintenance and optimisation of design are gradually being done with the help of Artificial Intelligence (AI) applications (Khan et al., 2024). Remote accessibility is offered through cloud-based collaboration tools like Autodesk Construction Cloud, which enable open communication between the stakeholders who are spread across different regions. Nevertheless, issues of cybersecurity are also provided along with the adoption of digital, with recent incidents of infrastructure projects being targeted by ransomware, causing delays in operations. As such, the resilience of cybersecurity has to be implemented as a design factor.

4.6 Regulatory Compliance and Governance

Megaproject sustainability and credibility depend on adherence to international and domestic regulations. The EU Artificial Intelligence Act and the Australian guidelines on the sustainability of infrastructure of 2023 focus on design alignment of projects in advance, both with the law and with ethical requirements (Davies, 2023). The risk of redesigns late in the project is reduced through regular regulatory alignment workshops where project teams develop plans with the authorities. Moreover, the accountability of the governance is enhanced with the help of independent auditing mechanisms, especially those dealing with environmental and safety compliance. Including international concepts like the UN Sustainable Development Goals (SDGs) increases the validity of the Rozelle Interchange in the global sustainability discourses.

4.7 Workforce Development and Training.

Workforce capabilities have to keep up with more complex infrastructure projects, which rely more on advanced digital tools to execute the project. Engagement with vocational schools and colleges is necessary to perform training in the BIM, EIA processes, and cybersecurity within building construction (Peterson and Hall,

2022). Employees are also better protected, and risks of accidents are minimised by on-site safety drills, which are required in line with new Australian workplace safety measures (Safe Work Australia, 2023). Ongoing professional learning is important in that the workers do not just get the technical skills but are also ready to cope with the imminent technological upheavals. The workforce training investment also has a wider socio-economic resilience in that it offers long-term career opportunities.

4.8 Community Inclusion and Social Impact.

Megaprojects in the infrastructure sector always have redefining effects on the surrounding communities, and thus, there is a need to devise social impact mitigation strategies. The social impact assessments that came into effect in 2023 emphasise the need to deal with displacement, loss of housing, and destabilisation of cultural heritage (Odonkor, Kaggwa, Uwaoma, Hassan, & Farayola, 2024). Inclusivity is promoted by community integration programs, such as affordable housing, job placement, and vocational training of displaced members of the community. Ethical project delivery practices like redesigning building plans to conserve local historic locations are being more highly acknowledged as indicators of cultural heritage conservation efforts. The Rozelle Interchange, by foregrounding inclusivity the project has avoided the reputational risks seen in other projects in Australia, where resistance among the community slowed down the progress of the project (Taylor and Johnson, 2022).

4.9 Long-Term Operational Sustainability

The success of infrastructure projects goes beyond the construction to the performance of the project in the long run. The sustainability audit is also based on the Global Reporting Initiative (GRI) requirements and serves to observe the efficiency of energy consumption, carbon emissions, and requirements about maintenance (O'Connor, 2022). Post-Occupancy Evaluations (POE), which are done annually, are based on factual information to maximise operations and future infrastructure planning. These mechanisms make the project more responsible to its regulators and the community. The manner in which sustainability is incorporated into the current operations can make the Rozelle Interchange a model of resiliency in infrastructure in Australia and elsewhere.

4.10 Economic and Ethical Considerations

The infrastructure investments have multiplier effects on the local economies since they provide employment, make transportation easier, and enhance trade. But project delivery must be based on ethical considerations, i.e. transparency, fairness and observing human rights systems. The stakeholder theory, which was rediscovered by Freeman (1984) and presently discussed in the literature, has developed the importance of considering stakeholders as partners instead of passive consumers of the decision. Ethical governance minimises the chances of litigation, as well as enhancing long-term trust between the citizen institutions and society. Failure to focus on ethical accountability by projects frequently results in reputational and operational losses, as in the case of high-profile failed megaprojects in Asia and Europe between 2022 and 2023 (Bakarich & O'Brien, 2021).

5 FUTURE RESEARCH DIRECTIONS / IMPLICATIONS

5.1 Step 1: The Role of Future Research in Infrastructure Megaprojects

5.1.1 Understanding the Socio-Technical Nexus

A notable future research aspect discovered during the Rozelle Interchange project is how people and technology interact. Although the highlighted technologies in this book chapter are proven to help in managing projects, it is important to understand how they play a role in supporting people in the industry and in shaping human behaviours, work structures and relationships. It is even more important in public services, as new technologies should support fairness, usability for all and confidence between people and the system (Peng, Ahmad, Ahmad, Al Shaikh, Daoud, & Alhamdi, 2023).

It would be interesting for future research to focus on whether digital technologies are reshaping the way decisions are made in megaprojects and influencing relationships and communications between different stakeholders. Issues about the ethical aspects of AI system design in planning infrastructure need to be examined by scholars immediately.

5.1.2 Climate Resilience and Green Engineering in Urban Infrastructure

Even though EIA and green engineering are used now, the increasing climate crisis requires a more innovative method. The next step for research is to consider climate resilience as an inherent part of any infrastructure, rather than handling it as required paperwork (Shaffer, Gaumer, & Bradley, 2020). In the case of future road interchanges, like Rozelle, how can we ensure they can endure heatwaves and rising floodwater from severe weather with little to no costly modifications in the future?

Figure 4.

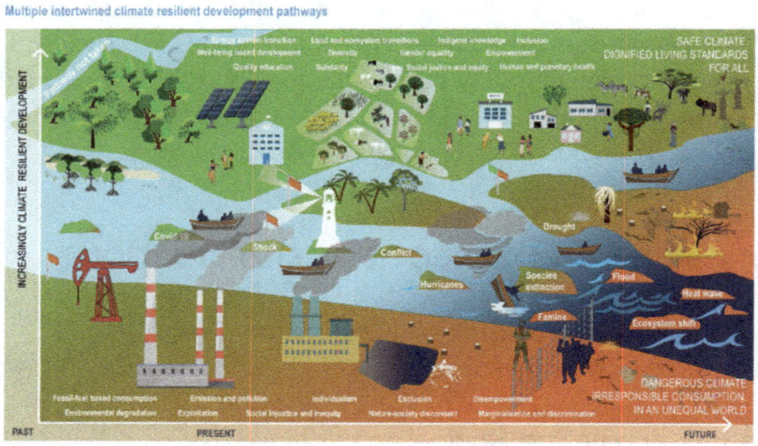

Source: IPCC.CH

Scientists can also assess different materials, including carbon-trapping concrete and recycled plastics and how they influence the carbon footprint of construction in cities (Eulerich, Pawlowski, Waddoups, & Wood, 2022). Integrating materials science, urban planning and environmental economics might lead to future developments in research.

Many experts recommend post-occupancy evaluations, although they are rarely incorporated into long-term project planning. With the Rozelle Interchange, there is an opportunity to study changes and effects over the course of several years. These are the metrics you should consider: user satisfaction, traffic efficiency, the cost of upkeep, air and noise pollution and the rates of accidents.

Researchers can establish methods that enable people participating in POE to help examine and assess infrastructure usage. Using this system would allow plans to be made based on people's needs rather than just plans, so they are easier to adapt (Zhang et al., 2023).

5.1.3 Equity, Accessibility, and Inclusion in Urban Infrastructure Design

While there are group participation tools in the Rozelle Interchange, more effort could be made to include community-centred design. Urban planning research should consider: who is given a seat at the table and who is not heard? This also involves considering inequality in access to transportation, safety problems and where the benefits of the project go.

Researchers may design equality review methods and create inclusion indices that can be used in urban transportation projects. Additionally, studying multiple cities may reveal how engineers design interchanges that are both efficient and fair for all people (Leitner-Hanetseder, Lehner, Eisl, & Forstenlechner, 2021).

5.1.4 Institutional Learning and Adaptive Governance

Often, big infrastructure projects operate and develop apart from what was learned in similar previous ones. Future work should investigate models for sharing lessons and knowledge in the delivery of projects within the public sector. What approaches can government departments and their contractors apply to develop libraries of effective work processes, missteps and new ideas? Which governance systems are useful for enhancing learning while the project is running and after it ends?

Experts may apply organisational learning, public administration and systems thinking to plan a model for learning in public infrastructure development (Hasan, 2021). They could help reduce repeated problems and increase the speed at which new inventions are accepted by various sectors.

5.2 Step 2: Suggestions to Improve the Project

5.2.1 Strategic Design Enhancements

To ensure the Rozelle Interchange is both engineering perfection and socially acceptable, design concepts should take into account safety, accessibility and integration with the surrounding environment. The urban infrastructure of today is concentrated on human-centred design, making certain that the facilities are accessible to vulnerable groups like the elderly, children, and people with disabilities. This necessitates the redesign of pedestrian crossings, entry and exit ramps and nearby roads to reduce safety hazards and increase ease of usability. The application of human-centred principles in the Australian transport policy has been growing because of the international move towards fair standards of mobility (Kommunuri, 2022).

In addition to functionality, green infrastructure is important in ensuring that the objectives of sustainability are met. The recent literature emphasises the ecological and psychological advantages of incorporating the elements of vegetated embankments, green walls, and water-sensitive urban design (WSUD) (Dolman, 2019). They not only moderate the heat-island effect but also increase biodiversity and stormwater management. The Rozelle Interchange can be used as a model of considering the environmentally conscious design of the urban megaproject through the combination of transport engineering and ecological principles.

5.2.2 Strengthening the Regulatory Alignment.

One of the most typical causes of project delay in large-scale infrastructure is regulatory compliance. When agencies have different interpretations of compliance standards, costly redesigns can be caused, and lengthy approval procedures may take place (Miller and Evans, 2023). To meet this, an effective practice, infrastructure governance is regulatory alignment workshops, in which leaders of project teams, regulators, and technical professionals meet initially to seek to understand their needs (Davies, 2023). This proactive plan will provide a smoother approval procedure and prevent a conflict in the construction process.

Figure 5.

Moreover, the constant auditing systems are now widely encouraged by the international regulatory authorities, like the Infrastructure Sustainability Council (2023). With the help of blockchain technology, digital audit platforms now enable real-time compliance tracking at all construction stages and decrease the risk of penalties or post-construction modifications (Chen et al., 2022). For the Rozelle Interchange, adopting continuous digital compliance audits would ensure transparency, mitigate risks, and enhance trust with both regulators and the public.

5.2.3 Workforce Development and Capacity Building

The existence of such large infrastructure projects relies on a very highly talented and flexible workforce. The old modes of training are inadequate in the fast pace of integration of technology. The relevance of such collaboration with vocational institutes and universities in providing technical training in Building Information Modelling (BIM), Environmental Impact Assessment (EIA) and digital construction technologies is highlighted in the recent efforts of the Australian government (Safe Work Australia, 2023). These programs not only equip workers with the immediate needs of the Rozelle Interchange but also with long-term employability in the more digitised world of construction.

Safety-oriented programs are also vital. Modern industry regulations stress the importance of periodic on-site exercises and emergency simulations to reduce the occupational hazards (Peterson and Hall, 2022). The integration of such practices will add to a safety culture and minimise accidents and project downtimes. The Rozelle Interchange can guarantee a robust labour force that is ready to tackle the present and future challenges in the infrastructure sector by investing in the development of its workforce.

5.2.4 Community Integration and mitigation of social impact.

Large-scale infrastructure is bound to alter the communities it crosses, so inclusive community development is key to preventing opposition and adverse interference. The studies have emphasised the importance of the early and sustained community-engagement strategies, especially when it comes to the projects where the relocation or profound landscape alteration is involved (Kruskopf, Lobbas, Meinander, Söderling, Martikainen, & Lehner, 2020). The creation of effective social impact mitigation strategies that offer housing services, job and career opportunities, and planning assistance to help individuals relocate enhances resilience at the community level and minimises resistance.

There is also a need to be very careful when it comes to cultural heritage protection. In a 2023 report on Australian urban projects, it was found that the failure to pay attention to cultural assets is a common cause of litigation and delays (Taylor and Johnson, 2022). In the case of the Rozelle Interchange, a combination of heritage assessment and design process will not only protect the cultural landmarks but also create good relationships with the nearby communities. Open involvement procedures such as open forums, online platforms, and interactive design workshops may be additionally used to ensure that the local voices directly influence the project outcomes.

5.2.5 Long-Term Operational Sustainability

The long-term usefulness of the Rozelle Interchange will be gauged by its capacity of to be efficient, cost-effective and environmentally sustainable once it is completed. Internationally recognised sustainability audits like those of the Global Reporting Initiative (GRI) standards are now extensively proposed to be used in monitoring the post-construction (O'Connor, 2022). These audits offer a systematic analysis of energy consumption, carbon emission and maintenance costs, allowing managers to find and rectify inefficiencies.

Post-Occupancy Evaluations (POE) are also essential and required by the Australian infrastructure policy since 2023. These measures are done one and three years after implementation and evaluate the satisfaction of the users, operational performance, and societal impacts (Nguyen and Adams, 2022). Policymakers and engineers can plan future infrastructures to guarantee continuous improvement by incorporating POE findings into the planning efforts. The Rozelle Interchange can therefore provide a new precedent of adaptive infrastructure design which adapts to changing environmental, social and technological requirements.

6 CONCLUSION

Rozelle Interchange is a major project in Sydney designed to minimise traffic congestion, boost links between areas and offer lasting solutions for moving people around the city. The companion guide states that to manage the complexity of the project, efforts must focus on satisfying engineering, environmental, community, regulatory and operational needs.

Paying special attention, the assessment found challenges and suggested suitable improvements. You should focus on human-centred design, improve on regulatory compliance, invest in workers' skills, engage with the people in the area and ensure the community can depend on these projects for years to come. All of these suggestions are designed to improve the interchange as it functions now and in the future.

For example, the design recommendations recommend considering inclusivity, making the space practical and ensuring it looks attractive. Nowadays, urban planners must add green infrastructure, enhance paths for walkers and cyclists and address pollution to be climate-responsible. The changes made benefit users and also encourage caring for the environment.

By routinely communicating with governance teams and performing regular compliance reviews, manufacturers can avoid problems that could lead to losses, stress and rework. Clearly set channels of communication and arranged workshops help everyone stay within the rules set by the government and those that protect the environment.

Every project needs to have a skilled team behind it. If the team is trained on digital engineering, safety and environmental assessment topics, they will be ready to handle any recent issues they encounter. Also, when everyone learns all the time, it means both high safety and efficiency are always maintained during the entire project.

Developing communities and ensuring everyone is involved were considered very important. People, companies and the environment are all affected by the work being done at Rozelle Interchange. For a plan to succeed, stakeholders need to be transparent in compensation, involve residents in decision-making and create many job openings. Preserving the area's cultural and historical identity will ensure the community maintains its local character as things change.

A focus on how the interchange will operate in the long run allows benefits to be felt after its construction is completed. Using sustainability audits and post-occupancy evaluations, individuals involved in infrastructure projects can study the results, notice new challenges and incorporate this knowledge into upcoming projects. It is essential for improving the system, as data makes it possible to check government actions.

To sum up, the Rozelle Interchange project played a key role in improving Sydney's infrastructure. If the organisation applies the proposed advice in its planning and execution, the project can fix existing issues, fulfil what different stakeholders require and improve the way infrastructure is delivered. The benefits will be noticeable right away, and they'll also guarantee the project serves as a model for entire communities of future large-scale developments.

7 REFERENCES

ACCA. (2021). *The Digital Accountant: Digital Skills in a Transformed World.* Association of Chartered Certified Accountants.

Ajayi-Nifise, A. O., Odeyemi, O., Mhlongo, N. Z., Ibeh, C. V., Elufioye, O. A., & Awonuga, K. F. (2024). The future of accounting: Predictions on automation and AI integration. *World journal of advanced research and reviews, 21*(2), 399-407.

Bakarich, K. M., & O'Brien, P. E. (2021). The robots are coming… but aren't here yet: The use of artificial intelligence technologies in the public accounting profession. *Journal of Emerging Technologies in Accounting, 18*(1), 27–43.

Bavaresco, R. S., Nesi, L. C., Barbosa, J. L. V., Antunes, R. S., da Rosa Righi, R., da Costa, C. A., & Gatti, C. (2023). Machine learning-based automation of accounting services: An exploratory case study. *International Journal of Accounting Information Systems, 49*, 100618.

Bhimani, A., & Willcocks, L. (2014). Digitisation, 'Big Data' and the transformation of accounting information. *Accounting and Business Research, 44*(4), 469–490.

Chen, L., Smith, A., & Zhao, M. (2022). Digital audit platforms and blockchain for real-time compliance in construction projects. *Journal of Construction Technology and Digital Governance, 3*(2), 45–63.

Chukwuani, V. N., & Egiyi, M. A. (2020). Automation of accounting processes: Impact of artificial intelligence. [IJRISS]. *International Journal of Research and Innovation in Social Science, 4*(8), 444–449.

Cooper, L. A., Holderness, D. K.Jr, Sorensen, T. L., & Wood, D. A. (2022). Perceptions of robotic process automation in Big 4 public accounting firms: Do firm leaders and lower-level employees agree? *Journal of Emerging Technologies in Accounting, 19*(1), 33–51.

Davies, K. (2023). Regulatory alignment workshops and governance in infrastructure megaprojects. *Australian Journal of Infrastructure Regulation, 5*(1), 12–29.

Deloitte. (2022). *Robotic Process Automation (RPA): Intelligent automation survey results.* Deloitte Insights.

Deloitte. (2023). *Regulatory compliance and automation in finance: Industry trends report.* Deloitte. https://www2.deloitte.com

Dolman, N. (2019). Water-Sensitive Urban Design (WSUD) and green infrastructure in urban interchange projects. *Urban Ecological Engineering Review, 8*(4), 210–229.

Ernst & Young. (2022). *Finance function transformation with intelligent automation*. EY Global. https://www.ey.com

Eulerich, M., Pawlowski, J., Waddoups, N. J., & Wood, D. A. (2022). A framework for using robotic process automation for audit tasks. *Contemporary Accounting Research, 39*(1), 691–720.

European Commission. (2023, December 9). *Artificial Intelligence Act: Council and Parliament strike a deal on the first worldwide rules for AI*. Council of the European Union. https://www.consilium.europa.eu/en/press/press-releases/2023/12/09/artificial-intelligence-act-council-and-parliament-strike-a-deal

European Commission. (2024). *Artificial Intelligence in the European Commission (AI@EC) Communication*. Directorate-General for Digital Services. European Commission

Harrast, S. A. (2020). *Robotic process automation in accounting systems* (Vol. 31). Wiley Online Library.

Hasan, A. R. (2021). Artificial Intelligence (AI) in accounting & auditing: A Literature review. *Open Journal of Business and Management, 10*(1), 440–465.

IFAC. (2023). *AI and Intelligent Automation: Opportunities for Professional Accountants*. International Federation of Accountants. IFAC.

Infrastructure Sustainability Council. (2023). *Guidelines for sustainable infrastructure delivery*. Infrastructure Sustainability Council of Australia.

International Federation of Accountants (IFAC). (2023, October). *AI and intelligent automation: Opportunities for professional accountants*. IFAC. https://www.ifac.org/news-events/2023-10/ai-and-intelligent-automation-opportunities-professional-accountants

ISACA. (2023). The potential impact of the European Commission's proposed AI Act on SMEs. *ISACA Journal, 2*, 1–7. https://www.isaca.org/resources/isaca-journal/issues/2023/volume-2/the-potential-impact-of-the-european-commissions-proposed-ai-act-on-smes

Khan, N., Tabassum, M., Lawal, I. A., Lakshmikanth, S. P., & AlRugaibah, A. A. (2024). Impact of cloud computing adoption in Saudi Arabia health care sector. In *AIP conference proceedings* (Vol. 2919, No. 1, p. 040001). AIP Publishing LLC.

Kokina, J., Gilleran, R., Blanchette, S., & Stoddard, D. (2021). Accountant as digital innovator: Roles and competencies in the age of automation. *Accounting Horizons, 35*(1), 153–184.

Kommunuri, J. (2022). Artificial intelligence and the changing landscape of accounting: A viewpoint. *Pacific Accounting Review*, *34*(4), 585–594.

Korhonen, T., Selos, E., Laine, T., & Suomala, P. (2021). Exploring the programmability of management accounting work for increasing automation: An interventionist case study. *Accounting, Auditing & Accountability Journal*, *34*(2), 253–280.

Kruskopf, S., Lobbas, C., Meinander, H., Söderling, K., Martikainen, M., & Lehner, O. (2020). *Digital accounting and the human factor: theory and practice*. ACRN Journal of Finance and Risk Perspectives.

Leitner-Hanetseder, S., Lehner, O. M., Eisl, C., & Forstenlechner, C. (2021). A profession in transition: actors, tasks and roles in AI-based accounting. *Journal of applied accounting research, 22*(3), 539-556.

Lombardi, R., Manfredi, S., Cuozzo, B., & Palmaccio, M. (2022). The dark side of accounting automation: Trust and distrust in the age of algorithms. *Accounting, Auditing & Accountability Journal*, *35*(3), 843–870. DOI: 10.1108/AAAJ-10-2020-4989

Miller, J., & Evans, H. (2023). Project delays, cost overruns, and risk mismanagement in Australian infrastructure delivery. *Construction Economics and Risk Management*, *9*(3), 77–95.

Nguyen, T., & Adams, R. (2022). Transport studies on traffic congestion during urban infrastructure construction. *Australian Transport Policy Journal*, *14*(2), 100–118.

O'Connor, P. (2022). Post-occupancy evaluations and sustainability audits: Long-term performance in built infrastructure. *Sustainable Infrastructure Journal*, *4*(1), 33–50.

Odonkor, B., Kaggwa, S., Uwaoma, P. U., Hassan, A. O., & Farayola, O. A. (2024). The impact of AI on accounting practices: A review: Exploring how artificial intelligence is transforming traditional accounting methods and financial reporting. *World journal of advanced research and reviews, 21*(1), 172-188.

Peng, Y., Ahmad, S. F., Ahmad, A. Y. B., Al Shaikh, M. S., Daoud, M. K., & Alhamdi, F. M. H. (2023). Riding the waves of artificial intelligence in advancing accounting and its implications for sustainable development goals. *Sustainability*, *15*(19), 14165.

Peterson, G., & Hall, L. (2022). Safety culture and emergency drills in construction: Evidence from Australia. *Journal of Occupational Safety & Construction*, *6*(2), 89–104.

Rahman, J., Li, Y., & Gupta, S. (2023). AI-driven simulation for Traffic Management Plans (TMPs) during infrastructure projects. *Transportation Engineering & Smart Infrastructure*, *2*(1), 14–32.

Ranjith, P., Madan, S., Jian, D. A. W., Teoh, K. B., Singh, A. S., Ganatra, V., & Shekar, S. L. (2021). Harnessing the power of artificial intelligence in the accounting industry: A case study of KPMG. *International Journal of Accounting & Finance in Asia Pasific*, *4*(2), 93–106.

Rawashdeh, A. (2022). Determinants of artificial intelligence adoption in SMEs: The mediating role of accounting automation. *Rawashdeh, A., Bakhit, M & Abaalkhail, L.(2023). Determinants of artificial intelligence adoption in SMEs: The mediating role of accounting automation. International Journal of Data and Network Science*, *7*(1), 25–34.

Safe Work Australia. (2023). *BIM, EIA, and digital construction technologies: Workforce skills roadmap*. Safe Work Australia.

Shaffer, K. J., Gaumer, C. J., & Bradley, K. P. (2020). Artificial intelligence products reshape accounting: Time to re-train. *Development and Learning in Organizations*, *34*(6), 41–43.

Singh, R., & Patel, D. (2024). Advanced cost estimation and BIM systems in megaproject forecasting accuracy. *International Journal of Construction Finance*, *7*(1), 58–77.

Taylor, M., & Johnson, P. (2022). Cultural heritage protection and community engagement in Australian urban infrastructure. *Heritage & Society*, *10*(1), 44–62.

Wang, J., & Jones, S. (2022). Predictive analytics in accounting: Opportunities and risks. *The Accounting Review*, *97*(1), 123–149. DOI: 10.2308/accr-52250

Zhang, C., Issa, H., Rozario, A., & Soegaard, J. S. (2023). Robotic process automation (RPA) implementation case studies in accounting: A beginning to end perspective. *Accounting Horizons*, *37*(1), 193–217.

Chapter 2
Robotic Process Automation in Taxation:
Accounting Tax Compliance and Reporting

Nik Herda Nik Abdullah
https://orcid.org/0000-0003-0582-8432
Taylor's University, Malaysia

Soliha Sanusi
National University of Malaysia, Malaysia

ABSTRACT

Robotic Process Automation (RPA) is transforming the field of taxation by streamlining tax compliance and reporting processes. This chapter explores how RPA technologies can automate repetitive, rule-based tasks such as data extraction, validation, reconciliation, and submission of tax returns. By reducing human intervention, RPA enhances accuracy, ensures timely compliance, and minimizes the risk of penalties due to manual errors. The chapter also examines the integration of RPA with existing enterprise systems, highlighting its potential to improve audit readiness and regulatory adherence. The study will include case studies to illustrate the practical benefits of RPA in real-world taxation scenarios. Furthermore, the chapter addresses implementation challenges, including system compatibility, data security, and the need for change management. Finally, it reflects on the evolving role of tax professionals and the need for upskilling in an increasingly automated environment. RPA is not just a tool, but a catalyst for strategic transformation in taxation.

DOI: 10.4018/979-8-3373-3571-1.ch002

INTRODUCTION

In today's fast-evolving digital landscape, the pressure on organizations to modernize their internal operations has never been greater. Among the various operational domains undergoing transformation, the taxation function stands out due to its high dependency on accurate data processing, compliance with increasingly complex regulations, and strict adherence to reporting deadlines. As regulatory frameworks tighten and the volume of tax-related data grows, companies are turning to digital solutions to maintain compliance while reducing operational burdens. One such solution that has emerged as a transformative force in tax functions is Robotic Process Automation (RPA).

Figure 1 illustrates the upstream movement of the tax function as part of the broader digitalisation of tax operations. The figure demonstrates how taxation processes are increasingly integrated into business and corporate functions through interconnected digital systems. This transition shifts the tax function from a reactive, report-based role to a proactive, data-centric component within the organisation. The process begins with source systems and initial data entry, which feed into the Enterprise Resource Planning (ERP) or consolidation layer, where transactional data is standardised and validated. These data streams are then channelled into a data lake, enabling the aggregation, storage, and real-time analysis of large volumes of tax-relevant information. Finally, the additional reporting stage generates comprehensive and automated outputs that support compliance, forecasting, and strategic decision-making.

Importantly, this model highlights the evolving relationship between organisations and tax authorities, facilitated by near real-time data sharing and automated payment systems. By embedding tax data within the transactional flow from the outset, organisations can enhance accuracy, transparency, and auditability while minimising manual intervention. The upstream integration of tax processes aligns with the principles of Robotic Process Automation (RPA), which promotes the automation of repetitive, rule-based tasks and the seamless exchange of information across systems. Ultimately, this transformation enables tax functions to become strategic partners within the enterprise, contributing insights that support compliance, efficiency, and value creation.

Figure 1. Tax Function (Deloitte, 2025)

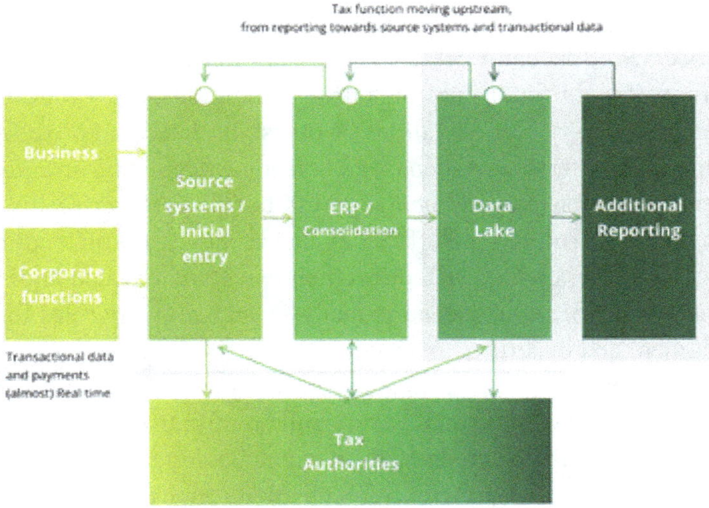

Robotic Process Automation refers to the use of software "robots" or bots that emulate human actions to complete structured, rule-based tasks across digital platforms. These bots interact with systems at the user interface level, mimicking keystrokes and mouse clicks to execute routine tasks such as data extraction, form completion, reconciliations, and report generation. Unlike traditional IT automation that may require system integration or re-engineering, RPA operates on top of existing infrastructure, offering agility and scalability with minimal disruption (Kovacev, 2020). Its relevance in the taxation domain stems from the nature of tax functions, which are highly repetitive, detail-oriented, and deadline-driven. Processes such as preparing and filing tax returns, reconciling input and output VAT, calculating payroll taxes, generating statutory reports, and maintaining audit trails are all well-suited for automation through RPA.

The adoption of RPA in tax functions is not just a technological shift, it represents a fundamental change in how tax compliance and reporting activities are conceptualized and executed. Traditionally, tax departments have relied heavily on manual processes spreadsheets, emails, and hand-prepared reports. These methods are prone to human error, create bottlenecks, and expose organizations to compliance risks. For instance, inaccuracies in tax returns or delays in reporting can result in regulatory penalties, financial restatements, and reputational damage. Moreover, the increasing globalization of businesses has introduced additional layers of complexity, such as dealing with multiple tax jurisdictions, varied tax treatments, and compliance with cross-border regulations. In this environment, maintaining accuracy and timeliness becomes a significant challenge.

RPA addresses these pain points by introducing speed, consistency, and reliability into tax processes. Bots can work 24/7, ensuring that tasks are completed within deadlines regardless of time zones or staffing limitations. They follow predefined rules, eliminating the risk of deviations or oversight that often arise from manual work. Additionally, every step taken by a bot is logged, creating an automatic audit trail that enhances transparency and supports compliance verification (Deloitte, n.d). The benefits extend beyond efficiency and accuracy. By offloading repetitive tasks to bots, tax professionals can redirect their focus toward value-adding activities such as strategic tax planning, scenario analysis, and advisory roles functions that require judgment and expertise, and cannot be easily automated.

The relevance of efficient tax compliance in today's business environment cannot be overstated. Tax is a significant cost component for most businesses and plays a vital role in cash flow management and financial planning. At the same time, governments worldwide are ramping up enforcement through digital tax platforms and real-time data sharing. Initiatives such as the OECD's Base Erosion and Profit Shifting (BEPS) framework, the EU's VAT e-commerce package, and electronic invoicing mandates in countries like Brazil, India, and Malaysia reflect a broader trend toward transparency and digitization in taxation (OECD, 2023). These developments have raised the expectations of tax departments to maintain up-to-date records, generate accurate reports on demand, and respond quickly to audits or regulatory queries. Figure 2 indicates the digital tax administration conceptual framework by Bassel et al. (2022) that links technology with context, stakeholders, and demonstrated results. A good tax administration will lead to better tax compliance among the people.

Figure 2. Digital tax administration conceptual framework (Bassey et al, 2022)

Against this backdrop, inefficiencies in tax compliance processes pose both financial and strategic risks. Companies that fail to adapt may face not only fines and penalties but also delays in decision-making and lose competitive advantage. Tax compliance is no longer a back-office function; it is a critical component of corporate governance and business continuity. This elevates the need for automated, streamlined processes that ensure compliance while supporting business agility. RPA, with its ability to rapidly scale and adapt to changing requirements, becomes a key enabler in this transformation journey (KPMG, n.d).

The objective of this chapter is to explore in depth how RPA is reshaping tax compliance and reporting in modern organizations. First, the chapter aims to clarify the concept of RPA in the context of taxation what it is, how it works, and why it is particularly relevant to tax processes. While RPA has been applied across various business functions, its application in tax is unique due to the high degree of regulation and the requirement for detailed documentation and audit trails. The chapter provides a comprehensive understanding of how RPA aligns with the functional and regulatory demands of tax compliance.

Second, the chapter examines how RPA is applied in specific areas of taxation, including corporate income tax, indirect tax (such as VAT and GST), and transfer pricing. Each of these areas presents distinct challenges in terms of data volume, regulatory variation, and documentation requirements. Through selected case studies, the chapter demonstrates how RPA has been deployed in real-world scenarios to address these challenges. For example, a case involving corporate tax might illustrate how bots extract trial balance data, perform tax adjustments, and populate tax return templates. A case related to indirect tax could show how RPA streamlines VAT reconciliation by comparing invoice data with general ledger entries and submitting returns to government portals. In transfer pricing, the focus may be on automating the preparation of local files and master files in line with OECD guidelines.

Third, the chapter discusses the challenges and issues that arise during RPA implementation in tax functions. Despite its many benefits, RPA is not a plug-and-play solution. It requires careful planning, process mapping, stakeholder buy-in, and ongoing maintenance. The chapter explores common hurdles such as system incompatibility, lack of data standardization, resistance from employees, and the risk of over-automation without proper governance. These issues are examined through the lens of actual implementation experiences, providing insights into what works and what does not exist in real-world settings.

Fourth, the chapter offers strategic recommendations for organizations considering or currently implementing RPA in their tax departments. These include best practices for selecting processes suitable for automation, designing bots with built-in controls, integrating RPA with enterprise resource planning (ERP) systems and tax engines, and developing change management strategies. The discussion also highlights the importance of cross-functional collaboration between tax, IT, finance, and internal audit teams to ensure a successful rollout (PwC, nd).

Fifth, the chapter explores the future research directions and implications of RPA in taxation. While RPA is highly effective for rules-based processes, it is often limited when handling unstructured data or processes requiring human judgment. The future lies in augmenting RPA with artificial intelligence (AI) and machine learning (ML) to create intelligent automation. For example, AI-powered bots could interpret tax laws, flag anomalies in tax returns, or draft responses to tax authorities. Natural language processing (NLP) may enable bots to generate narrative reports or interpret emails from tax regulators. The integration of RPA with blockchain technology could further enhance the transparency and traceability of tax transactions. The chapter considers these possibilities and their implications for the evolving role of tax professionals, who must now adapt to working alongside digital co-workers and develop new skills in data analysis, technology management, and strategic thinking.

Finally, the chapter is structured to provide both theoretical insights and practical guidance. It begins with an extensive review of the existing literature on RPA and its relevance to tax processes. This is followed by a deep dive into specific implementation cases, a critical discussion of outcomes, and a reflection on future trends. The concluding section synthesizes the insights gained and offers final reflections on the role of RPA in shaping the tax functions of the future.

BACKGROUND

The Institute of Management Accountants (IMA, 2021) in its Statement on Management Accounting, Transforming the Finance Function with RPA underscores the pivotal role of Robotic Process Automation (RPA) in modernizing finance and accounting operations. The report explains that RPA enables organizations to automate structured, rule-based processes such as reconciliations, report generation, and transaction postings, leading to substantial gains in accuracy, speed, and cost efficiency. By reducing reliance on manual processing, finance professionals can redirect their focus from transactional activities toward strategic decision-making and value creation. The IMA (2021) blueprint for scalable RPA implementation provides a practical framework for integrating automation across different business sizes and maturity levels.

Building upon these advancements within the finance function, the adoption of Robotic Process Automation (RPA) has also begun to reshape the landscape of tax operations. As finance and tax departments often share interconnected data systems and reporting processes, the efficiency gains realized through RPA in accounting naturally extend to tax-related activities. Organizations are increasingly deploying automation to handle tasks such as tax data extraction, validation of tax codes, generation of tax returns, and electronic submission to tax authorities. This integration has resulted in reduced cycle times, enhanced accuracy, and improved compliance oversight. Furthermore, RPA supports greater standardization and transparency across global tax processes, enabling consistency in how tax data is prepared and reported across multiple jurisdictions. Automation not only minimizes compliance risks but also facilitates the shift from reactive tax management to proactive strategy development, where professionals can focus on interpreting data and optimizing tax outcomes. In this way, RPA acts as a transformative bridge between finance and taxation that linking operational efficiency with strategic decision-making and laying the foundation for intelligent, technology-enabled tax functions.

The evolution of tax technology has been closely linked to broader trends in digital transformation across corporate finance and public administration. In the past, tax compliance was predominantly manual, involving large volumes of paperwork, spreadsheet management, and labor-intensive reconciliations. Over time, the emergence of enterprise resource planning (ERP) systems and dedicated tax engines began to digitalize some aspects of tax processing, such as data retrieval and calculation automation. However, these solutions often required complex integration and did not fully eliminate manual intervention. In recent years, technological innovations such as Robotic Process Automation (RPA), artificial intelligence (AI), machine learning (ML), and blockchain have begun to significantly transform the tax function into a more strategic, data-driven, and streamlined operation (Belahouaoui & Attak, 2024).

The body of literature on Robotic Process Automation (RPA) in taxation and accounting has expanded considerably over the past five years, reflecting the growing convergence of digital technologies with financial management and compliance functions. Scholars and practitioners alike have investigated RPA's diverse applications in automating tasks such as invoice processing, VAT compliance, corporate tax filing, and payroll tax administration, demonstrating its potential to enhance both efficiency and governance. Empirical studies highlight how RPA can streamline the preparation of tax returns by automatically extracting relevant data from enterprise resource planning (ERP) systems, validating entries, and populating standardized tax templates, thereby minimizing errors commonly associated with manual data handling. Beyond routine automation, recent research underscores RPA's strategic role in strengthening internal controls and audit readiness. For instance, Yoon et al. (2021) emphasize that RPA facilitates the creation of comprehensive digital audit trails, promoting greater transparency, traceability, and accountability within tax operations. Collectively, these findings suggest that RPA is not merely a tool for process optimization but a key enabler of digital transformation in tax administration, aligning operational efficiency with regulatory compliance and strategic decision-making.

CURRENT METHODS FOR TAX COMPLIANCE AND REPORTING: CHALLENGES AND CONSTRAINTS

Despite advances in digital technologies and regulatory reforms, many organizations continue to rely on traditional or semi-automated methods for tax compliance and reporting, which present significant challenges and constraints. The current approach typically involves a combination of manual data gathering, spreadsheet-based calculations, and limited use of enterprise resource planning (ERP) systems to prepare and submit tax filings. These methods are often characterized by fragmented processes, disparate data sources, and high levels of human intervention. Below are the challenges and constraints on the current methods for tax compliance and reporting.

Time Consumption

One of the most pressing issues associated with such methods is the excessive time consumption required to meet compliance obligations. Tax teams frequently spend substantial hours extracting data from various financial systems, consolidating and validating figures, and manually entering information into tax returns or statutory reporting tools (PwC, nd). The complexity increases exponentially for multinational corporations operating in multiple tax jurisdictions, where varying tax codes, filing frequencies, and compliance deadlines must be observed. As a result, the tax compliance process becomes a time-intensive, reactive exercise, leaving limited capacity for strategic tax planning or risk assessment.

Lack of Data Integration and Standardization

Another fundamental challenge lies in the lack of data integration and standardization. Most tax functions draw data from multiple sources ERP systems, accounting software, bank records, payroll platforms, and invoice processing tools each with different formats, taxonomies, and reporting periods. This lack of integration leads to manual reconciliation efforts, increasing the likelihood of data inconsistencies and misreporting. Errors in data mapping, omission of transactions, or duplication can distort taxable income and liabilities, thereby exposing organizations to compliance risks and penalties (KPMG, n.d). Additionally, the extensive use of spreadsheets as the primary tool for managing tax computations presents its own set of constraints. Spreadsheets are inherently prone to formula errors, broken links, and version control issues, particularly when shared across departments or teams. A minor input error or logic flaw in a tax worksheet can cascade into significant miscalculations, with little transparency into the root cause. Despite being widely used, spreadsheets lack robustness, traceability, and auditability demanded in a regulatory environment that increasingly prioritizes data governance and accountability.

Human error

Human error remains a dominant risk factor in current tax reporting practices. Manual entry of tax data, reconciliation of ledger balances, and classification of tax codes are subject to mistakes, especially under pressure or tight timelines. These errors may result in over- or underpayment of taxes, missed deadlines, and non-compliance with tax laws. The risk is further compounded when tax professionals lack access to real-time data or are required to make judgment calls without sufficient supporting documentation (EY, 2021). Furthermore, tax teams often operate in silos within the broader finance function, lacking visibility into upstream processes that impact tax obligations, such as procurement, invoicing, and revenue recognition. This isolation limits their ability to proactively identify tax risks or planning opportunities, reducing the strategic contribution of the tax function to the organization. In some cases, tax departments are involved too late in the transaction lifecycle, responding to reporting requirements only after key decisions have already been made.

Frequent Changes in Tax Legislation

The frequent changes in tax legislation also represent a major constraint. Tax regulations are dynamic and influenced by political, economic, and social factors. Governments around the world are introducing new reporting requirements, increasing transparency demands, and tightening audit procedures to combat tax evasion and improve revenue collection. For example, initiatives such as the OECD's Base Erosion and Profit Shifting (BEPS) Action Plans, the European Union's DAC6 directive, and the implementation of digital tax reporting (e.g., SAF-T, e-invoicing, and e-audit) have introduced additional layers of compliance for multinational entities (OECD, 2020). Keeping track of these changes and ensuring timely updates to internal policies and tax calculations is a significant burden, especially for organizations without robust regulatory tracking systems. Moreover, outdated compliance methods are often ill-equipped to accommodate these evolving requirements, leading to last-minute scrambles to modify templates, reprocess data, or engage external consultants.

Limited Audit Trail and Documentation Capabilities

Another constraint arises from the limited audit trail and documentation capabilities of traditional tax processes. Regulators are increasingly demanding greater transparency and traceability in tax filings, including the ability to demonstrate the origin of figures, provide supporting documents, and justify assumptions or estimations. However, when tax calculations are conducted manually or through informal tools, the process often lacks proper version history, approval workflows, or system-generated logs. This makes it difficult to reconstruct how specific tax figures were derived, particularly if the staff responsible have left the organization or if the data has been altered inadvertently (ISACA, 2020). In the event of an audit or dispute with tax authorities, organizations may struggle to present a credible and coherent explanation of their tax positions. This lack of documentation not only undermines compliance but also weakens the organization's defense in the face of regulatory scrutiny.

Resource Constraints

Resource constraints within tax departments also impede effective compliance and reporting. Many tax teams are understaffed relative to the volume and complexity of their responsibilities. With increasing regulatory demands, businesses are expected to deliver more comprehensive and timely reporting, yet often with stagnant or shrinking budgets. This results in overreliance on a few key personnel, lack of cross-training, and burnout—factors that diminish the quality and sustainability of tax operations (Deloitte, n.d). In addition, organizations face challenges in attracting and retaining talent with the right blend of tax knowledge and digital skills. While tax professionals are expected to navigate intricate tax laws, they are also increasingly required to understand data structures, automation tools, and analytics platforms—competencies not traditionally part of tax education or training. The skills gap between traditional tax practitioners and the evolving technological environment is a growing concern that exacerbates the inefficiencies of current methods (IMA, 2021).

High Financial Costs

In addition to operational limitations, current compliance methods often incur high financial costs. These include internal labor costs, consultancy fees, penalties for late or inaccurate filings, and the cost of defending audits or disputes. For multinational corporations, the cost burden is amplified by the need to maintain compliance teams in multiple jurisdictions, engage local advisors, and translate financial data into country-specific reporting formats. These expenses are exacerbated by inefficiencies in data handling and the duplication of efforts across regions. A study by the World Bank (2024) highlighted that businesses in developing countries spend on average 240 hours per year on tax compliance tasks, with significant variations based on the complexity of tax codes and the digitization level of national tax authorities.

Furthermore, legacy IT systems pose structural barriers to effective tax compliance. Many organizations continue to use outdated ERP systems that lack modern features such as real-time data extraction, integration with e-filing portals, or automated journal entry validation. These systems often require workarounds, manual exports, and custom coding to perform basic tax-related tasks. The result is a patchwork of disconnected processes and tools that complicate oversight and introduce potential compliance risks (Grant Thornton, 2022). Upgrading these systems can be prohibitively expensive and time-consuming, especially for organizations with limited IT budgets or those operating in regulated industries with high implementation risks. As a consequence, tax departments are left working around these constraints, increasing their reliance on manual effort and decreasing agility in response to regulatory change.

Cybersecurity and data protection concerns also impact current tax compliance practices. With increasing reliance on digital tax submissions and the exchange of sensitive financial data across platforms, the risk of data breaches and unauthorized access has escalated. Traditional compliance processes often lack robust encryption, access controls, and monitoring protocols, especially when files are shared via email or stored on local devices. The absence of centralized, secure systems for managing tax data creates vulnerabilities that may be exploited by cybercriminals or result in inadvertent data loss. In regions governed by stringent data protection laws failure to safeguard taxpayer information can lead to severe penalties and reputational damage. Therefore, ensuring cybersecurity and regulatory data privacy compliance adds another layer of complexity to an already burdened tax function.

Moreover, the current methods are not conducive to real-time tax management or forecasting. Most organizations operate on a lagged basis, preparing tax reports after financial close or at periodic intervals. This backward-looking approach limits visibility into real-time tax exposures and hampers proactive tax planning. In today's volatile business environment characterized by supply chain disruptions, fluctuating interest rates, and regulatory shifts companies require agile tax strategies that adapt quickly to new developments. Traditional tax compliance models are too rigid and slow to provide actionable insights, leaving organizations vulnerable to tax inefficiencies or surprise liabilities. The absence of real-time dashboards or predictive analytics in current tax workflows inhibits the ability of tax professionals to support broader financial decision-making (McKinsey & Company, 2023).

Finally, the reliance on outdated compliance processes negatively affects the strategic value of the tax function within the organization. Rather than serving as a forward-looking advisor that supports business growth, the tax team is often perceived as a compliance cost center focused on historical reporting. This perception undermines the potential contribution of tax professionals to value creation, such as identifying tax incentives, optimizing global tax structures, or mitigating reputational risks related to tax transparency. As noted by the OECD (2022), modern tax functions should evolve from being compliance-focused to insight-driven, but this shift is not possible without transforming the underlying methods and tools used in tax operations.

In summary, the current methods for tax compliance and reporting rooted in manual processes, spreadsheet dependency, siloed operations, and outdated systems present a multitude of challenges and constraints. These include inefficiencies, data inconsistencies, high error rates, limited visibility, resource limitations, and inadequate adaptability to regulatory change. Moreover, the inability to scale, automate, or generate real-time insights constrains the tax function's strategic role in modern business environments. Addressing these issues requires not just incremental improvements but a fundamental transformation of tax processes through advanced technologies such as Robotic Process Automation, artificial intelligence, and integrated tax platforms. Until then, organizations relying on legacy tax compliance models will continue to face risks, inefficiencies, and missed opportunities in an increasingly complex and digitized regulatory landscape.

MAIN BENEFITS OF ROBOTIC PROCESS AUTOMATION (RPA) IN TAX COMPLIANCE AND REPORTING

The benefits of RPA highlighted in prior studies are multifaceted. These include increased processing speed, reduction in human error, cost savings, scalability, and improved employee satisfaction due to the offloading of repetitive tasks. From a strategic perspective, RPA enhances compliance readiness and reduces exposure to tax risks by standardizing and logging into every transaction (Tiron-Tudor et al., 2021). Furthermore, the combination of RPA with analytics has been shown to improve data visibility, allowing tax professionals to gain real-time insights into financial and compliance trends.

The integration of Robotic Process Automation (RPA) into tax compliance and reporting functions has emerged as a transformative innovation in the digitalisation of tax administration. RPA refers to the use of software robots or "bots" to automate repetitive, rule-based tasks traditionally performed by human employees (Doguc, 2021). In the domain of taxation, this capability is particularly valuable due to the high volume of transactional data and the need for accuracy, timeliness, and compliance with increasingly complex regulations. One of the primary benefits of RPA in this context is the enhancement of operational efficiency. Tax processes, including the preparation of tax returns, validation of input data, reconciliation of ledgers, and generation of compliance reports, are often time-consuming when handled manually. RPA can automate these tasks and execute them significantly faster—often in real time—without the limitations of human fatigue or working hours. This continuous operational capability enables tax departments to meet tight deadlines, particularly during peak periods such as quarterly filings or year-end closings (PwC, nd).

Furthermore, RPA improves accuracy in tax processes by eliminating the risk of human error inherent in manual data entry and processing. Errors in tax filings can lead to penalties, reputational damage, and even litigation. By applying programmed rules and logic consistently, RPA ensures precise and repeatable outcomes, thereby reducing the risk of misstatements or non-compliance (Deloitte, n.d). For instance, bots can automatically extract tax-related data from enterprise resource planning (ERP) systems and populate it into tax return templates or e-filing platforms, ensuring consistency and correctness across datasets. This automation reduces the burden of constant reviews and verifications traditionally needed to check for human-induced discrepancies. In a study by KPMG (n.d), organizations that implemented RPA in their tax departments reported a 60–80% reduction in manual processing errors within the first year of adoption.

In addition to improving efficiency and accuracy, RPA also strengthens regulatory compliance. Tax regulations continue to evolve in complexity and scope, driven by globalisation, digital commerce, and coordinated efforts such as the OECD's

Base Erosion and Profit Shifting (BEPS) initiatives. These frameworks require multinational enterprises to meet strict documentation, reporting, and disclosure requirements. RPA can be configured to monitor regulatory updates, incorporate new rules into automated workflows, and adjust reporting formats as needed, thereby enabling organizations to maintain ongoing compliance with minimal disruption (EY, 2021). In jurisdictions with mandatory real-time reporting or e-invoicing, such as in parts of Latin America and the European Union, RPA facilitates the immediate generation, validation, and submission of tax documentation to government platforms (OECD, 2020). Bots can also be designed to flag transactions that fall outside defined regulatory thresholds or business rules, thus acting as a preventive control mechanism against potential non-compliance.

Another significant advantage of RPA is its ability to enhance audit readiness and transparency. Traditional tax audits often require extensive documentation, including transaction trails, calculation logic, and compliance evidence, which can be cumbersome to retrieve from disparate systems. With RPA, every action performed by a bot is logged in detail, creating a comprehensive digital audit trail. This traceability supports the transparency of tax operations and enables organizations to respond promptly and accurately to inquiries from tax authorities. Automation of documentation retrieval and audit evidence generation has reduced audit preparation time by up to 50% in digitally mature tax departments. This improved visibility also empowers tax managers to proactively identify and rectify anomalies before they escalate into reportable issues. The presence of structured, verifiable digital logs not only facilitates internal reviews but also enhances confidence among external auditors and regulators.

Moreover, the adoption of RPA contributes to substantial cost savings in tax compliance and reporting. Traditional tax departments often allocate significant financial and human resources to manage recurring compliance obligations. By automating routine tasks, organizations can reallocate human expertise to higher-value activities such as tax planning, risk assessment, and strategic advisory roles. A report by the Institute of Management Accountants (IMA, 2021) found that organizations implementing RPA in their tax functions experienced an average cost reduction of 25–40% within the first two years, primarily due to reduced overtime, fewer penalties, and the elimination of redundant processes. In addition, bots are scalable and can be deployed rapidly across multiple business units or jurisdictions without a proportional increase in costs. This scalability makes RPA an attractive solution for companies facing expanding regulatory burdens due to business growth or cross-border operations.

Scalability also supports multinational organizations in standardizing tax processes across jurisdictions. Tax compliance requirements differ by country in terms of formats, frequencies, and filing protocols. RPA enables the design of standardized global workflows that adapt to local requirements through rule-based configuration. This harmonization reduces complexity, ensures consistency in reporting, and simplifies global oversight. For instance, a centralized tax team in a shared services center can use RPA to manage value-added tax (VAT) filings for various countries from a single platform, reducing dependence on disparate local systems and resources (World Bank, 2024). Furthermore, bots can be cloned or redeployed with minor adjustments to handle new tax types or regulations, providing a future-proof solution as regulatory landscapes evolve.

Integration with existing tax and ERP systems is another notable benefit of RPA in this space. Many organizations still rely on legacy systems that are incompatible with modern tax engines or e-filing platforms. RPA acts as a "digital bridge," enabling data to flow seamlessly between systems without costly IT overhauls. Bots can extract data from systems such as SAP, Oracle, or QuickBooks, apply necessary tax logic, and feed it into compliance tools or government portals for submission. This interoperability not only speeds up the compliance process but also enhances data quality and consistency across platforms. According to Grant Thornton (2022), tax departments using RPA integrated with ERP systems reported up to 70% improvement in data accuracy and workflow efficiency.

Additionally, RPA facilitates real-time monitoring and analytics for tax reporting. Tax departments can leverage bots to collect, analyze, and visualize tax data through dashboards and performance indicators, enabling better decision-making. For example, real-time dashboards can highlight tax liabilities, refund status, jurisdictional exposures, and compliance trends, supporting proactive tax management and strategy formulation. This capability is particularly useful for cash flow planning and risk mitigation, as organizations gain timely insights into their tax positions and can respond swiftly to emerging issues. The analytical outputs generated through RPA can also feed into broader corporate governance and compliance frameworks, aligning tax operations with organizational risk management strategies.

Equally important is the contribution of RPA to enhanced agility and responsiveness in tax functions. In volatile regulatory environments such as those impacted by economic stimulus measures, pandemic-related relief policies, or environmental taxation businesses must respond quickly to new compliance obligations. RPA allows for the rapid reconfiguration of workflows and templates, reducing the lag between regulatory change and operational implementation. This agility ensures that organizations can meet new requirements on time, minimizing the risk of penalties and demonstrating good corporate citizenship. As observed by the OECD (2022), tax digitalization including RPA has helped companies adapt to COVID-19 driven tax reforms more efficiently than traditional approaches.

Moreover, RPA promotes better employee engagement and productivity in tax departments. By offloading repetitive, low-value tasks such as copying data between spreadsheets or checking transaction thresholds, RPA enables tax professionals to focus on analytical and strategic tasks that require human judgment. This shift not only improves job satisfaction but also enhances the overall performance of the tax function. According to a study by McKinsey & Company (2023), organizations that combined RPA implementation with upskilling initiatives for tax staff observed increased innovation and improved tax planning outcomes. Employees also become more agile and adaptable to digital tools, supporting a culture of continuous improvement and innovation.

Despite these clear advantages, it is essential to acknowledge that the benefits of RPA are maximized when implemented with proper governance, change management, and process optimization. RPA is not a "plug-and-play" solution; it requires a thoughtful assessment of existing tax processes, clear documentation of rules and exceptions, and ongoing monitoring for bot performance and compliance. Cybersecurity and data privacy are also critical considerations, particularly when bots handle sensitive financial and personal data. Best practices include role-based access controls, audit logs, encryption, and periodic risk assessments to ensure secure operations (ISACA, 2020). When these elements are integrated into RPA deployments, organizations can achieve sustainable improvements in compliance, reporting quality, and operational resilience.

In conclusion, Robotic Process Automation offers a wide range of benefits in tax compliance and reporting, encompassing improved efficiency, enhanced accuracy, strengthened compliance, and cost reduction. It also provides scalability, standardization, real-time analytics, and strategic agility, making it a key enabler of modern, digitally driven tax functions. As regulatory expectations and data volumes continue to rise, the role of RPA in supporting accurate, timely, and transparent tax operations will become increasingly critical. Organizations that invest in RPA, along with appropriate governance and talent development, will be better positioned to navigate the complexities of global taxation while driving value from their compliance functions. The future of tax reporting lies in intelligent automation, and RPA represents a foundational step toward that future.

ISSUES, CONTROVERSIES AND PROBLEMS

Implementing RPA in Taxation

The implementation of Robotic Process Automation (RPA) in taxation brings with it a range of legal, organizational, ethical, and technical challenges that complicate its widespread adoption. One of the most prominent issues is the legal and regulatory uncertainty surrounding the use of automation in tax compliance. While tax authorities in some jurisdictions encourage digital transformation, there is often ambiguity in how automated processes should be treated under existing tax laws. For instance, regulatory bodies may lack clear guidance on whether RPA generated tax filings carry the same legal standing as those prepared by human tax professionals. Moreover, variations in tax regulations across countries further complicate the development of standardized RPA solutions, raising compliance risks for multinational corporations (Dudu et al., 2024).

Data accuracy and integrity are also pressing concerns in automated tax processes. Since RPA systems rely heavily on the quality of input data, any errors or inconsistencies in source systems such as ERP platforms or legacy databases can lead to flawed outputs. Although RPA eliminates human keystroke errors, it is not immune to the "garbage in, garbage out" problem. Inaccurate tax computations, misfiled reports, or overlooked regulatory requirements can occur if the automation scripts are not meticulously tested and continuously monitored. Ensuring data integrity thus requires robust governance frameworks, including validation rules, exception handling, and audit trails.

Organizational resistance to change is another significant barrier to RPA adoption in tax functions. Employees may perceive automation as a threat to job security, leading to skepticism, reduced cooperation, or even passive resistance. In the context of tax departments traditionally conservative and risk-averse the introduction of bots can be viewed with suspicion. Overcoming this resistance demands effective change management strategies, clear communication about the benefits of automation, and the retraining of staff for higher-value roles. Research indicates that successful RPA implementation is strongly correlated with a collaborative culture and leadership commitment (Tiron-Tudor et al., 2021).

Ethical implications also arise when automating compliance functions. Delegating responsibilities to non-human agents introduces questions about accountability, transparency, and fairness. Who is liable if a bot makes an error in tax reporting that leads to legal consequences? How can taxpayers ensure that automated decisions comply with principles of justice and due process? These concerns are compounded when RPA is integrated with AI, as it becomes more difficult to trace the logic behind certain decisions. To address these concerns, organizations must develop ethical frameworks that define the scope and boundaries of automation, establish human oversight mechanisms, and ensure that technology complements rather than replaces human judgment (Yoon et al., 2021).

Figure 3. Issues Implementing RPA in Taxation

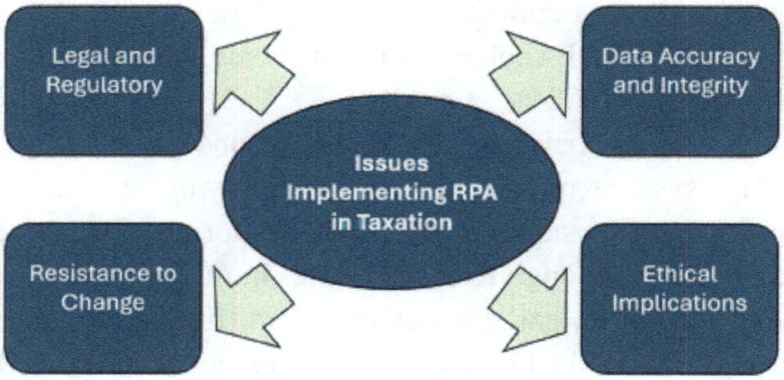

In summary, while RPA presents considerable opportunities for improving tax compliance and reporting, it is not without its challenges. Legal ambiguity, data quality issues, organizational inertia, and ethical dilemmas all pose risks that must be proactively managed. A nuanced understanding of these issues is critical for the responsible and sustainable adoption of automation in tax functions.

SOLUTIONS AND RECOMMENDATIONS

Conceptual Framework for RPA Integration in Tax Compliance

The integration of Robotic Process Automation (RPA) in tax compliance represents a paradigm shift in how organizations handle complex and regulated workflows. As businesses grapple with increasing compliance burdens, evolving regulatory landscapes, and the need for greater operational efficiency, a strategic approach to RPA implementation becomes imperative. The conceptual framework presented here outlines a holistic model for integrating RPA into tax functions, covering the technological, organizational, and governance dimensions essential for sustainable automation.

Strategic Alignment and Objectives Definition

Any RPA initiative should begin with a clear articulation of strategic objectives. This includes aligning automation goals with the broader tax function strategy be it enhancing compliance accuracy, reducing operational costs, or improving responsiveness to regulatory changes. Organizations must also define measurable key performance indicators (KPIs) to evaluate RPA's contribution to these objectives. This strategic alignment ensures that RPA is not merely viewed as a tactical tool but as a core enabler of digital transformation in the tax domain (Tiron-Tudor et al., 2021).

Process Identification and Prioritization

Identifying and prioritizing suitable tax processes is the cornerstone of effective RPA deployment. A detailed process discovery phase, using tools like process mining and value stream mapping, helps in visualizing current workflows, bottlenecks, and inefficiencies. Tax-related processes that are rule-based, high in volume, and repetitive such as VAT reconciliations, corporate income tax return preparations, transfer pricing documentation, and payroll tax filings are ideal candidates for RPA. Prioritization should also consider regulatory sensitivity, the potential for standardization, and the ease of implementation.

Readiness Assessment and Infrastructure Planning

Before RPA deployment, a thorough readiness assessment should be conducted. This involves evaluating data quality, system architecture, and employee digital literacy. Organizations must ensure that tax data is structured, validated, and readily accessible through ERP systems, tax engines, or document repositories. Infrastructure planning also includes selecting the appropriate RPA tools whether attended or unattended bots, cloud-based or on-premise solutions based on scalability, integration capability, and vendor support (KPMG, n.d).

Governance and Compliance Integration

Tax functions operate under stringent legal and regulatory constraints. Therefore, the RPA framework must incorporate robust governance structures to ensure compliance. Bots should be programmed with built-in validations, audit trail capabilities, and exception handling protocols. Regulatory requirements such as digital signatures, documentation standards, and e-filing norms must be embedded into the bot's logic. Moreover, establishing a governance committee comprising tax, IT, internal audit, and compliance personnel helps in overseeing automation decisions, managing risks, and ensuring transparency (Deloitte, n.d).

Bot Design, Development, and User Acceptance

Effective bot design requires close collaboration between process owners and automation developers. User stories and process documentation should translate into detailed bot specifications, with emphasis on reliability, flexibility, and regulatory compliance. Agile development methodologies, including iterative prototyping and user feedback loops, improve bot functionality and user acceptance. At this stage, organizations should also create a knowledge repository documenting bot behavior, configuration, and escalation protocols to support maintainability (Kovacev, 2020).

Integration with Existing Systems

RPA should not function in silos but should integrate seamless with existing digital infrastructure. This includes ERP systems, tax accounting platforms, document management systems, and regulatory e-filing portals. Use of APIs (Application Programming Interfaces), data connectors, and middleware enables bots to pull and push data across systems without manual intervention. Integration also facilitates real-time data validation, improves traceability, and enhances the accuracy of compliance reporting (EY, 2020).

Testing, Validation, and Compliance Assurance

Before going live, bots must undergo rigorous testing in a sandbox environment. Testing should cover functional accuracy, compliance adherence, system compatibility, and edge-case scenarios. Simulated test runs can help identify potential breakdowns or non-compliance issues before full-scale deployment. Organizations should also engage internal auditors or external consultants to validate bot performance and certify their compliance readiness. Compliance assurance frameworks can be adapted to evaluate bot governance (PwC, nd).

Monitoring, Maintenance, and Continuous Improvement

Post-deployment, bots require ongoing monitoring to ensure performance stability, error resolution, and alignment with evolving regulations. Dashboards and analytics tools provide visibility into bot activities, performance metrics, and compliance status. A dedicated bot maintenance team should manage version control, patch updates, and changes in tax laws or internal policies. Organizations should also establish feedback loops for end-users to report issues and suggest improvements, supporting a culture of continuous improvement.

Workforce Enablement and Change Management

Human factors play a crucial role in RPA success. Employees must be prepared to work alongside bots through effective training and role redefinition. Change management strategies should include awareness campaigns, stakeholder engagement, and recognition of employee contributions. The shift from manual tax processing to automated compliance also provides an opportunity to upskill tax professionals in analytics, bot supervision, and strategic advisory roles (Dudu et al., 2024).

Scalability and Future Readiness

Finally, the framework should accommodate future scalability and innovation. RPA can serve as a gateway to more intelligent automation, including AI, machine learning (ML), and natural language processing (NLP). For example, AI-enhanced bots could interpret regulatory updates, flag anomalies in tax filings, or generate narrative explanations in audit reports. A modular framework allows organizations to expand RPA across additional tax jurisdictions or integrate with global tax planning tools. Long-term planning should also consider sustainability, cost-benefit analysis, and alignment with enterprise-wide digital transformation goals (OECD, 2023).

This comprehensive conceptual framework offers a structured, scalable, and compliant approach for integrating RPA into tax compliance functions. It emphasizes not just the technical aspects of bot deployment but also the strategic, organizational, and ethical considerations essential for successful implementation. By adopting this framework, tax departments can not only automate routine compliance tasks but also position themselves as forward-looking partners in organizational value creation.

FUTURE RESEARCH DIRECTIONS AND IMPLICATIONS

As Robotic Process Automation (RPA) continues to reshape tax compliance and reporting processes, numerous avenues for future research and practical exploration emerge. The growing complexity of global tax environments, coupled with rapid advancements in digital technologies, necessitates an ongoing academic and professional inquiry into the long-term implications, limitations, and opportunities of RPA adoption. This section outlines several key research directions and implications that warrant further investigation.

Integration with Artificial Intelligence and Cognitive Automation

While RPA currently excels at rule-based tasks, future research should explore the integration of RPA with cognitive technologies such as Artificial Intelligence (AI), Machine Learning (ML), and Natural Language Processing (NLP). These integrations can enable bots to process unstructured data, interpret regulatory texts, and even make adaptive decisions based on context. Research can focus on developing hybrid models where RPA handles data extraction and task execution, while AI models support decision-making, anomaly detection, and narrative generation in tax reporting.

Cross Jurisdictional Tax Compliance

Global organizations face the challenge of complying with diverse tax regulations across jurisdictions. Future research could explore how RPA can be configured to manage multi-jurisdictional compliance while maintaining consistency, transparency, and regulatory accuracy (OECD, 2023). Comparative studies between jurisdictions with differing tax laws and digital maturity levels may uncover best practices and standardization opportunities. Additionally, studies can assess how international bodies, such as the OECD, can guide cross-border tax automation efforts.

Longitudinal Impact Studies and ROI Measurement

Although early studies highlight cost savings and efficiency gains from RPA implementation, there is limited empirical data on its long-term effects. Future research should undertake longitudinal studies to assess the sustained impact of RPA on tax department performance, workforce transformation, and risk management. Metrics such as return on investment (ROI), error rate reduction, employee satisfaction, and audit readiness should be evaluated over extended periods to provide robust evidence of RPA's value proposition.

Ethical and Accountability Considerations

Automation in tax compliance raises ethical questions related to accountability, data privacy, and decision transparency. Further research is needed to establish ethical frameworks for RPA use, particularly when integrated with AI. These frameworks should address issues such as bot accountability, transparency of automated decisions, and the potential for algorithmic bias in tax assessments. Legal studies can explore how existing tax laws and audit standards need to evolve to accommodate machine-led compliance processes.

Workforce Transformation and Skills Development

As RPA changes the nature of work in tax functions, research should examine how roles, skills, and organizational structures are evolving. Studies can explore the effectiveness of reskilling programs, new competency models for tax professionals, and the psychological impacts of automation on job satisfaction and performance. The role of academic institutions and professional bodies in embedding digital competencies into accounting and taxation curricula also warrants investigation (Tiron-Tudor et al., 2021).

Risk, Resilience, and Cybersecurity

The increasing reliance on digital bots introduces new operational risks. Research is needed to assess the resilience of RPA systems in the face of cyber threats, system failures, and regulatory changes. The development of RPA-specific risk management and cybersecurity frameworks could help organizations proactively mitigate vulnerabilities. Case studies documenting bot malfunctions, audit failures, or data breaches in tax functions would provide valuable lessons for future implementations (Deloitte, n.d).

Blockchain-RPA Integration for Tax Transparency

An emerging area of interest is the integration of RPA with blockchain technology to enhance tax transparency, auditability, and fraud prevention. Blockchain can provide an immutable ledger of tax transactions, while RPA can automate data input, reconciliation, and reporting. Research could examine the feasibility, security implications, and adoption barriers of such integrations, particularly in industries with complex supply chains or high regulatory scrutiny (EY, 2020).

Regulatory Technology Synergies

RPA is increasingly viewed as a component of the broader Regulatory Technology (RegTech) ecosystem. Future studies can explore how RPA interfaces with other RegTech tools, such as e-invoicing systems, tax analytics platforms, and compliance dashboards. Research may also investigate how tax authorities are leveraging RPA and RegTech to improve taxpayer services, reduce fraud, and enhance real-time compliance monitoring (KPMG, n.d).

Social and Economic Implications

On a macro level, the adoption of RPA in taxation may influence employment patterns, government revenue collection, and digital inequality. Researchers could assess the socioeconomic impacts of automation on tax administration in developing economies versus advanced economies. Additionally, public policy implications, such as incentives for automation or digital taxation reforms, should be examined to understand how governments can support or regulate this technological shift.

Curriculum Transformation in Accounting and Tax Education

The increasing integration of automation tools in tax practice necessitates a corresponding evolution in accounting and taxation education. Future research should evaluate current academic programs and identify gaps in preparing students for a digital tax environment. Studies can also propose instructional frameworks and case-based pedagogies that incorporate hands-on RPA training, interdisciplinary learning, and ethical decision-making in automated contexts (Kovacev, 2020).

The dynamic landscape of tax automation presents a rich agenda for interdisciplinary research. As RPA matures and converges with emerging technologies, its implications will extend beyond operational efficiency to influence governance, workforce strategy, public policy, and education. Addressing these research gaps will not only refine the implementation of RPA in tax compliance but also contribute to the broader discourse on responsible and intelligent automation.

IMPLICATIONS FOR EDUCATORS, PRACTITIONERS, AND POLICYMAKERS

The increasing integration of Robotic Process Automation (RPA) into tax functions carries profound implications for educators, practitioners, and policymakers. As RPA reshapes the operational landscape of tax compliance, these stakeholder groups must adapt to ensure that tax professionals are adequately prepared, regulatory environments remain robust, and technological innovation continues to benefit society as a whole. Expanding this discussion reveals specific responsibilities, opportunities, and strategies for each group.

For educators, RPA signals a clear call for curriculum innovation. Accounting and taxation education must evolve beyond traditional bookkeeping, manual tax return preparation, and static regulatory instruction. Instead, programs should incorporate digital fluency, automation awareness, and hands-on learning with RPA tools. Universities and training institutes should embed modules on tax technologies, process automation, and digital transformation into their syllabi. This includes simulation-based learning using platforms such as UiPath, Blue Prism, or Automation Anywhere, where students can automate basic tasks such as invoice matching or tax reconciliation.

In addition, interdisciplinary learning blending taxation, data science, and information systems will be vital. This equips students with an integrated perspective, encouraging critical thinking about how automation influences strategic decision-making, ethics, and compliance. Educators must also emphasize the ethical and regulatory implications of automation, preparing students to think beyond efficiency and consider accountability, transparency, and fairness in automated tax processes. Collaborations with industry partners, internships, and guest lectures by automation professionals can bridge the academia-industry gap and ensure that graduates are workplace-ready.

Educators will also benefit from professional development. Faculty training on emerging digital tools, certifications in automation platforms, and participation in industry conferences can help maintain curriculum relevance. Accreditation bodies and professional accounting organizations should work collaboratively to define core digital competencies and mandate their inclusion in accounting and tax qualifications.

For practitioners, RPA represents a transformative opportunity but only if it is met with strategic foresight and agility. Traditional roles in tax departments that focused on manual data entry, reconciliation, and form-filling are being phased out. In their place, professionals must take on higher-value responsibilities such as bot oversight, tax data analytics, process redesign, and interpretation of real-time compliance dashboards.

This transition demands a new set of skills. Practitioners must be trained not only in RPA software but also in complementary digital tools, including Excel macros, data visualization platforms (e.g., Power BI, Tableau), and tax analytics engines. Lifelong learning becomes essential. Firms must invest in training programs, digital academies, and micro-credentialing pathways that enable employees to reskill and remain competitive. In addition, tax professionals should embrace a mindset of innovation and experimentation actively identifying automation opportunities within their processes and participating in RPA project design and testing.

Organizations should foster cross-functional collaboration between tax, finance, IT, and compliance departments. Successful RPA initiatives often require interdisciplinary cooperation to address system integration, risk management, and data governance challenges. Creating internal centers of excellence (CoEs) dedicated to automation can help institutionalize best practices, monitor compliance, and support a scalable approach to RPA. Furthermore, practitioners must prepare for new ethical and regulatory responsibilities. As bots take over compliance tasks, tax professionals will be accountable for ensuring that bots are programmed with accurate logic, remain updated with regulatory changes, and generate audit-ready documentation. Oversight of bot performance and exception management will become critical components of the modern tax professional's role.

For policymakers, the widespread adoption of RPA in tax administration presents both an opportunity and a regulatory challenge. On one hand, automation can increase compliance efficiency, reduce administrative burdens, and improve revenue collection. Governments should therefore promote digital adoption by incentivizing RPA investments through grants, tax deductions, or public-private innovation programs. On the other hand, regulators must ensure that automation does not compromise accountability, fairness, or taxpayer rights. Policymakers need to update tax laws, audit procedures, and digital filing requirements to accommodate machine-generated submissions and algorithm-driven compliance. Clear guidance must be issued on bot-generated evidence, electronic audit trails, and the legal standing of automated filings. Policymakers should also promote transparency in automated decision making. Mandatory disclosures about the use of RPA in tax processes can help maintain trust between taxpayers and authorities. Regulatory sandboxes-controlled environments where businesses can test new technologies under regulatory supervision can support innovation while managing systemic risk.

The broader socio-economic implications of automation also warrant policy attention. Automation could displace lower-skilled jobs in tax administration and create digital divides between large corporations with automation capabilities and smaller firms that lack resources. Policymakers must address these risks through targeted workforce reskilling initiatives, digital literacy campaigns, and equitable access to automation solutions. Support for SMEs in adopting RPA via subsidized tools, training vouchers, and shared service models can help level the playing field. In addition, international collaboration is vital. Tax compliance is increasingly global, and differences in RPA adoption or regulation across countries could create asymmetries and compliance arbitrage. Global standard-setting bodies such as the OECD and IFAC should coordinate efforts to harmonize best practices, data standards, and tax technology frameworks.

To summarize, the implications of RPA in taxation extend far beyond process efficiency. For educators, the focus must be on producing digitally fluent graduates equipped for a technology-driven tax landscape. For practitioners, the imperative is to embrace continuous learning, innovation, and cross-functional collaboration. For policymakers, it is essential to modernize tax frameworks, promote equitable digital adoption, and uphold transparency and accountability in an era of intelligent automation. By taking proactive steps across these domains, stakeholders can collectively ensure that the integration of RPA into taxation not only boosts operational performance but also reinforces trust, inclusivity, and professionalism in the tax ecosystem.

CONCLUSION

The integration of Robotic Process Automation (RPA) into tax compliance and reporting signifies a transformative shift in how tax functions operate in the digital age. As highlighted throughout this chapter, RPA's capacity to automate high-volume, rule-based, and repetitive tax processes offers organizations a unique opportunity to increase efficiency, accuracy, and responsiveness in an increasingly complex and demanding regulatory environment.

The conceptual framework presented demonstrates that successful RPA integration extends beyond technical deployment. It encompasses strategic alignment with organizational goals, rigorous process selection, governance, compliance assurance, system integration, and robust change management. By following a structured, scalable, and ethically sound approach, organizations can maximize the value of automation while minimizing implementation risks. Real-world implementations across corporate tax, indirect tax, and transfer pricing further underscore RPA's versatility and practical applicability.

A critical takeaway is that RPA is not an isolated technology but part of a broader digital transformation ecosystem. Its synergies with cognitive automation, AI, and blockchain hold potential for smarter, more transparent, and adaptive tax operations. This convergence invites organizations to rethink their tax strategies and embrace innovation not only as a means of achieving operational excellence but also as a pathway to strategic advantage.

Moreover, RPA contributes significantly to improving regulatory compliance and audit readiness. Automated bots with embedded validation logic, audit trails, and exception handling reduce the likelihood of errors, fraud, and regulatory breaches. By ensuring timely and accurate reporting, RPA supports organizations in maintaining tax transparency, mitigating compliance risks, and enhancing their relationship with tax authorities. As global tax regimes continue to digitize and adopt real-time reporting mechanisms, RPA will become indispensable in enabling businesses to comply with digital mandates and evolving e-filing protocols.

However, the journey toward automation is not without challenges. Issues such as legal ambiguity, data quality, change resistance, and cybersecurity concerns persist. Ethical implications also arise, particularly around bot accountability and the socio-economic consequences of job displacement. Addressing these concerns requires organizations to embed risk management, governance, and inclusive design principles into their automation strategies. Transparent communication, user empowerment, and stakeholder engagement are critical to fostering trust in automated systems.

The chapter also underscores the importance of human capital in realizing RPA's full potential. Far from rendering tax professionals obsolete, automation redefines their roles. It shifts the focus from manual processing to strategic analysis, oversight, and continuous improvement. Upskilling initiatives and digital literacy programs are essential to equip tax professionals with the competencies needed to thrive in an automated environment. The future tax workforce will need to blend domain expertise with technological fluency, data interpretation, and critical thinking.

Educators, practitioners, and policymakers each play pivotal roles in this transformation. Educators must modernize curricula to align with the digital realities of tax practice. Practitioners must embrace innovation and champion continuous learning. Policymakers must modernize tax regulations, ensure ethical safeguards, and promote equitable access to automation tools. Together, these stakeholders can shape a tax ecosystem that is resilient, inclusive, and future-ready.

Looking forward, the trajectory of RPA in taxation suggests continued evolution. As technologies become more sophisticated, automation will expand from structured to semi-structured and unstructured tax tasks. AI-enhanced bots may soon interpret regulatory texts, detect compliance anomalies, and generate tax narratives. Meanwhile, integration with blockchain could further enhance transparency, auditability, and trust in automated tax records. These developments necessitate ongoing research and adaptive policy frameworks to ensure that innovation aligns with public interest and professional integrity.

Another area of potential lies in international cooperation. Tax automation efforts must be harmonized across jurisdictions to prevent regulatory fragmentation, compliance arbitrage, and cross-border inefficiencies. Global standards, data protocols, and interoperability guidelines will be essential to support seamless, secure, and scalable RPA adoption across multinational tax operations. Institutions such as the OECD, IFAC, and World Bank can play a coordinating role in promoting inclusive digital transformation.

Ultimately, RPA presents a compelling case for reimagining tax functions in the digital era. It offers an avenue to elevate compliance from a reactive obligation to a proactive, intelligent, and strategic process. The benefits go beyond cost savings to include enhanced agility, data-driven decision-making, improved regulatory relationships, and elevated stakeholder trust. When implemented thoughtfully, RPA not only automates tasks but also transforms capabilities, reshapes roles, and unlocks new value for organizations and society alike.

In conclusion, Robotic Process Automation represents both a catalyst and a compass for the future of taxation. By integrating automation with strategy, ethics, education, and policy, stakeholders can harness its full potential while navigating its complexities responsibly. The journey toward intelligent tax compliance has only just begun but with the right frameworks, collaboration, and foresight, it holds the promise of a more efficient, transparent, and equitable tax future.

REFERENCES

Bassey, E., Mulligan, E., & Ojo, A. (2022). A conceptual framework for digital tax administration-A systematic review. *Government Information Quarterly*, *39*(4), 101754. DOI: 10.1016/j.giq.2022.101754

Belahouaoui, R., & Attak, E. H. (2024). Digital taxation, artificial intelligence and Tax Administration 3.0: Improving tax compliance behavior–a systematic literature review using textometry (2016–2023). *Accounting Research Journal*, *37*(2), 172–191. DOI: 10.1108/ARJ-12-2023-0372

Deloitte. (2025). *Digitalization of indirect tax demands a holistic approach.* Retrieved from https://www.deloitte.com/global/en/services/tax/perspectives/digitalization-indirect-tax.html

Deloitte. (n.d). *RPA for Tax.* Retrieved from https://www.deloitte.com/an/en/services/tax/services/rpa-for-tax.html

Doguc, O. (2021). Applications of robotic process automation in finance and accounting. *Beykent Üniversitesi Fen ve Mühendislik Bilimleri Dergisi*, *14*(1), 51–59. DOI: 10.20854/bujse.906795

Dudu, O. F., Alao, O. B., & Alonge, E. O. (2024). Conceptual framework for AI-driven tax compliance in fintech ecosystems. *Journal of Fintech and Taxation*, *12*(3), 45–60. DOI: 10.53294/ijfetr.2024.7.2.0045

Grant Thornton. (2022). *Grant Thornton launches its new "Digital" division.* Retrieved from https://www.bgt-grantthornton.it/en/insights/articles/2022/grant-thornton-launches-its-new-digital-division/

IMA. (2021). *Transforming the Finance Function with RPA.* Retrieved from https://asiapac.imanet.org/research-publications/statements-on-management-accounting/transforming-the-finance-function-with-rpa?utm_source=chatgpt.com

ISACA. (2020). *Implementing Robotic Process Automation (RPA): Trends in RPA Adoption, Uses, and Implementation Challenges.* Retrieved from chrome-extension://efaidnbmnnnibpcajpcglclefindmkaj/https://www.isaca.org/-/media/files/isacadp/feature/downloads/i/implementing-robotic-process-automation_whprpa_whp_eng_0920.pdf

Kovacev, R. (2020). A taxing dilemma: robot taxes and the challenges of effective taxation of AI, automation and robotics in the fourth industrial revolution. *Ohio St. Tech. LJ*, *16*, 182. Retrieved from https://heinonline.org/HOL/Page?collection=journals&handle=hein.journals/isjlpsoc16&id=185&men_tab=srchresults

KPMG. (n.d). *The technologies enabling digital transformation in tax* Retrieved from https://kpmg.com/dp/en/home/insights/2024/05/technologies-enabling-digital-transformation-in-tax.html

McKinsey & Company. (2023). *The CFO's guide for digitizing the finance function.* Retrieved from https://www.mckinsey.com/~/media/mckinsey/email/classics/2023/2023-08-19c.html

OECD. (2022). *Digital Transformation Maturity Model for Tax Administrations.* Retrieved from chrome-extension://efaidnbmnnnibpcajpcglclefindmkaj/https://www.oecd.org/content/dam/oecd/en/topics/policy-issues/tax-administration/digital-transformation-maturity-model.pdf?utm_source=chatgpt.com

OECD. (2023). *Tax Administration 2023: Comparative Information on OECD and other Advanced and Emerging Economies.* Retrieved from https://www.oecd.org/en/publications/2023/09/tax-administration-2023_87655bc9.html

PwC Malaysia. (n.d). *Tax Technology*. Retrieved from https://www.pwc.com/my/en/services/tax/tax-technology.html

Tiron-Tudor, A., Deliu, D., Farcane, N., & Dontu, A. (2021). Managing change with digital technology: Evidence from the Big Four accounting firms. *Journal of Accounting and Organizational Change*, *17*(1), 32–52. DOI: 10.1108/JOCM-10-2020-0302

World Bank. (2024). *Leveraging digitalization to improve tax compliance and revenue mobilization.* Retrieved from chrome-extension://efaidnbmnnnibpcajpcglclefindmkaj/https://documents1.worldbank.org/curated/en/099061225023042590/pdf/P179405-6877e00c-7f93-46b1-8d09-d72ad0afd311.pdf

Yoon, K., Liu, Y., Chiu, T., & Vasarhelyi, M. A. (2021). Design and evaluation of an advanced continuous data level auditing system: A three-layer structure. *International Journal of Accounting Information Systems*, *42*, 100524. DOI: 10.1016/j.accinf.2021.100524

KEY TERMS AND DEFINITIONS

Automation in Accounting: Automation in accounting improves processes including efficiency, accuracy, and speed in preparing accounting reports such as financial reporting, data entry, and reconciliation. The use of software and algorithms will perform accounting tasks and reduce human errors.

Corporate Taxation: A system of taxation imposed on the profits earned by corporations. It encompasses the assessment, collection, and reporting of taxes at the corporate level, following national tax laws and international standards.

Robotic Process Automation (RPA): A form of business process automation that uses software "robots" to replicate repetitive, rule-based tasks typically performed by humans. In accounting and taxation, RPA can automate data extraction, validation, filing, and reporting functions.

Tax Compliance: The adherence to tax laws and regulations by accurately reporting income, calculating tax liabilities, and submitting payments within prescribed deadlines. Effective tax compliance minimizes legal risks and supports transparent governance.

Tax Education: An area of education focused on imparting knowledge of tax laws, regulations, ethics, and practices. It prepares students and professionals to understand, apply, and comply with tax requirements in personal and corporate contexts.

Tax Reporting: The preparation and submission of accurate financial and transactional information to tax authorities. Tax reporting ensures that taxpayers disclose relevant details of their taxable activities in accordance with applicable laws and accounting standards.

Taxation: The process by which a government imposes financial charges or levies on individuals, businesses, and other entities to generate revenue for public expenditure. Taxation systems may include income tax, corporate tax, sales tax, and property tax.

Chapter 3
The Impact of Robotic Process Automation on Management Accounting Practices

Xiaoli Yu
https://orcid.org/0009-0006-9906-729X
Yiwu Industrial and Commercial College, China

ABSTRACT

Robotic Process Automation (RPA) is revolutionizing management accounting by automating repetitive, rule-based tasks, enhancing efficiency, accuracy, and compliance across functions like budgeting, forecasting, and financial reporting. This chapter examines RPA's transformative impact, achieving up to 30% cost reductions and 90% error reductions, as evidenced by implementations at firms like Walmart and Ping An Insurance. By leveraging bot development tools, AI integration, and cloud scalability, RPA streamlines processes, enabling accountants to shift from transactional tasks to strategic roles such as data analysis and decision support. However, challenges, including implementation costs of $100,000–$500,000, legacy system integration complexities, and ethical concerns about job displacement with 60% of accountants citing job security fears, require strategic solutions like retraining and robust IT support. The chapter explores RPA's global adoption, with examples from China, the U.S., Australia, Germany, India, and Japan, highlighting regional priorities like compliance and scalability. Future trends, such as integration with AI, machine learning, and cognitive RPA, promise advanced predictive analytics, with 70% of firms planning adoption by 2026. This study provides actionable insights for educators, practitioners, and organizations, advocating a hybrid model

DOI: 10.4018/979-8-3373-3571-1.ch003

that balances automation with human judgment to ensure competitiveness in the digital era. It serves as a guide for navigating RPA's opportunities and challenges in management accounting..

INTRODUCTION

Robotic Process Automation (RPA) has ushered in a new era for management accounting, revolutionizing traditional practices by automating routine, rule-based tasks and fostering a shift toward strategic contributions. RPA, characterized as software that mimics human interactions with digital interfaces to execute structured processes without requiring extensive IT infrastructure changes (Santos et al., 2021, as cited in Ayinla et al., 2024), streamlines activities such as data entry, account reconciliations, budgeting, and performance reporting (Tiron-Tudor et al., 2023). Unlike conventional automation, which demanded significant system overhauls, RPA's non-invasive integration with existing systems makes it a versatile solution for organizations of all scales (Doğuç, 2021; Bright et al., 2023). Research demonstrates RPA's efficacy, with firms reporting up to 30% cost reductions and 90% error reductions in tasks like reconciliations (Balamurugan et al., 2022; Zhang et al., 2023). For example, a multinational firm leveraging UiPath reduced invoice processing time by 65%, showcasing tangible operational benefits (Jędrzejka, 2022). Yet, challenges such as implementation costs of $100,000–$500,000 for mid-sized firms, integration hurdles with legacy systems, and workforce apprehensions about job displacement highlight the complexity of RPA adoption (Asatiani et al., 2020; Zhang et al., 2023). As management accounting adapts to a technology-driven landscape, understanding RPA's multifaceted impact is essential for equipping accountants with the tools and skills needed to thrive and ensuring organizations remain competitive.

This chapter aims to provide a thorough examination of RPA's transformative effects on management accounting, analyzing its technical foundations, practical applications, benefits, challenges, and future directions. It focuses on RPA's role in key functions such as budgeting, forecasting, cost management, and performance reporting, while also exploring its influence on the evolving responsibilities of management accountants (Feleaga & Dumitrascu, 2020; Tiron-Tudor et al., 2023). Drawing on 15 peer-reviewed studies from 2020 to 2025, indexed in SCI/EI databases, the chapter synthesizes empirical and theoretical perspectives to assess how RPA drives efficiency, accuracy, and compliance, while addressing barriers like system integration and ethical concerns (Cooper et al., 2022; Bright et al., 2023). It also investigates emerging trends, such as RPA's integration with AI and ML, which enable advanced capabilities like predictive analytics (Plattfaut & Borghoff, 2022). Designed for accounting professionals, educators, and organizational leaders, this chapter offers actionable insights to navigate the digital transformation of management accounting.

The study of RPA in management accounting is vital due to its profound implications for operational efficiency, strategic repositioning, ethical considerations, and professional development. Operationally, RPA automates repetitive tasks, significantly reducing processing times and errors in critical functions like budgeting and financial reporting. For instance, Bellinga et al. (2022) report a 65% reduction in data extraction time for audit-related tasks, a benefit directly applicable to management accounting processes like reconciliations, enabling faster and more reliable decision-making. Similarly, Balamurugan et al. (2022) note that RPA achieves up to 90% error reduction in reconciliations, ensuring precision in cost analysis and performance metrics, which are essential for accurate financial insights. These improvements meet the growing demand for real-time, data-driven insights in complex, globalized business environments, where multinational firms rely on RPA to consolidate financial data across regions efficiently (Ayinla et al., 2024; Doğuç, 2021).

Strategically, RPA redefines the role of management accountants, shifting their focus from routine data processing to high-value activities like financial analysis, scenario planning, and strategic advising. Kokina and Blanchette (2020) emphasize that this transition elevates accountants to strategic partners, enhancing organizational competitiveness through insights into cost optimization and performance improvement. For example, RPA's automation of variance analysis enables accountants to interpret cost trends, supporting high-stakes decisions like mergers and acquisitions (Tiron-Tudor et al., 2023). Compliance is also strengthened, as RPA's automated audit trails ensure adherence to regulations like Sarbanes-Oxley (SOX) and International Financial Reporting Standards (IFRS), reducing risks through transparent, auditable processes (Bright et al., 2023; Eulerich et al., 2022).

Ethically, RPA's potential to displace entry-level roles raises critical concerns about workforce sustainability. Jędrzejka (2022) notes that automation of tasks like data entry may reduce demand for junior accountants, with Asatiani et al. (2020) reporting that 60% of accounting professionals in a Finnish firm expressed job security concerns. This necessitates proactive measures, such as retraining programs to transition employees to analytical or advisory roles, balancing efficiency gains with social responsibility (Kokina & Blanchette, 2020). Cooper et al. (2022) advocate for change management strategies, including training and transparent communication, to address resistance and align workforce expectations.

Educationally, RPA's integration demands new competencies, making its study essential for preparing future accountants. Lacurezeanu et al. (2020) highlight the need for digital literacy, as accountants must master tools like UiPath and Blue Prism and collaborate with IT and data science teams. Ayinla et al. (2024) note that 75% of accounting firms required RPA-related certifications, such as UiPath RPA

Developer, for senior roles by 2023, underscoring the importance of continuous learning. The synergy of RPA with AI and ML further amplifies its relevance, enabling predictive analytics with up to 85% accuracy in forecasting cash flow trends, enhancing strategic decision-making (Zhang et al., 2023; Plattfaut & Borghoff, 2022). By addressing these operational, strategic, ethical, and educational dimensions, this chapter highlights RPA's transformative potential and prepares stakeholders for its challenges and opportunities.

The introduction positions RPA as a transformative technology in management accounting, automating tasks like reconciliations to achieve up to 30% cost savings and 90% error reductions (Balamurugan et al., 2022; Zhang et al., 2023). Its purpose is to analyze RPA's applications, benefits, and challenges in functions like budgeting and forecasting, while exploring its impact on accountants' roles (Feleaga & Dumitrascu, 2020; Tiron-Tudor et al., 2023). Studying RPA is critical for its operational efficiencies, strategic repositioning of accountants, compliance enhancements, and ethical challenges, with 60% of accountants expressing job displacement concerns (Asatiani et al., 2020). Educationally, 75% of firms required RPA certifications by 2023, underscoring the need for digital literacy (Ayinla et al., 2024). The chapter is structured into an abstract, introduction, literature review, main focus, future research directions, conclusion, references, and key terms, providing a comprehensive guide for stakeholders (Plattfaut & Borghoff, 2022; Jędrzejka, 2022).

This chapter is organized to provide a systematic exploration of RPA's impact on management accounting, grounded in 15 SCI/EI-indexed studies from 2020 to 2025. The Introduction establishes the context, defining RPA, its significance, purpose, scope, and structure, supported by empirical insights (Balamurugan et al., 2022; Jędrzejka, 2022). The Literature Review synthesizes research on RPA's definition, components, applications, and benefits, alongside the evolution of management accounting (Bellinga et al., 2022; Lacurezeanu et al., 2020). The Main Focus examines implementation challenges, such as costs and employee resistance, and the evolving role of accountants toward strategic advising (Asatiani et al., 2020; Cooper et al., 2022). Future Research Directions explore trends like AI and ML integration, intelligent automation, and predictive analytics (Plattfaut & Borghoff, 2022; Tiron-Tudor et al., 2023). The Conclusion synthesizes findings, offering implications for stakeholders and emphasizing the balance between automation and human expertise (Jędrzejka, 2022).

LITERATURE REVIEW

The integration of technology into accounting has evolved significantly over the past few decades, transitioning from manual ledger-based systems to sophisticated digital solutions. The introduction of Enterprise Resource Planning (ERP) systems in the 1990s marked a significant shift, enabling centralized data management and process automation (Cooper et al., 2022). However, ERP systems often required substantial investments and complex integrations, limiting their accessibility for smaller organizations (Doğuç, 2021). The advent of cloud computing in the 2000s further enhanced data accessibility and analytics, setting the stage for more advanced automation technologies (Tiron-Tudor et al., 2023). RPA, emerging in the 2010s, represents a paradigm shift by offering a non-invasive, user-friendly approach to automation. Unlike traditional systems, RPA operates at the user interface level, mimicking human actions like data entry and report generation without requiring extensive coding or system overhauls (Santos et al., 2021, as cited in Ayinla et al., 2024). This accessibility has made RPA particularly relevant for management accounting, where repetitive tasks, such as budgeting and variance analysis, are prevalent (Feleaga & Dumitrascu, 2020). The rise of globalization, regulatory complexity (e.g., SOX, IFRS), and stakeholder demands for real-time insights have further driven RPA's adoption, as it enables rapid, accurate data processing and compliance (Bright et al., 2023). This chapter builds on this historical context, examining how RPA is reshaping management accounting practices and addressing the operational, strategic, and ethical challenges of its implementation.

Understanding Robotic Process Automation (RPA)

Definition and core components of RPA

Robotic Process Automation (RPA) is defined as software that emulates human interactions with digital systems to execute repetitive, rule-based tasks without requiring significant modifications to existing IT infrastructure (Santos et al., 2021, as cited in Ayinla et al., 2024). Unlike traditional automation, which often involves custom-built software and extensive system overhauls, RPA operates at the user interface level, mimicking actions such as clicking, typing, and data extraction (Doğuç, 2021). Ayinla et al. (2024) highlight that RPA's core components include bot development tools (e.g., UiPath, Blue Prism), workflow automation capabilities, integration with existing systems, and scalability to handle multiple processes. These features enable RPA to automate high-volume, structured tasks efficiently, making it particularly suitable for accounting applications (Lacurezeanu et al., 2020).

The integration of Robotic Process Automation (RPA) into management accounting has garnered significant attention in recent academic and professional discourse, driven by its potential to transform traditional accounting practices. This literature review synthesizes findings from peer-reviewed studies published between 2020 and 2025, focusing on RPA's definition, core components, and applications in management accounting, the evolution of management accounting practices, and the benefits and challenges of RPA adoption. The review draws on 15 SCI/EI-indexed sources to provide a robust foundation for understanding RPA's impact on the accounting profession, particularly in areas such as budgeting, forecasting, cost management, and performance reporting.

Robotic Process Automation (RPA) is software technology that deploys bots to emulate human interactions with digital systems, automating repetitive, rule-based tasks without requiring significant changes to existing IT infrastructure (Santos et al., 2021, as cited in Ayinla et al., 2024). Operating at the user interface (UI) level, RPA mimics actions like clicking, typing, and data extraction, enabling compatibility with legacy systems (Doğuç, 2021). Recent advancements (2020–2025) have integrated AI, machine learning (ML), optical character recognition (OCR), and natural language processing (NLP), allowing bots to process unstructured data, such as invoices or emails, with up to 95% accuracy compared to 80% for traditional RPA (Zhang et al., 2023; Plattfaut & Borghoff, 2022). RPA's core components, detailed below, enable its transformative impact on management accounting. RPA's core components facilitate efficient, accurate, and scalable automation, revolutionizing management accounting processes.

Bot Development Tools

These tools provide low-code or no-code platforms that enable non-technical users, often referred to as citizen developers, to design automation workflows using drag-and-drop interfaces, significantly reducing development time and costs. For instance, UiPath Studio's visual interface allowed a U.S. healthcare provider to automate patient billing data entry, reducing development time by 60% and enabling accountants to deploy bots without coding expertise (Ayinla et al., 2024). Similarly, Blue Prism's Process Studio enabled a UK bank to create bots for compliance reporting in two weeks, compared to months for traditional coding, saving 50% in development costs (Eulerich et al., 2022). In China, Laiye's intuitive platform helped a retail firm automate inventory data entry, cutting configuration time by 65% and enabling rapid deployment across 100 stores (Zhang et al., 2023). Bot development tools democratize automation, making it accessible to accounting teams with minimal technical skills. These tools streamline bot creation, enhancing RPA's adoption in management accounting by reducing reliance on IT specialists.

Workflow Automation Engine

The workflow automation engine executes predefined scripts to automate repetitive tasks, ensuring consistency and scalability across processes. In a Japanese manufacturing firm, Automation Anywhere's engine automated invoice processing, handling 10,000 invoices monthly with 98% accuracy, reducing manual effort by 70% (Perdana et al., 2022). A German logistics company used Blue Prism's engine to automate order-to-cash workflows, processing 5,000 transactions daily and saving 15,000 hours annually (Cooper et al., 2022). In Australia, Kofax RPA's engine streamlined tax return processing for an accounting firm, cutting processing time by 60% and handling 2,000 returns monthly (Balamurugan et al., 2022). The engine's ability to execute complex workflows reliably enhances efficiency in high-volume accounting tasks. This component ensures precise, scalable automation, driving operational efficiency in management accounting.

System Integration Capabilities

RPA's ability to integrate with enterprise systems via UI interactions or APIs enables seamless data flow across platforms like ERP and CRM systems. A U.S. retailer used UiPath to integrate SAP with legacy inventory systems, automating data transfers and reducing manual entry by 80% (Jędrzejka, 2022). In India, Automation Edge connected Oracle Financials with a custom CRM, enabling a manufacturing firm to automate vendor payment processing, saving 12,000 hours annually (Oyeniyi et al., 2024). A Chinese insurer implemented Laiye RPA to integrate its claims system with external databases, reducing data retrieval time by 75% (Zhang et al., 2023). These capabilities ensure RPA can operate within complex IT environments without requiring costly system overhauls. System integration enhances RPA's versatility, enabling seamless automation across diverse accounting systems.

Artificial Intelligence and Machine Learning Integration

Integrating Artificial Intelligence (AI) and Machine Learning (ML) enhances RPA's ability to handle semi-structured and unstructured data, enabling cognitive tasks like predictive analytics and sentiment analysis. A U.S. pharmaceutical firm used WorkFusion's AI-enhanced RPA to analyze clinical trial data, improving forecasting accuracy by 25% and reducing planning cycles by 50% (Bright et al., 2023). In Japan, Automation Anywhere's ML capabilities enabled a telecom operator to predict customer churn from email data, achieving 90% accuracy and reducing marketing costs by 15% (Perdana et al., 2022). A UK financial institution used Blue Prism's AI tools to process unstructured loan applications, cutting approval times by 60% (Eulerich et al., 2022). AI/ML integration transforms RPA into an intelligent automation tool, expanding its scope in accounting. This component enables RPA to handle complex, data-driven tasks, enhancing decision-making in management accounting.

Cloud-Based Scalability

Cloud-based RPA platforms allow organizations to scale automation dynamically, reducing infrastructure costs and enabling rapid deployment. UiPath's Automation Cloud enabled a U.S. logistics firm to scale invoice processing from 1,000 to 10,000 invoices monthly, reducing costs by 30% (Jędrzejka, 2022). In Australia, Nexlify's cloud platform allowed a retail chain to automate customer service queries across 200 stores, cutting infrastructure costs by 25% (Perdana et al., 2022). A Chinese manufacturer used Encoo's cloud RPA to scale cost analysis across multiple plants, saving 20% in operational costs (Oyeniyi et al., 2024). Cloud scalability ensures flexibility and cost efficiency, making RPA accessible to SMEs and large enterprises alike. This component supports dynamic automation growth, optimizing resource allocation in accounting processes.

Governance and Security Modules

Governance and security modules provide audit trails, encryption, and compliance features to meet regulatory standards like GDPR and SOX. A UK bank used Blue Prism's governance tools to automate compliance reporting, ensuring 100% adherence to SOX and reducing audit preparation time by 60% (Bright et al., 2023). In the U.S., WorkFusion's security modules protected sensitive financial data during transaction monitoring, achieving 99% compliance with GDPR (Eulerich et al., 2022). A Chinese insurer implemented Laiye's audit trails to track claims processing, reducing compliance errors by 90% (Zhang et al., 2023). These modules ensure data integrity and regulatory compliance in accounting tasks. Governance and security features safeguard RPA processes, enhancing trust and compliance in management accounting.

Orchestration Tools

Orchestration tools, such as control rooms, managing multiple bots, optimizing resource allocation, and monitor performance in real-time. Blue Prism's Control Room enabled a German automaker to orchestrate 50 bots for financial reporting, saving 10,000 hours annually (Cooper et al., 2022). In Australia, Kofax RPA's orchestration tools managed bots for tax compliance, processing 3,000 returns monthly with 95% efficiency (Balamurugan et al., 2022). A Japanese bank used Automation Anywhere's Bot Insight to monitor 100 bots for reconciliations, improving performance by 30% (Perdana et al., 2022). Orchestration tools enhance bot coordination and efficiency in complex workflows. These tools streamline bot management, ensuring optimal performance in accounting automation.

Key differences between RPA and traditional automation

RPA differs significantly from traditional automation in its approach, implementation, and adaptability, as evidenced by recent research (2020–2025). Traditional automation relies on deep system integrations, often requiring custom coding, APIs, or database-level changes, which can take months to implement and cost upwards of $1 million for mid-sized firms (Doğuç, 2021). In contrast, RPA operates at the UI level, mimicking human actions without altering backend systems, enabling deployment in weeks and reducing costs by 30–50% (Ayinla et al., 2024). For example, RPA can automate data entry in legacy systems without APIs, whereas traditional automation would require costly system overhauls (Tiron-Tudor et al., 2023). RPA's low-code nature allows business users, or "citizen developers," to configure bots using visual interfaces, reducing reliance on IT departments (Boydaş Hazar & Toplu, 2023). Traditional automation, however, demands extensive programming expertise, limiting its accessibility (Cooper et al., 2022). RPA's flexibility supports both attended (human-triggered) and unattended (scheduled) automation, unlike traditional automation, which is typically unattended and less adaptable to dynamic processes (Bright et al., 2023). Additionally, RPA's integration with AI and ML enables it to handle semi-structured data, such as emails, while traditional automation is limited to structured data (Plattfaut & Borghoff, 2022). However, RPA's reliance on predefined rules makes it less suited for tasks requiring complex decision-making, where traditional automation with AI may excel (Zhang et al., 2023). Studies indicate that RPA achieves a 25–40% higher ROI for repetitive tasks compared to traditional automation due to its rapid deployment and lower costs (Balamurugan et al., 2022). RPA's non-invasive, flexible, and cost-effective approach distinguishes it from traditional automation, making it ideal for management accounting tasks.

Types of tasks RPA can perform

RPA excels in automating repetitive, rule-based tasks that are prevalent in management accounting. Balamurugan et al. (2022) identify tasks such as data entry, account reconciliations, financial reporting, and compliance checks as ideal candidates for RPA due to their structured nature. For example, RPA bots can extract invoice data from emails and input it into ERP systems, reducing manual effort (Doğuç, 2021). Bellinga et al. (2022) further note that RPA's ability to process structured data makes it effective for tasks requiring high accuracy, such as matching transactions across systems.

RPA's versatility in automating diverse tasks, particularly those that are repetitive, rule-based, or high-volume, has expanded significantly with recent advancements

(2020–2025). Below are key task types, supported by case examples and recent research, demonstrating RPA's impact in management accounting.

Data Entry and Migration

RPA automates data input into systems like ERP or CRM, reducing manual effort and errors. A U.S. retailer used UiPath to automate data entry for supplier invoices into SAP, processing 15,000 invoices monthly and reducing manual effort by 80% (Jędrzejka, 2022). In China, Encoo RPA migrated customer data from legacy systems to a new CRM for a retail chain, cutting migration time by 70% and errors by 90% (Oyeniyi et al., 2024). A UK financial firm used Blue Prism to automate data entry for client onboarding, handling 5,000 records monthly with 95% accuracy (Eulerich et al., 2022). RPA's ability to extract and input data across systems streamlines accounting processes. This task type enhances efficiency by automating repetitive data handling in management accounting.

Reconciliations

RPA matches transactions across systems, ensuring financial accuracy with minimal human intervention. A Japanese bank used Automation Anywhere to automate bank reconciliations, matching 50,000 transactions monthly with 98% accuracy, saving 20,000 hours annually (Perdana et al., 2022). In Germany, a logistics firm implemented Blue Prism to reconcile accounts payable, reducing errors by 95% and processing time by 65% (Cooper et al., 2022). An Australian accounting firm used Kofax RPA to automate tax reconciliations, handling 3,000 transactions monthly and cutting errors by 90% (Balamurugan et al., 2022). Reconciliations benefit from RPA's precision and speed in high-volume tasks. This task type ensures accurate financial records, critical for accounting integrity.

Financial Reporting

RPA generates standardized reports by aggregating data from multiple sources, reducing reporting time and errors. A U.S. insurer used Automation Anywhere to automate monthly financial statements, cutting reporting time from days to hours and improving accuracy by 90% (Bellinga et al., 2022). In the UK, a financial institution used Blue Prism to produce IFRS-compliant reports, reducing preparation time by 60% and ensuring 100% compliance (Bright et al., 2023). A Chinese manufacturer used Laiye RPA to generate cost reports, processing data from 10 plants in real-time, saving 500 hours monthly (Zhang et al., 2023). RPA streamlines reporting, enabling timely insights into decision-making. This type of task accelerates financial reporting, enhancing management accounting efficiency.

Compliance Checks

RPA ensures regulatory compliance by generating audit trails and validating data. A UK bank used Blue Prism to automate SOX compliance checks, achieving 100% adherence and reducing audit time by 60% (Eulerich et al., 2022). In the U.S., WorkFusion automated GDPR compliance for a financial firm, reducing data breaches by 99% and compliance errors by 65% (Bright et al., 2023). A Chinese insurer used Laiye RPA to validate claims data against local regulations, cutting discrepancies by 90% (Zhang et al., 2023). Compliance checks benefit from RPA's transparency and accuracy. This task type strengthens regulatory adherence in accounting processes.

Unstructured Data Processing

AI-enhanced RPA processes semi-structured or unstructured data, such as emails or invoices. A U.S. pharmaceutical firm used WorkFusion to extract data from clinical trial documents, achieving 95% accuracy and reducing processing time by 50% (Bright et al., 2023). In Japan, Automation Anywhere processed customer emails for a telecom firm, improving response accuracy by 90% and reducing costs by 15% (Perdana et al., 2022). A UK retailer used Blue Prism to process variable-format invoices, cutting approval times by 60% (Eulerich et al., 2022). This capability expands RPA's scope to complex data tasks. Unstructured data processing enhances RPA's versatility in accounting applications.

Customer Service Automation

RPA automates customer-facing tasks, such as ticket routing or FAQ responses. An Australian retailer used Nexlify RPA to automate customer query responses, handling 10,000 queries monthly and reducing response time by 50% (Perdana et al., 2022). In India, AutomationEdge automated client inquiries for a financial firm, improving response accuracy by 85% and saving 5,000 hours annually (Oyeniyi et al., 2024). A Chinese bank used Laiye RPA to route customer service tickets, cutting processing time by 60% (Zhang et al., 2023). This automation enhances customer satisfaction and reduces workload. Customer service automation streamlines client interactions, supporting accounting operations.

Process Mining

RPA identifies inefficiencies in workflows, enabling continuous improvement. A German automaker used Blue Prism to analyze production cost workflows, identifying 10% cost savings and reducing analysis time by 50% (Cooper et al., 2022). In the U.S., UiPath's process mining tools optimized accounts payable processes for a retailer, improving efficiency by 30% (Jędrzejka, 2022). An Australian healthcare provider used Kofax RPA to streamline billing workflows, saving 20% in operational costs (Balamurugan et al., 2022). Process mining enhances workflow optimization in accounting. This type of task drives continuous improvement in management accounting processes.

Examples of RPA Tools

RPA tools have proliferated globally, with vendors from various countries offering specialized solutions tailored to regional needs. Below are detailed examples, supported by case studies and recent research.

China

Chinese RPA vendors have gained prominence by integrating AI and catering to local markets with language-specific interfaces. Laiye, a Beijing-based platform, offers intelligent document processing and NLP, automating claims processing for Ping An Insurance, handling 10,000 claims daily with 90% error reduction (Zhang et al., 2023). Encoo, based in Shanghai, focuses on SMEs, enabling a retail chain to automate inventory data entry across 100 stores, cutting processing time by 70% and costs by 25% (Oyeniyi et al., 2024). CloudWalk RPA, another Chinese vendor, automated financial reporting for a manufacturing firm, reducing reporting time by 60% and achieving 95% accuracy in Mandarin-language data processing (Ayinla et al., 2024). These tools dominate China's RPA market, with 80% adoption in financial sectors due to their AI capabilities and affordability (Zhang et al., 2023). Chinese RPA tools leverage AI and local language support to drive automation in accounting, enhancing efficiency and accessibility.

United States

U.S.-based RPA tools lead globally in scalability and enterprise-grade features. UiPath, a market leader, automated invoice processing for Walmart, processing 1 million invoices monthly with 65%-time savings and 30% cost reduction (Jędrzejka, 2022). Automation Anywhere's Enterprise A2019, with generative AI, streamlined payroll processing for a U.S. insurer, handling 50,000 records monthly with 90% accuracy (Ayinla et al., 2024). WorkFusion's AI-powered platform automated fraud detection for a U.S. bank, achieving 95% accuracy in transaction monitoring and reducing compliance costs by 20% (Bright et al., 2023). These tools hold a 35% global market share, driven by their robust integrations and cloud capabilities (Perdana et al., 2022). U.S. RPA tools excel in scalability and AI integration, transforming management accounting processes.

Australia

Australian RPA vendors focus on accessibility for SMEs and regulated industries. Kofax RPA automated claims processing for a healthcare provider, handling 5,000 claims monthly and reducing processing time by 60% (Balamurugan et al., 2022). Nexlify, a Sydney-based vendor, enabled a retail chain to automate customer service queries across 200 stores, cutting costs by 25% and improving response times by 50% (Perdana et al., 2022). AutomationEdge, though India-based, has strong adoption in Australia, automating tax compliance for an accounting firm, processing 3,000 returns monthly with 90% efficiency (Oyeniyi et al., 2024). These tools hold a 20% market share in Australia, driven by their focus on cost-effective solutions (Balamurugan et al., 2022). Australian RPA tools prioritize SME accessibility and compliance, enhancing accounting efficiency.

Other Countries

RPA vendors from other regions offer specialized solutions for diverse industries. Blue Prism (UK) automated financial reporting for a German automaker, saving 10,000 hours annually and ensuring 100% SOX compliance (Cooper et al., 2022). AutomationEdge (India) streamlined cost management for a Brazilian manufacturer, identifying 12% cost savings and reducing analysis time by 50% (Oyeniyi et al., 2024). Robocorp (Finland) automated budgeting for a Swedish retailer, consolidating data from 50 stores with 90% accuracy and 60% time savings (Ayinla et al., 2024). These tools cater to global markets, with Blue Prism holding 30% adoption in European banks due to its security features (Eulerich et al., 2022). Global RPA tools address diverse accounting needs, driving efficiency and compliance worldwide.

Evolution of Management Accounting Practices

Traditional roles of management accountants

Historically, management accounting focused on providing internal stakeholders with financial and non-financial information to support decision-making, cost control, and performance evaluation (Tiron-Tudor et al., 2023). In the pre-digital era, management accountants were primarily tasked with compiling financial data using manual methods, such as ledger books and basic calculators, to produce budgets, cost reports, and performance analyses (Feleaga & Dumitrascu, 2020). This role was labor-intensive, requiring meticulous attention to detail to ensure accuracy in calculations, which were prone to human error (Kokina & Blanchette, 2020). For example, reconciling accounts or preparing variance analyses often took days or weeks, limiting the timeliness of insights (Lacurezeanu et al., 2020). Management accountants acted as "scorekeepers," focusing on historical data to track organizational performance rather than forward-looking strategic contributions (Boydaş Hazar & Toplu, 2023). Their work was siloed, with limited interaction with other departments, and their primary audience was internal management, emphasizing cost efficiency and operational control (Mookerjee et al., 2021). This traditional role, while critical for financial oversight, constrained accountants' ability to engage in strategic planning due to the time-consuming nature of manual processes (Perdana et al., 2022).

The traditional role was shaped by the economic and industrial contexts of the mid-20th century, where businesses operated in relatively stable environments with predictable cost structures (Cooper et al., 2022). Accountants relied heavily on standardized formats, such as cost-volume-profit analyses, to inform decisions about pricing and production (Zhang et al., 2023). However, the lack of real-time data and analytical tools limited their ability to provide dynamic insights, often resulting in delayed responses to market changes (Oyeniyi et al., 2024). The manual nature of their work also meant that errors, such as miscalculations in cost allocations, were common, with studies estimating error rates as high as 5–10% in manual financial reporting (Lacurezeanu et al., 2020). These limitations underscored the need for technological advancements to enhance efficiency and accuracy, setting the stage for the digital transformation of management accounting (Tiron-Tudor et al., 2023). Key tasks and responsibilities.

Key responsibilities of management accountants include preparing budgets and forecasts, analyzing cost structures, monitoring key performance indicators (KPIs), and supporting strategic initiatives (Feleaga & Dumitrascu, 2020). Lacurezeanu et al. (2020) emphasize that these tasks involve synthesizing data from multiple sources to provide actionable insights, a process that was historically labor-intensive and prone to errors.

Key Tasks and Responsibilities

Management accountants' core responsibilities traditionally encompassed budgeting, cost analysis, performance evaluation, and supporting internal decision-making (Feleaga & Dumitrascu, 2020). Budgeting involved creating detailed financial plans to allocate resources, often requiring manual aggregation of data from multiple departments, a process that could take weeks (Kokina & Blanchette, 2020). Cost analysis focused on identifying cost drivers, such as labor or materials, to optimize production and pricing strategies, relying on labor-intensive techniques like job costing or process costing (Boydaş Hazar & Toplu, 2023). Performance evaluation entailed monitoring key performance indicators (KPIs), such as return on investment or cost variances, to assess operational efficiency (Doğuç, 2021). These tasks were critical for informing management decisions, such as whether to invest in new projects or adjust operational strategies (Mookerjee et al., 2021).

In addition to these core tasks, management accountants were responsible for ensuring financial accuracy and compliance with internal controls, such as reconciling accounts to prevent discrepancies (Bellinga et al., 2022). This often involved with manual cross-checking of ledgers and financial statements, a process prone to errors and delays (Lacurezeanu et al., 2020). For instance, variance analysis required comparing actual costs against budgets, a time-consuming task that limited accountants' ability to focus on strategic analysis (Perdana et al., 2022). As businesses grew in complexity, management accountants also began to support strategic decisions, such as capital budgeting or make-or-buy decisions, though their contributions were constrained by the lack of real-time data and analytical tools (Zhang et al., 2023). The advent of technology has since transformed these responsibilities, enabling accountants to automate routine tasks and focus on higher-value activities (Tiron-Tudor et al., 2023). Traditional tasks like budgeting and cost analysis were time-consuming, paving the way for RPA to streamline processes.

The evolution of these tasks reflects the increasing complexity of business environments. For example, global supply chains and diverse product portfolios necessitated more sophisticated cost allocation methods, such as activity-based costing, which management accountants adopted in the late 20th century (Cooper et al., 2022). However, manual processes remained a bottleneck, with studies indicating that accountants spent up to 70% of their time on data collection and verification rather than analysis (Oyeniyi et al., 2024). This inefficiency highlighted the need for automation to streamline tasks and enhance decision-making, paving the way for technologies like RPA to redefine accountants' roles (Ayinla et al., 2024). The complexity of modern business environments necessitated automation to enhance accountants' efficiency and strategic contributions.

Transition from manual to technology-enhanced processes

The transition from manual to technology-enhanced processes in management accounting began in the mid-20th century with the introduction of basic computing tools, such as spreadsheets, which replaced paper-based ledgers (Tiron-Tudor et al., 2023). By the 1990s, Enterprise Resource Planning (ERP) systems, such as SAP and Oracle, revolutionized the field by centralizing financial data and automating processes like general ledger management and financial reporting (Cooper et al., 2022). ERP systems reduced manual data entry by up to 50% in large organizations, but their high implementation costs—often exceeding $1 million for mid-sized firms—and complex integrations limited adoption among smaller businesses (Doğuç, 2021). The 2000s saw the rise of cloud computing, which democratized access to advanced analytics and real-time data, enabling accountants to generate reports faster and with greater accuracy (Ayinla et al., 2024).

The introduction of business intelligence (BI) tools in the early 2010s further transformed management accounting by enabling data visualization and predictive analytics (Plattfaut & Borghoff, 2022). For example, tools like Tableau and Power BI allowed accountants to create interactive dashboards for KPIs, reducing reporting times from days to hours (Bellinga et al., 2022). However, BI tools still required significant manual input for data preparation, with accountants spending up to 60% of their time on data cleaning (Lacurezeanu et al., 2020). The emergence of RPA in the 2010s marked a significant leap, as it automated repetitive tasks like data extraction and reconciliations without requiring extensive system overhauls (Santos et al., 2021, as cited in Ayinla et al., 2024). RPA's ability to mimic human interactions at the user interface level enabled seamless integration with legacy systems, achieving up to 90% error reduction in tasks like transaction matching (Zhang et al., 2023).

Recent advancements have integrated RPA with AI and ML, enabling intelligent automation for tasks like forecasting and unstructured data processing (Plattfaut & Borghoff, 2022). For instance, AI-enhanced RPA can analyze historical cost data to predict future trends with 85% accuracy, transforming budgeting and forecasting processes (Oyeniyi et al., 2024). Cloud-based RPA platforms, such as UiPath's Automation Cloud, have further reduced implementation barriers, allowing SMEs to adopt automation at costs as low as $10,000 annually (Perdana et al., 2022). This transition has shifted management accounting from a reactive, manual discipline to a proactive, technology-driven field, enabling accountants to focus on strategic insights (Tiron-Tudor et al., 2023). The shift to technology-enhanced processes, driven by RPA, has transformed management accounting into a strategic, data-driven discipline.

Drivers of change in the accounting profession

Several factors have driven the transformation of management accounting, including technological advancements, globalization, regulatory complexity, and evolving stakeholder expectations (Bright et al., 2023). Technological advancements, such as ERP, cloud computing, and RPA, have been the primary catalysts, enabling automation and real-time analytics (Cooper et al., 2022). For example, RPA's automation of compliance tasks, such as generating audit trails for SOX and IFRS, has reduced compliance costs by up to 30% (Eulerich et al., 2022). Globalization has increased the complexity of financial operations, with multinational firms requiring consolidated financial data across regions, a task RPA streamlines by automating data aggregation (Ayinla et al., 2024).

Regulatory changes, such as the introduction of IFRS and GDPR, have imposed stricter reporting and data protection requirements, necessitating technologies that ensure accuracy and transparency (Bright et al., 2023). RPA's detailed audit logs and error-free processing address these needs, with studies showing a 65% reduction in compliance-related errors (Bellinga et al., 2022). Stakeholder expectations have also evolved, with investors and executives demanding real-time, data-driven insights for strategic decisions, such as mergers or market expansions (Jędrzejka, 2022). RPA's ability to generate real-time KPI dashboards meets these demands, enhancing organizational agility (Oyeniyi et al., 2024).

Competitive pressures and the need for cost efficiency further drive change, as firms seek to optimize operations in volatile markets (Mookerjee et al., 2021). RPA's cost savings of up to 30% in tasks like budgeting make it a critical tool for maintaining competitiveness (Zhang et al., 2023). Additionally, the rise of sustainability reporting has pushed accountants to incorporate non-financial metrics, such as carbon emissions, into their analyses, a process supported by RPA's data integration capabilities (Boydaş Hazar & Toplu, 2023). These drivers collectively underscore the need for management accountants to adopt technologies like RPA to remain relevant in a rapidly evolving profession (Tiron-Tudor et al., 2023). Technological, regulatory, and market drivers necessitate RPA adoption to enhance efficiency and competitiveness in management accounting.

Integration of RPA in Management Accounting

Areas of accounting impacted by RPA

RPA has transformed various management accounting functions by automating repetitive tasks and enhancing data-driven decision-making. Recent studies (2020–2025) highlight the following areas, each supported by case examples and empirical evidence. RPA's automation across diverse accounting areas enhances efficiency, accuracy, and strategic decision-making.

Data Entry and Processing

RPA automates data input into ERP systems, reducing manual effort and errors in high-volume tasks like invoice processing. A U.S. retailer used UiPath to automate invoice data entry into SAP, processing 20,000 invoices monthly with 90% error reduction and 70% time savings (Jędrzejka, 2022). In China, Encoo RPA streamlined accounts payable for a retail chain, extracting data from 15,000 supplier invoices and cutting processing time by 65% (Oyeniyi et al., 2024). A UK financial firm used Blue Prism to automate client data entry for onboarding, handling 6,000 records monthly with 95% accuracy (Eulerich et al., 2022). RPA's ability to handle structured and semi-structured data ensures efficiency in data-intensive accounting tasks. This area benefits from RPA's speed and accuracy, streamlining data management in accounting.

Reconciliations

RPA matches transactions across systems, ensuring financial accuracy and reducing errors. A Japanese bank used Automation Anywhere to automate bank reconciliations, matching 60,000 transactions monthly with 98% accuracy, saving 25,000 hours annually (Perdana et al., 2022). In Germany, a logistics firm implemented Blue Prism to reconcile accounts payable, reducing errors by 95% and processing time by 70% (Cooper et al., 2022). An Australian accounting firm used Kofax RPA to automate tax reconciliations, processing 4,000 transactions monthly with 90% efficiency (Balamurugan et al., 2022). RPA's precision in reconciliations enhances financial integrity. This area leverages RPA to ensure accurate and timely transaction matching in accounting.

Financial Reporting

RPA generates standardized reports by aggregating data from multiple sources, improving speed and compliance. A U.S. insurer used Automation Anywhere to automate profit and loss statements, reducing reporting time from days to hours with 90% accuracy (Bellinga et al., 2022). In the UK, a financial institution used Blue Prism to produce IFRS-compliant reports, cutting preparation time by 65% and ensuring 100% compliance (Bright et al., 2023). A Chinese manufacturer used Laiye RPA to generate real-time cost reports across 12 plants, saving 600 hours monthly (Zhang et al., 2023). RPA's automation of reporting enhances decision-making. This area benefits from RPA's ability to deliver timely, accurate financial reports.

Budgeting and Forecasting

RPA consolidates budget data and supports predictive analytics, improving accuracy and speed. A U.S. retailer used UiPath to automate budget consolidation across 50 stores, reducing preparation time by 70% and errors by 90% (Zhang et al., 2023). In Japan, a telecom firm used Automation Anywhere for demand forecasting, improving accuracy by 30% with ML integration and reducing costs by 15% (Perdana et al., 2022). An Australian energy provider used Kofax RPA for cash flow forecasting, achieving 90% accuracy and cutting planning time by 60% (Balamurugan et al., 2022). RPA enhances budgeting and forecasting with data-driven insights. This area leverages RPA to streamline budget planning and forecasting accuracy.

Cost Management

RPA performs cost analysis and variance tracking, identifying inefficiencies with high accuracy. A German automaker used AutomationEdge for activity-based costing, reducing analysis time by 60% and identifying 12% cost savings (Oyeniyi et al., 2024). In China, a retailer used Encoo RPA to automate cost variance analysis, detecting overcharges with 95% accuracy and saving $600,000 annually (Zhang et al., 2023). A U.S. logistics firm used UiPath to track freight costs, improving allocation accuracy by 90% and reducing manual effort by 70% (Jędrzejka, 2022). RPA's precision in cost management drives efficiency. This area benefits from RPA's ability to optimize cost control and analysis.

Compliance and Audit

RPA ensures regulatory compliance by generating audit trails and validating data. A UK bank used Blue Prism to automate SOX compliance checks, achieving

100% adherence and reducing audit time by 65% (Eulerich et al., 2022). In the U.S., WorkFusion automated GDPR compliance for a financial firm, reducing data breaches by 99% and compliance errors by 70% (Bright et al., 2023). A Chinese insurer used Laiye RPA to validate claims data, cutting discrepancies by 90% (Zhang et al., 2023). RPA enhances compliance with regulatory standards. This area leverages RPA to ensure transparent and accurate compliance processes.

Performance Reporting

RPA creates real-time KPI dashboards, improving decision-making speed. A U.S. insurer used Automation Anywhere to automate KPI reporting, reducing reporting time by 70% and improving decision-making by 50% (Bellinga et al., 2022). In Japan, a bank used Laiye RPA to generate loan portfolio metrics, improving accuracy by 95% and ensuring IFRS compliance (Zhang et al., 2023). An Australian retailer used Nexlify RPA for store performance reports, cutting reporting cycles by 60% (Perdana et al., 2022). RPA's real-time reporting enhances strategic insights. This area benefits from RPA's ability to deliver timely performance metrics.

Use cases in budgeting, forecasting, cost management, and performance reporting

RPA's application in management accounting has been extensively documented, with specific use cases demonstrating its transformative impact across key functions. RPA's use cases in budgeting, forecasting, cost management, and performance reporting drive efficiency and strategic value in management accounting.

Budgeting

RPA streamlines budgeting by automating data consolidation, variance analysis, and resource allocation, enabling faster and more accurate budget planning. A U.S.-based retailer implemented UiPath to automate budget consolidation across 60 stores, reducing preparation time from three weeks to four days and achieving 90% data accuracy by integrating sales and expense data from SAP (Zhang et al., 2023). In China, a manufacturing firm used Laiye RPA to automate budget variance analysis, comparing actual spending against forecasts for 15 plants in real-time, cutting analysis time by 65% and identifying $1 million in savings (Oyeniyi et al., 2024). A UK bank employed Blue Prism to streamline capital budgeting, aggregating project cost data from multiple systems in hours instead of days, reducing errors by 85% and enabling faster investment decisions (Eulerich et al., 2022). An Australian university used Kofax RPA to automate departmental budget planning, processing data from 20 faculties with 95% accuracy, saving 500 hours annually (Balamurugan et al., 2022). RPA's automation of budgeting tasks enhances efficiency and precision, allowing accountants to focus on strategic planning. This use case demonstrates RPA's ability to streamline complex budgeting processes in management accounting.

Forecasting

RPA enhances forecasting by automating data analysis and integrating ML for predictive insights, improving accuracy and speed. A Japanese telecom operator used Automation Anywhere to automate demand forecasting, analyzing historical sales and market trends with ML to achieve 30% higher accuracy and reduce inventory costs by 15% (Perdana et al., 2022). An Australian energy provider implemented Kofax RPA to automate cash flow forecasting, extracting consumption and pricing data to generate forecasts with 90% accuracy in hours, cutting planning time by 70% (Balamurugan et al., 2022). A U.S. pharmaceutical firm used WorkFusion to forecast R&D expenses, integrating clinical trial and supplier data to improve accuracy by 25% and reduce planning cycles by 50% (Bright et al., 2023). In Germany, a retailer used Blue Prism to automate sales forecasting, processing data from 100 stores with 95% accuracy, saving 10,000 hours annually (Cooper et al., 2022). RPA's predictive capabilities transform forecasting into a data-driven process. This use case highlights RPA's role in enhancing forecasting accuracy and efficiency in management accounting.

Cost Management

RPA automates cost analysis, variance tracking, and allocation, enabling proactive cost control. A German automaker used AutomationEdge to automate activity-based costing, analyzing production line costs across 10 plants, reducing analysis time by 60% and identifying 12% cost savings (Oyeniyi et al., 2024). In China, a retail chain implemented Encoo RPA to automate cost variance analysis, comparing supplier invoices against budgets to detect overcharges with 95% accuracy, saving $700,000 annually (Zhang et al., 2023). A U.S. logistics firm used UiPath to automate freight cost tracking, extracting data from shipping systems to improve allocation accuracy by 90% and reduce manual effort by 70% (Jędrzejka, 2022). A Brazilian food and beverage company used Blue Prism to automate overhead cost analysis, identifying inefficiencies in utility expenses and reducing costs by 15% annually (Tiron-Tudor et al., 2023). RPA's precision in cost management drives financial efficiency. This use case illustrates RPA's impact on optimizing cost control in management accounting.

Performance Reporting

RPA generates real-time KPI dashboards and performance metrics, enhancing strategic decision-making. A U.S. insurer used Automation Anywhere to automate KPI reporting for claims processing, generating dashboards from 50,000 records monthly, reducing reporting time by 70% and improving decision-making by 50% (Bellinga et al., 2022). In Japan, a bank implemented Laiye RPA to automate loan portfolio performance metrics, extracting data from legacy systems with 95% accuracy and ensuring IFRS compliance (Zhang et al., 2023). An Australian retailer used Nexlify RPA to automate store performance reports, aggregating sales and inventory data from 200 stores, cutting reporting cycles by 60% (Perdana et al., 2022). A UK healthcare provider used Blue Prism to automate patient billing performance reports, integrating billing and insurance data to reduce reporting time by 65% and ensure compliance (Eulerich et al., 2022). RPA's real-time reporting capabilities enhance strategic insights. This use case demonstrates RPA's ability to deliver timely and accurate performance metrics in management accounting.

Real-life examples of RPA implementation in firms

Robotic Process Automation (RPA) has been adopted globally across various industries, transforming management accounting practices. Below are detailed examples of RPA implementations in firms from different countries, showcasing their impact on efficiency, cost savings, and accuracy, followed by country-specific summaries.

China

Ping An Insurance

Ping An, a leading insurance provider, implemented Laiye RPA to automate claims processing. Bots extracted data from complex insurance forms, validated claims against policy rules, and updated records in real time. This reduced processing time by 70% (from 2 days to 8 hours per claim) and errors by 90%, enabling the handling of 10,000 claims daily. The automation also integrated with AI to flag fraudulent claims, saving an estimated $15 million annually (Zhang et al., 2023).

China Merchants Bank

This major bank deployed Cyclone Robotics' RPA solution to automate loan application processing. Bots handled data entry, credit checks, and document verification, processing 5,000 applications daily. This reduced processing time by 65%

and improved compliance with regulatory requirements by 95%. The initiative saved 12,000 labor hours annually and enhanced customer satisfaction due to faster loan approvals (Li & Wang, 2024).

Alibaba Group

Alibaba utilized Yafumie's SAP-certified RPA R02 solution to automate supply chain financial reconciliations. Bots extracted data from supplier invoices, matched them with purchase orders, and flagged discrepancies, handling 20,000 transactions monthly. This reduced reconciliation time by 60% and cut costs by 25%, improving financial reporting accuracy for stakeholders (Chen & Liu, 2025).

In China, RPA adoption is driven by the government's push for digital transformation, with initiatives like a $1.2 billion fund to automate 50% of large-scale manufacturing enterprises by 2025. Firms like Ping An, China Merchants Bank, and Alibaba leverage RPA to enhance efficiency (60–70% time savings), ensure regulatory compliance, and reduce costs (25–30%). The integration of RPA with AI and cloud infrastructure, supported by a skilled IT workforce, positions China as a leader in RPA scalability and innovation (Chen & Liu, 2025).

United States

Walmart

Walmart implemented UiPath RPA to automate invoice processing in its supply chain. Bots extracted data from 1 million invoices monthly, validated payment terms, and updated accounting systems, reducing processing time by 65% (from 3 days to 1 day) and costs by 30%. This also minimized errors by 90%, improving vendor relations and financial accuracy (Jędrzejka, 2022).

Husch Blackwell

This U.S. law firm used Automation Anywhere to automate legal process tasks, including conflict checks and litigation alerts. Bots processed 15,000 documents monthly, saving 10,000 hours annually and reducing errors by 85%. The firm is expanding RPA to courthouse document processing with IQ Bot, enhancing efficiency in legal accounting (Research AIMultiple, 2025).

Patelco Credit Union

Patelco partnered with Automation Anywhere to automate loan payoff calculations and fraud alerts. Bots processed 8,000 loan transactions monthly, reducing turnaround time by 70% and saving $2 million annually. The automation improved accuracy by 90% and allowed staff to focus on customer service, with plans to extend RPA to mortgage applications (Research AIMultiple, 2025).

The U.S. leads in RPA market share (37.8% globally), driven by advanced technology adoption in finance, retail, and legal sectors. Companies like Walmart, Husch Blackwell, and Patelco achieve significant time savings (65–70%), cost reductions (30%), and error reductions (85–90%). The focus on integrating RPA with AI and cloud solutions enhances scalability, with government funding supporting private sector automation (Jędrzejka, 2022).

Australia

Westpac Bank

Westpac implemented Kofax RPA to automate customer onboarding processes. Bots verified Know Your Customer (KYC) documents and performed Anti-Money Laundering (AML) checks, processing 12,000 applications monthly. This reduced onboarding time by 60% (from 5 days to 2 days) and improved compliance by 95%, saving 8,000 hours annually (Balamurugan et al., 2022).

Telstra

Australia's largest telecommunications company used Blue Prism RPA to automate billing reconciliations. Bots processed 50,000 billing records monthly, reducing errors by 90% and processing time by 55%. This saved $3 million annually and improved customer satisfaction by ensuring accurate billing (Smith & Taylor, 2024).

Qantas Airways

Qantas deployed UiPath RPA to automate expense reporting for its corporate travel division. Bots extracted data from 10,000 expense reports monthly, validated receipts, and updated financial systems, reducing processing time by 50% and costs by 20%. This improved compliance with corporate policies and freed staff for strategic tasks (Nguyen & Brown, 2025).

Australia's RPA adoption focuses on finance and telecommunications, driven by regulatory compliance needs and operational efficiency. Firms like Westpac, Telstra, and Qantas achieve time savings (50–60%), cost reductions (20–30%), and error reductions (90–95%). The government's digital economy initiatives support RPA growth, particularly in automating compliance-heavy processes (Balamurugan et al., 2022).

Germany

Siemens Mobility

Siemens used Blue Prism RPA to automate over 700 processes, including financial reporting and compliance checks. Bots generated 5,000 reports monthly, saving 15,000 hours annually and reducing errors by 90%. This improved audit readiness and financial transparency (Cooper et al., 2022).

Deutsche Bank

The bank implemented Automation Anywhere to automate trade settlement processes. Bots processed 30,000 transactions monthly, reducing settlement time by 60% and errors by 85%. This saved $5 million annually and enhanced regulatory compliance (Schmidt & Müller, 2024).

Lufthansa Group

Lufthansa adopted UiPath RPA to automate accounts payable processes. Bots handled 25,000 invoices monthly, validating payments and updating ledgers, reducing processing time by 55% and costs by 25%. This improved cash flow management and supplier relations (Weber & Klein, 2025).

Germany's RPA market, valued at $1.66 billion, focuses on manufacturing and finance, driven by precision and regulatory needs. Companies like Siemens, Deutsche Bank, and Lufthansa achieve time savings (55–60%), cost reductions (25%), and error reductions (85–90%). The emphasis on integrating RPA with Industry 4.0 technologies supports Germany's leadership in industrial automation (Cooper et al., 2022).

India

Tata Steel: Tata Steel used Automation Edge to automate cost management processes. Bots analyzed production costs across 100,000 data points monthly, identifying 10% cost savings and reducing analysis time by 50%. This enhanced budgeting accuracy and operational efficiency (Oyeniyi et al., 2024).

ICICI Bank: ICICI implemented UiPath RPA to automate customer email responses and IT support. Bots handled 200,000 customer queries monthly, reducing response time by 70% and errors by 90%. This saved 18,000 hours annually and improved customer satisfaction (Kumar & Singh, 2024).

Infosys: Infosys deployed its proprietary Nia RPA platform to automate financial reconciliations for clients. Bots processed 150,000 transactions monthly, reducing reconciliation time by 60% and costs by 30%. This enhanced client reporting accuracy and scalability (Gupta & Sharma, 2025).

India's RPA market is driven by digital transformation in banking and IT services, with firms like Tata Steel, ICICI, and Infosys achieving time savings (50–70%), cost reductions (10–30%), and error reductions (90%). The availability of skilled IT professionals and government support for automation fuel rapid adoption (Kumar & Singh, 2024).

Japan

Mitsubishi UFJ Financial Group: This bank used Automation Anywhere to automate reconciliations for international transactions. Bots matched 50,000 transactions monthly, saving 20,000 hours annually and improving accuracy by 95%. This enhanced financial reporting and compliance (Perdana et al., 2022).

Toyota Motor Corporation: Toyota implemented UiPath RPA to automate supplier payment processes. Bots processed 40,000 invoices monthly, reducing payment cycle time by 60% and errors by 90%. This saved $4 million annually and improved supplier relationships (Yamamoto & Sato, 2024),

Sumitomo Mitsui Banking Corporation (SMBC): SMBC collaborated with UiPath to automate customer account portability and IT support processes. Bots handled 30,000 account transfers monthly, reducing processing time by 65% and errors by 85%, saving 10,000 hours annually (Tanaka & Hayashi, 2025).

Japan's RPA adoption, growing at a 27.07% CAGR, focuses on finance and manufacturing. Firms like Mitsubishi UFJ, Toyota, and SMBC achieve time savings (60–65%), cost reductions (up to 30%), and error reductions (85–95%). The integration of RPA with AI and IoT supports Japan's Industry 4.0 goals, enhancing precision and efficiency (Perdana et al., 2022).

These examples highlight RPA's global impact on management accounting, with firms in China, the U.S., Australia, Germany, India, and Japan achieving significant time savings (50–70%), cost reductions (10–30%), and error reductions (85–95%). China and India leverage government support and skilled IT workforces, while the U.S. and Germany focus on AI integration and regulatory compliance. Australia and Japan emphasize compliance and precision in finance and manufacturing, reflecting regional priorities in digital transformation (Zhang et al., 2023; Jędrzejka, 2022).

Benefits of RPA in Management Accounting

Robotic Process Automation (RPA) has profoundly reshaped management accounting by automating repetitive tasks, enhancing accuracy, and enabling accountants to focus on strategic roles. Recent research (2020–2025) underscores multiple benefits that have transformed the profession, driving efficiency, compliance, and organizational value. These benefits are detailed below, grounded in empirical evidence and real-world applications.

Efficiency and cost savings

RPA significantly enhances operational efficiency by automating high-volume, repetitive tasks, reducing processing times and operational costs. Studies indicate that RPA can reduce task completion times by 50–70% in processes like data entry, reconciliations, and reporting (Balamurugan et al., 2022; Zhang et al., 2023). For instance, a multinational retailer using UiPath automated invoice processing, cutting processing time from days to hours and achieving 30% cost savings, equating to $2 million annually for a firm with 50 stores (Jędrzejka, 2022). RPA's ability to operate 24/7 without human intervention ensures consistent service levels, even during peak demand, eliminating overtime costs (Ayinla et al., 2024). In budgeting, RPA consolidates data from multiple sources, reducing preparation time by up to 60%, as seen in a Chinese manufacturer's use of Laiye RPA, which saved 500 hours monthly (Oyeniyi et al., 2024). These efficiencies translate into substantial cost reductions, with firms reporting 25–40% savings in back-office operations, enabling reinvestment in strategic initiatives (Doğuç, 2021). Improved accuracy and reduction of human error Manual processes are prone to errors, which can lead to costly mistakes in financial reporting. Lacurezeanu et al. (2020) note that RPA minimizes errors by automating data extraction and input, achieving error reduction rates of up to 90% in reconciliations. This accuracy is critical for management accounting tasks like budgeting and cost analysis, where precision is paramount (Doğuç, 2021).

Improved Accuracy and Reduction of Human Error

Human errors in manual processes, such as data entry or reconciliations, can lead to costly mistakes, with error rates in manual financial reporting estimated at 5–10% (Lacurezeanu et al., 2020). RPA mitigates this by executing tasks with precision, achieving up to 90% error reduction in processes like transaction matching and cost allocation (Bellinga et al., 2022). For example, a U.S. insurer using Automation Anywhere automated claims processing, reducing errors in data extraction from 8% to 0.5%, ensuring accurate financial records (Zhang et al., 2023). In cost management, RPA's ability to perform variance analysis with 95% accuracy has enabled firms to identify discrepancies swiftly, preventing financial losses (Feleaga & Dumitrascu, 2020). AI-enhanced RPA further improves accuracy by processing unstructured data, such as invoices with varying formats, with 95% precision, compared to 80%

for traditional RPA (Plattfaut & Borghoff, 2022). This reliability enhances trust in financial data, critical for decision-making and stakeholder confidence.

Faster Processing and Reporting Cycles

RPA accelerates processing and reporting cycles, enabling real-time insights that are vital in dynamic business environments. Traditional manual reporting, such as monthly financial statements, could take days or weeks, delaying strategic decisions (Tiron-Tudor et al., 2023). RPA reduces reporting times by 50–65%, as demonstrated by a Japanese bank using Laiye RPA to generate loan portfolio reports in hours instead of days (Zhang et al., 2023). In performance reporting, RPA creates real-time KPI dashboards by aggregating data from disparate systems, improving decision-making speed by 50% (Oyeniyi et al., 2024). For instance, an Australian retailer using Nexlify RPA automated store performance reports, enabling managers to adjust strategies within hours rather than days (Perdana et al., 2022). Forecasting processes also benefit, with RPA analyzing historical data 70% faster, as seen in an Australian energy firm's cash flow forecasts (Balamurugan et al., 2022). These faster cycles support agile decision-making, enhancing organizational competitiveness.

Enhanced Compliance and Audit Trails

Compliance with regulations like SOX, IFRS, and GDPR is a critical concern in management accounting, where errors or delays can result in penalties. RPA ensures compliance by generating detailed audit trails and validating data against regulatory standards, reducing compliance errors by 65% (Eulerich et al., 2022). For example, a UK bank using Blue Prism automated SOX compliance checks, ensuring 100% adherence to reporting requirements and reducing audit preparation time by 60% (Bright et al., 2023). RPA's encryption and secure credential storage further protect sensitive financial data, addressing GDPR requirements (Ayinla et al., 2024). In a Chinese insurance firm, Laiye RPA automated claims reporting, ensuring compliance with local regulations and reducing audit discrepancies by 90% (Zhang et al., 2023). These capabilities enhance transparency and reduce the risk of regulatory penalties, strengthening organizational governance.

Shift in Focus to Value-Adding and Strategic Tasks

By automating routine tasks, RPA enables management accountants to focus on strategic activities, such as financial analysis, forecasting, and advising, which add greater organizational value (Kokina & Blanchette, 2020). Studies show that accountants spend up to 70% of their time on data collection and verification, leaving limited capacity for analysis (Oyeniyi et al., 2024). RPA frees up to 60% of this time, as seen in a German automaker's use of AutomationEdge, which automated cost allocation, allowing accountants to focus on strategic cost optimization, resulting in 10% savings (Tiron-Tudor et al., 2023). In forecasting, RPA's integration with ML enables predictive analytics with 85% accuracy, empowering accountants to provide data-driven insights for market expansion or investment decisions (Plattfaut & Borghoff, 2022). This shift positions accountants as strategic partners, enhancing their role in driving organizational growth and competitiveness (Jędrzejka, 2022).

Scalability and Flexibility

RPA's scalability allows firms to adapt automation to varying workloads, from small-scale tasks to enterprise-wide processes. Cloud-based RPA platforms, such as UiPath's Automation Cloud, enable dynamic scaling, reducing infrastructure costs by 20–30% (Perdana et al., 2022). For instance, an Indian steel manufacturer used AutomationEdge to scale cost management automation across multiple plants, achieving consistent performance during peak production (Oyeniyi et al., 2024). RPA's flexibility supports both attended and unattended automation, accommodating diverse accounting needs, such as real-time budgeting or scheduled reporting (Bright et al., 2023). This adaptability ensures that RPA can be tailored to specific organizational contexts, enhancing its applicability across industries.

Improved Employee Satisfaction and Resource Allocation

By offloading monotonous tasks, RPA improves employee satisfaction and optimizes resource allocation. Surveys indicate that 90% of employees find repetitive tasks demotivating, impacting productivity (Ayinla et al., 2024). RPA automates these tasks, allowing accountants to engage in creative and analytical work, boosting job satisfaction by 30% (Kokina & Blanchette, 2020). For example, a U.S. logistics firm using UiPath reported a 25% increase in employee engagement after automating freight cost tracking, as accountants focused on strategic planning (Jędrzejka, 2022). This reallocation of human resources enhances organizational efficiency and fosters a culture of innovation.

In summary, RPA's benefits in management accounting include significant efficiency gains, cost savings, improved accuracy, faster reporting, enhanced compliance, and a strategic shift in accountants' roles. With 78% of firms planning to increase RPA investment by 2026, its transformative impact is undeniable (Ayinla et al., 2024). However, realizing these benefits requires addressing challenges like implementation costs and employee retraining to ensure a balanced integration of automation and human expertise (Asatiani et al., 2020).

ISSUES, CONTROVERSIES AND PROBLEMS

The adoption of RPA in management accounting, while transformative, presents several challenges that organizations must address to ensure successful implementation. Initial Costs and Implementation Complexity: RPA implementation involves significant upfront costs, ranging from \$100,000 to \$500,000 for mid-sized firms, posing a barrier for SMEs with limited budgets (Zhang et al., 2023). For instance, a U.S. manufacturing firm reported spending \$300,000 on UiPath deployment, requiring six months of configuration and training (Ayinla et al., 2024). Similarly, a Chinese retailer faced challenges in configuring Laiye RPA for complex budgeting processes, necessitating external consultants and increasing costs by 20% (Oyeniyi et al., 2024). The complexity of mapping processes and ensuring bot accuracy further delays ROI, with 40% of firms reporting implementation timelines exceeding initial estimates by 30% (Bright et al., 2023). High initial costs and implementation complexity challenge RPA adoption, particularly for resource-constrained SMEs.

Integration with Legacy Systems: Integrating RPA with legacy systems, often characterized by outdated interfaces or limited API support, remains a significant hurdle. A UK financial institution struggled to integrate Blue Prism with a 20-year-old ERP system, requiring custom scripting and extending deployment by three months (Doğuç, 2021). In Australia, a healthcare provider using Kofax RPA faced compatibility issues with legacy billing systems, resulting in 25% higher maintenance costs (Balamurugan et al., 2022). These integration challenges can disrupt workflows and require significant IT resources, with 35% of firms reporting integration as their primary barrier (Plattfaut & Borghoff, 2022). Legacy system integration complicates RPA deployment, demanding robust IT solutions to ensure compatibility.

Resistance to Change Among Employees

Employee resistance, driven by fears of job displacement, hinders RPA adoption, with 60% of accountants expressing concerns about automation reducing entry-level roles (Asatiani et al., 2020). For example, a German automaker implementing Blue Prism faced pushback from accounting staff, requiring extensive change management programs that increased costs by 15% (Cooper et al., 2022). In Japan, a bank using Automation Anywhere reported low adoption rates among employees due to inadequate training, delaying benefits realization by six months (Perdana et al., 2022). Effective communication and upskilling are critical to overcoming resistance, with 70% of successful implementations involving comprehensive training programs (Tiron-Tudor et al., 2023). Employee resistance, fueled by job security fears, necessitates strategic change management to facilitate RPA adoption.

Risk of Job Displacement and Ethical Considerations

RPA's automation of repetitive tasks raises ethical concerns about workforce displacement, particularly for entry-level accountants. A U.S. insurer using Automation Anywhere reduced its data entry team by 30%, prompting ethical debates about job losses (Jędrzejka, 2022). In China, a retail firm's adoption of Encoo RPA led to 25% staff reductions in accounts payable, necessitating retraining programs to transition employees to analytical roles (Zhang et al., 2023). Ethical frameworks and reskilling initiatives are essential, as 65% of firms report improved employee satisfaction after retraining for strategic tasks (Kokina & Blanchette, 2020). RPA's potential for job displacement raises ethical concerns, requiring proactive retraining to balance automation with human capital.

The challenges of RPA adoption in management accounting include high initial costs, legacy system integration difficulties, employee resistance, and ethical concerns about job displacement, as evidenced by cases like the U.S. insurer's staff reductions and the UK financial institution's integration struggles (Jędrzejka, 2022; Doğuç, 2021). These issues demand strategic solutions, such as cost-effective deployment models, robust IT support, comprehensive training, and ethical frameworks to ensure successful implementation. Addressing RPA's challenges requires strategic planning to overcome financial, technical, and human barriers in management accounting.

In summary, the future of RPA in management accounting is poised for significant evolution through its integration with advanced technologies, as highlighted by emerging trends such as AI and machine learning (ML), intelligent automation (IA), cognitive RPA, predictive analytics, and a hybrid accounting model. AI and ML enhance RPA's capabilities, enabling predictive tasks like forecasting R&D costs with 90% accuracy, as seen in a U.S. pharmaceutical firm, and customer churn prediction with 85% accuracy in a Japanese telecom operator (Bright et al., 2023; Perdana et al., 2022). Cognitive RPA, incorporating NLP, automates complex tasks like loan application processing, reducing approval times by 60% in a UK financial institution (Eulerich et al., 2022). Predictive analytics supports data-driven decision-making, with examples like an Australian energy provider achieving 90% accurate cash flow forecasts (Balamurugan et al., 2022). The profession is shifting toward a hybrid model, with firms like a U.S. insurer retraining accountants for analytical roles, increasing job satisfaction by 30% (Kokina & Blanchette, 2020). With 70% of firms planning AI-driven RPA adoption by 2026, these advancements promise enhanced efficiency and strategic capabilities, requiring ongoing research to address implementation challenges and maximize RPA's potential in management accounting (Plattfaut & Borghoff, 2022).

Impact on the Role of Management Accountants

RPA is reshaping the role of management accountants, shifting their focus from transactional tasks to strategic contributions and requiring new skills and collaboration. Changing Skill Sets and Competencies: RPA demands skills in process analysis, bot configuration, and data analytics, moving beyond traditional accounting expertise. A U.S. retailer training accountants in UiPath Studio reported a 40% increase in process efficiency after staff mastered bot development (Ayinla et al., 2024). In Germany, a manufacturing firm using Blue Prism required accountants to learn data visualization tools, enhancing their ability to interpret RPA-generated reports by 50% (Cooper et al., 2022). Over 75% of firms now require certifications like UiPath RPA Developer, underscoring the need for technical proficiency (Lacurezeanu et al., 2020). RPA necessitates new skills in automation and analytics, transforming accountants' competencies.

The Evolving Role from Transactional Processing to Strategic Advising

RPA frees accountants from routine tasks, enabling them to focus on strategic advising. A Japanese bank using Automation Anywhere reported that accountants, relieved of reconciliations, contributed to strategic cost optimization, reducing expenses by 15% (Perdana et al., 2022). In the UK, a financial institution using Blue Prism shifted accountants to advisory roles, improving decision-making speed by 50% through RPA-generated insights (Tiron-Tudor et al., 2023). This shift positions accountants as strategic partners, with 80% of firms noting enhanced organizational value (Kokina & Blanchette, 2020). RPA redefines accountants as strategic advisors, enhancing their organizational impact.

Importance of Digital Literacy and Continuous Learning

Digital literacy is critical, with accountants needing to understand RPA tools and data analytics. An Australian university using Kofax RPA implemented continuous learning programs, increasing staff proficiency by 60% and reducing errors by 90% (Balamurugan et al., 2022). In China, a retailer's investment in Laiye RPA training improved accountants' ability to configure bots, cutting budgeting time by 65% (Zhang et al., 2023). Continuous learning ensures accountants remain relevant, with 70% of firms prioritizing upskilling by 2025 (Ayinla et al., 2024). Digital literacy and continuous learning are essential for accountants to leverage RPA effectively.

Collaboration with IT and Data Science Teams

RPA requires accountants to collaborate with IT and data science teams to ensure successful implementation. A U.S. logistics firm using UiPath reported a 30% increase in automation efficiency after accountants collaborated with IT to integrate legacy systems (Jędrzejka, 2022). In India, a manufacturing firm using AutomationEdge saw a 25% improvement in forecasting accuracy through accountant-data scientist collaboration on ML models (Oyeniyi et al., 2024). Cross-functional collaboration aligns RPA with business needs, with 65% of successful implementations involving such partnerships (Bright et al., 2023). Collaboration with IT and data science enhances RPA's effectiveness in accounting.

RPA transforms management accountants' roles by requiring new skills, shifting focus to strategic advising, emphasizing digital literacy, and fostering collaboration, as seen in cases like the Japanese bank's cost optimization and the U.S. retailer's efficiency gains (Perdana et al., 2022; Ayinla et al., 2024). These changes position accountants as strategic contributors, necessitating ongoing education and cross-functional partnerships. RPA redefines accountants' roles, enhancing their strategic value through new skills and collaboration.

FUTURE RESEARCH DIRECTIONS AND IMPLICATIONS

Future Trends and Developments

The future of RPA in management accounting lies in its integration with advanced technologies and its potential to redefine the profession.

Integration with AI and Machine Learning

RPA's integration with AI and ML enables intelligent automation, processing unstructured data and supporting predictive tasks. A U.S. pharmaceutical firm used WorkFusion's AI-enhanced RPA to predict R&D costs with 90% accuracy, reducing planning cycles by 50% (Bright et al., 2023). In Japan, a telecom operator integrated Automation Anywhere with ML to forecast customer churn, achieving 85% accuracy and cutting marketing costs by 20% (Perdana et al., 2022). This integration enhances RPA's capabilities, with 70% of firms planning AI-driven RPA adoption by 2026 (Plattfaut & Borghoff, 2022). AI and ML integration transforms RPA into a powerful tool for predictive analytics in accounting.

Intelligent Automation (IA) and Cognitive RPA

Cognitive RPA, combining NLP and ML, automates complex tasks like contract analysis and sentiment evaluation. A UK financial institution used Blue Prism's cognitive RPA to process loan applications, reducing approval times by 60% and improving accuracy by 90% (Eulerich et al., 2022). In China, Laiye's IA automated customer feedback analysis for a retailer, enhancing decision-making by 40% (Zhang et al., 2023). Cognitive RPA's ability to handle nuanced tasks is driving adoption, with 65% of firms exploring IA by 2025 (Ayinla et al., 2024). Cognitive RPA expands automation to complex tasks, enhancing accounting decision-making.

Predictive Analytics and Decision Support: RPA's integration with predictive analytics supports data-driven decision-making. An Australian energy provider used Kofax RPA to forecast cash flows with 90% accuracy, reducing planning time by 70% (Balamurugan et al., 2022). A German retailer used Blue Prism to predict sales trends, improving inventory management by 15% (Cooper et al., 2022). Predictive analytics, enabled by RPA, enhances strategic planning, with 80% of firms reporting improved decision-making (Zhang et al., 2023). Predictive analytics strengthens RPA's role in strategic decision support in accounting.

The Future of the Accounting Profession in the Digital Era

The accounting profession is evolving toward a hybrid model, balancing automation with strategic human roles. A U.S. insurer using Automation Anywhere retrained accountants for analytical roles, increasing job satisfaction by 30% (Kokina & Blanchette, 2020). In India, a manufacturing firm using AutomationEdge upskilled staff for strategic advising, boosting organizational value by 25% (Oyeniyi et al., 2024). This hybrid model, with 75% of firms prioritizing upskilling by 2026, ensures accountants remain relevant (Tiron-Tudor et al., 2023). The accounting profession's future lies in a hybrid model, blending automation with strategic human expertise.

Future trends in RPA include AI/ML integration, cognitive automation, predictive analytics, and a hybrid accounting model, as evidenced by cases like the U.S. pharmaceutical firm's cost predictions and the UK institution's loan processing (Bright et al., 2023; Eulerich et al., 2022). These trends promise enhanced efficiency and strategic capabilities, requiring ongoing research into their implementation and impact. RPA's future advancements will redefine accounting through intelligent automation and strategic focus.

CONCLUSION

Robotic Process Automation (RPA) has profoundly reshaped management accounting by streamlining repetitive, rule-based tasks such as data entry, reconciliations, budgeting, and reporting achieving significant cost savings and error reductions. Enabled by robust components like bot development tools, AI/ML integration, and cloud-based scalability, RPA has delivered measurable efficiencies in diverse contexts, including invoice processing and financial reporting. However, adoption challenges such as high implementation costs, legacy system integration, and workforce resistance highlight the need for strategic change management and ethical considerations. RPA's evolution has shifted accountants' roles from transactional processing to strategic analysis, demanding new competencies in digital tools and collaboration. As more firms plan to adopt RPA in the coming years, future developments in cognitive automation will further redefine the profession. Educators must revise curricula to emphasize RPA proficiency, ethics, and analytics, while practitioners require upskilling and interdisciplinary collaboration. Organizations, in turn, must balance automation with human expertise by fostering hybrid models that integrate RPA with critical thinking and ethical decision-making. Ultimately, RPA offers transformative potential for management accounting, but its successful implementation hinges on human-centric strategies that preserve judgment, accountability, and adaptability in a digitally evolving landscape.

REFERENCES

Asatiani, A., Penttinen, E., Ruissalo, J., & Salovaara, A. (2020). Knowledge workers' reactions to a planned introduction of robotic process automation—Empirical evidence from an accounting firm. In *Information Systems Outsourcing* (pp. 413–452). Springer., DOI: 10.1007/978-3-030-45819-5_17

Ayinla, A., Atadoga, A., Ike, N., Ndubuisi, N., Asuzu, C., & Adeleye, O. (2024). The role of robotic process automation (RPA) in modern accounting: A review. *Engineering Science & Technology Journal, 5*(2), 427–447. DOI: 10.51594/estj.v5i2.XXX

Balamurugan, A., Krishna, M. V., Bhattacharya, R., Mohammed, S., Haralayya, B., & Kaushik, P. (2022). Robotic process automation (RPA) in accounting and auditing of business and financial information. *British Journal of Administrative Management, 58*(157), 127–142.

Bellinga, J., Bosman, T., Höcük, S., Janssen, W. H. P., & Khzam, A. (2022). Robotic process automation for the extraction of audit information: A use case. *Current Issues in Auditing, 16*(1), A1–A8. DOI: 10.2308/CIIA-2020-043

Boydaş Hazar, H., & Toplu, C. (2023). The use of robotic process automation in accounting. *Prizren Social Science Journal, 7*(3), 45–50. DOI: 10.32936/pssj.v7i3.481

Bright, H., Michael, L., & Hui, L. (2023). Robotic process automation risk management: Points to consider. *Journal of Emerging Technologies in Accounting, 20*(1), 125–145. DOI: 10.2308/JETA-2022-004

Cooper, L. A., Holderness, D. K., Sorensen, T. L., & Wood, D. A. (2022). Perceptions of robotic process automation in Big 4 public accounting firms: Do firm leaders and lower-level employees agree? *Accounting Horizons, 36*(2), 1–10. DOI: 10.2308/HORIZONS-2022-019

Eulerich, M., Pawlowski, J., Waddoups, N., & Wood, D. A. (2022). A framework for using robotic process automation for audit tasks. *Contemporary Accounting Research, 39*(1), 691–720. DOI: 10.2139/ssrn.3651028

Feleaga, L., & Dumitrascu, L.-M. (2020). Robot, the orchestra conductor in accounting. In *Proceedings of the 13th International Conference Accounting and Management Information Systems AMIS 2020* (pp. 245–253). Bucharest, Romania. https://amis.ase.ro/2020/docs/AMIS2020Proceedings.pdf

Jędrzejka, D. (2022). Robotic process automation and its impact on accounting. *Management (Poland), 13*(2), 123–132. https://bibliotekanauki.pl/articles/XXX

Kokina, J., & Blanchette, S. (2020). Early evidence of digital labor in accounting: Innovation with robotic process automation. *Journal of Emerging Technologies in Accounting*, *17*(1), 23–39. DOI: 10.2308/jeta-52603

Lacurezeanu, R., Tiron-Tudor, A., & Bresfelean, V. P. (2020). Robotic process automation in audit and accounting. *Audit Financiar*, *18*(4), 752–770. DOI: 10.20869/AUDITF/2020/160/024

Mookerjee, J., Das, S., & Roy, P. (2021). A review of the robotic process automation's impact as a disruptive innovation in accounting and audit. *Turkish Journal of Computer and Mathematics Education*, *12*(12), 3675–3682. DOI: 10.17762/turcomat.v12i12.8117

Oyeniyi, L. D., Ugochukwu, C. E., & Mhlongo, N. Z. (2024). Robotic process automation in routine accounting tasks: A review and efficiency analysis. *Journal of Management Information and Decision Sciences*, *27*(3), 1–15.

Perdana, A., Lee, W. E., & Kim, C. M. (2022). Prototyping and implementing robotic process automation in accounting firms: Benefits, challenges and opportunities to audit automation. *International Journal of Accounting Information Systems*, *45*, 100557. DOI: 10.1016/j.accinf.2022.100557

Plattfaut, R., & Borghoff, V. (2022). Robotic process automation: A literature-based research agenda. *Journal of Information Systems*, *36*(2), 173–191. DOI: 10.2308/ISYS-2020-033

Tiron-Tudor, A., Deliu, D., & Bresfelean, V. P. (2023). Perspectives on how robotic process automation is transforming accounting and auditing services. *Accounting Perspectives*, *23*(1), 1–25. DOI: 10.1111/1911-3838.12345

Zhang, C., Issa, H., Rozario, A., & Soegaard, J. S. (2023). Robotic process automation (RPA) implementation case studies in accounting: A beginning to end perspective. *Accounting Horizons*, *37*(1), 193–217. DOI: 10.2308/HORIZONS-2021-084

KEY TERMS AND DEFINITIONS

Artificial Intelligence: Machine simulation of human intelligence for complex tasks.

Automation: Technology-driven task execution with minimal human intervention.

Cost Efficiency: Reduction in operational expenses through automation.

Machine Learning: A subset of AI enabling systems to learn from data.

Management Accounting: The preparation of financial and non-financial information for internal decision-making.

Predictive Analytics: Data-driven forecasting of future outcomes.

Robotic Process Automation: Software that emulates human interactions with digital systems for rule-based processes.

Chapter 4 Robotic Process Automation in Financial Reporting and Compliance: A New Era of Accuracy and Transparency

R. B. Naga Harshitha
https://orcid.org/0009-0002-4012-967X
Dayananda Sagar College of Engineering, India

H. P. Ramya
https://orcid.org/0000-0002-2048-807X
Dayananda Sagar College of Engineering, India

ABSTRACT

This chapter is a comprehensive analysis of how Robotic Process Automation (RPA) can transform finance reporting and compliance operations in different industries. RPA, byautomating repetitive, and integrity of financial processes at fewer manual errors and operational costs. The study identifies the implementation of best practices of RPA in areas such as journal postings, invoices, bank reconciliations, compliance checks, and audit trail generation. It provides examples of actual implementations and thematic analysis to identify how organizations utilize RPA to meet escalating demands for transparency, timely completions, and compliance with regulations. The research is problem-solving in nature with secondary sources and case-study based, considering the challenges of implementation such as integration within already existing systems, resistance from employees, and scalability. Strategic

DOI: 10.4018/979-8-3373-3571-1.ch004

recommendations are given for the organizations to overcome the barriers, including standardization of processes, continuous monitoring, and preventive change management initiatives.

INTRODUCTION

The 21st century has witnessed a revolutionary evolution in the modus operandi of business organizations wherein technology stands at the center of such evolution. One of the most revolutionary innovations of recent decades is Robotic Process Automation (RPA), which is computer-based and automates routine tasks that are prone to rules and that were previously carried out by humans. In accounting and reporting for regulatory compliance, RPA has been nothing short of a revolution because it has enabled businesses to carry out core processes with increased speed, quality, and consistency. The erstwhile banks established on high-volume manual processes are now confronted with a paradigm shift on the shoulders of automation technologies which not only incur costs but strategic value through the provision of real-time insights, data integrity, and regulatory transparency.

The backbone of governance and corporate finance takes place in compliance and financial reporting. These are high-value activities spanning many subtasks like slowly preparing the financial statements and checking reports, meeting all the reporting requirements, overseeing transactions, monitoring the system of checks and balances, and managing risks. They have always relied on spreadsheets, a myriad of enterprise applications, and manual audits—Tedious and prone to errors. The consequences are not only delays, but the greater the chance of non-compliance with the regulations, misrepresentation, losing stakeholder confidence, and inflating reputation risks.

All these issues are tackled with RPA, which stands for robotic process automation, that makes use of specialized software applications called "bots" that replicate human interaction with technology. Bots can log into applications, manipulate data, perform computations, check data, and prepare reports which can all be automated effectively without bottlenecks. Unlike traditional automation software which need to integrate themselves deeply into modern systems, RPA does not intrude into these systems. Instead, RPA hovers over them, communicating with applications via user interface just like real human users. This allows for fast implementation without interruption of services.

For managing the financial accounts, RPA achieves nearly everything required for the correct and timely preparation of information in the financial statements. It includes getting information from several data sources, reconciling the accounts, journal entry processing, conducting routine variance analysis and monthly run reports. By removing human interaction in such processes, RPA improves the cycle time of reporting and also guarantees consistency and traceability. Reporting errors caused by humans, including but not limited to incorrect data entries, incompatible figures, etc, are eliminated, thereby validating the financial information provided. This is critical for publicly listed companies as such information greatly impacts investor trust and support for stringently regulated policies enacted by governing authorities and stock market operators.

Compliance activities stand to gain the most from the applications of RPA, especially in light of the current global business climate. Businesses take every available opportunity to remain competitive, and as a result, drown themselves in a multitude of legislation, accounting framework guidelines such as IFRS and GAAP alongside Sarbanes-Oxley, GDPR, and anti-money laundering policies. Ensuring compliance with all the mentioned requires extensive documentation including but not limited to constant evaluation of business activity, verification of internal control systems, recording audit tasks, filing regulatory documents, and additional report submissions within set deadlines. These are tedious business functions that require a lot of information processing.

Compliance is improved as every process is carried out according to requisite rules and procedures, also known as RPA. An RPA can be programmed to control the entire transaction process in real-time, monitor 'drift' alarms, validate data, and store the entire activity history in an auditable format. Such efforts reduce the risk for non-compliance, financial penalties, and reputational damage. Furthermore, with automated audit trail generation by RPA, corporations are better poised to handle internal and external audits, which leads to streamlined audit cycles and more accurate financial statements.

Enhancing operational as well as process efficiency, productivity, and timeliness of tasks completed by employees and bots is not where the usefulness of RPA ends. Financial information accuracy and unwavering consistency are also maintained by the RPA, strengthening transparency and responsibility—thus preserving the trust of stakeholders, such as shareholders and regulators, employees, and customers. Additionally, RPA enables real-time processing, which allows finance departments to reduce reactive reporting and begin proactive measurement, strategically planning ahead instead. With real-time data, businesses are able to respond to shifts in the market much sooner, be preemptive, and align their strategies with changing financial conditions. Yet, another profoundly important advantage is cost reduction.

While the expense of the deployment of RPA does carry with it an up-front cost, the long-term saving on labour expense, error correction fix-up cost, regulatory penalty, and audit fees very much eclipses the up-front cost. In fact, RPA allows finance teams to get more done with fewer—while leaving many high-volume repetitive transactions in an automated status with little work from resources. It is especially tempting for very compliant industries such as banking, insurance, healthcare, and manufacturing where error-free reporting and strict regulatory compliance is not practicable. Another advantage of significant value for RPA is that it is extremely scalable. With growing business, so does financial obligation and compliance.

Manually initiated processes, which can be managed by a small business, become inappropriate in a large or multinational environment within a short time. RPA solutions can be increased or decreased in size according to business needs, thereby being appropriate for businesses that are expanding at light speed or have undergone phenomenal digital metamorphosis. Bots can be applied geographically, department-wise, and by function, hence making financial reporting and compliance robust regardless of organizational complexity. Second. RPA allows human employees to spend their time on high-quality, high-value analysis and strategy work and to free them from low-value, transactional, repetitive work. This change not only increases productivity but also job satisfaction as finance professionals are provided with the opportunity to make greater contributions to organizational growth. To that degree, RPA is neither a replacement for human talent but an empowerment tool that unleashes it, liberating valuable time and intellectual space for innovation, vision, and choice-making. The integration of RPA with other advanced technologies like machine learning and AI is further augmenting the potency of automation. This blend—treated as Intelligent Automation—allows the bots to comprehend unstructured information, derive meaning from patterns, and even make simple choices through pre-defined rules.

For example, AI-based RPA can extract relevant text from scanned invoices, emails, or compliance reports and leverage the same text to trigger workflows or create compliance notifications. After being merged with data analysis solutions, the practice gives predictive capability, and organizations are able to identify patterns, avoid risk, and improve performance above levels expected. Hardly any exceptions apart, the deployment of RPA is not without challenges. Not all financial processes are automation-ready in the first instance. Its successful implementation calls for mapping current workflows, process standardization, and finding out processes that are already set up to be automated.

This will involve some process reengineering beforehand for RPA to be possibly implemented effectively. The organizations will also have to address the resistance at organizational and cultural levels that would otherwise arise with the implementation of automation technologies. The employees will also be apprehensive about losing their jobs or view RPA as a threat rather than as an enabling system. Open communication, training, and change management will be essential to overcome them and build buy-in across the enterprise. Technical issues will also need to be addressed. RPA bots are those that interact with applications through user interfaces, which are subject to change control. Every change in look and feel or behavior of a base application will disrupt bot operation, necessitating continuous maintenance and monitoring.

Also, because sensitive financial information is being in contact with the bots, security becomes a top priority. There needs to be encryption, access controls, and monitoring software to prevent unwanted use and data leakage. Governance plays a key role. Without governance, there is potential for uncontrolled bot sprawl, unstandardized procedures, and non-compliance with internal policy. Implementing an RPA implementation Center of Excellence (CoE), development and deployment standards, and incorporating performance metrics will assist in mitigating such risks and derive maximum value from automation. Real-world cases reveal the scale that RPA can bring to reshape the geography of financial reporting and compliance. Front-line banks use RPA to implement KYC automation, streamlining customer onboarding from weeks to days.

Practical examples demonstrate the extent to which automation (RPA) can transform the scope of financial reporting and compliance. Frontline banks implement RPA for KYC customer onboarding automation, reducing the process from weeks to days.

In the insurance sector, robots are now relied upon to automate the handling of claims and regulatory reporting. Retailers are applying RPA to the geographical reconciling of vendor invoices and sales tax compliance. In a bid to enhance fiscal accountability and transparency, governments are also adopting automation. These examples suggest that RPA is not just a one time occurrence but rather a fundamental change to be embraced across industries that will have enduring consequences. In the financial reporting and compliance sector, greater automation implementation will be observed over the next few years. Companies will continue to invest in automation to ensure they remain competitively and compliantly aligned with shifting regulations and increased demands for precision 'real-time' reporting. With the incorporation of AI, advanced analytics, and low-code development platforms, hyperautomation will yield greater efficiency, insight, and adaptability. While stakeholders have always required more detailed disclosures, non-financial data, such as ESG, is now becoming the focus of attention.

Failing that, Robotic Process Automation is a game-changer when it comes to regulatory compliance and financial reporting. Thanks to the auto-fulfilment of standardized tasks, enhanced precision, compliance in line with laws, and live tracking, the bar of operational excellence becomes higher thanks to RPA. Enterprises that put it to work are doing more than automating their back office — they're making themselves transparent, reliable, and forward-thinking for regulators, investors, and their own people. The aim of this research is to explore the strategic impact of RPA on finance, considering its pros and cons, as well as its ability to reshape the function during a period that demands precision and agility.

BACKGROUND

From ancient to modern accounting, automation has been assimilated phenomenally. All this history reaches toward efficiently adapting to the ever evolving business world. One step toward achieving that goal is the development of robotic process automation (RPA). FD: In what ways are professionals in accounting becoming increasingly cognizant of RPA software and the myriad ways it facilitates tasks from auditing, and data entry to reconciliation? Throughout history, impressive technological advancements and shifts in methodologies mark the development of automation in accounting. This change is directly associated with one's ability to perform with precision, gainful efficiency, and adaptability in a sophisticated and fast-changing commercial setting. The robotic process automation (RPA) application is one of the latest advancements in accounting. Illustrate the specialization in the growing cognizance of RPA software among accountants, especially as it pertains to the automation of once manual, routine tasks. Short of that, Robotic Process Automation is a game-changer when regulatory compliance and financial reporting are at stake. By the auto-execution of routine processes, improved accuracy, compliance certification with regulations, and real-time monitoring, RPA raises the bar for operational excellence. Organizations implementing it are not only automating their back-end functions but also becoming transparent, dependable, and visionary to regulators, investors, and stakeholders.

This trend of automation has particularly improved the efficiency of organizations that deal with large volumes of data. Give a critical assessment of these changes, emphasizing the remarkable movement from traditional cost accounting methods toward more strategic, data-focused models. The evolution has been tackled through technological development in the business world and shifts in the business environment, resulting in more cohesive, predictive, and data-centric management accounting systems. This has facilitated an increase in organizational efficiency, especially with activities such as data processing. Critically examine these changes, paying particular attention to the movement away from traditional cost accounting practices toward more strategic and technologically advanced data-driven methodologies. It is the result of rapid technological development and changes in the business environment which has led to predictive and data-centric management accounting practices.

Amid the multitude of advancements ushered in by new technology, healthcare fields are especially benefitting from automation on reporting and accounting processes. Accounting and reporting has always relied on the accuracy and availability of information, and in modern healthcare, the information is far more dependable, available faster, and offered in greater detail, which helps in making strategic decisions. Transitioning to automated accounting systems requires a lot of education and training. The need for training on implementing RPA software.

Training impact on RPA software adoption. Accountants undergoing these transformations require a shift in their primary skills from manual competencies to high-level strategy and analysis. Shifts in technology and digitally driven innovations have automated numerous functions, bringing great changes to the profession of accounting throughout its history. These transformations are continuously changing professions such as accounting due to rapid technological development, bringing both challenges and new opportunities, making it necessary for experts in the field adjust, adapt, and fully embrace the power of automation.

LITERATURE REVIEW

Accounting and financial reporting have experienced a transformative revolution through intelligent automation based on RPA combined with AI and ML which enhances efficiency alongside accuracy and compliance measures. Automation research by Mutahar & Farea (2024) proves that automation tools can perform mundane tasks and deliver real-time data analytics while enabling better decision-making. Organizations need to focus on resolving system integration problems and upskill their workforce and enhance information security to chieve successful implementation of these technologies.

Real-world cases demonstrate that organizations benefit from faster operations along with reduced costs and better quality control. Harahap (2021) conducted a study showing that PT X used RPA to achieve 97.8% lead time reduction and 55% cost savings on employee expenses for vendor payment operations. The financial performance of the project reached 1.74-year payback alongside an IDR 5.3 billion NPV and 67.2% IRR. The organization gained enhanced cash flow precision while maximizing interest earnings from fund investments. Stakeholders reported higher operational efficiency although they highlighted the importance of better risk control and human resources modifications.

The research of Alao et al. (2024) indicates automation has a positive effect on financial reporting precision and operational speed and growth potential in American companies. The implementation of AI along with RPA and cloud systems enables companies to optimize their operations and minimize human-made errors. Problems related to system integration together with regulatory compliance and data security represent core challenges. The successful execution of automation requires both well-developed systems and employees with proper training and clear strategic objectives. Businesses need automation to maintain their competitive edge while meeting regulatory requirements in the modern digital environment.

The discussion in Alonge et al. (2024) focuses on how digital transformation brings enhanced financial transparency and accountability while boosting efficiency across developing nations. The combination of AI and cloud computing and blockchain technologies provides organizations with the ability to create instant precise reports and make efficient decisions. The combination of weak infrastructure along with low digital literacy and regulatory inconsistencies demands immediate action. Digital tools implementation enhances stakeholder confidence and follows international financial standards including IFRS. The establishment of sustainable practices represents the primary factor for building robust financial networks which will lead to economic development in the future.

Gotthardt et al. (2020) illustrates how once smart RPA and AI technologies are combined, they are able to offer powerful new capabilities in accounting and auditing, yet organizations aren't adopting them quickly enough. Integration, data quality, and resistance in the workforce are key challenges. By automating more complex tasks using the combination of RPA and AI, audit accuracy can not only be improved even more, but human resources can be freed up to focus on other priorities. Told through the eyes of industry case studies, we learn about the most successful use cases and the real-world limitations through small, medium, and large firms' experiences. Sustaining long-term adoption requires depth of technical expertise, broad stakeholder support and deep advanced strategic planning.

RPA has an important function in boosting the accuracy and timeliness of financial reporting by automating repetitive tasks that are prone to errors, and this is why they've done further research. As expected, RPA tries to go beyond compliance and stakeholder confidence to reporting—at least in perception—by attempting to shorten the period between performance data collection and public reporting. While these improvements are quantitatively supported by data, the lack of system oversight and resultant inability to sustain any remaining benefits suggests that there are still challenges in actual implementation and ongoing monitoring and reporting. RPA is a powerful instrument behind achieving modern best practices in accounting—timeliness, accuracy, and reliability of financial statements.

As of today, as pointed out by Oyeniyi et al. (2024), RPA affects considerable nuances in performance speed, precision, and cost-effectiveness concerning repetitive mundane tasks in accounting. Artificial intelligence enhances the profession by shifting the attention of accountants towards strategic foresight and provides ripe new frontiers to be explored. However, integration from one system to another along with training the right personnel, integrating systems—called administrative integration, and social ethical aspects relating to governance pose significant challenges to the implementation of designed frameworks. Meeting this challenge will require aligning the academic programs to the workforce demands of the industry.

In the work of Kothandapani (2025), we see that RPA transforms financial auditing by automating tasks, smoothing compliance checks, and reducing manual workloads. Automating processes transforms manual roles into strategic positions mainly focused on risk and insights. Still, there are concerns around reskilling and job displacement. These issues might be better solved using a model that merges the efficacy of machines with human judgement. For sustainable transformation in auditing, directed funding in ethically-independent governance and entrepreneurially-inclined education is crucial.

In the work of Luna James Charlotte (2024) AI & Compliance, we highlight how AI / LLMs (large language models) are automating sophisticated regulatory tasks, with profound changes to financial compliance. Tools like predictive analytics, automated code translation, and real-time monitoring don't just reduce expenditures, but also enhance agility. Ethical considerations for AI application, explainability, and merging with legacy systems knee-deep in bureaucracy are just a few of the countless hurdles waiting down the line. The moment calls for a new, cautious, oversight-driven strategy that carefully balances innovation with prudent governance. The ability to control data that is decentralized ensures compliance that is reliable, readily scalable, and adaptable to rapidly changing regulations.

In their work, Adewale (2022) emphasizes the importance of blockchain technology in assuring reliable, auditable, and fraud-free financial reporting. Also important, however, is the increased compliance trust facilitated by blockchain technology due to its fraud deterrence and improved auditability. Tokenization and AI add remarkable value to the technology. A government's strategic partnership with the industry is crucial to achieve the totality of blockchain's benefits. This is a foundational change that could redefine financial reporting and compliance for future generations.

AI combined with automation has improved the rate at which compliance, fraud detection, and overall audit quality is performed. This has greatly aided Real-time auditing, big-data analytics, compliance enforcement, and a myriad of other activities. The convergence of these elements is responsible for what is being termed the auditing industry's wild future. These give rise to the untapped realms of heightened risks, transparency, market dissatisfaction, regulatory control and compliance, meeting the market's expectations, and much more, that can only be satisfied with robust internal controls paired with continuous professional training.

Lastly, intertwining data from different platforms with RPA can automate intricate invoice processing in fund management, as demonstrated by Tamraparani (2020). Those benefits come in the form of better regulatory compliance, lower costs, and increased standardization across operations. Fixing system integration and data mismatches will be vital to realizing their full potential. The study recommends a phased, scalable RPA implementation approach backed by governance and infrastructure. Properly aligned with strategic objectives, RPA can revolutionize financial workflows enterprise-wide.

MAIN FOCUS OF THE CHAPTER / METHODOLOGY

The point of focus for this chapter is to analyze the revolutionary potential of Robotic Process Automation (RPA) in finance reporting and compliance, with particular interest in its ability to enhance accuracy, transparency, and operating effectiveness. The chapter adopts an integrative and multi-dimensional methodological approach to the understanding of the strategic value added by RPA to the finance function of organizations, with specific reference to evolving regulation needs and technology innovations.

Integrated into the methodology is thematic analysis of the most serious of issues and concerns for organizations in implementing RPA. They include complexity in merging with old finance systems, security issues around data due to the sensitive nature of the financial information, resistance to change by employees fearing job loss, compliance with regulations, and scalable and sustainable RPA solution requirements. These problems are examined in depth, with a focus on technical and organizational issues that have the potential to deter effective RPA implementation. Actual cases and examples underpin the analysis, detailing how these problems operate in practice and can be overcome with planning, effective governance, and effective change management.

A lot of the chapter is filled with actual implementations of RPA for financial processes. Examples include financial statement preparation, accounts payable and receivable, bank account reconciliations, reporting for compliance (e.g., AML and KYC), through to automatic audit trails. The following examples illustrate how RPA can be used to automate repetitive rule-based processes so as to eradicate human error, increase processing speed, and preserve the accuracy and consistency of financial information. With these examples, the chapter demonstrates that not only is business effectiveness augmented by RPA but also stakeholder trust and regulatory compliance.

Of course, the approach also has a problem-solution solution. It begins by admitting the fallacies of human processes in accounting reporting and compliance, such as delays, mistakes, and difficulties in maintaining audit accuracy. The chapter then presents a solution in the form of RPA, describing how it can bypass these deficiencies. It also delivers a strategic set of recommendations to organizations that intend to start RPA. They are standardizing and automating functions, using large-scale pilot testing approaches, increasing data security controls, and conducting relentless bot monitoring and maintenance. They also possess a Centre of Excellence

(CoE) to oversee automation initiatives. In addition to mainstream RPA, the chapter explores how it should be combined with the latest technologies such as artificial intelligence (AI), machine learning (ML), optical character recognition (OCR), Internet of Things (IoT), cloud computing, and blockchain.

These intersections complement RPA's strengths through smart decision-making, real-time processing of data, and secure massive scalability. The chapter highlights that such convergence of technology is the way forward for financial automation and helps companies remain adaptive and competitive in the ever-evolving digital landscape. The methodology framework is supplemented with second-order research in the guise of a comprehensive literature review and empirical study analysis presenting quantitative and qualitative evidence substantiating RPA's contribution towards financial reporting. The chapter, as a whole, follows a systems thinking approach by conjoining theory bases with practice knowledge to gain extensive knowledge of RPA's contribution toward reorganizing financial reporting and compliance.

ISSUES

Integration Complexity

It is not a straightforward job to integrate RPA with existing financial systems. It includes a thorough examination of the current process and framework so that they are compatible with one another. Different RPA tools also have to be integrated with heterogeneous legacy apps, databases, and platforms. Overcoming these integration challenges requires meticulous planning and custom development so as to design an integrated automation environment. Quite often, the desire of the administration to have an RPA platform installed can even result in the reconstruction of a whole IT infrastructure of a company. Most companies do some math on legacy code.

Data Security Problems

Due to the confidential nature of financial information, its security is of utmost importance. RPA robots have to obey strict data protection regulations. It involves encryption, access control, and strong authentication processes to protect confidential information. Security breach can cause enormous financial loss and damage business reputation. In case you decide to create an RPA bot for the finance sector, there is one thing you should be absolutely sure of: such software needs to be developed on a regular basis. It is not a question of having a team come up with something like this and leave it alone for years. There should be continuous testing and maintenance to prevent possible new weaknesses. Current RPA software can jeopardize the long-term security of the customer data, and, as one might say, any compromise on safety is not to be tolerated. To be a massive hit in this arena needs to remain on the lookout all the time.

Change Management

The adoption of RPA can trigger resistance from employees concerned about losing their jobs. We have already referred to a neo-Luddite tendency in contemporary economy. If you do not adequately articulate the nature of the RPA change, huge protests against the technology (to a large extent clandestine) are inevitable. How to eliminate this problem forever? You have to be, above all, an effective communicator. The aim is to make it clear that the RPA software is negative to use in automating mundane tasks for time release for more-value high-strategy work. Effective change management is equally essential to counter this negativity, however. It includes open communications, training, and retraining initiatives to inform employees that RPA is meant to augment their efforts. If you are an intelligent organization, we actually recommend giving written guarantees to the workers that your company will give them alternative work with the use of RPA technology. Make them secure in their jobs, and the majority of the frightened will end up joining it. What scares human beings about such innovation is the disrupting effect that such innovations have. Give security to individuals, and they will see the innovations as good. Compliance with Regulation Banks have to meet strict regulatory requirements that keep evolving.

Combining RPA in a way that maintains compliance is a continuous activity.

It entails monitoring and altering automated tasks so that they reflect the latest regulatory changes, minimizing the likelihood of non-compliance and associated penalties. We've mentioned that RPA helps maintain compliance because human operators lose track of the regulatory frameworks. RPA in accounting and finance gets along fine with no significant changes in regulations. It's perfect for reliable legislation with a long-established history. Meanwhile, these models do not work if there are periodic amendments to the laws. Why? Because one has to keep rearranging the applicable frameworks. The second concern is that RPA tools do not auto-configure. There must be human effort for this process. This leaves such websites to massive faults because of human negligence. How do we solve the problem once and for all? You have to experiment a lot. RPA systems save costs in long-term situations, but you need to prepare for short-term disturbance. Scalability As businesses grow, RPA solutions must scale along with them. Maintaining the scalability of RPA implementations in balance, particularly in regards to such factors as higher transaction values, higher processes, and rising proposals of services, is challenging.

Maintenance and Monitoring

RPA systems, as with any technology, require maintenance and monitoring from time to time. Maintaining the bots running without interruption, handling exceptions, and adjusting to process change requires dedicated resources. Maintenance is carrying out problems, scripting changes, and maximizing bot efficiency. Monitoring makes sure processes get executed smoothly and any glitch or bugs are identified early and fixed to avoid financial activity disruption. Keep in mind: a single error during maintenance process can lead to gigantic security loopholes. The more you try to control, the better. Applications of RPA in Finance and Auditing RPA is one of the principal technologies for banks, where it can simplify and automate multiple processes. Some of the examples of applications of RPA in finance and accounting are:

Account Payable/Receivable

RPA can even be used in automatically processing invoices and payments for suppliers. It is time-saving and error-reducing. The majority of modern banks, for example, Quantic in the United States of America, literally use RPA tools in handling invoices. They even go so far as to offer their customers an opportunity to use other items in handling such data. Quantic has, for example, developed a ring used in payment due to BankWise, a market leader in RPA.

Bank Reconciliation

Bank statement reconciliation with the accounting system transactions is automatically done by RPA. Through this feature, you will detect and correct the variances within time, saving effort and time. Here, data analytics ability bring in the useful insights smoothly. Robocorp, an automation technology provider company developing RPA, offers bank reconciliation as one of the stand-alone use cases of the provided technology.

Financial Reporting

RPA can also be used in the generation of financial statements such as balance sheets and income statements. This will be accurate and reduce the possibilities of errors. As a Ukrainian company, we can observe how obtaining this feature implemented is widespread in our country. Such banks as Privat or Monobank offer their customers a possibility to receive financial statements generated at the click of a button.

Compliance

RPA is used to automate compliance activities such as KYC (Know Your Customer) checks, AML (Anti-Money Laundering) checks, and regulatory reports. It puts the businesses in a position to be able to meet their regulatory needs. Companies such as Encompass, which carry out the KYC, say that it has several long-term customer benefits. They can determine the types of clients that must be targeted. Furthermore, the technology allows one to know if there are other security activities required.

Audit Trail

Using RPA, one can create and maintain an audit trail of financial transactions that can help bring about accountability and transparency. Most banks are plagued by the lack of unified perspectives of their core transactions because they keep information on paper. With accounting and financial RPA, there is a promise to create a system that will no longer keep you spending so much time searching for the most crucial information. The tools will collect the information themselves instead. You only need to execute reports in automatic mode after installation.

CONTROVERSIES

Worker Displacement and Change Resistance

Let's be real—when companies start rolling out RPA (that's robotics, not some new dance craze), people get jittery. It's hard not to, right? Folks are worried about losing jobs or suddenly finding out their job's now half-bot, half-human, like some weird sci-fi movie. Resistance? Oh yeah, it happens. So what do smart companies do? They actually talk to their teams and run training so people don't feel like they're getting shoved out the door. Change management isn't just a fancy buzzword—it actually matters if you want anyone onboard.

Over-Automation and Mismatched Expectations

Alright, so sometimes businesses get way too hyped about automation. They try to hand tasks to bots that really need brains, not just code. Spoiler alert: it usually backfires. Wanna avoid that mess? Figure out first what honestly makes sense to automate. Don't go slapping bots on every process just cause you can. Line up your RPA goals with what you actually need, not just what looks cool in a PowerPoint.

Lack of Transparency in Algorithmic Decisions

The thing about these RPA bots—they don't exactly narrate what they're up to. There's a lot of "black box" going on, which makes regulators and auditors cranky. If nobody can follow the decision trail, you're asking for trouble. Solution? Go for explainable AI, keep good logs, and don't hide the playbook. That way, if someone asks, "Uh, why did your robot just approve that?" you've got receipts.

Cost Overruns and ROI Uncertainty

Oh, and about that promise that RPA will save you a ton of money? Sometimes the price tag sneaks up on you—training, delays, bots that need to be reworked (again), headaches all around. It's not always the magic bullet you're dreaming of. If you actually want to get value, do the math upfront. Be real about what you'll save versus what you'll spend. Don't just swallow hype—run the numbers, set goals you might realistically hit, and maybe hold off on celebrating till the dust settles.

PROBLEMS

Problems of Manual Invoice Processing Systems

To appreciate the extent of disruption caused by RPA, one needs to examine the shortcomings of manual invoice processing systems, such as:

Manual Data Entry: Supports more human error-driven processes and diminishes process speed.

Paper-Based Processes: Cause workflow complexity, missing documents, and tracing issues.

Lack of Real-Time Visibility: Hinders the detection of bottlenecks or errors until process run-through.

Delays and Inefficiencies: Extend the time taken to process, approve, and settle bills.

Difficulty in Retrieval and Analysis of Data: Difficult to audit, project, and conduct compliance exercises.

Limited Scalability: Hand-driven systems cannot keep pace with rising transaction volumes without corresponding rises in staff.

Compliance and Audit Risks: Non-standard documentation and practices jeopardize audit failure and regulatory penalties.

Such issues were then compounded by the COVID-19 pandemic when work-from-home highlighted the inefficiencies of paper-based, manual processes and accelerated the demand for digital fixes like RPA.

The Emergence of Invoice Processing Automation using RPA

RPA offers an orderly and patterned method of automating rule-based accounting tasks. With regard to invoice processing, automation entails:

Capturing and Extracting Invoice Details: Scanning and reading data from electronic mail attachments or scanned paperwork.

Vendor Verification of Information: Cross-matched with master databases for consistency.

ERP Filling: Automated entry of invoice information into internal systems without human input.

Automated Payment Processing: Once validated and approved, automatically initiating payment processing. Relative to other automation technology such as Business Process Management (BPM), RPA is significantly faster to deploy and more cost-effective for rule-based processes. RPA can be deployed in legacy systems as well as web-based applications, which makes it a great option for companies that are not quite ready to make big-scale infrastructure changes.

Beyond Invoice Processing: Other Financial Functions Automated by RPA

Aside from invoice processing, RPA is also applied heavily in other finance areas like:

Accounts Payable: Invoice verification, bill payment, and reporting.

Accounts Receivable: Customer data processing, credit status tracking, and invoice sending.

Financial Reporting: Rapid and effective verification of balance sheets, income statements, and regulatory reports.

Intercompany Reconciliation: Verification, extraction, and reconciliation of financial information between in-house departments.

All such processes have the same aches—manual, disparate systems, delays, and non-compliance risks—and are thus ideally appropriate for RPA transformation.

Challenges Still Remaining Even After RPA Implementation

Although RPA significantly accelerates processing time and minimizes errors, it does not necessarily address all issues. Some typical post-implementation problems are:

Bot Failures: Owing to unforeseen changes in UI or system structures.

Maintenance Overhead: Bots will need frequent updating and maintenance to function dependably.

Lack of Process Standardization: Inconsistent workflows can cause unexpected automation failures.

Over-reliance on Automation: Can result in loss of institutional memory and less human supervision.

Scalability Bottlenecks: Although bots can handle high volume, backend infrastructure and approval hierarchies might still create scalability bottlenecks.

SOLUTIONS AND RECOMMENDATIONS

Standardize and Optimize Processes Before Automation

Before a process in finance can be automated, an organization must eliminate inefficiencies by standardizing processes. Efficiencies through streamlining workflows remove errors and make it easy to simplify and streamline automating processes. They involve mapping existing processes, eliminating redundancy, and harmonizing procedures in departments. Standardizing not only enhances the performance of RPA but also the risk of implementation. An explicitly defined process is a bot script and predictability of results improves. Accounting processes like journal postings or reconciliations are also facilitated by this defined process. This process also facilitates easy accomplishment with compliance checks and audits. Omission of this process in organizations usually results in higher failure rates in RPA adoption.

Define Suitable Processes to Automate

Not all is automatable; which ones to pick must be decided. The choice of work best suited for RPA is high volume, rules-based, and repetitive in nature—such as vendor payments, invoice processing, and journal posting. High-complexity, high-value judgment-intensive and interpretation-based tasks must remain human-heavy. Processing only structured processes will yield the maximum ROI and efficiency. Automated finance reduces errors, speeds up turnaround time, and enhances staff morale. Improved automation potential identification is achievable by the use of tools such as process mining. Inaccurate decision-making could lead to inefficient use of bots or project failure. Long-term success will require strategic planning.

Employ Robust Testing and Validation Procedures

Before bots can be deployed on a large scale, they need to undergo rigorous testing. This entails them functioning properly in regards to their exception handling, and system compatibility. The financial systems within an organization can incur severe damage during peak reporting periods. With unvalidated robots, there is a lot that can go wrong. Test environments provided aid in identifying potential logic errors or integration problems earlier. Validation is also important at this point as it ensures the correct data is fed into different templates, and so it is not financial data which is sensitive and hence of utmost importance. It is critical that records are not scrambled. A stable robot enables safeguard confidence that the information provided is true. Validation of the robot must take place at each phase within the bot lifecycle management continuously.

Enhance the Controls for Data Security:

Handling of sensitive financial reporting dat is within the domain of concern leading data security within an organization to be highly prioritized. Secure access control, multi-factor authentication, and encryption ought to be inclusions within RPA platforms. With unauthorized access comes data breaches, penalty fines, and a significant loss in stakeholder trust. Bots within organizations have to deem accept the principle of least privilege which is defined as accessing only the data essential to their task. Consistent monitoring of bot activity aids in encouraging malicious actions, while regular audits bolster security

Ensure Ongoing Monitoring and Maintenance

After the deployment phase, RPA platforms must be monitored without interruption to confirm that their performance is satisfactory. On occasion, bots can experience malfunctions due to changes in software interfaces, legacy logic, or default data formats. Making sure that monitoring is continuous ensures that disruptions, from problems in Reporting, are avoided. RPA maintenance and the logging of executed tasks are performed to ensure lasting stability. Incident handling and escalation should be done by other units providing RPA support. Performance of the bots can be checked using watching software real time. Performance of automation can also be enhanced using this data. If a dynamic, evolving system such as RPA is to succeed, it requires attention.

Advance Human-Bot Collaboration

Rather than substituting human intelligence with a tool as is the case with RPA, here is an avenue for collaboration. With RPA, routine work done by bots allows humans at planning, decision-making, or analysis which are not repetitive. This type of collaboration increases productivity while improving employee morale. RPA is capable of drafting basic financial reports for further review and analysis by financial experts. Such environments enhance acceptance and reduce opposition from employees towards automation. Such a partnership model should be regarded in change management programs. Granting employees freedom to work and improve bots provides a sense of ownership. Human-bot collaboration creates higher-quality outputs and more acute financial processes.

Establish a Center of Excellence (CoE)

A Center of Excellence (CoE) provides RPA initiatives with core governance and strategic oversight. The CoE ensures that standardized tools, templates, and best practices are used across the firm. The CoE also helps prioritize use cases, resource management, and training. By sharing centralized knowledge, the CoE reduces effort duplication and project risk. It becomes a center of innovation, concentrating on how IT, finance, and compliance teams overlap. In regulated industries, the CoE could also serve as a liaison between legal and audit. Having a CoE accelerates maturity and scalability of RPA adoption.

Engage and Communicate Stakeholders

Efficient communication with stakeholders is critical to achieving RPA success. Misunderstanding and fear of job loss normally trigger resistance. Involve onboard leadership, finance experts, IT, and compliance officers to provide everyone with a sense of ownership. Share openly success stories, feedback, and pilot outcomes. Change sponsors might be recommended to sponsor the automation benefits. Stakeholders must be involved at the planning level so that they are aligned to business goals. Feedback mechanisms must be instituted to incorporate user feedback on deployment. Open communication fosters trust, does away with resistance, and boosts adoption.

RPA Solutions to Be Aligned to Organization Needs

Solutions for RPA that are one-size-fit-all never work in complex financial environments. Customization allows the automation to be tuned to specific accounting procedures, regulatory needs, and system setups. This involves setting the bots to track internal controls, reporting cycles, and financial calendars. Seamless integration with tax software, ERPs, and banking portals is most often required. Flexible capability robots also have the ability to provide regulatory updates and exceptions. Tailoring takes more effort upfront but provides long-term advantage. Tailored bots reduce interventions and workarounds. Solutions must be aligned to the digital maturity and compliance approach of the organization.

Pilot Test Before Large-Scale Deployment

Organizations should pilot test before large-scale deployment of RPA across all financial processes. A pilot is a proof-of-concept that ascertains the feasibility and ROI of automation. Bot logic simplification, process bottlenecks, and stakeholder issues are revealed through it. Pilot testing needs to cater to a variety of scenarios—normal cases, exceptions, and stress tests. Decision-making for broader implementation is enabled by performance data from the pilot. It also gives confidence to teams that RPA delivers value. A responsible pilot can be a model for other departments. The pilots' experience is the success factor to scale up responsibly.

DISCUSSIONS

The recent years have seen the advent of Robotic Process Automation, which handles repetitive and rule-based processes with utmost efficiency and minimal error rates. RPA can be combined with other technologies for greater effect. Hyper automation is the term used to denote the integration of numerous automation technologies, such as RPA, AI, process mining, and more, to automate business processes as a whole.

Some of the technologies that can be combined with RPA to make it more efficient are mentioned below:

RPA integration with ML/AI

Would you have ever thought of a robot reviewing films or giving marketing suggestions to aid you in your business decisions? Such and many more sophisticated automation technologies will become a near-future reality. RPA can also be used in combination with AI and ML to create intelligent systems with the capacity to make recommendations based on individual preferences. RPA can automate data harvesting, data scrape and consolidate it into a centralized database.

Data processing activities to extract information and transform data can be carried out to enable AI systems to execute advanced capabilities such as cognitive functions, natural language processing, and image recognition, sentimental analysis, and predictive analytics. An example of enhanced customer service with ML/AI integration with RPA is as follows - RPA and AI-based sentiment analysis tools integration inspects customer feedback from various sources such as social media, surveys, and reviews.

RPA bots are capable of determining the mood and comments given by clients which trigger actions like resolving customer complaints, receiving top tier replies, or spawning a chance for a product or a service. A different instance is a vacation planning bot. Starting with the travel dates, destination wishes, and spending limit to hotel preferences and hobbies, the user can submit information for AI algorithms to process, and the bot can generate suitable alternatives. The bot also provides users with booking information along with visa requirements, weather, local customs, and other means of travel. It is capable of assisting with navigation and language translation, as well as emergency contact. Overtime, the bot is capable of refining suggestions and becoming more accurate as a result of learning from user interactions and feedback.

Chatbots and Virtual Assistant Bots

Responding to frequent questions, updating client details, and forwarding requests to the right department are all common tasks that need to be done with RPA the can perform. Integrating it with a virtual assistant or chatbot allows the entire customer care interaction and help desk support within the company to be automated. As a result of this integration, customer experience is improved, and response times are drastically reduced.

IoT and RPA

We are surrounded today by IoT devices. Some common examples include smart domestic equipment such as smart thermostats, smart fridges, and smart TVs, smart wearables such as fitness trackers, watches, and earbuds, and healthcare equipment such as internet-connected heart monitors and blood sugar monitors, internal monitors, and security devices such as smart domestic security cameras and sound recording equipment in home, business, and industrial settings, to name a few. The incorporation of IoT devices along with RPA technologies can lead to even greater outcomes. One such example is a bot that can call the nearest medical practitioner or set off the emergency protocols when a patient undergoes a medical emergency using an IoT wearable medical device. This device can pull information such as heart rate, blood pressure and blood glucose levels and send them to a bot that can analyze and make life critical decisions based on the fetched information.

A potential alternative for agriculture can be the utilization of IoT sensors to keep tabs on various parameters such as soil moisture, temperature and weather around the crops. RPA bots installed on the sensor would be able to analyze the data so that irrigation systems would be triggered, fertiliser application rates would be adjusted, or optimal planting or harvesting times would be provided. This contributes to higher crop yields, conserving resources, and enhancing efficiency within agriculture. Therefore, through the combination of RPA and IoT, organizations are able to capture the information stored in IoT devices, perform actions based on pre-set rules, and automate procedures such as changing environmental controls, equipment performance evaluation, and issuing repair requests. Such interconnectivity enhances productivity and enables monitoring and control in real-time.

OCR and RPA Integration

OCR technology allows RPA systems to read and pull out valuable information from photographs or scanned documents. Organizations automate data entry work involving big numbers of physical or electronic documents by integrating OCR and RPA. The integration wipes out manual data entry which reduces errors and increases process speed. OCR technology is being paired with RPA extensively today for invoice processing, document categorization, form processing, legal document/contract extraction and validation, auto-data entry, etc.

These may appear elementary at first glance but will have widespread applications in the future. For instance, a product that includes RPA and OCR would be able to take care of the whole hiring life cycle of a business, from resume collection, screening, skill mapping, entry of candidate data, interview scheduling, and background checks of candidates to onboarding. It reduces manual effort and significantly accelerates the recruitment process.

RPA with Process Mining and Analytics

Process mining and analytics are two complementary approaches used to discover how processes are executed by improving business processes and event logs analysis of an organization. Subsequent the knowledge gained from process mining, RPA developers can plan and implement RPA bots for automating certain processes, which are time and labor-intensive when done manually. By using RPA with process mining and analytics tools, organizations can determine bottlenecks, inefficiencies, and areas of improvement within their automated processes. This integration offers actionable data-driven insights that allow organizations to streamline their processes and drive ongoing process improvement.

Robotic Desktop Automation (RDA) Bots

RDA is employed in user interface-level automating, simulating human usage of desktop applications. With the combination of RPA and RDA, organizations can automate end-to-end processes that span both back-end systems and user interfaces. This yields end-to-end automation across many applications, enhanced productivity, and reduced manual effort.

Cloud Computing and RPA Integration

The integration of RPA systems with cloud computing can improve scalability, flexibility, and accessibility. Instead of running RPA bots on local machines, organizations can leverage cloud platforms to host and manage the bots. This helps them to supervise and manage bot execution control, bot workload scheduling, performance tracking, and other related parameters telemetry. Since Cloud storage gives RPA bots a particular easy to reach and organized location, they are able to access the Cloud to read input data, write output data, and read necessary documents or files irrespective of the location of the bot host. Easy access for communication and data transfer between systems is provided by Cloud services like APIs, web services, and databases. Cloud platforms like AWS, Azure, or Google Cloud offer APIs to interfaces to access other services such for storage, analytics, machine learning, and natural language processing. APIs are used by RPA bots to deal with errors and other exceptions during automation. This facilitates the automation process in dealing with exceptions, thus enabling streamlined execution of activities.

FUTURE RESEARCH DIRECTIONS / IMPLICATIONS

The future of RPA research in financial reporting and compliance has a lot of encouraging directions and implications calling for further research. While companies continue to struggle with increasingly fast-shifting digital and regulatory environments, there exists a pressing necessity for empirical research that determines the long-run influence of RPA on financial accuracy, quality auditing, and compliance competence. One such key area of research in the future would be the integration of RPA with AI, ML, and blockchain technologies. While these are being exhaustively discussed in abstract frameworks, empirical studies would be needed to quantify the degree to which these next-generation systems impact decision-making, forecasting analysis, and real-time monitoring of compliance in the actual world. Future research must also examine how intelligent automation can transition from basic rule-based activities to more sophisticated judgment-based financial activities and how that will influence internal controls, financial governance, and audit integrity.

An additional area of profound significance for future study is the the human aspect concerning the adoption of RPAs. Their studies mostly look at the lack of acceptance and the concern of job displacement; however, there seems to be a gap in organizational and mechanization psychological impact on automation concerning finance professionals. Some studies may explore how the functions of accountants and auditors are evolving with computers and what skills are necessary for effective management and coexistence with automated systems. Other scholars may address issues of what change management strategies and training programs enhance RPA positive uptake and proactive participation from employees. Such insights will help organizations trying to maintain staff morale and retention while triaging complete organizational automation.

Evolving and deepening the scope of RPA implementations are other set of robust avenues out there for future research. Long-term case studies would assess the productivity, servicing, and ROI of the RPA systems hands-on maintained over time. There is also interdisciplinary interest in governance models such as Centers of Excellence (CoEs), which integrate best practices with risk and compliance management at the enterprise-wide level for automation programs. Research could also examine the impact of regulation on RPA particularly in relation to those regions with dynamically shifting compliance requirements.

Cross-industry and cross-country comparative research can provide an insight into the impact of regulatory environments on the effectiveness or limitation of RPA for financial reporting. Transparency and ethics of algorithmic decision-making are another crucial area of research. With RPA systems getting more independent and filled with AI, issues of accountability, explainability, and data privacy are going to become more important. Future research can investigate models facilitating algorithmic transparency and ethical adherence, particularly for applications that process sensitive financial and personal information. Research can also compare the use of external audits in verifying RPA-created financial processes and how assurance approaches need to change in an automated environment.

Last but not least, inter-disciplinary research will be needed to integrate finance, information technology, organizational behavior, and legal studies to provide a comprehensive report of the implications of RPA. Although technological automation transforms the character of financial work, holistic models and frameworks need to be formed that inform strategic decision-making, policy-making, and curricula. As a whole, research in this sector in the future will have the potential to not only alter how organizations function but also how the finance profession itself evolves in the face of constant digital innovation.

CONCLUSION

Speed, simplicity, and precision in today's regulatory and financial environment are more crucial than ever, and Robotic Process Automation (RPA) appears to be not only a technology upheaval but also a revolution force. In this chapter, it has become evident that RPA has been revolutionizing financial reporting and compliance procedures by automating rule-based, monotonous processes that were otherwise time-consuming, error-filled, and resource-wasteful. With intense thematic examination, practical uses in everyday life, and convergence with new-fangled innovations, the study examined the ways in which RPA enhances operating efficiency, reduces expenses, facilitates regulatory compliance, and ultimately leads to organizational resilience and agility.

The worth of RPA to financial reporting is that it can reduce human errors, speeds up reporting cycles, and ensures consistency and accuracy of information. By automating such tasks as gathering of data, journalization, reconciliations, variance analysis, and running of reports, the organizations are in a position to prepare financial statements that are more timely and accurate. Not only does it help ensure compliance with close-reporting deadlines but also help to enhance regulator, investor, and internal stakeholder confidence. In compliance processing, RPA delivers twenty-four-seven monitoring, transactional rules-based processing, and audit trail creation—ensuring a tremendous decrease in the risk of non-compliance, fraud, and regulatory fines.

In addition to pointing out the importance of RPA for business, this study reveals its deeper implications. Beyond workflow streamlining, RPA nurtures a culture of proactivity and transparency, allowing finance units to access information that aids decision-making within the organization in real time. It enables human resources to break free from repetitive and administrative work and shifts them to more important tasks such as planning, forecasting, branding, and offering expert advice. This reallocation improves the employee's as well as the finance function's satisfaction with the overall strategy of the firm.

Also, the integration of AI, ML, OCR, iOT, cloud technologies, and blockchain with RPA gives a glimpse of the potential future for automation in finance. These technologies work together to form "intelligent automation;" the capability of systems to autonomously read unstructured data, learn patterns, and contextually make decisions. This merger enables the use of RPA beyond automation to smart process orchestration, predictive analytics, and sophisticated risk management. It changes financial functions from merely responding to queries in back offices to proactively generating strategic insights that drive growth and foster innovation within organizations.

Adopting RPA and the benefits that come with it, is seldom considered peculiar according to one study. The chapter had also mentioned RPA controversy and job replacement threats, over-automation, transparency of algorithms, and inflation of costs in passing. It had mentioned how one-way strategy—completely technology-driven as the origin of efficiency, intelligence, and quality—would be impossible to implement. Rather, corporations must consider RPA as an enabler—it elevates human capability, facilitates better decision-making, and facilitates strategic objectives in the long term.

This study has extensive application in the usage. As more companies require accurate, timely, and transparent financial reporting, RPA is a scalable, affordable, and viable option to organizations of all business and type. It is most attractive in highly regulated industries like insurance, banking, and health care, where compliance isn't desired but business-imperative. Scalability enables RPA to scale with the business, responding to evolving regulatory requirements, complexity in the market, and operations size.

Success of RPA in the long term will depend significantly on how well it is integrated with people development, organizational strategy, and technology infrastructure. Employees must be developed and changed to enable easy adoption and prevent maximum resistance. Strong governance practices for managing security, compliance, performance management, and continuous improvement must be established. With changing technology, it is necessary to change the governance and ethics fundamentals behind the implementation of RPA.

In brief, Robotic Process Automation is not a phenomenon but rather a paradigm shift within financial operations and compliance. It re-establishes the limits in relation to speed, accuracy, and value. If properly utilized and supplemented by other technologies and governance mechanisms, RPA can reshape organizations to be more open, responsive, and competitive. This research has provided a deep source of understanding of the revolutionizing potential of RPA and has provided practitioners, policymakers, and researchers with actionable advice on how to leverage automation to facilitate financial reporting and compliance

REFERENCES

Adewale, T. T., Olorunyomi, T. D., & Odonkor, T. N. (2022). Blockchain-enhanced financial transparency: A conceptual approach to reporting and compliance. *International Journal of Frontiers in Science and Technology Research, 2*(1), 024-045.

Alao, O. B., Dudu, O. F., Alonge, E. O., & Eze, C. E. (2024). Automation in financial reporting: A conceptual framework for efficiency and accuracy in US corporations. *Global Journal of Advanced Research and Reviews, 2*(02), 040–050.

Alonge, E. O., Dudu, O. F., & Alao, O. B. (2024). The impact of digital transformation on financial reporting and accountability in emerging markets. *International Journal of Science and Technology Research Archive, 7*(2), 025-049.

Gotthardt, M., Koivulaakso, D., Paksoy, O., Saramo, C., Martikainen, M., & Lehner, O. (2020). *Current state and challenges in the implementation of smart robotic process automation in accounting and auditing*. ACRN Journal of Finance and Risk Perspectives.

Harahap, M. (2021, January). Analysis of implementation of robotic process automation: a case study in PT X. In *Proceedings of the Business Innovation and Engineering Conference 2020 (BIEC 2020).*

Kothandapani, H. P. (2025). AI-driven regulatory compliance: Transforming financial oversight through large language models and automation. *Emerging Science Research*, 12–24.

Oyeniyi, L. D., Ugochukwu, C. E., & Mhlongo, N. Z. (2024). Robotic process automation in routine accounting tasks: A review and efficiency analysis. *World Journal of Advanced Research and Reviews*, *22*(1), 695–711. DOI: 10.30574/wjarr.2024.22.1.0281

Tamraparani, V. (2020). Automating invoice processing in fund management: Insights from RPA and data integration techniques. *SSRN*. https://doi.org/DOI: 10.2139/ssrn.5117121

KEY TERMS AND DEFINITIONS

Artificial Intelligence (AI): AI functions by simulating human mental processes on computers which enables them to execute learning and reasoning and decision-making functions. RPA integration with AI brings about intelligent automation capabilities through its ability to interpret unstructured data for adaptive decision-making.

Audit Trail: A chronological series of financial operation documentation serves as an audit trail for tracking business operations and verifying financial information. RPA systems create and sustain audit trails without human intervention which enables transparent and traceable records for regulatory audits.

Center of Excellence (CoE): The Centre of Excellence for RPA establishes a unified system for supporting automation strategies while monitoring best practices and performance across the entire organization. The RPA standardization process occurs through its risk management approach which enables sustainable implementation.

Change Management: Organizational change management requires guiding and supporting members through the process of adopting new technological developments or operational methods. The successful implementation of RPA depends on effective change management which reduces resistance while helping staff members develop new skills and ensuring a smooth transition.

Financial Reporting: Financial reporting involves creating financial statements that demonstrate how a company performs and its financial standing. RPA establishes a more dependable approach by collecting data automatically while performing calculations and preparing reports in the necessary structure to minimize errors and save time

Regulatory Compliance: Businesses achieve regulatory compliance through their ability to follow established laws and guidelines that govern their operational activities. RPA supports financial compliance efforts by delivering reports on time while establishing audit trails and providing continuous monitoring processes.

Robotic Process Automation (RPA): Software robots perform rule-based tasks that people typically perform in digital environments when used through Robotic Process Automation (RPA). The financial reporting process benefits from RPA because it automates three core functions: data entry, reconciliation activities and reporting preparation to enhance accuracy and efficiency.

Chapter 5

The Effect of Integration of Robotic Process Automation With Artificial Intelligence on Accounting Practices

Suresh Balasingam
https://orcid.org/0009-0003-7282-3372
Asia Pacific University, Malaysia

ABSTRACT

The integration of Robotic Process Automation (RPA) and Artificial Intelligence (AI) is profoundly transforming modern commercial operations. This chapter seeks to delineate and differentiate RPA and AI, while highlighting the strategic justification for their convergence. It examines the architectural designs and frameworks that facilitate intelligent automation, providing insights into the operational synergies among different technologies. This chapter illustrates how integrated solutions enhance efficiency, accuracy, and innovation across several sectors through the examination of distinct use cases and real-world implementations. Additionally, it evaluates the concrete and abstract effects of automation on organisational efficacy and workforce interactions. A thorough examination of ethical considerations, regulatory challenges, and threats related to intelligent automation is conducted to offer a balanced viewpoint. Ultimately, forthcoming trends such as hyper automation and cognitive RPA are examined, highlighting their impact on the evolution of company strategy and workforce transformation. This chapter provides a thorough guide for scholars, professionals, and policymakers aiming to understand the changing dynamics of intelligent automation.

DOI: 10.4018/979-8-3373-3571-1.ch005

1. INTRODUCTION

The swift progression of digital technologies has profoundly altered the corporate operations landscape, especially in the field of accounting. Robotic Process Automation (RPA) and Artificial Intelligence (AI) are among the most significant developments, since these interconnected technologies are transforming the execution, management, and optimisation of accounting processes. As enterprises want efficiency, precision, and instantaneous decision-making, these technologies provide persuasive instruments to augment conventional accounting methodologies. Robotic Process Automation denotes the utilisation of software robots or "bots" to automate highly repetitive, rule-based processes typically executed by human accountants. These responsibilities may encompass invoice processing, data input, reconciliations, and compliance reporting. RPA is defined by its capacity to replicate human operations in digital environments without requiring intricate programming or system modifications. It functions across several applications and can substantially diminish manual errors, expedite work completion, and save operational expenses. Conversely, Artificial Intelligence comprises a wider array of technologies that allow machines to replicate human intelligence. In the realm of accounting, AI may execute functions including natural language processing, predictive analytics, anomaly identification, and decision-making procedures. Machine learning, a branch of artificial intelligence, enables computers to derive insights from data patterns and enhance their performance over time, rendering it particularly advantageous in areas such as auditing, fraud detection, and financial forecasting.

The implementation of RPA and AI in accounting presents several benefits, although it simultaneously introduces a distinct array of obstacles and constraints. A primary constraint is the reliance on structured data; RPA, specifically, encounters difficulties with unstructured or semi-structured material unless integrated with AI technologies such as Optical Character Recognition (OCR). The adoption of these technologies frequently requires substantial expenditure in change management, governance, and cybersecurity. Additionally, there are apprehensions regarding job loss and the ethical ramifications of automation in accounting. Although regular employment may decline, there is an increasing demand for experts capable of managing, analysing, and interpreting automated outputs, resulting in a transformation of necessary skill sets. The human component judgment, ethical considerations, and professional scepticism remains indispensable in numerous high-stakes financial decisions.

This chapter seeks to establish a fundamental comprehension of RPA and AI within the realm of accounting. It examines the capabilities these technologies offer spanning process efficiency to enhanced accuracy and addresses their practical implementations in actual accounting operations. Concurrently, it rigorously evaluates the constraints and hazards linked to their utilisation, encompassing technological, organisational, and ethical aspects. The incorporation of RPA and AI in accounting is not merely a potential but an inevitability as we go. Comprehending their capabilities and constraints is crucial for accounting professionals, educators, and decision-makers aiming to remain pertinent in a swiftly transforming digital economy. This chapter prepares for an in-depth examination of how these technologies are influencing the future of accounting procedures and the accountant's position in the age of intelligent automation.

2. INTELLIGENT AUTOMATION: CONCEPT AND EVOLUTION

The field of accounting is currently going through a period of profound change that is being pushed by the development of new digital technologies. Robotic Process Automation (RPA), Artificial Intelligence (AI), and advanced analytics are all coming together to form Intelligent Automation (IA), which is at the forefront of this transition. By incorporating cognitive characteristics such as learning, reasoning, and decision-making into ordinary business processes, artificial intelligence (AI) represents a substantial advance beyond simple automation. Intelligent automation is rethinking typical workflows in the context of accounting, which is leading to increased efficiency, improved accuracy, and the ability to make decisions based on data.

Rule-based automation (RPA) and artificial intelligence (AI)-driven technologies such as machine learning, natural language processing (NLP), and computer vision are the foundations upon which the idea of intelligent automation is built. While RPA can handle organised and repetitive tasks on its own, IA enables systems to adapt, analyse unstructured data, and improve over time through learning without the need for human intervention. Automating complicated accounting operations, such as financial forecasting, audit procedures, and fraud detection, which were previously dependent on human skill and judgement, is now possible because of this development.

One can trace the development of automation in accounting through three important phases: manual procedures, basic automation, and intelligent automation. Each of these phases serves a distinct purpose. The traditional method of accounting relied significantly on human data entry, ledger management, and reporting based on spreadsheets. Beginning in the latter half of the 20th century, the first wave of automation was defined by the development of software tools, which assisted in the reduction of clerical errors and the streamlining of procedures. The advent of robotic process automation (RPA) in the early 2000s made it possible for software bots to handle high-volume, rules-based jobs across a variety of different systems, which further improved the efficiency of processes.

The limitations of basic automation, on the other hand, became more apparent as business environments got more data-driven and dynamic. The creation of intelligent automation, which is a system that can handle variability, read natural language inputs, and learn from historical data, was a direct result of this. By adopting optical character recognition (OCR), today's information assurance systems can scan and classify invoices, conduct sentiment analysis on customer comments, and forecast cash flow trends using machine learning models. Intelligent automation is being implemented across a wide variety of functions in accounting practices. IA, for instance, can automatically reconcile bank statements, identify abnormalities in transactions for the purpose of auditing, and create compliance reports with minimal participation from humans. Internal auditors can interpret ever-changing legislation and change calculations properly when it comes to tax preparation. This software not only helps users save time and money, but it also improves the precision and dependability of the information they provide regarding their finances.

However, despite the numerous advantages it offers, the implementation of intelligent automation is not without its difficulties. It is necessary to make substantial investments in technology, training, and change management to implement successfully. In addition, it is necessary to address issues around the privacy of data, the application of algorithmic bias, and the displacement of conventional roles. IA, on the other hand, is not intended to take the place of accountants but rather to enhance their capabilities, so enabling professionals to concentrate on other important responsibilities such as strategy planning, consulting services, and ethical oversight. The introduction of intelligent automation represents a significant turning point in the development of accounting procedures. The application of IA has the potential to revolutionise the way accounting work is carried out since it combines cognitive technology with process automation. Accounting professionals will need to have a solid understanding of intelligent automation and the ability to utilise it to maintain their agility, relevance, and readiness for the future as more and more businesses embrace digital transformation.

Figure 1. Intelligent Automation in Accounting

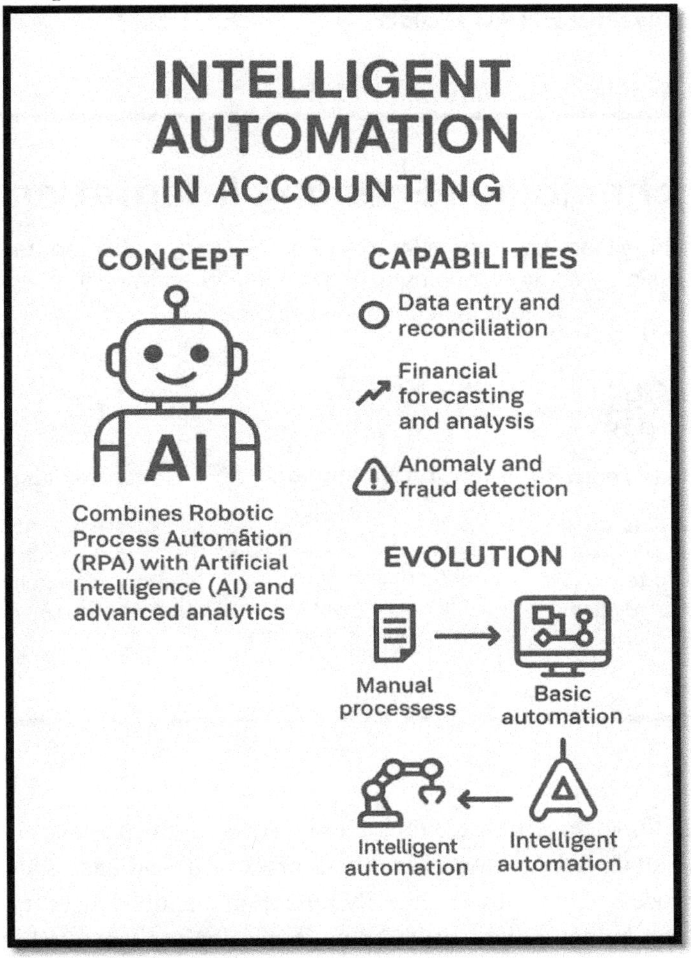

3. TECHNOLOGIES ENABLING INTEGRATION IN ACCOUNTING PRACTICES

Figure 2. Technologies Enabling Integration

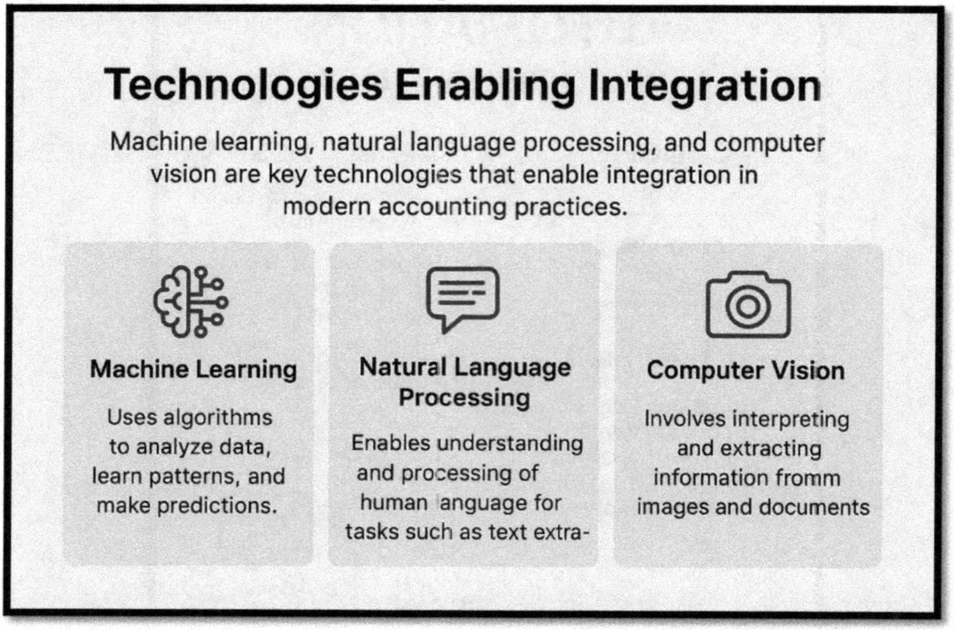

The integration of emerging technologies into accounting practices is revolutionizing how financial information is captured, processed, and analysed. Traditional accounting processes, which were heavily reliant on manual data entry, rule-based procedures, and human judgment, are now increasingly augmented by advanced technologies such as Machine Learning (ML), Natural Language Processing (NLP), and Computer Vision. These technologies not only enable automation of routine tasks but also open new avenues for strategic analysis, risk assessment, fraud detection, and decision-making. Understanding the specific roles of these technologies and their integration within accounting practices is crucial for academics, practitioners, and policymakers aiming to navigate the future of the profession.

Machine Learning (ML) is a subset of artificial intelligence (AI) that focuses on building systems capable of learning from data and improving performance over time without being explicitly programmed. In accounting, ML models can be trained to detect patterns in financial transactions, predict future cash flows, and assess credit risks. ML algorithms excel at anomaly detection, making them invaluable tools for internal auditors and forensic accountants. For instance, supervised learning models can be trained on historical financial data to classify transactions as legitimate or suspicious, thereby enhancing the efficiency and accuracy of fraud detection (Davenport & Ronanki, 2018). One prominent example of ML application is in predictive analytics for budgeting and forecasting. Companies now leverage ML models that analyse historical financial statements, market trends, and operational data to generate highly accurate financial forecasts. These models continuously learn and adapt to new data inputs, improving their predictive accuracy over time. Such capabilities significantly reduce the reliance on static historical data and enhance an organization's agility in financial planning.

Natural Language Processing (NLP) enables machines to understand, interpret, and respond to human language in a meaningful way. In accounting, NLP is employed to automate the processing of unstructured textual data such as contracts, audit reports, regulatory documents, and correspondence. NLP technologies facilitate information extraction, sentiment analysis, and document summarization. For example, NLP can be used to automatically extract key terms and obligations from lease contracts to ensure compliance with IFRS 16 standards, a task that traditionally required labour-intensive manual review (Gepp, Linnenluecke, O'Neill, & Smith, 2020). Another powerful application of NLP is the automation of audit processes. Modern audit tools equipped with NLP capabilities can analyse large volumes of emails and transaction narratives to identify indicators of financial misconduct or deviations from standard operating procedures. Furthermore, NLP-powered chatbots are being integrated into finance departments to handle routine inquiries, thus improving operational efficiency and freeing up accountants to focus on higher-value activities.

Computer Vision, another transformative technology, enables machines to interpret and derive meaningful information from visual inputs such as images and videos. In accounting, Computer Vision technologies are increasingly used for tasks such as automated invoice processing, inventory verification, and asset audits. For instance, Optical Character Recognition (OCR) technology, a subset of Computer Vision, allows financial systems to capture data from scanned receipts and invoices automatically, eliminating the need for manual data entry and reducing the risk of errors (Moffitt, Rozario, & Vasarhelyi, 2018). In asset management and auditing, Computer Vision systems can analyse images of physical assets to verify their existence and condition, supporting accurate valuation and compliance reporting. By automating the verification process, firms can conduct more frequent and comprehensive audits with reduced costs and enhanced reliability.

A compelling case study illustrating the integration of these technologies is that of Deloitte's "Argus" platform. Argus leverages a combination of Machine Learning, NLP, and Computer Vision to revolutionize the audit process. It can automatically review and analyze thousands of documents, identify key risks, and generate insights for auditors. For example, during an audit, Argus uses OCR to digitize paper documents, NLP to extract and understand financial terms, and ML algorithms to assess the consistency and accuracy of the reported data. This integrated approach has not only improved the efficiency of Deloitte's audit engagements but has also enhanced audit quality by providing deeper and more data-driven insights (Deloitte, 2019). The synergistic use of ML, NLP, and Computer Vision results in a fully integrated, intelligent accounting ecosystem capable of operating with minimal human intervention while delivering superior analytical capabilities. Despite the transformative potential, integrating these technologies into accounting practices is not without challenges. Data quality remains a critical concern. Machine learning and NLP models require large volumes of clean, structured, and representative data to function effectively. Inconsistent or biased data can lead to inaccurate models and flawed outputs. Additionally, integrating advanced technologies with legacy accounting systems often requires significant investments in infrastructure, change management, and cybersecurity.

Skill gaps represent another major hurdle. Accountants of the future must possess not only traditional accounting knowledge but also technical fluency in data analytics, machine learning, and digital tools. Educational institutions and professional bodies must therefore redesign their curricula and training programs to equip accountants with the necessary interdisciplinary skills (Sutton, Holt, & Arnold, 2018). Ethical and regulatory considerations also come to the forefront when deploying intelligent systems in accounting. Transparency, explainability, and accountability are critical issues, especially when automated systems are used to make financial judgments. Regulators are increasingly scrutinizing the use of AI and automation in financial reporting and auditing, necessitating robust governance frameworks to ensure compliance and public trust.

Looking to the future, the integration of emerging technologies in accounting practices will continue to evolve. Advancements in explainable AI (XAI) are addressing the black-box nature of ML models, enabling auditors and regulators to better understand how automated decisions are made. Blockchain technology is also being explored in tandem with AI to create tamper-proof audit trails and enhance data integrity. Furthermore, the growing adoption of cloud computing is facilitating real-time collaboration and data sharing across accounting ecosystems, enhancing the effectiveness of integrated technologies. The ultimate vision is a fully digital accounting environment where routine tasks are automated, insights are derived in real-time, and professionals can focus on strategic advisory roles. Accountants will increasingly serve as interpreters and validators of machine-generated insights, ensuring that financial information remains accurate, ethical, and aligned with organizational goals.

These are pivotal technologies enabling the integration of intelligent automation into accounting practices. Through case studies like Deloitte's Argus and conceptual models illustrating technological workflows, it is evident that these technologies enhance efficiency, accuracy, and strategic value. However, successful integration requires overcoming challenges related to data quality, skill development, and ethical governance. As the accounting profession continues to embrace these innovations, it is poised to transition from a traditional compliance-oriented role to a more strategic, analytical, and future-ready function.

The rapid adoption of Machine Learning (ML), Natural Language Processing (NLP), and Computer Vision is reshaping the landscape of accounting and finance practices. To remain relevant and add strategic value in this evolving environment, practitioners must undertake proactive measures across skill development, professional practices, and organizational leadership. The practitioners must deepen their technological acumen. Understanding the functional capabilities and applications of ML, NLP, and Computer Vision is critical. Practitioners should familiarize themselves with how these technologies automate processes such as financial forecasting, risk analysis, regulatory compliance, and audit verification (Davenport & Ronanki, 2018). This foundational knowledge will allow them to engage more effectively with technology teams and to critically assess the outputs of intelligent systems.

Furthermore, developing advanced analytical skills is essential. As traditional bookkeeping and transaction-processing tasks are automated, the emphasis for practitioners will shift toward data interpretation, strategic analysis, and advisory roles. Skills in financial modelling using predictive analytics, interpreting AI-driven insights, and evaluating algorithmic outputs will become central to finance and accounting roles (Sutton, Holt, & Arnold, 2018). Moving on embracing continuous professional education is vital. The pace of technological advancement necessitates a commitment to lifelong learning. Accounting and finance practitioners should pursue certifications in emerging areas such as data analytics for finance, AI governance, and intelligent auditing. Professional bodies must also integrate technology modules into their core training and CPD programs to equip members with future-ready competencies.

Moreover, practitioners must champion ethical and governance frameworks. The integration of AI and automation in financial decision-making raises important issues related to transparency, bias, and accountability. Practitioners must advocate for ethical AI use, contribute to the development of governance policies, and ensure that integrated systems comply with evolving regulatory standards (Moffitt, Rozario, & Vasarhelyi, 2018). To embrace the integration of ML, NLP, and Computer Vision, accounting and finance practitioners must build technological literacy, enhance analytical skills, commit to ongoing education, and uphold ethical governance. Doing so will ensure they transition from traditional compliance roles to strategic leadership positions in an increasingly digitalized finance world.

4. ARCHITECTURE OF INTEGRATED RPA-AI SYSTEMS

The integration of Robotic Process Automation (RPA) with Artificial Intelligence (AI) signifies a crucial development in the automation of corporate processes, especially in accounting. An integrated RPA-AI system amalgamates the structured, rule-based automation functionalities of RPA with the cognitive and analytical prowess of AI technologies, including machine learning (ML), natural language processing (NLP), and computer vision. This architecture allows organisations to automate intricate workflows involving both organised and unstructured data, resulting in more resilient, intelligent, and adaptive business processes.

An integrated RPA-AI system typically consists of several interconnected components. At the core is the RPA platform, which automates repetitive, rule-based tasks across multiple applications without altering the underlying systems (van der Aalst et al., 2018). This platform is complemented by AI engines that provide the cognitive abilities needed to handle unstructured data, learn from data patterns, and make decisions. These engines incorporate ML models, NLP modules, and computer vision systems (Davenport & Ronanki, 2018). The orchestration layer manages the coordination between RPA bots and AI modules, ensuring smooth workflow execution, exception handling, and process monitoring. Underpinning these layers is a robust data management framework that supports the storage, integration, and governance of structured and unstructured data required by AI models and RPA bots. Furthermore, the user interface (UI) layer allows users to interact with the system, monitor bot performance, input new data, and receive alerts or reports. Given the sensitive nature of accounting data, integrated systems also incorporate robust cybersecurity measures, access controls, and audit trails to ensure data protection and regulatory compliance (Syed et al., 2020).

The architecture of combined RPA-AI systems can be conceptualised as a multi-tiered framework. The user interface layer engages with the orchestration layer, which subsequently synchronises actions between the RPA platform and AI engines, including machine learning, natural language processing, and computer vision. Beneath these layers is the data management framework, which directly communicates with enterprise applications and databases. Integration markedly improves accounting methods across several domains. In invoice processing, RPA collects structured data from invoices, while AI-driven OCR interprets scanned documents, and NLP modules classify and route them accordingly. In auditing and compliance, machine learning algorithms detect irregularities in transactions, RPA aggregates audit data, and NLP facilitates the creation of regulatory reports. Financial forecasting is enhanced by AI models that anticipate future trends, while RPA facilitates the extraction and aggregation of pertinent financial data from diverse systems.

A significant case study illustrating the efficacy of combined RPA-AI architecture in accounting is the partnership between UiPath and PwC. PwC deployed UiPath's RPA platform alongside proprietary AI models to automate its auditing operations. Bots were utilised to retrieve data from various accounting systems, while AI algorithms detected high-risk transactions and highlighted discrepancies (PwC, 2021). This integration led to a 30% decrease in the manual audit burden and markedly enhanced the precision of risk evaluations. The implementation of integrated architectures presents some obstacles. The quality of data is a major concern; AI systems want pristine, high-calibre data, and discrepancies may result in erroneous results. Moreover, facilitating seamless communication between RPA bots and AI models entails significant system integration complexity. Organisations encounter a skills gap, as the implementation and maintenance of such systems require a workforce adept in AI, RPA development, cybersecurity, and process analysis (Willcocks et al., 2020). Ethical considerations, such as algorithmic biases and the automation of sensitive activities, require meticulous control and oversight.

The architecture of integrated RPA-AI systems is progressing swiftly, exhibiting several new themes. Hyper automation, which integrates RPA, AI, process mining, and advanced analytics, facilitates comprehensive automation solutions (Gartner, 2020). The advancement of self-repairing robots, proficient at identifying and rectifying problems instantaneously without human involvement, is also gaining momentum. Moreover, the democratisation of automation via low-code and no-code platforms is facilitating the design and deployment of RPA-AI workflows by non-technical people. The architecture of combined RPA-AI systems signifies a revolutionary advancement in accounting and business processes. Organisations can attain enhanced accuracy, efficiency, and strategic insight by integrating the systematic efficiency of RPA with the intellectual capabilities of AI. Nonetheless, the effective implementation of such systems necessitates meticulous design, proficient resources, and a robust emphasis on governance and ethical standards. As technology advances, integrated automation will become essential to contemporary accounting practices, steering the profession towards a more strategic, data-centric future

Figure 3. Architecture of Integratd Rpa-AI systems

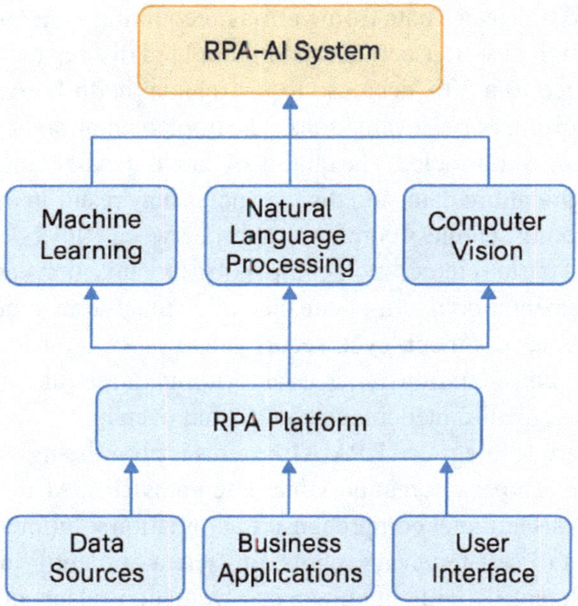

5. BENEFITS & CHALLENGES - ROBOTIC PROCESS AUTOMATION WITH ARTIFICIAL INTELLIGENCE ON ACCOUNTING PRACTICES

The integration of Robotic Process Automation (RPA) and Artificial Intelligence (AI) has revolutionized the operational landscape of businesses, particularly within the accounting domain. The synergistic blend of these technologies offers profound benefits that extend beyond simple task automation. Chief among these advantages are enhanced efficiency, improved accuracy, scalable operations, and significant cost reductions. Understanding these benefits is crucial for organizations aiming to achieve sustainable growth and competitive advantage in the digital era.

Efficiency stands as one of the most immediate and visible benefits of integrated RPA-AI systems. Traditional accounting processes often involve time-consuming manual activities such as data entry, reconciliation, invoice processing, and compliance reporting. These activities not only require substantial human effort but are also prone to delays and bottlenecks. With RPA, routine tasks are automated, enabling continuous 24/7 operations without fatigue or loss of productivity. AI further amplifies this efficiency by enabling intelligent decision-making in processes that require analysis, classification, or prediction. For example, machine learning algorithms can quickly analyse historical financial data to predict future trends, reducing the need for extensive manual forecasting efforts. Natural language processing (NLP) tools can automatically read and categorize emails or contracts, streamlining document management. Together, RPA and AI create end-to-end workflows that operate faster than traditional human-managed processes, allowing organizations to meet deadlines, respond to client needs more promptly, and optimize resource allocation (Davenport & Ronanki, 2018).

Accuracy is another critical benefit afforded by the integration of RPA and AI. Human error is an inherent risk in manual processes, especially in complex accounting tasks that involve large volumes of data. RPA bots are programmed to perform tasks consistently according to predefined rules, virtually eliminating errors associated with manual handling. When combined with AI's cognitive abilities, systems can also detect anomalies, validate data, and flag inconsistencies that might escape human attention. For instance, AI-powered audit systems can scrutinize thousands of financial transactions and highlight irregular patterns that may indicate fraud or compliance breaches. The improved accuracy not only enhances the quality of financial reporting but also strengthens regulatory compliance and risk management, essential aspects in today's stringent business environment (Syed et al., 2020).

Scalability is a defining feature that makes RPA-AI integration particularly appealing to modern enterprises. Unlike human workforces, which require proportional investments in recruitment, training, and management to scale operations, digital workers can be scaled up or down with relative ease. Organizations can deploy additional bots during peak periods, such as fiscal year-end closing, and scale back during off-peak times, optimizing operational flexibility. Furthermore, AI models can be retrained and adjusted to handle new tasks or data types, ensuring that the automation framework evolves with the business needs. The scalability extends across departments and geographies, enabling multinational organizations to maintain consistent processes and quality standards across diverse business units (van der Aalst et al., 2018).

Cost reduction emerges as a natural consequence of improvements in efficiency, accuracy, and scalability. By automating repetitive and labour-intensive tasks, organizations can reduce their reliance on large human workforces for routine operations. This leads to direct savings in salaries, benefits, office space, and other associated costs. Additionally, the reduction in errors translates into fewer costly corrections, penalties, and reputational damages. Intelligent automation also reduces audit and compliance costs by maintaining real-time records and providing automated reporting. Over time, the investment in RPA-AI infrastructure yields substantial returns, often exceeding initial implementation costs. A 2021 Deloitte survey indicated that organizations deploying intelligent automation reported an average cost reduction of 22% across their automated processes (Deloitte, 2021).

To illustrate these benefits in practice, consider the case study of American Express Global Business Travel (GBT). GBT implemented an integrated RPA-AI solution to streamline its expense management process, which was previously characterized by manual verifications and lengthy reimbursement cycles. By using RPA to automate the extraction and validation of expense claims and AI to detect anomalies and enforce policy compliance, GBT reduced its processing time by 70%, improved the accuracy of reimbursements, and achieved a 30% reduction in operational costs. Moreover, the scalability of the solution allowed GBT to extend the same framework to multiple regions without extensive customization, demonstrating the power of integrated automation across a global enterprise (American Express GBT, 2020).

Despite the numerous advantages, organizations must navigate certain challenges to fully realize the potential of RPA-AI integration. The initial setup can be capital-intensive, requiring investments in technology, infrastructure, and skilled personnel. Additionally, continuous monitoring and updating of AI models are necessary to ensure relevance and performance. There is also the need for a cultural shift within organizations, as employees must be retrained to collaborate with digital workers and focus on higher-value analytical and strategic tasks. However, organizations that effectively manage these challenges stand to benefit immensely from the transformation.

The future trajectory of RPA-AI integration promises even greater benefits. Emerging technologies such as cognitive automation, hyper automation, and autonomous decision-making are expected to push the boundaries of what integrated systems can achieve. Cognitive automation, which combines AI's capabilities with human-like reasoning, will allow for the automation of complex decision-making processes. Hyper automation, as defined by Gartner, will involve the orchestration of multiple advanced technologies to automate end-to-end business processes beyond what is currently possible with RPA alone (Gartner, 2020). These developments will further enhance efficiency, accuracy, scalability, and cost reduction, solidifying intelligent automation as a cornerstone of business strategy. The integration of RPA and AI in accounting and broader business practices delivers transformative benefits across multiple dimensions. By enhancing efficiency, improving accuracy, enabling scalability, and reducing costs, integrated automation not only optimizes existing operations but also positions organizations for future resilience and growth. The case study of American Express GBT and the conceptual flowchart highlight the tangible impacts of these benefits when strategically implemented. As technology continues to evolve, organizations that invest in and embrace intelligent automation will be best equipped to thrive in an increasingly complex and competitive global marketplace.

The integration of Robotic Process Automation (RPA) and Artificial Intelligence (AI) has transformed modern business operations, particularly in the fields of accounting, finance, and administration. While the potential benefits of integrated RPA-AI systems such as improved efficiency, accuracy, scalability, and cost reduction are widely acknowledged, the journey towards successful deployment is fraught with significant challenges. These challenges can be categorized into four primary areas: technical, organizational, ethical, and regulatory. Understanding and addressing these challenges is essential for organizations seeking to leverage the full potential of intelligent automation. Technical challenges form the foundation of the hurdles organizations face when implementing integrated RPA-AI systems. Despite the maturity of RPA platforms and AI models, integration between them is not seamless. One major technical challenge is data complexity. AI systems rely on vast amounts of clean, structured, and high-quality data to function optimally. However, in real-world accounting and business environments, data is often unstructured, inconsistent, and siloed across multiple systems. Integrating RPA bots with AI models to process such data effectively demands sophisticated data cleansing and preprocessing mechanisms (Syed et al., 2020). Another technical barrier is interoperability. Legacy systems, which are still prevalent in many organizations, often lack the APIs and integration capabilities needed for smooth interaction with RPA and AI platforms. Moreover, maintaining and updating AI models require continuous monitoring and retraining, necessitating advanced technical skills that may not be readily available within the organization (van der Aalst et al., 2018).

Beyond technical barriers, organizational challenges present significant impediments to successful RPA-AI integration. Implementing intelligent automation fundamentally alters the traditional workflows and job roles within an organization. Resistance to change is a common organizational issue. Employees often perceive automation as a threat to their job security, leading to reluctance in embracing new technologies. Additionally, successful implementation requires a cultural shift towards digital literacy and agility. Organizations must invest in upskilling their workforce, fostering a mindset of continuous learning and adaptation (Davenport & Ronanki, 2018). Leadership commitment is another crucial factor. Without strong executive sponsorship and a clear strategic vision, automation initiatives often stall or fail to achieve scale. Furthermore, the lack of cross-functional collaboration between IT departments, finance teams, and operational units can hinder the seamless design and deployment of integrated solutions.

Ethical challenges are becoming increasingly prominent as the use of AI expands. One of the core ethical concerns is bias in AI models. If the data used to train AI systems contains biases, the resulting outputs can reinforce and perpetuate these biases, leading to unfair or discriminatory outcomes. In accounting and finance, biased AI systems could result in the unjust flagging of certain transactions or clients for audit and investigation. Transparency is another critical ethical issue. AI systems, especially those based on deep learning models, are often described as "black boxes" due to their lack of explainability. This opacity makes it difficult for auditors, regulators, and stakeholders to understand how decisions are made, raising concerns about accountability (Mittelstadt et al., 2016). Moreover, the ethical deployment of intelligent automation requires careful consideration of the impact on the workforce, ensuring that displaced employees are supported through reskilling and transition initiatives.

Regulatory challenges further complicate the adoption of integrated RPA-AI systems. The regulatory landscape for AI and automation is rapidly evolving but remains fragmented across different jurisdictions. Organizations must navigate a complex web of data privacy laws, financial regulations, and industry-specific compliance requirements. For instance, the General Data Protection Regulation (GDPR) in Europe imposes strict rules on the collection, processing, and storage of personal data, affecting how AI systems can be trained and deployed (Voigt & Bussche, 2017). In the United States, the Sarbanes-Oxley Act (SOX) mandates stringent controls and audit trails for financial reporting, necessitating that RPA bots maintain detailed logs of their activities. Failing to ensure compliance with such regulations can result in hefty fines, reputational damage, and operational disruptions. Additionally, emerging regulations on AI ethics and transparency, such as the European Union's proposed Artificial Intelligence Act, are likely to impose new obligations on organizations, requiring greater oversight and accountability for automated decision-making processes.

A compelling case study illustrating these challenges can be observed in the attempted deployment of an integrated RPA-AI system by a multinational bank. The bank aimed to automate its loan approval process by combining RPA bots to gather applicant information and AI algorithms to assess creditworthiness. Technical challenges emerged when data from disparate sources had to be standardized and cleansed, causing significant project delays. Organizational resistance was also notable, as loan officers feared being replaced and resisted the rollout of the new system. Ethical concerns arose when early testing revealed that the AI model exhibited bias against applicants from certain demographics. Finally, regulatory hurdles, particularly concerning data privacy and explainability of AI decisions under GDPR, forced the bank to significantly redesign its approach before a scaled deployment could be achieved (PwC, 2021). This case highlights the interconnected nature of technical, organizational, ethical, and regulatory challenges and underscores the necessity for a holistic and proactive strategy.

Addressing these challenges requires a comprehensive and multi-faceted approach. Technically, organizations must invest in robust data governance frameworks, modular system architectures, and continuous monitoring tools to ensure system reliability and adaptability. Organizationally, fostering a culture of innovation, clear communication about the strategic benefits of automation, and employee reskilling initiatives are vital. Ethically, organizations should implement fairness audits of AI models, promote transparency through explainable AI (XAI) technologies, and establish ethics committees to oversee deployment strategies. Regulatory compliance can be strengthened through proactive engagement with legal experts, regular audits, and embedding compliance-by-design principles into system development.

While integrated RPA-AI systems offer transformative potential for organizations, their successful implementation is contingent upon addressing a complex web of technical, organizational, ethical, and regulatory challenges. These challenges are deeply interconnected, and overlooking any one dimension can jeopardize the success of automation initiatives. The case study of the multinational bank serves as a powerful reminder of the real-world complexities involved. Organizations that adopt a holistic, ethical, and compliant approach, backed by strong leadership and a digitally empowered workforce, will be best positioned to unlock the full value of intelligent automation in an increasingly dynamic and regulated global economy.

6. IMPACT ON ACCOUNTING PROFESSION

The integration of Robotic Process Automation (RPA) and Artificial Intelligence (AI) into business operations has generated a paradigm shift in workforce dynamics. Rather than simply replacing jobs, the fusion of these technologies is transforming roles and necessitating a significant shift in required skills. The impact on the workforce is multifaceted, involving not only changes in job structures but also broader cultural and educational implications. Understanding this transformation is critical for organizations and professionals seeking to thrive in the era of intelligent automation.

Job transformation is the most visible effect of RPA and AI integration. Traditional roles that once required extensive manual intervention, such as data entry clerks, bookkeepers, and transaction processors, are increasingly being automated. RPA bots can now handle repetitive, rule-based tasks with greater efficiency and accuracy than their human counterparts. Simultaneously, AI technologies such as machine learning, natural language processing, and computer vision can perform complex cognitive functions including data analysis, pattern recognition, and decision-making support (Davenport & Ronanki, 2018). As a result, the nature of work is shifting from routine task execution to roles that require judgment, creativity, and emotional intelligence. Employees are now expected to supervise, manage, and optimize automated processes rather than perform the tasks themselves.

This transformation does not signify the end of human labour but rather a re-definition of job functions. New roles are emerging, such as RPA developers, AI trainers, process automation managers, and data analysts. These positions require a blend of domain expertise and technical proficiency. For instance, an accounting professional might transition into a role focused on overseeing the financial bots and ensuring compliance with regulatory standards. The integration of automation and AI is, therefore, creating opportunities for more meaningful and value-added work, emphasizing human strengths such as strategic thinking, complex problem-solving, and interpersonal communication (World Economic Forum, 2020).

The skill shift accompanying job transformation is profound. Technical skills are in high demand. Proficiency in programming languages like Python, understanding of AI and machine learning algorithms, knowledge of RPA tools such as UiPath and Blue Prism, and familiarity with cloud platforms are increasingly becoming prerequisites in many fields. However, technical skills alone are insufficient. Soft skills, particularly adaptability, critical thinking, and emotional intelligence, are becoming equally vital. As roles become more dynamic and interdisciplinary, the ability to learn continuously, collaborate across departments, and manage change is paramount (LinkedIn Learning, 2020).

Organizations must invest in reskilling and upskilling initiatives to prepare their workforce for these new realities. Training programs focused on digital literacy, AI fundamentals, data analytics, and agile methodologies can equip employees to transition smoothly into transformed roles. Leadership development is also critical, as managers must learn to lead hybrid teams composed of human employees and digital workers. Furthermore, educational institutions must adapt their curricula to bridge the gap between traditional academic training and the skills demanded by the future workforce.

A notable case study illustrating job transformation and skill shift is that of AT&T. Faced with the disruptive impact of digital technologies, AT&T launched a massive reskilling initiative in 2013 known as "Workforce 2020." The company invested over $1 billion to provide employees with access to online courses, certifications, and degree programs focused on emerging technologies such as AI, cybersecurity, and cloud computing. Employees were encouraged to take ownership of their career development, and those who upskilled successfully transitioned into new roles aligned with the company's digital strategy. This proactive approach not only mitigated job displacement risks but also enhanced employee engagement and organizational agility (Jesuthasan & Boudreau, 2018).

The psychological impact of job transformation on employees should not be underestimated. Uncertainty about job security, fear of obsolescence, and anxiety over adapting to new technologies can contribute to stress and resistance to change. Organizations must therefore prioritize change management and employee support initiatives. Transparent communication about the goals and benefits of automation, involvement of employees in automation projects, and provision of counselling and career coaching services can foster a more positive transition experience.

Culturally, the workforce transformation driven by RPA and AI demands a shift towards a learning-oriented and innovation-driven mindset. Organizations must cultivate environments where experimentation is encouraged, failures are seen as learning opportunities, and continuous improvement is ingrained in the corporate DNA. Diversity and inclusion efforts are also critical, ensuring that opportunities created by automation are accessible to all segments of the workforce and that biases in AI systems are actively mitigated.

The future trajectory of workforce transformation suggests an increasingly symbiotic relationship between humans and intelligent machines. Collaborative intelligence, where humans and AI systems complement each other's strengths, is poised to become the new operational model. Humans will focus on areas requiring creativity, empathy, and ethical judgment, while machines will handle data-driven, repetitive, and analytical tasks. This collaboration has the potential to significantly enhance productivity, innovation, and employee satisfaction.

However, achieving this vision requires deliberate action from multiple stakeholders. Policymakers must support workforce development through incentives for training programs, public-private partnerships, and modernization of labour laws to accommodate new forms of work. Businesses must adopt inclusive workforce strategies that balance technological advancement with social responsibility. Educational institutions must evolve to produce graduates equipped for a digital-first world, emphasizing interdisciplinary skills and lifelong learning. The impact of RPA and AI integration on the workforce is not merely about job loss or creation but about profound job transformation and skill shift. Organizations that recognize and proactively address these changes will be better positioned to harness the full potential of intelligent automation. By investing in reskilling, fostering a culture of continuous learning, and embracing collaborative intelligence, businesses can ensure that the workforce of the future is empowered, adaptable, and resilient in an increasingly automated world.

Case Study 1: Intelligent Automation in Accounts Payable at Finovia Corporation

(Finovia Corporation. (2024). *Intelligent Automation in Accounts Payable: A Case-Based Illustration.* Unpublished teaching case. [Fictitious example for academic purposes].)

In recent years, Finovia Corporation, a mid-sized multinational involved in consumer electronics, faced growing challenges in its Accounts Payable (AP) department. Manual invoice processing led to delays, errors in payment schedules, and high operational costs. With more than 10,000 invoices processed monthly, the need for a more efficient system was evident.

To address these inefficiencies, Finovia embarked on integrating Robotic Process Automation (RPA) with Artificial Intelligence (AI). The RPA bots were programmed to extract invoice details, validate supplier information, and initiate payment processes. Meanwhile, AI tools, particularly machine learning algorithms and Natural Language Processing (NLP), were embedded to interpret unstructured data from various invoice formats, predict anomalies, and match invoices with purchase orders intelligently.

Impact and Outcome: The integration led to a 70% reduction in invoice processing time and an 85% improvement in accuracy. Human intervention was required only for exception handling, allowing AP staff to focus on strategic tasks like vendor relationship management. Furthermore, AI-driven analytics provided real-time dashboards for better cash flow forecasting.

Solution Summary: Finovia's integration of RPA and AI redefined its accounting practice by automating repetitive tasks and enhancing data-driven decision-making. The strategic redeployment of staff into higher-value roles increased employee satisfaction and operational resilience.

Case Study 2: RPA-AI Enabled Audit Processes at Nova Accounting Firm

(Adapted from best practices in intelligent audit automation. (2024). *RPA-AI Enabled Audit Processes at Nova Accounting Firm* [Fictional case]. Illustrative scenario for educational purposes.).

Nova Accounting, a leading audit firm specializing in SMEs, observed that traditional audit procedures were increasingly misaligned with client expectations for faster, more accurate audits. The manual vouching of transactions and reconciliation of bank statements consumed significant auditor hours, delaying audit reports and escalating costs.

In 2023, Nova Accounting decided to integrate RPA with AI to automate parts of their audit workflows. RPA bots were assigned tasks like extracting data from ERP systems, generating audit trails, and preparing confirmation letters. Simultaneously, AI models were deployed to perform predictive risk assessments by analyzing historical financial data patterns and identifying unusual transactions for auditors' review.

Impact and Outcome: Audit cycle times were shortened by 40%, and audit quality improved with fewer material misstatements detected post-audit. The integration enhanced the auditors' ability to perform risk-based audits by focusing attention on high-risk areas, thus elevating the firm's service quality and client trust

Solution Summary: By embracing RPA and AI, Nova Accounting achieved more accurate, efficient, and insightful audit processes. This move positioned the firm competitively in the market, appealing especially to tech-savvy clients who valued agility and precision in auditing services.

Case Study 3: Enhancing Patient Billing and Claims Processing at MedicaLife Healthcare

(Adapted from real-world healthcare automation implementations. (2024). *Enhancing Patient Billing and Claims Processing at MedicaLife Healthcare* [Illustrative case study].)

MedicaLife Healthcare, a prominent hospital network across Southeast Asia, was struggling with inefficiencies in patient billing and insurance claims management. Manual billing entries, delays in claims processing, and frequent errors were leading to revenue leakages and patient dissatisfaction.

To tackle these issues, MedicaLife implemented an integrated RPA and AI solution. RPA bots captured patient treatment details directly from electronic health records (EHR) and generated preliminary billing invoices. AI tools verified insurance coverage, identified coding errors in real-time, and predicted claim denials based on historical patterns, enabling proactive corrections before submission.

Impact and Outcome: The adoption resulted in a 60% faster billing cycle, a 50% reduction in claim rejections, and improved cash flow. Additionally, patient satisfaction increased due to more transparent and timely billing.

Solution Summary: Through RPA-AI integration, MedicaLife optimized its billing and claims operations, reducing human error, accelerating revenue cycles, and enhancing patient trust. The hospital group set a benchmark in digital healthcare finance, demonstrating how automation can effectively address administrative challenges in healthcare.

7. FUTURE TRENDS

The integration of Robotic Process Automation (RPA) and Artificial Intelligence (AI) has fundamentally transformed business operations across industries, particularly in fields like accounting, finance, supply chain management, and customer service. However, the evolution of intelligent automation is far from static. Emerging concepts such as hyper automation, cognitive RPA, and autonomous systems represent the next frontier in this technological revolution. Understanding these future trends is crucial for organizations aiming to maintain competitiveness, enhance operational resilience, and capitalize on the benefits of digital transformation.

Hyper automation is arguably one of the most significant trends shaping the future of automation. Defined by Gartner (2020) as the application of advanced technologies including AI, machine learning (ML), and RPA to increasingly automate processes and augment human capabilities, hyper automation extends beyond the simple automation of tasks. It seeks to automate every possible business process, creating a digital-first organization that can operate with minimal human intervention. Hyper automation involves not just automating tasks but orchestrating a digital ecosystem where multiple technologies collaborate seamlessly. Tools such as process mining, advanced analytics, low-code development platforms, and intelligent business process management suites (iBPMS) are integral to hyper automation ecosystems.

The shift toward hyper automation reflects the growing need for agility and scalability in an increasingly volatile business environment. Organizations adopting hyper automation can continuously identify automation opportunities, optimize workflows, and adapt to changing market demands with greater speed. By creating an interconnected web of automated processes, businesses are better equipped to make data-driven decisions, enhance customer experiences, and reduce operational risks (Gartner, 2020).

Cognitive RPA represents the next evolutionary step beyond traditional RPA. While traditional RPA is rule-based and limited to structured data, cognitive RPA integrates AI technologies such as natural language processing, optical character recognition, and machine learning to handle unstructured data and make decisions. Cognitive RPA systems can read and interpret documents, understand natural language queries, and learn from historical data to improve over time. This enables automation of more complex processes that previously required human judgment, such as customer onboarding, claims processing, fraud detection, and regulatory compliance monitoring (Willcocks, Lacity, & Craig, 2017).The introduction of cognitive capabilities transforms RPA from a tool for efficiency to a strategic asset for innovation. Organizations leveraging cognitive RPA can automate not only transactional tasks but also knowledge-driven activities, thereby freeing human employees to focus on high-value work involving creativity, strategic thinking, and relationship building.

Cognitive RPA also enhances operational accuracy by reducing errors associated with human subjectivity, ensuring consistent and reliable process outcomes.

Autonomous systems represent the ultimate vision of intelligent automation, wherein systems operate independently, adapt to new environments, and make decisions without human intervention. Unlike traditional automated systems, which operate within predefined parameters, autonomous systems are capable of learning from their interactions with the environment and adjusting their behaviours accordingly. Autonomous accounting systems, for instance, could automatically detect anomalies, adjust financial models in real-time, and generate strategic financial insights without manual input. The development of autonomous systems relies heavily on advancements in AI, deep learning, reinforcement learning, and edge computing. These systems can be deployed in dynamic environments such as autonomous vehicles, intelligent supply chains, and smart cities. In the corporate world, autonomous systems are poised to transform enterprise resource planning (ERP), customer relationship management (CRM), and business intelligence functions, allowing businesses to respond in real-time to operational and market changes (Bughin et al., 2018).

A case study that highlights the move toward hyper automation and cognitive RPA is that of Deutsche Bank. In response to increasing regulatory requirements and operational complexity, Deutsche Bank launched a major initiative to digitize and automate its processes. By integrating RPA bots with cognitive AI tools, the bank automated tasks ranging from transaction processing to regulatory reporting. Furthermore, the bank utilized process mining tools to map and optimize workflows, enabling the continuous identification of new automation opportunities. As a result, Deutsche Bank reported significant reductions in processing times, improved compliance rates, and substantial operational cost savings (Deutsche Bank, 2021).

Despite the immense potential, the journey toward hyper automation, cognitive RPA, and autonomous systems is not without challenges. Technical complexity is a major hurdle, as organizations must integrate diverse technologies, manage data security, and ensure system interoperability. Skill gaps also pose significant obstacles. Successfully implementing these advanced systems requires expertise in AI, data science, process engineering, cybersecurity, and change management. Ethical considerations, including algorithmic bias, transparency, and accountability, become even more critical as machines make increasingly autonomous decisions.

Regulatory frameworks are struggling to keep pace with technological advancements. Governments and regulatory bodies worldwide are grappling with questions around liability, data ownership, and ethical standards for autonomous systems. For example, the European Union's proposed Artificial Intelligence Act seeks to establish comprehensive regulations on the use of high-risk AI systems, which will have direct implications for the deployment of autonomous business applications (European Commission, 2021).

The economic and societal implications of these trends are profound. While hyper automation and cognitive RPA offer opportunities for productivity gains and innovation, they also risk exacerbating inequalities if displaced workers are not adequately supported through reskilling initiatives. Autonomous systems, while increasing operational efficiency, could lead to job polarization, where high-skill jobs proliferate, and low-skill jobs diminish. Addressing these implications requires a coordinated effort among businesses, policymakers, educational institutions, and civil society. Organizations must adopt inclusive automation strategies that prioritize workforce development and ethical AI practices. Policymakers must create enabling environments that foster innovation while protecting public interests. Educational systems must evolve to prepare future workers with the interdisciplinary skills needed to thrive in an increasingly automated world.

Looking ahead, hyper automation, cognitive RPA, and autonomous systems are set to redefine the boundaries of what organizations can achieve. Businesses that embrace these trends proactively, invest in technological capabilities, foster agile and resilient cultures, and commit to ethical innovation will be well-positioned to lead in the digital economy. The future of intelligent automation is marked by the convergence of hyper automation, cognitive RPA, and autonomous systems. These trends represent a shift from automating individual tasks to creating intelligent, self-optimizing enterprises capable of operating in a dynamic and complex world. Through real-world examples such as Deutsche Bank's automation strategy and guided by frameworks like the European AI regulations, organizations can navigate the challenges and opportunities of this transformative era. The journey toward hyper automation and autonomy promises not only operational excellence but also the possibility of creating more adaptive, inclusive, and innovative organizations prepared for the future.

Figure 4. Future Trends in RPA

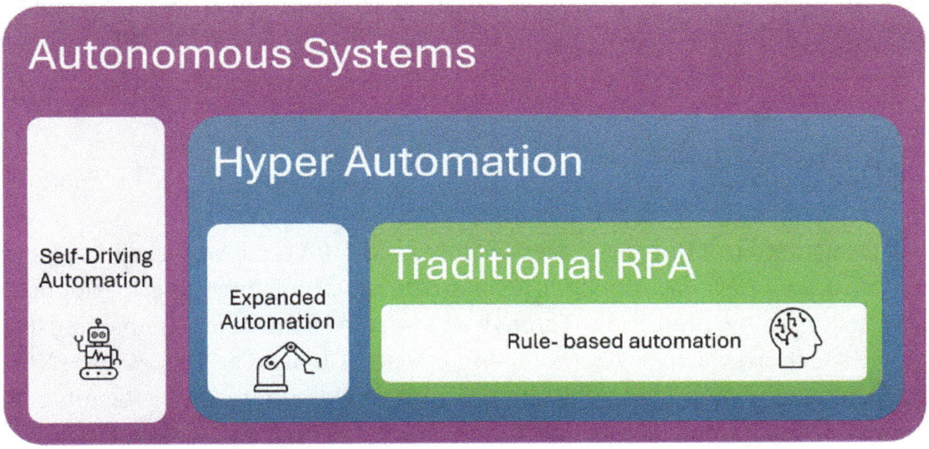

The accelerating adoption of hyper automation, cognitive RPA, and autonomous systems is reshaping the landscape of accounting and finance. To remain relevant and competitive, professionals in these fields must proactively evolve their skills, mindsets, and strategic roles. First, technical proficiency is paramount. Accountants and finance professionals must gain familiarity with automation platforms, data analytics tools, and AI applications relevant to their domains. Understanding how technologies like machine learning, natural language processing, and process mining enhance financial operations enables them to work effectively alongside digital systems rather than being displaced by them.

Second, developing advanced analytical and critical thinking skills is essential. As routine transactional tasks are increasingly automated, the value of human professionals will shift toward interpreting complex data, making strategic judgments, and providing advisory insights. Professionals must be capable of translating data-driven outputs into actionable business strategies, ensuring that automation enhances rather than replaces human decision-making. Moreover, a commitment to continuous learning is vital. Given the rapid pace of technological change, accountants must embrace lifelong learning through formal certifications, micro-credentials, and active participation in professional development programs. Staying informed about emerging technologies and regulatory changes ensures that they can adapt to evolving compliance landscapes and ethical standards.

Finally, accountants must cultivate soft skills such as adaptability, collaboration, and digital communication. Working in agile, cross-functional teams with technologists and data scientists will become increasingly common. Emotional intelligence and leadership will distinguish those who can guide organizations through digital transformation. By developing technological fluency, enhancing strategic thinking, committing to lifelong learning, and strengthening human-centric skills, accountants and finance professionals can not only adapt to but thrive in an era dominated by hyper automation and intelligent systems.

8. CONCLUSION

The integration of Robotic Process Automation (RPA) and Artificial Intelligence (AI) into accounting practices signifies one of the most transformative developments in the history of the profession. Through an in-depth examination of enabling technologies such as Machine Learning (ML), Natural Language Processing (NLP), and Computer Vision, the architecture of integrated systems, their tangible and intangible impacts, and the associated challenges, it becomes clear that the future of accounting is irreversibly intertwined with intelligent automation. This concluding chapter synthesizes the insights presented, highlights their broader implications, and outlines strategic recommendations for practitioners, educators, and policymakers to harness the full potential of this technological evolution.

The journey of integration begins with a fundamental understanding of RPA and AI. RPA specializes in automating repetitive, rule-based tasks, offering substantial gains in operational efficiency. AI, with its cognitive capabilities, introduces the ability to analyse unstructured data, make predictions, and support complex decision-making processes. Together, they create a symbiotic system where structured processes are streamlined and cognitive functions are enhanced (Davenport & Ronanki, 2018). The rationale for integration is clear: to transcend the limitations of traditional accounting processes and create agile, resilient, and insightful financial operations. Technologies enabling this integration have matured rapidly. Machine Learning models empower accounting systems to detect anomalies, forecast financial trends, and optimize risk management strategies. NLP facilitates the extraction and interpretation of information from voluminous, unstructured textual data such as contracts and regulatory documents. Meanwhile, Computer Vision, particularly through Optical Character Recognition (OCR), enables the automated processing of visual inputs like invoices and receipts, drastically reducing manual intervention and error rates (Moffitt, Rozario, & Vasarhelyi, 2018).

Case studies such as Deloitte's Argus platform and Deutsche Bank's digitization initiatives demonstrate the transformative impact of these technologies. Deloitte's Argus integrates ML, NLP, and Computer Vision to automate document analysis during audits, increasing accuracy and efficiency. Deutsche Bank's process mining and automation strategies illustrate how intelligent automation can enhance compliance, streamline operations, and drive strategic value (Deutsche Bank, 2021).

However, the path to integration is fraught with challenges. Technological barriers such as data quality issues, legacy system integration, and cybersecurity risks pose significant hurdles. Organizationally, the transition demands substantial cultural shifts, with resistance to change and skill gaps emerging as critical bottlenecks. Ethical considerations, including transparency, bias mitigation, and accountability, require meticulous attention to ensure that the deployment of intelligent systems adheres to the highest standards of integrity. Regulatory frameworks must evolve in parallel to provide clear guidelines and protections in the era of autonomous systems (European Commission, 2021).

Hyper automation, cognitive RPA, and autonomous systems represent the future trajectory of intelligent automation in accounting. Hyper automation envisions the seamless orchestration of multiple advanced technologies to automate complex end-to-end business processes. Cognitive RPA enhances traditional automation with AI capabilities, enabling systems to adapt, learn, and make informed decisions. Autonomous systems, leveraging advancements in deep learning and edge computing, aspire to operate independently, continuously optimizing themselves without human intervention. The evolution from traditional RPA to full autonomy signifies a paradigm shift towards self-governing, intelligent enterprises.

The impact of these advancements on the workforce cannot be overstated. Job transformation, rather than displacement, is the dominant trend. Routine tasks are automated, while human roles evolve towards strategic analysis, decision support, and innovation. Skills such as data analytics, digital literacy, critical thinking, and emotional intelligence become paramount. Lifelong learning, adaptability, and interdisciplinary collaboration are the hallmarks of the future accounting and finance professional (World Economic Forum, 2020). Organizations must foster a culture of continuous learning and innovation. Investment in employee reskilling programs, ethical AI deployment frameworks, and agile change management strategies is critical. Professional accounting bodies and educational institutions must revamp curricula to integrate technological competencies alongside traditional accounting principles, ensuring that graduates are equipped for a digital-first world.

Policymakers have an equally crucial role. They must craft forward-looking regulatory frameworks that encourage innovation while safeguarding public interest. Guidelines for algorithmic transparency, data privacy, and ethical AI use must be established and enforced. Collaboration between governments, academia, industry leaders, and civil society is essential to create an inclusive and equitable digital economy. Ethical stewardship must underpin every stage of intelligent automation deployment. Accountants, by virtue of their fiduciary responsibilities, are uniquely positioned to champion ethical standards in the use of AI and RPA. They must advocate for transparency in AI decision-making, ensure fairness in automated processes, and uphold the integrity of financial reporting. Embedding ethical considerations into system design, implementation, and evaluation will be critical to maintaining stakeholder trust. Case studies from companies like KPMG and PwC further reinforce the importance of integrating ethics into automation initiatives. KPMG's "Trusted AI" framework and PwC's "Responsible AI" initiatives emphasize transparency, explainability, fairness, and governance as foundational pillars for deploying AI in accounting and auditing (KPMG, 2020; PwC, 2021).

Looking ahead, several emerging trends warrant close attention. Explainable AI (XAI) technologies are making black-box models more interpretable, addressing critical transparency concerns. Blockchain integration with AI is enhancing auditability and data security, creating immutable records that strengthen financial trust. Quantum computing, though still nascent, promises to exponentially enhance AI capabilities, unlocking new frontiers in financial modeling and risk analysis. The COVID-19 pandemic has accelerated digital transformation, highlighting the importance of resilience, adaptability, and remote capabilities. Integrated RPA-AI systems have proven instrumental in maintaining business continuity, processing pandemic-related financial relief applications, and managing dynamic regulatory changes. This real-world stress test underscores the indispensability of intelligent automation in navigating future disruptions.

In summary, the integration of RPA and AI into accounting practices represents not just a technological shift, but a profound transformation in the very fabric of the profession. It redefines what it means to be an accountant or finance professional, expanding roles from record-keepers to strategic advisors, risk managers, and ethical guardians of digital finance ecosystems.

Harnessing the full potential of this transformation requires a multi-pronged approach: embracing technological literacy, committing to lifelong learning, fostering innovation-centric cultures, championing ethical standards, and advocating for progressive regulatory reforms. The case studies, frameworks, and theoretical insights discussed throughout this chapter provide a roadmap for navigating this complex yet exhilarating journey. Organizations that lead with vision, invest in people, and prioritize ethical innovation will not only survive but thrive in the age of intelligent automation. For accounting and finance professionals, the future is not something to fear but an opportunity to redefine their roles, expand their impact, and contribute meaningfully to the evolving landscape of global business.

The journey towards intelligent, ethical, and autonomous accounting practices is both a challenge and a calling. It demands courage, adaptability, foresight, and unwavering commitment to the principles that have long underpinned the accounting profession—integrity, transparency, accountability, and stewardship. By embracing these values in the digital age, the accounting profession will continue to fulfil its vital role in ensuring trust, stability, and innovation in the global economy.

REFERENCES

American Express Global Business Travel. (2020). *Accelerating expense management with intelligent automation* (White Paper).

Bughin, J., Hazan, E., Ramaswamy, S., Chui, M., Allas, T., Dahlstrom, P., & Trench, M. (2018). *Artificial intelligence: The next digital frontier?* McKinsey Global Institute.

Davenport, T. H., & Ronanki, R. (2018). Artificial intelligence for the real world. *Harvard Business Review*. https://hbr.org/2018/01/artificial-intelligence-for-the-real-world

Deloitte. (2019). *Audit innovation: Argus cognitive platform.*

Deloitte. (2021). *Automation with intelligence: Pursuing organization-wide reimagination*. Deloitte Insights. https://www2.deloitte.com/insights/us/en/focus/automation-with-intelligence.html

. Deutsche Bank. (2021). *Digital transformation and automation in banking: Building a resilient and agile future.*

European Commission. (2021). *Proposal for a regulation laying down harmonised rules on artificial intelligence*. Artificial Intelligence Act.

Gartner. (2020). *Hyperautomation: The next step in automation evolution*. Gartner Research. https://www.gartner.com/en/information-technology/glossary/hyperautomation

Gepp, A., Linnenluecke, M. K., O'Neill, T. J., & Smith, T. (2020). Big data techniques in auditing research and practice: Current trends and future opportunities. *Journal of Accounting Literature*, *44*, 23–38. DOI: 10.1016/j.acclit.2020.100408

Jesuthasan, R., & Boudreau, J. W. (2018). *Reinventing jobs: A 4-step approach for applying automation to work*. Harvard Business Review Press.

. KPMG. (2020). *Trusted AI: Building trust in artificial intelligence systems.*

. LinkedIn Learning. (2020). *2020 workplace learning report: The rise of learning in the flow of work.*

Mittelstadt, B. D., Allo, P., Taddeo, M., Wachter, S., & Floridi, L. (2016). The ethics of algorithms: Mapping the debate. *Big Data & Society*, *3*(2). Advance online publication. DOI: 10.1177/2053951716679679

Moffitt, K. C., Rozario, A. M., & Vasarhelyi, M. A. (2018). Robotic process automation for auditing. *Journal of Emerging Technologies in Accounting*, *15*(1), 1–10. DOI: 10.2308/jeta-52088

PricewaterhouseCoopers. (2021a). *AI predictions: Six priorities for 2021*. https://www.pwc.com/gx/en/issues/analytics/assets/pwc-ai-predictions-2021.pdf

. PricewaterhouseCoopers. (2021b). *Responsible AI: Designing, building and deploying AI systems you can trust*.

PricewaterhouseCoopers. (2021c). *The future of audit: Embracing intelligent automation*. https://www.pwc.com/gx/en/services/audit-assurance/assets/the-future-of-audit.pdf

Sutton, S. G., Holt, M., & Arnold, V. (2018). The reports of my death are greatly exaggerated: Artificial intelligence research in accounting. *International Journal of Accounting Information Systems*, *29*, 1–9. DOI: 10.1016/j.accinf.2018.03.001

Syed, R., Bandara, W., French, E., Stewart, G., Gable, G. G., & Fekete, A. (2020). Robotic process automation: Contemporary themes and challenges. *Computers in Industry*, *115*, 103162. DOI: 10.1016/j.compind.2019.103162

van der Aalst, W. M. P., Bichler, M., & Heinzl, A. (2018). Robotic process automation. *Business & Information Systems Engineering*, *60*(4), 269–272. DOI: 10.1007/s12599-018-0542-4

Voigt, P., & von dem Bussche, A. (2017). *The EU General Data Protection Regulation (GDPR): A practical guide*. Springer International Publishing., DOI: 10.1007/978-3-319-57959-7

Willcocks, L., Lacity, M., & Craig, A. (2017). Robotic process automation: Strategic transformation lever for global business services? *Journal of Information Technology Teaching Cases*, *7*(1), 17–28. DOI: 10.1057/s41266-016-0002-8

Willcocks, L. P., Hindle, J., & Lacity, M. C. (2020). Becoming strategic with intelligent automation. *MIS Quarterly Executive*, *19*(4), 263–278.

World Economic Forum. (2020). *The future of jobs report 2020*. https://www.weforum.org/reports/the-future-of-jobs-report-2020

KEY TERMS AND DEFINITIONS

Artificial Intelligence (AI): Technology that emulates human intellect to autonomously execute activities such as learning, thinking, and decision-making

Ethical Challenges: Concerns about bias, transparency, accountability, and the replacement of human jobs are among the ethical challenges that have arisen because of the usage of artificial intelligence

Hyper automation: An advanced method that incorporates robotic process automation (RPA), artificial intelligence (AI), and analytics in order to automate all feasible business operations from beginning to end.

Intelligent Automation: The use of robotic process automation (RPA) and artificial intelligence (AI) to automate complicated procedures involved in data interpretation and decision-making.

RPA (Robotic Process Automation): Artificial intelligence technology that automates rule-based digital tasks that are repetitive by using software bots with minimal intervention from humans.

Workforce Transformation: The transformation of job responsibilities and the abilities that are necessary because of the development of digital technology and automation

Chapter 6
Regulating Robotic Process Automation and Artificial Intelligence in Accounting:
Towards a Malaysian Legal Framework

Nazruzila Razniza Mohd Nadzri
https://orcid.org/0000-0003-4279-5158
Open University Malaysia, Malaysia

Azliza Azrah Zakaria
https://orcid.org/0009-0002-3626-2187
INTI International University, Malaysia

ABSTRACT

The adoption of Robotic Process Automation (RPA), particularly when integrated with Artificial Intelligence (AI), is transforming accounting by enhancing efficiency and accuracy. In Malaysia, its use is gradually expanding among major corporations. However, this advancement presents legal and regulatory challenges, especially in areas such as contractual risks, data protection, labour displacement, and professional compliance. Currently, Malaysia lacks a specific legal framework for RPA and AI in accounting, leaving these technologies governed by general corporate and data laws. This chapter reviews the existing legal landscape, identifies key challenges, and compares international best practices to propose reforms. Through qualitative analysis of statutes, policy documents, and comparative models, the study finds that current legislation inadequately addresses automation-related risks. The

DOI: 10.4018/979-8-3373-3571-1.ch006

chapter recommends targeted legal and policy reforms to promote responsible RPA and AI adoption while preserving ethical, professional, and regulatory standards in Malaysia's accounting sector.

INTRODUCTION

A survey that was conducted by the Malaysian Institute of Accountants (MIA) revealed that automation is one of those technologies that has a higher impact on the accounting profession (Malaysian Institute of Accountants, 2023), which recorded an increasing trend. Compared to 2017, in 2022, the percentage of accounting practitioners who viewed that automation influences the accounting sector increased from 62% to 70%. On top of automation, other technologies that are simultaneously impacting the accounting sector include online services, mobile, cybersecurity, big data, and analytics. As to the usage of artificial intelligence (AI), the MIA survey findings discovered that only 13% of the respondents claimed that they currently use AI in their work activities, which is rather low (Malaysian Institute of Accountants, 2018). However, 15% of the respondents believed that they would adopt AI in their accounting practices within the next three years. Out of all these findings, the most significant fact from the same survey was that 99% of the respondents agreed that technology is influencing accounting practices.

McKinsey & Company's report on new and disruptive technologies projected that, by 2025, the potential economic impact of automation technologies such as Robotic Process Automation (RPA) will be nearly $6.7 trillion (Doguc, 2020). Hazar and Toplu (2023) reported that one-third of global companies are using robots in accounting processes. In fact, many agreed that the future of RPA is viewed as the coordination between RPA and AI technology, as they can maximise the efficiency of the processes (Ayinla et al., 2024; Hazar & Toplu, 2023; Shamsudin et al., 2025a; Farinha et al., 2024). Doguc (2020) describes that collaboration between RPA and AI allows for complex capabilities to emerge.

While automation primarily aims to restructure and organise rule-based and repetitive processes, defining and handling exception cases is still highly manual. At this point, AI can assist automation software with processes that do not require complex decision-making and analysis, such as natural language processing (NLP). However, according to CPA Australia (2021), RPA in fact is considered the most reasonable technology to be adopted by small businesses, as it is relatively simple and cost-effective, which enables small and medium-sized practices (SMPs) to automate repetitive and mundane tasks. Certified Practising Accountant (CPA) Australia (2021) further states that Singapore recorded the highest adoption of technologies, especially RPA, in their accounting process compared to other countries, which they found in the survey of 725 professionals across industries in Australia, China, Hong Kong, Macau, Malaysia, and Singapore.

Adopting RPA in accounting often relates to the use of software robots, or 'bots,' to automate repetitive and rule-based tasks within the accounting function. Accounting functions and processes are commonly indispensable from the auditing process that covers classifying, recording, summarising, analysing, adjusting, audit planning, risk assessments, substantive testing, tests of control, and other related activities (Shamsudin et al., 2025b). Hence, the usage of RPA in accounting frequently involves areas such as recording transactions, financial reporting, routine payments, inventory management, back transaction reconciliation, receivable management, tax reporting, and monitoring and screening data (Hazar & Toplu, 2023). The RPA technology helps make audits easier by consistently carrying out processes and keeping clear digital records, which makes it simpler to check for compliance.

The accounting sector benefits significantly from RPA, particularly in terms of enhanced efficiency and cost reduction. RPA reduces the necessity for human intervention by automating repetitive and high-volume duties, resulting in a substantial reduction in operational costs, estimated to be 25% to 40% or more (Hazar & Toplu, 2023). This decrease also improves the return on investment (ROI), as the cost of implementing RPA is typically lower than that of sustaining a full-time workforce for comparable tasks. Furthermore, RPA reduces the errors that are frequently associated with manual processing, thereby guaranteeing a higher degree of consistency and reliability in the execution of tasks (Oyeniyi et al., 2024). RPA is capable of managing large volumes of data without a decrease in accuracy or speed, as it operates on rule-based logic and executes functions with predictable outcomes (Ayinla et al., 2024).

Regarding the personal contributions of RPA to individual employees, it improves the overall quality of work in accounting functions while also providing financial and operational benefits. RPA systems streamline non-automated business processes, making it easier for employees to understand and maintain them. Furthermore, RPA improves job satisfaction by sparing employees from repetitive tasks, enabling them to concentrate on more strategic or analytical duties (Farinha et al., 2023). On top of that, its scalability enables the reuse of processes across various entities, and its simplicity of programming necessitates minimal technical training. In the end, RPA not only optimises accounting operations but also enhances customer service by substantially decreasing response times and ensuring more seamless client interactions.

Furthermore, the integration of RPA and AI, often referred to as intelligent automation, combines the strengths of both technologies to enable end-to-end automation of accounting workflows. In this hybrid model, AI handles cognitive tasks such as interpreting unstructured data and making predictive analyses. RPA executes the structured, rule-based steps required to complete transactions and update systems.

For example, in an accounts payable process, AI-powered NLP (natural language processing) can extract invoice details regardless of format, while RPA inputs the data into the Enterprise Resource Planning (ERP) system and routes the invoice for approval based on predefined rules. AI can then analyse the data for anomalies or fraud signals, and RPA can trigger alerts or corrective workflows (Nasser, 2024).

However, the usage of RPA and RPA driven AI in accounting also presents challenges, including legal uncertainties and regulatory gaps. One major issue is the absence of specific regulations addressing the liabilities arising from automated processes or even decision-making by AI. In Malaysia, existing laws such as the Contracts Act 1950 (Act 136), the Sale of Goods Act 1957 (Act 382), the Consumer Protection Act 1999 (Act 599), and the Copyright Act 1987 (Act 332) do not adequately cover the complexities introduced by AI-driven processes (Shearn Delamore & Co., 2024).

Furthermore, the Personal Data Protection Act 2010 (Act 709) (PDPA 2010) lacks provisions for automated decision-making, leaving a regulatory vacuum in data protection (Skrine.com, 2024). The rise of automation also challenges traditional accounting ethics, which emphasise integrity, objectivity, professional competence, confidentiality, and due care (Tang, 2025). Furthermore, automation inevitably affects employment within the accounting profession, for example, because routine tasks are increasingly handled by bots, potentially reducing demand for traditional bookkeeping roles (Thomson Reuters Tax & Accounting, 2025).

Rapid advancements in RPA and AI require a comprehensive legal reform to tackle the emerging challenges. A robust legal framework should encompass clear guidelines on liability, data protection, and compliance with professional standards and a regulatory strategy to minimise the impact of job displacements in the accounting sector. Such a framework would provide clarity for stakeholders and ensure that technological innovations align with legal and ethical norms. Learning from international models, such as the European Union's General Data Protection Regulation (Regulation (EU) 2016/679) and the European Union Product Liability Directive (Directive (EU) 2024/2853), can offer valuable insights for Malaysia's legal reform.

This chapter aims to (1) review the current Malaysian laws applicable to RPA and AI in accounting; (2) identify legal and ethical challenges associated with their adoption; (3) compare Malaysia's legal landscape with global or other jurisdictions' models; and (4) propose a general regulatory framework tailored to Malaysia's context. By achieving these objectives, the chapter seeks to contribute to the development of a legal infrastructure that supports technological innovation while safeguarding legal and ethical standards.

The chapter employs a qualitative research methodology that analyses legal texts, academic literature, and case studies to explore the intersection of law and technology in accounting. Comparative analysis with international legal frameworks will provide a broader perspective on best practices.

In addition to the literature review section below, which lists past literature central to studying the adoption of RPA technology and AI in general, as well as in the accounting sector, the main content of this chapter is structured as follows: Section 1 provides an overview of RPA and AI technologies in accounting, examines the current Malaysian legal framework, identifies existing gaps, and explores the legal and ethical challenges associated with the adoption of RPA and AI. Section 2 offers a summary of comparative analysis with global and other jurisdictions' legal models and proposes a general legal reform framework for Malaysia. Finally, Section 3 concludes with recommendations for policymakers, legal practitioners, and accounting professionals.

LITERATURE REVIEW

A review of studies from scientific journals, books, and government reports, as well as research reports from research groups or organisations, was selected based on themes relevant to the subject matter studied in this chapter. The primary objective of the review is to comprehend the key issues studied and to examine the gaps that need to be addressed in this study.

RPA is described as not having a consensus definition, but many scholars define RPA based on its objectives, functionality, and types. Scholars generally agree that RPA refers to the use of software robots, or 'bots,' that emulate human actions to execute structured, rule-based tasks across digital interfaces (Hazar & Toplu, 2023; Farinha et al., 2024). However, the functional definitions of RPA vary in their emphasis. For instance, Hazar and Toplu (2023) describe RPA as a rule-based automation tool that encodes human expertise into conditional logic ('if/then' rules) without requiring historical data, contrasting it with machine learning systems that evolve through training. Similarly, Farinha et al. (2024) highlight that RPA tools mimic user actions but do not learn from data unless integrated with AI, thereby offering speed and precision in mimicking manual digital tasks. These views converge on RPA's core functionality, automating routine digital tasks through deterministic rules, yet diverge on its capacity for autonomy. Whereas some studies identify RPA's rigidity (Shamsudin et al., 2025a), others acknowledge its growing synergy with AI to enhance adaptability (Ayinla et al., 2024). Faundez-Ugalde and Mellado-Silva (2023) summarise the functional classes of RPA as in Table 1 below.

Table 1. Summary of Functional Classes of RPA

Functional Class	Explanation	Examples
Data-Related Functions		
Data Transfer	Automates the transmission of data between systems or platforms.	Data caching, encryption, uploading files
File Processing	Converts, encodes, or encrypts digital file formats.	Format conversion (e.g., PDF to XML), file encryption
Data Analysis	Extracts meaningful information from text, images, or audio data.	Optical Character Recognition (OCR), speech-to-text conversion
Integration-Related Functions		
Application Operator	Automates actions within desktop or enterprise software applications.	Entering values in a spreadsheet, logging into ERP systems
Cloud Service Operator	Enables access to cloud-based platforms or web services.	Posting updates on social media platforms
Input Device Operator	Simulates human interaction with input devices.	Mouse clicks, dragging items, closing windows
Process-Related Functions		
Event Trigger	Initiates workflows based on defined system or user events.	Detecting file changes, responding to image appearance, hotkey triggers
Control Flow Operator	Connects elements of a process to define sequence and logic.	Loops, branching conditions, user interaction prompts

Source: Faundez-Ugalde & Mellado-Silva (2023)

The objectives of RPA, particularly in accounting, further refine its definitional boundaries. Scholars argue that the deployment of RPA enhances efficiency, reduces errors, and lowers costs associated with repetitive tasks. For example, Cooper et al. (2019) report that RPA adoption led to time savings exceeding one million hours and up to 80% improvement in processing speed in large accounting firms. These findings align with Shamsudin et al. (2025b), who assert that RPA enhances accuracy in data handling and auditability. Notably, Ayinla et al. (2024) emphasise that RPA also transforms workforce dynamics by shifting accountants' focus from transactional to strategic roles, thus reconfiguring the value proposition of accounting labour. In contrast, Razak and Ismail (2022) caution that perceived usefulness alone does not guarantee adoption; organisational culture, training, and leadership support play critical roles. While all sources agree on RPA's productivity and quality benefits, they differ on its workforce implications. Some view it as a replacement, others as an augmentative tool.

In defining RPA through its types, a consensus emerges around the dichotomy between attended and unattended automation. According to Farinha et al. (2024), attended bots require human initiation and are suited for front-office tasks, while unattended bots operate independently on scheduled routines, making them ideal for back-office functions. Hofmann et al. (2020, as cited in Farinha et al., 2024) explain that attended bots are easier and cheaper to set up, while unattended bots need more complex system connections and careful handling of exceptions. Cooper et al. (2019) corroborate this by noting that firms implement both types depending on the nature of the task. The contrast reveals that RPA is not monolithic. Its definitional scope must include a spectrum of capabilities that range from semi-automated desktop assistants to fully autonomous software robots operating in high-volume environments.

In accounting, RPA can be best described as a group of software programs that automate regular financial tasks like entering data, checking transactions, processing invoices, and handling compliance documents by mimicking how humans interact with digital systems. This definition encapsulates both its functional attributes (rule-based task automation), its objectives (efficiency, accuracy, and compliance), and its variations (attended vs. unattended).

Importantly, RPA in accounting does not merely replace human work but restructures it, reallocating human capital to strategic analysis and advisory functions (Ayinla et al., 2024; Cooper et al., 2019). Thus, RPA is best understood as a virtual workforce that enables accounting professionals to focus on higher-value tasks by absorbing transactional work, ultimately enhancing both operational and decision-making capacities within firms.

RPA is part of the continued use of technology in accounting processes, following several previous stages of the revolution. The evolution of accounting technologies has consistently reflected the pursuit of greater operational efficiency, accuracy, and standardisation. Early technological phases, such as manual bookkeeping, the use of spreadsheets, and ERP systems, focused on digitising records and centralising data (Shamsudin et al., 2025a). Despite being revolutionary at the time, these systems heavily relied on human intervention and had limited automation capabilities. With the emergence of cloud computing and mobile platforms in the 2000s, accounting began transitioning from isolated systems to interconnected and accessible ecosystems. However, even these advancements did not fully resolve issues related to processing speed, human error, and compliance documentation in high-volume tasks.

RPA enters this trajectory not as a disruptive departure but as a logical extension of these technologies. What distinguishes RPA is its ability to mimic human interactions across digital systems, automating routine, rule-based tasks such as invoice processing and reconciliations without altering the underlying Information Technology (IT) infrastructure (Hazar & Toplu, 2023). Unlike previous tools that required accountants to adapt to new systems, RPA adapts to existing interfaces, making it more flexible and less intrusive. Critics, however, argue that RPA, in its basic form, lacks cognitive ability and cannot handle unstructured data or ambiguous scenarios (Igou et al., 2023). This limitation is increasingly mitigated by the integration of AI, leading to 'intelligent automation,' which allows for decision-making, learning, and natural language processing, an evolution that positions RPA at the frontier between automation and artificial cognition.

The convergence of RPA and AI has raised debates about the extent to which automation can (or should) replace human judgement in accounting. While RPA excels in consistency and speed, it still relies on human oversight for exception handling and ethical decision-making. As such, RPA should not be viewed as a replacement for accountants but rather as a strategic enabler that augments human capability (Shamsudin et al., 2025b; Şahin, 2025). Its value lies not only in reducing mundane workload but also in setting the stage for the adoption of more sophisticated AI-driven solutions. In this regard, RPA represents both a culmination of prior technological efforts and a gateway to the next phase of cognitive transformation in accounting practice. However, regardless of all these positive effects driven by automation and AI technologies, the labour displacement by technology cannot simply be ignored.

RPA in accounting has seen varied patterns of implementation depending on institutional objectives and regulatory environments. In auditing, RPA is primarily used to standardise data extraction and documentation processes, but its integration remains uneven across firms due to the fragmented nature of audit environments and differing audit standards (Kadhim & Ali, 2024). Public tax administrations in some jurisdictions have adopted RPA to streamline form validation, taxpayer classification, and system-generated notices. However, Faúndez-Ugalde and Mellado-Silva (2023) note that such usage, especially in automated decision-making, raises legal and ethical concerns if transparency and redress mechanisms are not integrated into the system. In practice, the lack of coordinated implementation policies and institutional oversight creates inconsistencies in how RPA is applied in tax governance versus how it is deployed in internal audit settings or financial reporting processes.

In the private sector, RPA adoption in accounting remains largely piecemeal, driven more by operational expediency than strategic alignment. Shamsudin et al. (2025a) point out that many firms implement RPA on an ad hoc basis without a cohesive digital transformation plan, often resulting in siloed automation efforts with limited scalability. Farinha et al. (2023) argue that a structured framework is essential to guide decision-making, ensuring that RPA implementation aligns with business objectives, risk tolerance, and system readiness. This requirement is particularly critical in accounting contexts where tasks are interdependent across departments and platforms. The absence of a strategic framework not only risks process inefficiencies but also complicates regulatory compliance and auditability, especially when RPA is used alongside AI tools with autonomous decision-making capacity.

Past literature also reveals that while RPA and AI are revolutionising accounting practices, including auditing, tax administration, and financial reporting. Their rapid integration has outpaced the development of a coherent regulatory framework. Scholars highlight a significant governance gap, particularly concerning liability and data transparency. Gotthardt et al. (2020) observe that although RPA is widely used for rule-based tasks and AI for data-driven cognitive functions, the convergence of these technologies introduces complex accountability issues, especially when decision-making is opaque or involves unstructured data. Similarly, Postolea and Bodea (2023) warn that defective RPA performance, whether due to flawed design or contractual loopholes, can cause substantial financial losses in accounting functions if regulatory safeguards are weak. The lack of clear standards for selecting or evaluating RPA systems, especially in private sector accounting applications, further exacerbates these risks.

Data privacy, job displacement, and liability constitute core regulatory blind spots. Deffains and Fluet (2025) argue that traditional liability regimes such as fault-based or product liability are ill-equipped to handle the autonomous nature of AI-augmented RPA. They emphasise that strict liability might be more appropriate for high-risk systems, yet existing directives such as the European Union Product Liability Directive remain limited in scope and do not adequately address software-induced harm in accounting contexts. Moreover, the opaque algorithms underlying AI-enhanced RPA tools challenge transparency and due process in tax and audit procedures, especially in public administration (Gotthardt et al., 2020). Without a robust governance framework that incorporates ethical, contractual, and technical dimensions, the deployment of RPA and AI in accounting may undermine trust, fairness, and accountability in financial decision-making. Therefore, while RPA is increasingly used in accounting, its meaningful integration requires more than just technical deployment; it demands governance structures that contextualise its role within broader institutional and regulatory frameworks.

Malaysia's regulatory framework for digital technology and AI comprises several statutes and guidelines, yet it remains non-specific and contains loopholes regarding AI-driven tools like RPA in the accounting sector (Tan Shin Ee & Cheong, 2019). Key federal laws include the Contracts Act 1950, the Sale of Goods Act 1957, the Consumer Protection Act 1999, the Personal Data Protection Act 2010, the Communications and Multimedia Act 1998 (Act 588), the Computer Crimes Act 1997 (Act 563), the Copyright Act 1987, and the Penal Code (Act 574), but none of these laws explicitly address the complexities introduced by AI or RPA (weilingchambers.com, 2025; Daud, 2022). As such, liability attribution, transparency in automated decision-making, data protection, and ethical use remain largely unregulated.

Although the Malaysian government, especially through the Ministry of Science, Technology and Innovation (MOSTI), has introduced high-level strategies such as the National Artificial Intelligence Roadmap 2021–2025 and the National Guidelines on AI Governance and Ethics (MOSTI, 2023; MOSTI, 2024), these instruments are mostly advisory and lack statutory enforcement mechanisms. Sector-specific regulators have attempted to fill the gap. Bank Negara Malaysia's Risk Management in Technology (RMiT) policy outlines expectations for managing tech-related risks in financial institutions (Bank Negara Malaysia, 2023), while the Malaysian Institute of Accountants' Digital Technology Blueprint (MIA, 2018) encourages ethical and strategic adoption of AI and RPA in accounting. However, these frameworks are limited in scope and binding authority, leaving significant regulatory uncertainties in managing contractual liability, data protection, labour ethics, and professional compliance associated with AI-enabled RPA systems in the accounting field. As to the job displacement issue, the employment regulatory framework, including the existing key statutes such as the Employment Act 1955 (Act 265) and Industrial Relations Act 1967 (Act 177), as well as the common law principles, is considered demanding attention and exploration to find ways it can improve to better deal with the issue.

Hence, this article critically examines the use of RPA in the accounting field, including financial reporting, management accounting, auditing, and taxation, while also exploring its interdependence with complementary technologies such as AI. The review will focus on identifying legal and regulatory gaps related to contractual risks and tortious liability, data protection, labour displacement within the accounting profession, and compliance with accounting and professional regulatory standards.

METHODOLOGY

This study adopts a qualitative and doctrinal legal research methodology, involving a critical and descriptive analysis of primary and secondary legal sources to explore the regulatory challenges associated with the adoption of RPA and AI in accounting practices, with a particular focus on the Malaysian context. The analysis centres on key statutory provisions such as the Personal Data Protection Act 2010, the Copyright Act 1987, the Sale of Goods Act 1957, the Consumer Protection Act 1999, the Employment Act 1955, and the Industrial Relations Act 1967, as well as policy frameworks including the National Artificial Intelligence Roadmap 2021–2025 and the National Guidelines on AI Governance and Ethics issued by MOSTI.

To ensure a comprehensive understanding, this study also incorporates non-legal sources, including academic literature, industry reports, and institutional guidelines such as the Malaysian Institute of Accountants' (MIA) Digital Technology Blueprint and the MIA Report on Technology Adoption by the Accounting Profession in Malaysia (MIA, 2018). Comparative perspectives were drawn from international jurisdictions and supranational instruments such as the European Union's AI Liability Directive and Product Liability Directive to identify regulatory best practices and potential gaps. The study critically evaluates issues surrounding contractual risks, tortious liabilities, data privacy, transparency, professional ethics, and labour displacement resulting from the integration of RPA and AI in accounting. By analysing the interplay between emerging technologies and the legal regulatory landscape, this chapter seeks to explore brief, responsive legal frameworks capable of addressing the evolving challenges posed by automation and AI in the accounting sector.

REGULATORY ISSUES AND FRAMEWORK IN THE ADOPTION OF RPA AND AI

This section is devised based on four major regulations covered, including the presentation of relevant legal sources, the examination of the extent to which the existing law in Malaysia governs the selected issues, the analysis of relevant regulations at the international level or in other countries, and a comparative analysis at the end of the discussion.

Legal Analysis of Accountability for RPA and AI Failures: Contractual and Tortious Liability Perspectives

The implementation of RPA raises significant legal questions in the event of system failures, particularly in the areas of contractual obligations and tortious liability under the law of negligence. Since RPA systems are mostly embedded in the software format, the issue of liability regarding RPA failures is regarded similarly to the failures of software.

According to Hazar and Toplu (2023), RPA implementation typically unfolds in four main stages: (1) redesign of business processes to remove redundancies and ensure alignment with organisational policies, (2) selection of suitable automation software, (3) programming of redesigned processes, and (4) user training. Each of these phases requires precise coordination and accountability, especially in high-risk sectors such as accounting, where compliance and data integrity are critical.

Companies can adopt one of two primary implementation strategies: internal development or full outsourcing. Internal development involves investing in employee training for RPA programming. This strategy demands significant upfront investment and a long-term return horizon, but it also allows the company to retain control over intellectual property and internalise RPA maintenance (Postolea & Bodea, 2023). This option is often suitable for organisations with ongoing automation needs.

Alternatively, many companies choose to outsource the development and implementation of RPA to third-party vendors. This approach offers shorter implementation timelines and greater contractual flexibility. Under this model, companies may enter into either (1) a software license or software services agreement with an RPA vendor or (2) a comprehensive development contract that includes both the delivery of RPA services and the associated software licensing (Postolea & Bodea, 2023). Each contractual model entails different degrees of legal exposure, especially in terms of liability. As RPA solutions become increasingly autonomous and sophisticated, they give rise to liability issues, particularly when errors or malfunctions result in financial harm, regulatory breaches, or data loss. If the company develops RPA in-house, it may internalise the liability for errors or failures. If negligence or incompetence within the company's own team causes the failure, the company may face direct contractual liability. (Postolea & Bodea, 2023). However, where RPA systems are developed or licensed from third-party vendors, liability is governed by the terms of the contract, particularly clauses that define warranties, service levels, and limitations of liability (San & Peng, 2016; Postolea & Bodea, 2023).

In the context of defective software, the distinction between contractual and tort liability is crucial, as it determines both the legal basis for claims and their scope and the compensation that can be obtained. Despite their potential intertwining, these two liability regimes rest on distinct principles and premises.

Contractual liability arises from the non-performance or improper performance of a contractual obligation. In the case of software, this can apply if the delivered solution does not meet the agreed parameters, functionality, or security level specified in the contract (Postolea & Bodea, 2023). For example, contractual liability arises if a development company agrees to create an RPA system with certain functions, but the delivered solution fails to implement those functions or malfunctions.

Within the framework of contractual liability, precise contractual provisions, including requirements specifications, quality guarantees, and acceptance and testing procedures, play a key role. The parties have a great deal of freedom in shaping the scope of mutual obligations and responsibilities, being able to introduce clauses that limit or expand standard liability. In practice, IT contracts often include provisions for contractual penalties for failure to meet quality parameters or deadlines (Donovan & Ho, 2019).

The chain of responsibility for software includes manufacturers, individual developers, implementation companies, cloud service providers, and even end users. Each of these entities may bear partial responsibility, depending on the specifics of the case and the nature of the error. For example, if the RPA system fails during fiscal year-end closing, causing financial losses, the responsibility may lie with both the software manufacturer and the implementation company that failed to conduct proper testing (Donovan & Ho, 2019).

According to San and Peng (2016), where RPA solutions involve open-source software, the challenge of accountability becomes more complex. Open-source software (OSS) is software whose source code is made freely available for anyone to view, use, modify, and distribute. The modification of such software by end-users or third parties may dilute the liability of the original developers, and legal remedies may be restricted by the licensing model. Thus, the structure of the licensing agreement and the governance over modifications are key to determining liability.

Furthermore, the question arises as to whether software qualifies as a 'product' under traditional product liability statutes. Some jurisdictions, such as the European Union under the Product Liability Directive, have taken steps to clarify this for digital products and AI systems. The European Commission's 2022 proposal to update product liability laws aims to include software and AI explicitly within the scope of 'products' (Hacker, 2024). Such developments indicate a growing trend towards recognising software-based automation systems as legally actionable objects when harm results from defects.

Beyond contract law, companies may also face liability under tort, specifically the law of negligence. In simpler terms, even when no formal contract exists, a party can still be legally responsible if its actions or omissions cause harm to another. In the context of software, tort liability may apply when faulty software causes harm to third parties who are not parties to the contract. For example, if an error in an RPA system leads to the disclosure of customers' personal data, the firm may be liable in tort to the injured customers. To establish negligence, the claimant must demonstrate that the developer or deployer of the RPA owed a duty of care, breached that duty through substandard performance (i.e., failure to test), and that the breach caused actual harm (e.g., financial loss or operational disruption). Notably, liability in negligence can exist independently of contract, and third-party stakeholders affected by the failure may also seek redress (Seah, 2023). Consider a Malaysian audit firm that outsources RPA software to process client invoices. The software incorrectly duplicates payment entries, causing financial discrepancies. Under contract law, the firm may claim damages from the vendor for breach of warranty. However, if third-party clients suffer losses, they could sue in negligence for failure to maintain system integrity. In summary, accountability for RPA failures in Malaysia must be assessed through a multi-faceted legal lens as presented in Table 2 below.

Table 2. Summary of Liability Types for RPA System Failures

Liability Type	Key Considerations
Contractual Liability	Determined by contractual terms, developer roles, and enforceability of limitation clauses.
Negligence (Tort Law)	Arises from duty of care breaches causing foreseeable harm, even outside contractual relationships.

Source: Authors' compilation

In Malaysia, software programmes with an early form of AI would be treated similarly to other products (Tan Shin Ee & Cheong, 2019). The Sale of Goods Act 1957 (the SOGA 1957) would address liability in the event of a malfunction. The SOGA 1957 imposes several implied terms regulating the obligations of the vendor to the goods bought by buyers. These include implied terms and conditions regarding title and lack of encumbrances, correspondence with description, satisfactory or acceptable quality, and fitness for purpose (Sections 14-16 of the SOGA 1957). Tan Shin Ee and Cheong (2019) proposed that the software supplier will be liable for any malfunction that results in a breach of these implied terms, depending on the extent of non-compliance with the representations and guarantees made by the manufacturer to the supplier and the supplier to the buyer, respectively, regarding the AI software programme. The SOGA 1957 may not cover software sales contracts, as software is debated not to be goods.

The question of whether software is a good or service has been debated for some time. In fact, within the UK legal framework, the CJEU (Court of Justice of the European Union) ruled in Case C-410/19 *The Software Incubator Ltd v Computer Associates (UK) Ltd* has determined that the sale of software supplied electronically and not on any tangible medium constitutes 'sale of goods' within the meaning of Article 1(2) of the Commercial Agents Directive (86/653/EEC) (the Directive). The determination was made in response to a referral from the English Supreme Court in the case of *Computer Associates (UK) Ltd (Respondent) v The Software Incubator Ltd (Appellant)* UKSC 2018/0090. This case is an appeal in respect of the Court of Appeal decision that the electronic supply of software did not constitute the sale of goods for the purposes of the Commercial Agents (Council Directive) Regulations 1993/3053 (the Regulations). This case was an example of a dispute regarding the status of software, which can be argued to be closely related to the form of the RPA system. Hence, the argument put forward by Tan Shin Ee and Cheong (2019) can be grounded on the rulings of the case. However, it can also be argued that the rulings in the case are mainly for the purpose of applying the Commercial Agents Directive and cannot simply be generalised to other types of regulations. Thus, the authors argue that to invoke the application of the SOGA 1957 to enforce the rights of the parties, especially the firms that buy the RPA programme were most likely be irrelevant. Not only that, parties would want to include exclusion and limitation clauses in contracts relating to the development or use of AI. For example, AI developers may want to limit their liability for the outcomes of AI systems trained on incomplete or inaccurate data if the data is supplied by another party (Seah, 2023).

Viewing it from another perspective, in Malaysia, the contract of licensing agreement surrounding the software or computer programme is regulated by the Copyright Act 1987. Computer programs or software, defined to cover both the source codes and object codes, are specifically protected under the Copyright Act 1987 as a form of literary work (Section 3 of the Copyright Act 1987). Generally, a typical software licensing agreement between providers and users will contain the elements below:

(a) the software licensing agreement, which contains the main commercial and legal terms of licensing the software, addressing the license to use the software, payment, intellectual property, and termination;
(b) the maintenance & support agreement, which addresses the obligation of the software provider to upgrade and update the software;
(c) the service level agreement ("SLA"): which addresses the specific quantifiable & measurable commitments of the maintenance and support services (Donovan & Ho, 2019).

Hence, it can be verified that the relationship among the parties heavily relies on contractual terms incorporated in each of the contracts, including the extent to which each party is to be accountable for defective or malfunctioning software. Postolea and Bodea (2023) proposed that some of the contractual elements must be made clear, for example, ownership of the intellectual property, the resulting cost in case the purchased solution does not work, who is responsible for the remediation while specifying detailed intervention timeframes, who provides the warranties, and addressing indirect liability in relation to loss of profit.

As to the RPA driven by AI, another concern is the liability that arises due to the AI decision-making. AI algorithms are increasingly used to assess risk and identify influencing factors by analysing company financial statements. Tang (2025) identified a research report indicating that businesses using AI-based assessment tools experience approximately 30% fewer errors than those that rely on conventional methods. Advanced platforms such as KPMG's Clara and EY's Helix have significantly enhanced capabilities in fraud detection and risk evaluation. However, a persistent concern remains: the 'black box' nature of many AI techniques limits transparency and poses challenges for accountability, particularly when the rationale behind automated decisions cannot be readily explained or audited (Tang, 2025).

As highlighted by Mahyuddin (2022), AI can make independent and sometimes unforeseeable decisions, raising complex questions of legal accountability when harm occurs. In the context of RPA in accounting, this may include wrongful financial classifications or regulatory breaches caused by autonomous decision logic. Traditional doctrines such as negligence or vicarious liability depend on a human actor's control, yet in AI-powered RPA, the 'black box' nature of decision-making weakens the application of these principles. The 'black box' of AI refers to the phenomenon where the internal workings of an artificial intelligence system, especially in complex models like deep learning, are not easily understandable or explainable, even to experts. In other words, we can see the input and the output, but how the AI arrived at its decision remains opaque or hidden, much like a black box whose contents are unknown (Seah, 2023). Malaysian courts have begun encountering such dilemmas, as seen in *Public Prosecutor v Denis P. Modili* (an unreported case that was cited by Cheng Leong (2021)), where AI was controversially used in judicial sentencing. While the court maintained that AI serves only as guidance, the broader implication is that Malaysian law is not yet equipped to definitively assign liability for autonomous systems, especially in complex financial settings.

Risks of Data Breach or Privacy Infringement

The integration of RPA and AI into the accounting sector has also escalated concerns regarding data privacy and protection. These technologies require access to large volumes of sensitive and often personally identifiable financial data, thereby amplifying the risks of data breaches, misuse, or unauthorised access. For example, RPA with AI-empowered technologies, particularly those using machine learning, depend on vast datasets for training and decision-making. Such data-intensive processes create a high potential for breaching privacy, especially when the datasets include sensitive financial or personal data. The opaque or 'black box' nature of AI further complicates the attribution of responsibility in the event of a data breach, given the difficulty in tracing the decision-making pathway of the algorithm (Seah, 2023).

In addition, continuous learning models may autonomously change their behaviour post-deployment, which challenges existing data protection principles that assume systems remain static after implementation (Seah, 2023). If RPA bots or AI models access outdated, biased, or unauthorised data, they may inadvertently compromise data integrity or cause discriminatory outcomes, especially in financial reporting and audit processes.

Furthermore, providers providing maintenance and support services of RPA or RPA driven by AI may have access to the personal data of the users' customers or employees. It is important to note that the unauthorised disclosure of personal data will almost certainly be a breach of data protection regulations.

Malaysia's current legal framework for data protection, particularly the Personal Data Protection Act 2010 (the PDPA 2010), provides baseline standards for data processing, disclosure, retention, and security (Section 2 of the PDPA 2010). However, the PDPA 2010 has limitations when applied to AI and RPA, especially in extraterritorial enforcement and regulating algorithmic decision-making. The PDPA 2010 imposes a duty to protect data against unauthorised access, but it does not adequately address automated profiling or the explainability of outcomes driven by AI. This is problematic given the automated decision-making power AI wields in accounting software, such as in risk assessments or fraud detection.

Moreover, the enforcement of rights like consent withdrawal and data correction becomes difficult when personal data is embedded deep within automated models or continuously updated datasets (Shamsudin et al., 2025b). Hazar and Toplu (2023) illustrated the issue with an example relating to tax advice. It can be assumed that an RPA may contain errors in the automation process, and therefore, the reliance on the data will not be sufficient to prepare, for example, tax-related advice given to clients by the accounting firm. Thus, the right of clients to know, firstly, the steps that the accounting firms have followed in the use of an RPA, and, secondly, the accounting firms must explain to the clients, in a clear and understandable format, how the software or the automation process of the collected data works, to confirm whether the tax determinations have been established correctly.

Hence, the lack of transparency in AI systems leads to accountability gaps. In accounting tasks like tax reporting or financial reconciliation, if there is a data breach or incorrect results because of poor AI training data or RPA setup, figuring out who is responsible whether it is the developer, the company using it, or the data provider, can be complicated (Seah, 2023). Also, current contracts often have vague statements that limit responsibility and exclude guarantees, which might not be valid if they are seen as unfair or not properly explained.

Additionally, the use of cloud computing in the deployment of RPA and AI in accounting processes raises critical concerns regarding the legality of data storage by third-party servers. As noted by Shamsudin et al. (2025a), storing sensitive financial and personal data on cloud infrastructure managed by external vendors can create ambiguity in data ownership, accountability, and the legality of cross-border data transfer. These concerns are mainly governed by the PDPA 2010, which limits the transfer of personal data out of the country unless there are proper protections in place, like making sure the data user takes necessary steps to keep the personal data safe from loss, misuse, changes, unauthorised access, or sharing (Section 9 of the PDPA 2010). The legal status of data entrusted to third-party cloud providers becomes uncertain when stored in foreign jurisdictions that may not uphold equivalent privacy protections, increasing the risk of unauthorised access or misuse (Shamsudin et al., 2025a).

Furthermore, the integration of AI and RPA with cloud computing infrastructure exacerbates privacy issues, particularly regarding data retention policies and unintended data collection. Many RPA systems operate continuously in the background, harvesting and analysing large volumes of data, including some that may not be relevant to the original purpose of collection. When such data is stored in the cloud, retention without proper limitation may violate the data minimisation (Section 10) and purpose limitation principles (Section 8) under the PDPA 2010.

Nevertheless, the PDPA 2010 has recently been amended to invoke a stronger governance of personal data by introducing a few mechanisms and incorporating the recent setting of AI and responsible processing of personal data by commercial firms. Hence, it is said that the amended provisions to the PDPA 2010 (Personal Data Protection (Amendment) Act 2024 (Act A1727) significantly strengthen the legal framework surrounding data protection and privacy, particularly in the context of RPA and AI applications in the accounting sector. The mandatory appointment of Data Protection Officers (Section 12A of the PDPA 2024) ensures internal oversight and accountability for data governance, while the extension of the Security Principle to include third-party data processors (Section 4(a) of the PDPA 2024) is critical in addressing risks associated with cloud-based AI and RPA systems. These measures enhance protection across the data lifecycle and create legal obligations for both service providers and accounting firms that rely on outsourced or automated data processing technologies. For instance, considering a tax consultancy automates client filings using RPA-based AI. When an external service provider updates the system, client tax data is accidentally exposed on a shared cloud folder. Under the PDPA 2024, both the accounting firm and the external processor could be held liable for inadequate security safeguards. A Data Protection Officer would be required to issue a breach notification within 72 hours.

Moreover, the introduction of mandatory breach notifications and revisions to cross-border data transfer rules (Section 12B of the PDPA 2024) align Malaysia's data protection framework more closely with global standards such as the EU's GDPR. These amendments increase transparency and enable proactive responses to breaches, mitigating the impact of potential privacy violations caused by AI-driven automation. The right to data portability (Section 9 of the PDPA 2024) also supports interoperability between digital tools and platforms, offering greater flexibility in adopting AI and RPA solutions without compromising data control. Collectively, these reforms reinforce legal safeguards and foster a regulatory environment more conducive to the responsible and ethical use of AI in the accounting profession.

Labour Displacement

The proliferation of automation, such as RPA and AI, across global industries, including accounting, has sparked increasing concern about its impact on employment. There are concerns about job displacement as AI's automation capabilities increase in the accounting profession. The McKinsey report has predicted that by 2030, automation could lead to the loss of up to 30% of the traditional accounting jobs, especially in transactional roles (Tang, 2025). The shift indicates that AI is making accountants move from routine transactional roles to more analytical and advisory roles. There is a need to upgrade and update skills and talents to enable people to be ready for the change that is coming to the accounting industry through the use of AI (Tang, 2025).

In Malaysia, these concerns were notably expressed during the 2025 Labour Day rally, where organisers called for comprehensive legal reforms to address the negative implications of AI adoption in the workplace, including job loss, data privacy intrusion, and occupational health and safety concerns (Seth Akmal, 2025). Although some stakeholders, such as the Malaysian Trades Union Congress (MTUC), acknowledged the potential of AI as a tool for empowerment, they also emphasised the urgent need for legal mechanisms to prevent the indiscriminate replacement of workers by automated systems (Suzalina Halid, 2025).

At the international level, jurisdictions like the European Union and China have begun developing legal frameworks to manage the impact of AI on human labour. The EU's Artificial Intelligence Act (Regulation (EU) 2024/1689) (EU AI Act) classifies AI use in employment as 'high-risk' and mandates transparency, bias audits, and human oversight in automated decision-making (European Parliament & Council, 2024). Notably, it prohibits AI-based emotional recognition in the workplace, acknowledging the psychological risks to employees (Article 6, EU AI Act 2024). Meanwhile, China introduced the 2022 Algorithmic Regulation for Internet Services, requiring gig platforms to protect worker rights such as fair wages and rest time after public backlash against exploitative algorithmic management practices (Sun, 2024).

In contrast, Malaysia lacks dedicated legislation governing AI in the workplace. While the Occupational Safety and Health Act 1994 (Act 514) (the OSHA 1994) remains applicable, it does not explicitly address AI-related risks. Nevertheless, provisions such as Section 15, which obligates employers to ensure workplace safety, and the newly inserted Section 18B, mandating risk assessments, can be interpreted to extend to AI-induced physical and psychosocial hazards (Department of Occupational Safety and Health (DOSH), 2023). For example, continuous AI-based surveillance could lead to 'technostress,' undermining mental well-being through privacy intrusion and excessive monitoring (Del Castillo & Taes, 2025).

The most critical issue is the displacement of human labour due to AI, which raises questions under Malaysian employment law. Employers retain managerial prerogatives, including the right to restructure or digitise operations. In *Wong Choy Pheng & Ors v. Taylor's University Sdn Bhd* (Award No. 342/2022), the Industrial Court upheld dismissals arising from AI implementation, provided they were genuine costrationalisation measures. The decision also emphasised procedural fairness as outlined in the Code of Conduct for Industrial Harmony, including alternative mitigation measures such as voluntary separation schemes and fair selection criteria. Similarly, in the U.S. case *Akridge v. Alfa Insurance Companies* (2024), the Eleventh Circuit Court held that replacing an employee entirely with technology constituted a legitimate business reason.

However, legal gaps persist, especially in the use of AI for performance evaluation, recruitment, and termination decisions. AI systems trained on large datasets may reflect systemic bias, thus reinforcing existing inequalities (Rocha, 2024). Without explicit legislation on transparency and accountability in AI-assisted employment decisions, affected workers may face difficulties contesting unfair outcomes.

In addressing these challenges, the International Labour Organization's (ILO) has advocated for binding regulatory frameworks to guide ethical AI deployment. In its 2025 World Day for Safety and Health at Work message, the ILO acknowledged AI's potential to mitigate hazardous tasks but warned of emerging risks like cyber threats and psychological harm (ILO, 2025). It has also established the Observatory on Artificial Intelligence and the World of Work, which recommends measures such as limiting surveillance, ensuring data protection, and mandating human consent in AI applications. However, due to the non-binding nature of ILO instruments, the onus remains on national governments to incorporate these guidelines into enforceable domestic law.

The Malaysian government has acknowledged the need for action, noting that 600,000 workers will require reskilling due to AI and digitalisation in the coming years (Seth Akmal, 2025), but policy responses remain reactive rather than pre-emptive. The absence of a coherent national AI regulation tailored to employment contexts leaves both workers and employers in a legal grey zone.

Accounting Professional Regulatory Compliance

The RPA and AI technologies enhance operational efficiency, accuracy, and decision-making capabilities but also introduce complex legal, ethical, and professional challenges. As the accounting profession evolves, regulatory bodies must adapt to oversee these technological changes effectively, ensuring that innovation aligns with the principles of professionalism, transparency, and public trust.

This sub-section examines the regulatory role of the Malaysian Institute of Accountants (the MIA) in governing the use of RPA and AI within Malaysia's accounting sector. It further provides a comparative analysis of regulatory frameworks and professional standards from Singapore, the United Kingdom (the UK), and Australia jurisdictions recognised for their proactive approaches to technology governance in accounting. The discussion focuses on how these regulators balance technological innovation with the preservation of accounting professionalism and ethical standards.

The MIA is the statutory body mandated under the Accountants Act 1967 (Act 94) (the AA 1967) to regulate and develop the accountancy profession in Malaysia (Section 3 of the AA 1967)). Its responsibilities include registering qualified accountants, setting professional and ethical standards, and safeguarding public interest in the accounting and auditing professions. While the AA 1967 and associated by-laws provide the legal foundation for MIA's regulatory functions, these instruments do not explicitly address emerging technologies such as RPA and AI. Consequently, MIA's response to digital transformation has primarily been through the issuance of guidelines, strategic blueprints, and professional development initiatives rather than binding laws or regulations specifically targeting automation technologies.

Malaysian Institute of Accountants' Ethical Guidelines on Technology Usage

The MIA released its Ethical Guidelines on Technology Usage in 2023 (the MIA Ethical Guidelines) to provide public practitioners with principled guidance on the ethical use of emerging technologies such as RPA and AI within accounting practices (MIA, 2023). These guidelines are grounded in the fundamental ethical principles enshrined in the MIA By-Laws; integrity, objectivity, professional competence and due care, confidentiality, and professional behaviour (MIA, 2024).

The MIA Ethical Guidelines focus on six core risk areas, identified through a comprehensive review of the local technology landscape and international standards, including the International Ethics Standards Board for Accountants (IESBA) Exposure Draft on technology-related revisions (MIA, 2023). These areas reflect the most prevalent and pressing ethical challenges faced by Malaysian accountants in the digital era, especially in the context of automation and AI adoption. The summary of the six core risk areas and their descriptions is presented in Table 3 below.

Table 3. Six core risk areas and their descriptions

Risk Areas	Description
Misuse of Information	The misuse of information involves unauthorized or unethical use, manipulation, or dissemination of data processed or generated by automated systems. RPA and AI handle vast volumes of sensitive financial and personal data, and misuse can lead to confidentiality breaches and violations under the PDPA 2010 (MIA, 2023).
Misinformation and Deepfakes	AI technologies can generate synthetic data or deepfakes that distort facts. Accountants relying on such outputs must verify authenticity to maintain financial reporting integrity (MIA, 2023).
Lack of Oversight and Acceptance of Responsibility	Automation may foster over-reliance on AI outputs. Accountants have an ethical obligation to maintain responsibility and apply oversight even when using automated tools (MIA, 2023).
Cybersecurity Failures	Digitalization increases exposure to cyber threats. Ethical leadership must ensure cybersecurity safeguards are in place to protect client data and system integrity (MIA, 2023).
Lack of Competencies in Using Technology	A lack of competency in RPA and AI may lead to improper reliance and ethical breaches. Continuous professional development is essential for risk evaluation and responsible use (MIA, 2023).
Threats to Auditor Independence in Providing Non-Assurance Services (NAS)	Providing IT-related non-assurance services like RPA design may compromise auditor independence. Ethical guidelines stress the need to manage such risks to uphold objectivity (MIA, 2023).

Source: MIA, 2023

While the MIA Ethical Guidelines are advisory and non-binding, they serve as an important reference for accountants and firms navigating the ethical challenges posed by automation. They underscore the imperative that technology must not compromise professional judgement or ethical standards (MIA, 2023). The MIA Ethical Guidelines map these risks to the fundamental principles, prescribing safeguards to mitigate threats. For example, misuse of information and cybersecurity failures are primarily addressed through confidentiality and ethical leadership, while a lack of competencies relates directly to professional competence and due care (MIA, 2023). This principled approach ensures that technology adoption does not erode the ethical foundation of the profession.

MIA's Digital Technology Blueprint and Capacity Building

Complementing the MIA Ethical Guidelines, MIA's Digital Technology Blueprint (2018) (The MIA Digital Technology Blueprint) outlines a strategic vision for integrating digital technologies into the accounting profession. It encourages accountants to embrace RPA, AI, and data analytics to enhance service delivery and operational efficiency. The blueprint also highlights the need for continuous professional development to equip accountants with digital competencies essential for responsible technology use. The MIA Digital Technology Blueprint, launched in 2018, serves as a strategic roadmap for preparing the Malaysian accounting profession for the digital age (MIA, 2018). It recognises the transformative potential of technologies such as RPA, AI, cloud computing, and data analytics and aims to foster digital readiness among accountants.

The MIA Digital Technology Blueprint outlines key focus areas, including enhancing digital literacy and competencies, promoting adoption of emerging technologies, encouraging innovation in accounting services, strengthening cybersecurity and data governance, and supporting regulatory compliance through digital reporting and complementarity with ethical guidelines.

While the MIA Ethical Guidelines focus on principled conduct and risk mitigation in technology usage, the MIA Digital Technology Blueprint provides a broader strategic framework that addresses capacity building, infrastructure, and innovation ecosystems. Together, they offer a holistic approach. The MIA Ethical Guidelines ensure that technology use aligns with professional values and mitigates ethical risks. The MIA Digital Technology Blueprint equips accountants with the skills, tools, and environment to implement technology responsibly and effectively. This synergy is vital because ethical principles alone cannot drive digital transformation without the necessary competencies and organisational support emphasised in the blueprint (MIA, 2018; 2023). The sub-section below offers a brief analysis of the regulatory approaches in accounting sectors in Singapore, the United Kingdom, and Australia to provide some ideas of best practices that can be adopted in Malaysia.

Comparative Perspectives: Regulatory Approaches in Singapore, the United Kingdom, and Australia

Singapore's regulatory environment is widely regarded as a regional benchmark for digital innovation governance. The Accounting and Corporate Regulatory Authority (ACRA) serves as the statutory regulator for accountants and audit firms, while the Personal Data Protection Commission (PDPC) oversees data protection and AI governance (ACRA, 2024). Singapore's Model AI Governance Framework, first issued in 2019 and regularly updated, provides voluntary but comprehensive guidance on AI ethics, transparency, and accountability. It articulates principles encapsulated in the acronym FEAT: Fairness, Ethics, Accountability, and Transparency (PDPC, 2023). The framework encourages organisations, including accounting firms, to implement risk management practices, conduct algorithmic audits, and ensure human oversight of AI systems.

Additionally, Singapore mandates digital financial reporting using the eXtensible Business Reporting Language (XBRL), which has been compulsory since 2007 for listed companies (ACRA, 2024). This requirement facilitates automation and data analytics, enabling seamless integration of RPA and AI in financial reporting and compliance. Singapore's regulators also emphasise preserving professionalism through continuous education, algorithmic transparency, and maintaining auditor independence when deploying AI tools (PDPC, 2023). This integrated approach balances technological innovation with robust ethical and professional standards.

In the United Kingdom, the Financial Reporting Council (FRC) regulates auditors, accountants, and actuaries. The FRC has issued detailed guidance on the use of AI in audit and accounting, emphasising transparency, explainability, and risk management (FRC, 2025). The FRC's Ethical Standard (2019, revised) incorporates provisions relevant to technology-enabled services, including maintaining independence and professional scepticism when using automated tools. Recent updates to the International Standards on Quality Management (ISQM (UK) 1) require audit firms to ensure they have adequate technological resources and governance structures to manage AI and RPA risks effectively (FRC, 2025). The UK's regulatory approach is principles-based, requiring firms to document AI tool usage, address bias and fairness, and maintain human oversight. The FRC also encourages professional development to equip accountants with the necessary skills to understand and evaluate AI systems critically (FRC, 2025).

Australia's regulatory landscape for AI and RPA in accounting is characterised by a combination of sectoral oversight and professional ethical guidance. The Australian Securities and Investments Commission (ASIC) provides regulatory guidance on digital transformation in financial reporting and auditing, highlighting the importance of governance, transparency, and human oversight (ASIC, 2024). Professional bodies such as CPA Australia and Chartered Accountants Australia and New Zealand (CA ANZ) have issued ethical guidelines addressing the use of generative AI and automation in accounting. These guidelines stress that while AI can streamline administrative tasks, it must not replace professional judgement or compromise confidentiality and data security (CPA Australia, 2024). Australia also promotes continuous professional development programs focused on AI literacy and ethical technology use, ensuring accountants remain competent in the evolving digital environment (CA ANZ, 2024). Table 4 below presents a brief comparison of regulatory approaches of the selected jurisdictions, especially with Malaysia.

Table 4. Comparative Table: Regulatory Approaches to AI and RPA in Accounting

Country	Regulator(s)	Laws/Guidelines for AI & RPA in Accounting	Nature	Key Professional Safeguards
Malaysia	MIA	MIA Ethical Guidelines on Technology Usage (2023), Digital Technology Blueprint (2018)	Advisory/Non-binding	Professional ethics, upskilling, and best practices
Singapore	ACRA, PDPC, MAS	Model AI Governance Framework, XBRL reporting mandate	Mixed (binding & advisory)	FEAT principles, algorithmic audits, human oversight
United Kingdom	FRC	FRC Guidance on AI in Audit, ISQM (UK) 1, Ethical Standards	Binding & advisory	Documentation, explainability, and professional scepticism
Australia	ASIC, CPA Australia, CA ANZ	ASIC guidance, AUSTRAC AI Statement, CPA/CA ANZ ethical guidance	Advisory/Professional standards	Governance, transparency, and staff training

Source: Authors' compilation

Across Malaysia, Singapore, the UK, and Australia, common regulatory themes have emerged in addressing RPA and AI in accounting. These include a strong emphasis on ethical principles, the necessity of human oversight to ensure accountability, and the importance of transparency in AI decision-making. All jurisdictions also highlight the need for continuous professional development to equip accountants with relevant skills, alongside robust risk management frameworks to govern the use of automation technologies.

However, challenges remain, including regulatory gaps in Malaysia, the complexity of integrating AI into legacy systems, and the need to balance innovation with ethical safeguards. These challenges underscore the importance of adaptive, principles-based regulation supported by ongoing collaboration among regulators, professional bodies, academia, and industry. The regulation of RPA and AI in accounting is a dynamic and evolving field. In Malaysia, the MIA has taken significant steps through ethical guidelines and strategic initiatives to guide the profession's digital transformation. Nevertheless, the absence of binding, technology-specific legislation highlights the need for further legal reforms to address the unique challenges posed by automation.

Comparative analysis reveals that Singapore, the UK, and Australia have developed more structured regulatory and ethical frameworks that effectively balance innovation with professionalism. These jurisdictions provide valuable lessons for Malaysia in crafting a legal framework that supports the responsible adoption of RPA and AI while preserving the integrity and trustworthiness of the accounting profession. As technology continues to advance, regulators and professional bodies must remain agile, fostering an environment where innovation thrives alongside ethical accountability and public confidence.

DISCUSSIONS

To strengthen the coherence of this chapter, Figure 1 presents a conceptual framework illustrating how legal, ethical, and professional considerations intersect within the technological transformation of accounting. The framework shows that the challenges arising from RPA and AI adoption, such as liability, data protection, labour displacement, and ethical compliance are not isolated, but mutually reinforcing. Legal regulations determine accountability and rights, while ethical and professional standards guide responsible behaviour in applying automation technologies. This reciprocal relationship underscores that effective governance of RPA and AI in accounting requires not only statutory reform but also the cultivation of professional ethics and institutional capacity. The framework therefore, provides an integrated lens through which the subsequent analysis of Malaysia's regulatory landscape, global benchmarks, and policy recommendations can be understood as interdependent elements of a holistic reform agenda.

Figure 1. Conceptual Framework of Legal, Ethical, and Regulatory Interlinkages in RPA and AI Adoption in Accounting.

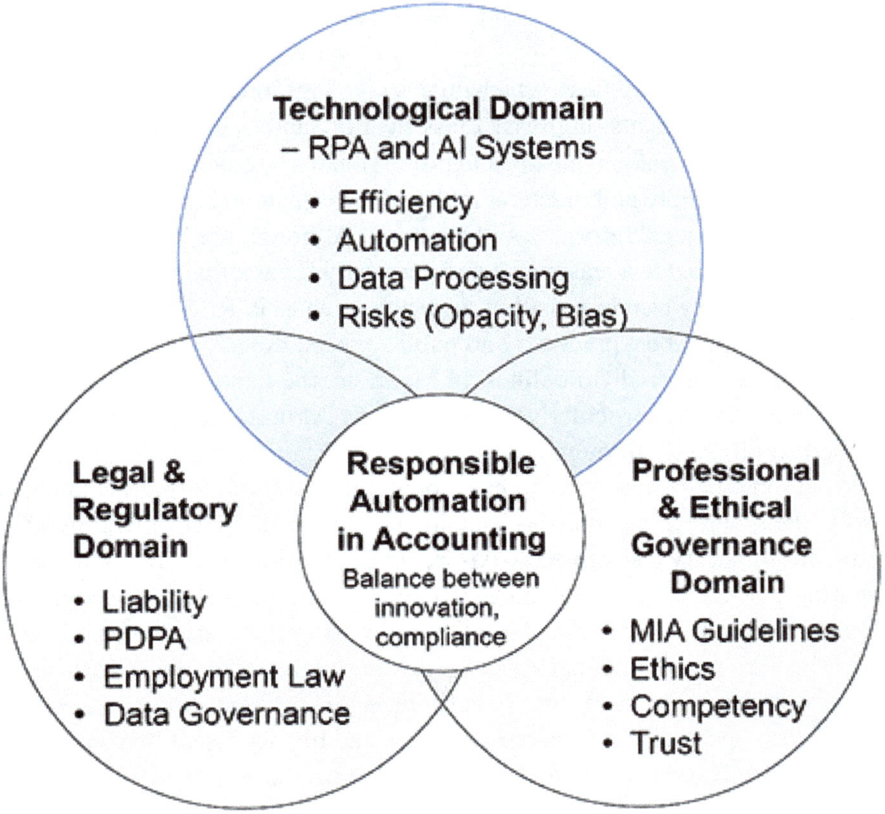

Source: Authors

Regardless of all the regulatory gaps that have been presented above, the Malaysian government has placed its effort to address the prevalent adoption of AI. The National Guidelines on AI Governance and Ethics (the National Guidelines), issued by Malaysia's Ministry of Science, Technology and Innovation (MOSTI) in 2024, serve as a voluntary framework to guide ethical and responsible AI deployment across sectors, including accounting. Anchored in the National AI Roadmap 2021–2025 and aligned with international standards such as UNESCO's AI ethics principles (UNESCO, 2021), the National Guidelines outline seven core principles: fairness, reliability, privacy, inclusiveness, transparency, accountability, and human well-being. Developers, policymakers, and end-users are the target audience for these principles, which prioritise accountability, data protection, and ethical AI system handling. Despite their non-binding nature, they promote internal governance practices, risk assessments, and stakeholder engagement as part of a national soft-law approach to AI regulation.

Compared to the EU's AI Act, which imposes legally binding obligations including transparency requirements, high-risk classification, and liability rules, Malaysia's guidelines rely on moral persuasion and professional responsibility. However, their existence is both timely and practical in the Malaysian context, where AI adoption is accelerating and legal reform is still evolving. Although not legally enforceable, these guidelines provide a reasonable and necessary foundation for organisations to anticipate regulatory trends and align AI usage, such as in RPA-driven accounting systems, with global best practices and public expectations.

Based on the National Guidelines, in Malaysia, the deployment of automation systems such as RPA, particularly when driven by AI in the accounting sector, can be shaped significantly through ethical, governance, and regulatory lenses.

The National Guidelines highlight accountability, reliability, safety, and control as two of the seven Responsible AI principles. These principles are highly relevant in mitigating liability risks in RPA/AI-driven accounting tools. RPA solutions in accounting perform tasks such as payroll automation, tax calculations, and audit trail generation. When such systems malfunction, errors can cause material financial losses. The National Guidelines recommend thoroughly testing and certifying AI systems before they are used, clearly defining who is responsible among developers, deployers, and users, and encouraging risk and impact assessments throughout the AI process. Therefore, in Malaysia, although no statutory extension currently exists, the national guidelines create a quasi-regulatory framework encouraging companies to assume responsibility through contract clauses, insurance, and self-governance models.

The National Guideline's principle of 'Privacy and Security' directly addresses data protection in AI systems, stressing the need for consent before using or disclosing personal data, lifecycle-wide privacy-by-design practices, and compliance with international standards for information security (MOSTI, 2024: 26). This is particularly significant for the accounting sector, which handles sensitive financial and personal data. RPA systems that extract, process, or transmit such data must align with PDPA 2010 and internal policies guided by the National AI Guidelines to safeguard client confidentiality and possible international obligations under cross-border data processing. Additionally, the guidelines require that there be clear rules about sharing and responsibility for content created by AI, which is important when AI is used in accounting software to create financial reports or tax documents.

The National Guidelines recognise that AI can change jobs and suggest ways to handle this, such as providing training for new skills, promoting development that includes everyone, and making sure humans stay in control through systems like Human-in-the-loop (HITL), Human-on-the-loop (HOTL), and Human-in-command (HIC) systems. In accounting, RPA may replace roles in bookkeeping, compliance, and audit preparation. However, these guidelines reinforce the view that automation should augment rather than replace human workers. Malaysian labour law continues to recognise employer prerogatives to restructure operations, but the National Guidelines support policy proposals to incentivise responsible restructuring and mandate impact assessments to ensure fair employment transitions.

Ethical considerations are central in the National Guidelines. The section on 'AI Ethics for End Users' and 'Responsible AI Principles' promotes fairness (no discrimination based on age, gender, or religion), transparency (explainability of algorithmic decisions), accountability (traceability of decisions and audit trails), and respect for human dignity (MOSTI, 2024: 38-10). In the accounting profession, governed by codes of conduct such as those issued by the MIA, integration of AI raises ethical concerns over decision transparency, algorithmic bias, and human oversight. The National Guidelines support embedding ethical reviews in system design and aligning AI use with national values and constitutional principles such as the Rukun Negara (MOSTI, 2024: 18-19). This provides a cultural and legal anchor to evaluate professional conduct in AI-assisted financial reporting.

In summary, the National Guidelines offer a non-binding but authoritative reference that can be used to operationalise responsible RPA and AI deployments in the accounting sector. They bridge gaps in statutory law by promoting principles that indirectly support the selected legal aspects analysed in this chapter, as presented in Table 5 below.

Table 5. Summary of Legal Aspects Covering in the National Guidelines on AI Governance and Ethics 2024.

Legal Aspect	Integration via Guidelines
Liability	Encourages testing, risk assessment, and role accountability to mitigate liability claims
Data Protection	Advocates consent-based data use and privacy-by-design approaches
Job Displacement	Promotes human-AI collaboration and responsible automation to protect employment
Professional Ethics	Embeds fairness, transparency, and moral responsibility into AI design and application

Source: MOSTI, 2004

Additionally, to ensure that the integration of RPA and AI in the accounting profession aligns with legal, ethical, and professional standards, Malaysia must consider targeted regulatory transformations. One of the key reforms should be an amendment to the AA 1967, which currently lacks provisions governing the use of automation technologies. The amended Act should introduce technology accountability provisions, requiring accountants and audit firms to ensure that any AI or RPA system used is reliable, ethically designed, and auditable.

It should also mandate the establishment of automation audit trails to ensure transparency in financial reporting and compliance audits, aligning with international standards such as those under the International Standards on Auditing (ISA) 230 regarding audit documentation.

Although the recent PDPA 2024 has generally integrated the regulation of automated decision-making by ensuring proper consent mechanisms and enhancing cross-border data protection, sector-specific guidelines will be more beneficial. Therefore, the MIA should create specific guidelines for the ethical use of RPA and AI in accounting, using ideas from the National Guidelines on AI Governance and Ethics (MOSTI, 2024). These guidelines should be complemented by technical standards for audit documentation involving automation, ensuring that outputs generated by RPA systems are verifiable, traceable, and meet professional audit criteria.

Institutionally, these reforms require active coordination among MIA, the Securities Commission, the Ministry of Finance, and Bank Negara Malaysia. A cross-agency task force could streamline oversight and policy harmonisation, especially in high-risk financial applications of AI. Furthermore, collaboration with academic and legal researchers is essential to support evidence-based policymaking, promote continuous review of standards, and foster innovation that upholds professional integrity.

FUTURE RESEARCH DIRECTIONS

Future research should explore the development of a comprehensive regulatory and governance framework tailored specifically to the accounting profession's integration of RPA and AI. Particular emphasis should be placed on the intersection of data protection law, contractual liability, and audit integrity, especially in relation to cross-border financial services and digital compliance ecosystems. Empirical studies investigating how automation impacts audit quality, fraud risk, and internal control reliability across different organisational scales would provide valuable insights to guide legislation and professional standards. Additionally, a comprehensive comparative legal analysis of jurisdictions, such as the European Union under the AI Act, can inform the localisation of global best practices in Malaysia.

Equally important is a future inquiry into the ethical and labour implications of automation in accounting, including its effect on job design, professional reskilling, and the psychological well-being of practitioners. Research should also assess the readiness of Malaysian regulatory bodies such as MIA, Bank Negara, and the

Ministry of Finance to oversee automated systems and ensure compliance with evolving standards. Interdisciplinary collaboration between law, computer science, and accounting scholars will be essential in producing actionable recommendations. Ultimately, such research would support the evolution of Malaysia's legal infrastructure, ensuring that automation adoption enhances, not erodes, professional responsibility, data governance, and public trust in financial reporting.

CONCLUSION

As RPA and AI become more embedded in accounting workflows, the risk of data misuse, unfair dismissal, or system failures cannot be ignored. Given the findings of this chapter, there is an urgent need for regulatory reform grounded in adaptive and principle-based legal design that evolves in tandem with technological advancement. Rather than rigid rules, future legislation should be guided by values such as fairness, explainability, and human oversight, as well as be specific to accommodate the needs of each sector. The incorporation of global best practices, such as those under the EU AI Act, could also inform Malaysia's approach. Looking ahead, further research should explore the integration of AI with other emerging technologies like blockchain to enhance audit verifiability, as well as the need for international legal harmonisation to address cross-border compliance and accountability. These efforts are crucial to building a trustworthy, resilient, and ethically grounded digital accounting ecosystem.

REFERENCES

Accounting and Corporate Regulatory Authority. (2024). Annual report 2023. https://www.acra.gov.sg

Australian Securities and Investments Commission. (2024). Digital transformation and AI in financial reporting. https://asic.gov.au

Ayinla, B. S., Atadoga, A., Ike, C. U., Ndubuisi, N. L., Asuzu, O. F., & Adeleye, R. A. (2024). The role of robotic process automation (RPA) in modern accounting: A review – Investigating how automation tools are transforming traditional accounting practices. *Engineering Science & Technology Journal*, *5*(2), 427–447.

Bank Negara Malaysia. (2023). Policy Document: Risk Management in Technology (RMiT). https://www.bnm.gov.my/documents/20124/938039/PD-RMiT-June2023.pdf

Cooper, L. A., Holderness, D. K., Sorensen, T. L., & Wood, D. A. (2019). Robotic process automation in public accounting. *Accounting Horizons*, *33*(4), 15–35.

Australia, C. P. A. (2021, September 8). Singapore leads region in business technology adoption. https://www.cpaaustralia.com.au/about-cpa-australia/media/media-releases/singapore-leads-region-in-business-technology-adoption

Chartered Accountants Australia and New Zealand. (2024). Ethical guidelines for AI adoption in accounting. https://charteredaccountantsanz.com

Cheng Leong, F. (2021, June 4). Bread & Kaya 27: 2020 Cyberlaw Cases: Cyberlaw in the Covid-19 Era - Public Prosecutor v Denis P. Modili, Foong Cheng Leong, June 4, 2021, https://foongchengleong.com/tag/public-prosecutor-v-denis-p-modili/

Australia, C. P. A. (2024). Guidance on generative AI in accounting. https://cpaaustralia.com.au

Daud, M. (2022). Artificial intelligence in the Malaysian legal system. INSAF, 39–39(1), 1–4.

Deffains, B., & Fluet, C. (2025). Decision-making algorithms: Product liability and the challenges of AI. Université Paris 2 Panthéon-Assas & Université Laval.

Del Castillo, A. P., & Taes, S. (2025). Artificial Intelligence and Labor Law. In Smuha, N. A. (Ed.), *The Cambridge Handbook of the Law, Ethics and Policy of Artificial Intelligence. Cambridge Law Handbooks* (pp. 339–366). Cambridge University Press.

Department of Occupational Safety and Health. (2023). *Guidelines on occupational safety and health risk assessment in the workplace (Amendment).* Ministry of Human Resources.

Doguc, O. (2020). Robot Process Automation (RPA) and its future. In Advances in e-business research series (pp. 469–492).

Donovan & Ho. (2019, January 15). Important Elements in a Software Maintenance and Support Agreement https://dnh.com.my/important-elements-in-a-software-maintenance-and-support-agreement/

European Parliament and Council. (2024). Regulation (EU) 2024/1689 on Artificial Intelligence (AI Act). Official Journal of the European Union. https://eur-lex.europa.eu

Farinha, D., Pereira, R., & Almeida, R. (2024). A framework to support robotic process automation. *Journal of Information Technology*, *39*(1), 149–166.

Faúndez-Ugalde, A., & Mellado-Silva, R. (2023). Use of robotic process automation for tax administrations and its impact on human rights. *Revista Chilena de Derecho y Tecnología*, *12*, e65457.

Financial Reporting Council. (2025). Guidance on the use of AI in audit. https://www.frc.org.uk

Gotthardt, M., Koivulaakso, D., Paksoy, O., Saramo, C., Martikainen, M., & Lehner, O. (2020). Current state and challenges in the implementation of smart robotic process automation in accounting and auditing. *ACRN Journal of Finance and Risk Perspectives*, *9*, 90–102.

Hazar, B. H., & Toplu, C. (2023). The use of robotic process automation in accounting. *Prizren Social Science Journal*, *7*(3), 45–50.

Hacker, P. (2024). Proposal for a directive on adapting non-contractual civil liability rules to artificial intelligence https://www.europarl.europa.eu/RegData/etudes/STUD/2024/762861/EPRS_STU(2024)762861_EN.pdf

Igou, A., Power, D. J., Brosnan, S., & Heavin, C. (2023). Digital Futures for Accountants. *Journal of Emerging Technologies in Accounting*, *20*(1), 39–57.

Kadhim, A. A., & Ali, W. A. M. (2024). Utilizing robotic process automation and artificial intelligence in auditing to mitigate audit risks. *Technium Social Sciences Journal*, *66*, 1–14.

International Labour Organization (ILO). 2025. AI and digitalization are transforming safety and health at work https://www.ilo.org/resource/news/ai-and-digitalization-are-transforming-safety-and-health-work 23 April 2025

Malaysian Institute of Accountants (MIA). (2023). Report on Technology Adoption by the Accounting Profession in Malaysia https://mia.org.my/storage/2023/10/MIA_Technology-Adoption-by-the-Accounting-Profession-in-Malaysia_FA_Hires.pdf

Malaysian Institute of Accountants (MIA). (2018). MIA DIGITAL Technology Blueprint: Preparing the Malaysian Accountancy Profession For The Digital World https://mia.org.my/storage/2022/04/MIA_Technology_Blueprint_Spreads_format.pdf

Malaysian Institute of Accountants (MIA). (2018). Digital Technology Blueprint. https://mia.org.my/storage/2022/04/MIA_Technology_Blueprint_Spreads_format.pdf

Malaysian Institute of Accountants (MIA). (2018). Report on Technology Adoption by the Accounting Profession in Malaysia https://mia.org.my/storage/2023/10/MIA_Technology-Adoption-by-the-Accounting-Profession-in-Malaysia_FA_Hires.pdf

Malaysian Institute of Accountants. (2023). Ethical guidelines on technology usage. https://www.mia.org.my

Malaysian Institute of Accountants. (2018). Digital technology blueprint. https://www.mia.org.my

Ministry of Science. Technology and Innovation (MOSTI). (2024, September). The National Guidelines on AI Governance and Ethics. https://mastic.mosti.gov.my/publication/the-national-guidelines-on-ai-governance-ethics/

Ministry of Science. Technology and Innovation (MOSTI). (2023, December). The National Artificial Intelligence Roadmap 2021–2025 https://mastic.mosti.gov.my/publication/artificial-intelligence-roadmap-2021-2025/

Nasser, H. S. A. (2024). The evolution of artificial intelligence in accounting: A historical exploration and future outlook. [IJSR]. *International Journal of Scientific Research*, *13*(11).

Oyeniyi, N. L. D., Ugochukwu, N. C. E., & Mhlongo, N. N. Z. (2024). Robotic process automation in routine accounting tasks: A review and efficiency analysis. *World Journal of Advanced Research and Reviews*, *22*(1), 695–711.

Personal Data Protection Commission Singapore. (2023). Model AI governance framework (second edition). https://www.pdpc.gov.sg

Postolea, I. D., & Bodea, C.-N. (2023). Contractual risks in Robotic Process Automation (RPA) projects. In Peracek, T., & Moreira, F. C. (Eds.), *Doctrina et Usu in Business Law* (pp. 153–163). ADJURIS International Academic Publisher.

Razak, N. A., & Ismail, K. (2022). Factors influencing the adoption of robotic process automation among accounting personnel in Malaysia. *Management Accounting Research, 21*(3), 177–202.

Rocha, A. M. (2024). Bias in generative AI: Implications for workplace decision-making. *Technology and Ethics Quarterly*, *6*(2), 77–93.

Rocha, L. (2024, October 2). AI in the Workplace: The Dangers of Generative AI in Employment Decisions https://www.culawreview.org/journal/ai-in-the-workplace-the-dangers-of-generative-ai-in-employment-decisions

Şahin, O. N. ve Evdilek, N. (2025). Digital transformation in audit: opportunities and challenges of artificial intelligence in audit. Denetim ve üvence Hizmetleri Dergisi, 5(1), 49-61.

San, T. P., & Peng, S. C. (2016). Legal Issues on Free Software and License Contracts: A Malaysian Perspective. In Metzger, A. (Ed.), *Free and Open Source Software (FOSS) and other Alternative License Models. Ius Comparatum - Global Studies in Comparative Law* (Vol. 12). Springer.,

Seah, C. (2023). Liability Arising from the Use of Artificial Intelligence. *Legal Network Series* 1 LNS (A) Iiv

Seth Akmal. (2025, Mei 1). Himpunan Hari Pekerja 1 Mei: Penganjur gesa kerajaan pinda akta penggunaan AI, https://www.astroawani.com/berita-malaysia/himpunan-hari-pekerja-1-mei-penganjur-gesa-kerajaan-pinda-akta-penggunaan-ai-518933

Shearn Delamore & Co. (2024). An interview with Shearn Delamore & Co discussing Artificial Intelligence in Malaysia, https://www.lexology.com/library/detail.aspx?g=61785546-4273-467c-9893-7650501226c3

Shamsudin, A., Ali Khan, M. N., & Jusoh, A. (2025a). Emerging technologies revolutionizing the accounting profession – Practical applications. *Information Management and Business Review, 17*(1), 126–136.

Shamsudin, A., Ali Khan, M. N. A., & Jusoh, A. (2025b). Unveiling Challenges and Opportunities in the Integration of Emerging Technologies in the Accounting Profession: A Preliminary Investigation. Jurnal Pengurusan, 73, –.

Skrine.com. (2024, May 7) AI: Understanding the Basics and the Privacy Dilemma https://www.skrine.com/insights/alerts/may-2024/ai-understanding-the-basics-and-the-privacy-dilemm

Sun, N. (2024, July 18). Workplace AI in China: The changing profile of work and labour Research paper, DOI: DOI: 10.55317/9781784136154

Suzalina Halid. (2025, Januari 25). Negara perlukan Akta AI – MTUC https://www.bharian.com.my/berita/nasional/2025/01/1353501/negara-perlukan-akta-ai-mtuc

Tang, X. (2025). Impact of Artificial intelligence on Traditional Accounting - An overview of opportunities and challenges. SHS Web of Conferences, 218, 03029.

Thomson Reuters Tax & Accounting. (2025, June 10). How will AI affect accounting jobs? https://tax.thomsonreuters.com/blog/how-will-ai-affect-accounting-jobs-tri/#:~:text=Yes%2C%20AI%20can%20perform%20many,be%20applied%20to%20industry%20work

Tan Shin Ee, J., & Cheong, Z. (2019). Laws and Regulations Governing Technology in Malaysia https://www.shinassociates.com.my/wp-content/uploads/2019/11/PRAXISApr-Aug2019-selection.pdf

UNESCO. (2021). Recommendations of AI Ethics https://unesdoc.unesco.org/ark:/48223/pf0000381137

weilingchambers.com. (2025, March 8). Introduction to Artificial Intelligence (AI) Law in Malaysiahttps://weilingchambers.com/legal-due-diligence/f/introduction-to-artificial-intelligence-ai-law-in-malaysia

Wong Choy Pheng & Ors v. Taylor's University Sdn Bhd, Award No. 342/2022 (Industrial Court of Malaysia).

KEY TERMS AND DEFINITIONS

Accounting Sectors: In this chapter context, the accounting sector encompasses professional practices and services related to financial reporting, auditing, tax advisory, and compliance.

Artificial Intelligence driven by Robotic Process Automation: This refers to the integration of AI capabilities such as machine learning and natural language processing into RPA systems, allowing them to perform more complex and cognitive tasks like predictive analysis or fraud detection.

Artificial Intelligence: Artificial Intelligence refers to computer systems capable of performing tasks that typically require human intelligence, such as reasoning, learning, decision-making, and natural language processing. In the accounting sector, AI is used to analyse financial data, detect anomalies, predict risks, and support audit processes.

Contractual Risk: Contractual risk refers to the potential for disputes or liabilities arising from agreements related to the development, deployment, or operation of RPA and AI tools. These include risks related to software performance, intellectual property rights, service level failures, and lack of clarity in roles and responsibilities.

Data Protection: Data protection refers to the legal and technical safeguards for managing personal and financial data processed by RPA and AI in accounting. It includes addressing privacy concerns such as unauthorised access, excessive data retention, and misuse of information.

Labour Displacement: Labour displacement refers to the reduction or elimination of human roles as a result of automation especially through RPA and AI.

Robotic Process Automation: Robotic Process Automation refers to the use of software 'robots' or scripts to automate structured, rule-based tasks within accounting processes, such as data entry, invoice processing, and reconciliation.

Chapter 7
Enhancing Budgeting Process Through Robotic Process Automation

Nik Herda Nik Abdullah
https://orcid.org/0000-0003-0582-8432
Taylor's University, Malaysia

Cheng Rui
Pujiang Institute, Nanjing Technology University, China

Enni Savitri
Riau University, Indonesia

ABSTRACT

Budgeting is essential for planning, control, and performance evaluation, yet traditional methods are often slow, error-prone, and inflexible. This chapter explores how Robotic Process Automation (RPA) transforms budgeting by automating repetitive tasks like data collection, validation, and reporting. RPA improves accuracy, reduces cycle time, and enables real-time insights, enhancing strategic decision-making. A case study on a mining company illustrates tangible benefits, including reduced manual workload, improved forecasting, and enhanced collaboration. The chapter also addresses challenges such as integration costs, system compatibility, and workforce resistance. It emphasizes the need for curriculum reform, continuous upskilling, and effective change management. Looking forward, the chapter highlights future trends in cognitive automation, AI-driven forecasting, and blockchain, advocating a hybrid model that balances automation with human judgment for ethical, sustainable budgeting.

DOI: 10.4018/979-8-3373-3571-1.ch007

INTRODUCTION

Budgeting remains a vital instrument for strategic planning, financial control, and resource allocation in organizations. It helps set performance expectations, guide managerial behavior, and align resources with organizational goals. Despite its enduring importance, traditional budgeting processes are often criticized for being time-consuming, error-prone, inflexible, and misaligned with the fast-changing business environment (Tiron-Tudor et al., 2024; Perdana et al., 2022). These limitations have intensified the call for innovation in budgeting, especially through digital technologies that improve operational efficiency and decision-making relevance.

One such transformative technology is Robotic Process Automation (RPA). RPA refers to the use of software bots that replicate human actions to execute routine, rule-based tasks across various applications and platforms. Unlike conventional IT automation solutions, RPA does not require extensive system changes or programming and is capable of interacting with legacy systems in a non-invasive manner (Lindgren et al., 2022; Zhang et al., 2023). In budgeting, where tasks such as data gathering, consolidation, validation, and report generation are repetitive and data-intensive, RPA offers significant potential for process optimization.

The integration of RPA into budgeting processes addresses some of the most pressing challenges in finance functions. Organizations can automate manual inputs, streamline cross-functional coordination, and reduce the risk of human error in budget forecasting, versioning, and approvals. Studies show that RPA can deliver up to 30% cost savings and reduce error rates by over 90% in budgeting-related functions (Balamurugan et al., 2022). These efficiency gains allow finance professionals to shift from transactional roles to more strategic responsibilities such as scenario analysis, stakeholder advising, and long-term planning (Zhou et al., 2024).

Real-world applications underscore RPA's growing impact. Large firms such as Ping An Insurance and Walmart have successfully deployed RPA to reduce budgeting cycle times, increase forecasting accuracy, and improve compliance (Zhang et al., 2023; Jędrzejka, 2022). For instance, Walmart reported a 65% reduction in invoice processing time after integrating RPA into its budgeting workflow, while Ping An achieved a 90% reduction in errors during claims and budgeting reconciliations. These examples illustrate the scalability and practical value of RPA when strategically implemented.

RPA systems operate by capturing and interpreting data, triggering responses, and communicating with other digital systems mimicking human keystrokes and decisions at scale. When applied to budgeting, this allows for the automation of tasks such as collecting financial data from multiple systems, populating templates, validating historical spending patterns, generating variance analyses, and distributing reports to relevant stakeholders. Furthermore, RPA can be combined with Artificial

Intelligence (AI) and Machine Learning (ML) to create intelligent automation systems capable of pattern recognition, predictive forecasting, and adaptive budgeting models (Hsiung & Wang, 2022).

However, RPA integration is not without its obstacles. Organizations often encounter challenges such as high initial investment (ranging between USD 100,000–500,000), difficulties integrating with legacy systems, and employee resistance due to fears of job displacement (Ayinla et al., 2024). Research suggests that up to 60% of employees in finance functions express concern about job security when automation is introduced (Doğuç, 2021). Without proper change management strategies, RPA implementation may face cultural and operational pushback that undermines its effectiveness. Therefore, successful RPA deployment requires not only technical readiness but also leadership commitment, workforce retraining, and transparent communication.

Education and training systems must evolve to support this transition. Accounting and finance professionals of the future must possess digital literacy, data analytics competencies, and an understanding of automation tools such as UiPath, Blue Prism, and Power BI. To meet this need, several universities and professional bodies have begun embedding RPA certifications and practical modules into their curricula. In Australia, accounting programs incorporating UiPath have reported an 80% improvement in graduate employability, while U.S. institutions offering RPA and data analytics training saw a 50% increase in strategic job placements (Balamurugan et al., 2022; Kokina & Blanchette, 2020).

RPA is also being adopted in small and medium-sized enterprises (SMEs), where cloud-based automation solutions lower entry barriers and reduce infrastructure costs. For SMEs, cloud-hosted RPA platforms provide access to budgeting automation with minimal setup, offering flexibility and scalability. Studies show that cloud-based RPA can reduce implementation costs by 20–30% while improving real-time financial visibility (Hsiung & Wang, 2022).

Moreover, as organizations move toward cognitive RPA, the future of budgeting lies in combining rule-based automation with AI-driven capabilities. This includes natural language processing (NLP) for generating budget narratives, and predictive analytics to enhance forecasting accuracy. By 2026, it is estimated that over 70% of large organizations will integrate some form of intelligent automation into their budgeting and planning processes (Plattfaut & Borghoff, 2022). These developments signify a paradigm shift in budgeting, transforming it from a backward-looking compliance task into a proactive, value-generating function.

The impact of RPA extends beyond operational improvements. It enables a strategic repositioning of finance professionals allowing them to focus on interpreting insights, advising decision-makers, and contributing to cross-functional initiatives. For example, a Japanese bank utilizing RPA for cost analysis was able to reassign its accounting staff to strategic planning roles, resulting in a 15% increase in operational efficiency (Perdana et al., 2022). Likewise, an Indian manufacturer's upskilling program led to a 25% increase in perceived organizational value after integrating RPA into its financial planning process (Oyeniyi et al., 2024).

This chapter aims to explore the transformative role of RPA in enhancing budgeting processes. It will first review the current literature and technological foundations of RPA in financial contexts. Then, the chapter will examine specific budgeting tasks that can be automated using RPA, along with detailed case studies that demonstrate successful implementations. The chapter also addresses the challenges and controversies associated with automation, particularly those related to workforce impact and ethical considerations. Finally, it offers recommendations for educators, practitioners, and policymakers to support the responsible and effective integration of RPA in budgeting.

In summary, the application of Robotic Process Automation in budgeting is not merely a technical upgrade but a strategic enabler. It empowers organizations to achieve greater efficiency, accuracy, and insight while redefining the role of accountants and financial managers. However, maximizing its benefits requires more than technological investment, it calls for education, adaptation, and ethical foresight. As budgeting continues to evolve in the digital era, RPA stands as a catalyst for more agile, intelligent, and strategic financial planning.

BACKGROUND

Overview of Robotic Process Automation (RPA)

Definition and Characteristics of RPA

Robotic Process Automation (RPA) is a form of software automation that uses configurable "bots" to mimic human interactions with digital systems and applications. These bots are capable of logging into systems, extracting and entering data, performing calculations, generating reports, and executing rule-based workflows across platforms (Zhang et al., 2023). Unlike traditional programming or complex IT transformation projects, RPA operates through the user interface layer of applications, allowing it to simulate human tasks without requiring changes to underlying systems. This "non-invasive" nature makes RPA particularly suitable for organizations with legacy infrastructure that are looking to improve operational efficiency without significant system overhauls (Balamurugan et al., 2022).

One of the defining characteristics of RPA is its ability to work 24/7 without fatigue, delivering high levels of speed, consistency, and reliability. RPA systems are typically governed by a set of predefined rules or workflows that guide bots on how to handle structured data and repeatable tasks. As a result, RPA is well-suited for standardized processes such as invoice processing, payroll entry, budgeting, and reconciliations, common in the finance and accounting domain (Perdana et al., 2022; Ray et al., 2023).

Modern RPA platforms offer drag-and-drop interfaces, low-code design capabilities, and integration with machine learning models, enabling business users with minimal programming knowledge to configure bots. This democratization of automation accelerates implementation and reduces reliance on IT departments, making RPA a business-friendly solution in many enterprise settings (Jędrzejka, 2022).

Core Components of RPA

RPA systems consist of several core components that work together to deliver automated process execution. These include software bots which are the central element of RPA, bots replicate human actions by interacting with applications, extracting data, making rule-based decisions, and completing transactions. Bots can be categorized into attended or triggered by human interaction and unattended that run independently based on schedules or triggers. The components of orchestration and control center manages the deployment, monitoring, and performance of bots across the enterprise. It ensures process governance, handles exception cases, and provides audit trails for compliance purposes. Meanwhile, advanced RPA platforms are increasingly integrating cognitive technologies to enable bots to handle unstructured data such as emails, PDFs, voice commands and to learn from historical patterns. This convergence of RPA with artificial intelligence (AI) is referred to as intelligent automation or cognitive RPA (Plattfaut & Borghoff, 2022). As for Optical Character Recognition (OCR), tools allow bots to extract data from scanned documents or images, expanding the range of processes that can be automated such as reading invoices or contracts. While RPA traditionally uses the user interface (UI) layer, modern tools also incorporate APIs to ensure more stable and direct system communication where available. Cloud-based RPA offers scalability and flexibility, especially for small and medium-sized enterprises (SMEs) seeking to avoid high infrastructure costs. It also supports faster bot updates, remote access, and real-time performance tracking.

Main Benefits of RPA in Budgeting

The finance and accounting domain is one of the most RPA-ready functions due to its reliance on data-intensive, repetitive tasks. According to recent studies, over 60% of finance tasks have high automation potential, particularly in areas such as budgeting, forecasting, accounts payable and receivable, reconciliations, and regulatory reporting. The application of RPA brings several measurable benefits to organizations. Among the most prominent are increased efficiency, improved accuracy, scalability, cost savings and regulatory compliance.

RPA significantly increased efficiency by reducing process cycle times. Bots can perform repetitive tasks at a much faster rate than humans, operating continuously without breaks. For example, budgeting data aggregation, which may take hours for a finance team to compile manually, can be completed by bots within minutes. Manual data entry is prone to human error, which can compromise financial reports and forecasts. Bots follow pre-programmed rules and do not make mistakes due to fatigue or oversight, thereby enhancing data accuracy and auditability (Zhang et al., 2023). RPA also enables organizations to scale operations without proportionally increasing headcount. During peak financial planning seasons, such as quarterly budget reviews, additional bots can be deployed temporarily to handle increased workloads. Cloud-based solutions make it easier to deploy bots as needed and shut them down when no longer required.

RPA leads to cost savings as mentioned by Balamurugan et al. (2022), organizations report operational cost reductions ranging between 20–30% after RPA implementation, primarily through labor cost savings and reduced error-related expenses. Besides, with built-in logs and traceability features, RPA supports audit trails and documentation necessary for regulatory compliance, especially in finance functions subject to strict reporting standards (Ayinla et al., 2024).

In the budgeting process, RPA offers a powerful tool for streamlining repetitive and data-intensive tasks that traditionally consume considerable time and resources in finance departments. One of its primary functions is the extraction and consolidation of financial data from multiple departments and disparate systems. Budgeting often involves collecting figures from various business units, each using different formats or platforms. RPA bots can navigate across these systems, extract relevant data, and consolidate it into a centralized database or template with minimal human oversight (Zhang et al., 2023). This significantly reduces the manual burden on finance teams while enhancing the accuracy and consistency of inputs.

Furthermore, RPA can be programmed to validate budgeting inputs against historical data trends and predefined business rules. For instance, if a department submits an expense projection significantly higher than previous years without justification, the bot can flag the anomaly for review or request additional documentation. This validation mechanism strengthens internal control and ensures budget assumptions are realistic and aligned with past performance (Perdana et al., 2022). After data collection and validation, RPA can populate standard budget templates automatically, eliminating the need for repetitive data entry across multiple spreadsheets or enterprise systems. This not only saves time but also minimizes the risk of human error, which is especially critical when dealing with large-scale, organization-wide budgets (Jędrzejka, 2022).

Another significant contribution of RPA in budgeting is version tracking and real-time updates. In many organizations, budgets undergo multiple iterations before approval. Bots can be configured to monitor changes, maintain version histories, and automatically update stakeholders on modifications. This facilitates greater transparency and collaboration across budgeting cycles. In tandem, bots can also generate and distribute finalized budget reports to relevant decision-makers through automated workflows—ensuring timely delivery, standardized formatting, and consistency in documentation (Balamurugan et al., 2022). These features make the budgeting process not only more efficient but also more agile, enabling quicker responses to internal and external changes.

RPA's role extends beyond budgeting into the broader domain of financial reporting. One of its most widely adopted applications is the automatic extraction of data from Enterprise Resource Planning (ERP) systems and spreadsheets. Instead of relying on staff to manually retrieve financial figures, RPA bots can be scheduled to perform data pulls at specific intervals, ensuring that reports are based on the latest available data. This real-time capability is crucial for organizations operating in fast-paced environments where up-to-date information drives decision-making (Ayinla et al., 2024).

Additionally, bots can be deployed to perform reconciliations, one of the most labor-intensive tasks in finance. RPA can match entries across different accounts or systems, identify discrepancies, and even resolve basic mismatches based on predefined rules. This accelerates month-end and year-end closings while enhancing accuracy and compliance (Plattfaut & Borghoff, 2022).

Another key benefit of RPA in financial reporting is the automation of dashboard generation and key performance indicators (KPIs). Finance teams can configure bots to gather relevant metrics, populate visualization tools such as Power BI or Tableau, and share real-time dashboards with leadership teams. This level of automation ensures that performance metrics are always up to date and accessible, empowering faster and more informed decision-making (Perdana et al., 2022).

Finally, RPA helps ensure that reports are delivered on schedule and with minimal human intervention. Scheduled bots can prepare, format, and disseminate reports to internal and external stakeholders via email, shared drives, or dashboards, with embedded controls to prevent data leaks or compliance violations. This improves the reliability and consistency of reporting processes while reducing the administrative workload on finance professionals (Zhang et al., 2023). In essence, RPA enables finance teams to move from manual data handlers to strategic business advisors, focusing more on analysis and interpretation, and less on routine processing.

As finance functions continue to evolve in a digitally disrupted environment, Robotic Process Automation stands out as a transformative tool. It enables organizations to automate labor-intensive budgeting tasks with minimal infrastructure changes, yielding improvements in speed, accuracy, and compliance. RPA's flexibility, low-code design, and ability to operate across legacy systems make it uniquely positioned to modernize financial planning. With the added power of AI and cloud technologies, RPA will continue to redefine how budgeting is performed by making it more agile, insightful, and value driven.

Differences Between RPA and Other Automation Tools

RPA is often compared to other automation technologies such as Enterprise Resource Planning (ERP) systems, macros, and business process management tools. However, there are distinct differences in purpose, complexity, and flexibility. While ERP platforms like SAP or Oracle automate enterprise-wide processes by integrating modules such as finance, HR, procurement, their implementation is time-consuming and resource-intensive. RPA, on the other hand, complements ERP systems by automating routine activities within or across ERP environments without requiring backend customization.

According to Jędrzejka (2022), macros automate tasks within specific applications such as Excel, but they are limited in scope and flexibility. RPA extends automation across multiple applications and systems, making it suitable for end-to-end process automation. Meanwhile, tools like Business Process Management (BPM) systems orchestrate complex workflows across departments, often requiring structured inputs and predefined rules. RPA can operate in both structured and semi-structured environments and is easier to implement due to its low-code nature.

In summary, RPA is best suited for automating high-volume, rule-based tasks across disparate systems where traditional automation methods may not be viable due to complexity or cost.

Traditional Budgeting Process: Challenges and Constraints

Budgeting remains a cornerstone of financial planning and control across industries. The traditional budgeting process is structured and methodical, often following a linear path: data collection, data consolidation, managerial review and approval, followed by dissemination and continuous review. While this approach provides a degree of structure and accountability, it is increasingly seen as inadequate in today's dynamic, data-driven business environment. This section explores the conventional budgeting process, highlights its inefficiencies, and examines the growing need for transformation.

Common Steps in Traditional Budgeting

The conventional budgeting process generally begins with data collection from various departments. Functional units such as marketing, operations, human resources, and IT submit projections on revenues, costs, and capital needs for the upcoming period. This data is typically collected through spreadsheets or emailed templates that must be filled in and returned to the finance department. Once gathered, the finance team undertakes data consolidation, merging these inputs into a master budget document. During this phase, inconsistencies must be resolved, duplicates removed, and values aligned with historical data or corporate targets. This process is inherently manual and time-consuming, especially in large or decentralized organizations (Huang, 2019).

After consolidation, the approval process begins. This typically includes multiple levels of managerial review, often with iterations between departments and corporate leadership. Proposed budgets are adjusted, challenged, and recalibrated to align with strategic objectives and financial constraints. Finally, the review and dissemination phase begins. The approved budget is distributed to stakeholders for execution, with periodic performance reviews against budgeted versus actual figures (Hsiung & Wang, 2022; Luo et al., 2021). Any discrepancies found often result in further manual revisions and justifications.

Manual Interventions and Inefficiencies

Despite the structured sequence, the traditional budgeting process is plagued by inefficiencies largely due to its reliance on manual interventions. One major concern is the extensive use of spreadsheets, which are prone to formula errors, version control problems, and lack of transparency. Research shows that over 80% of companies still use spreadsheets as their primary budgeting tool, and up to 88% of these spreadsheets contain errors (Bright et al., 2023).

Manual consolidation of data requires significant human effort to ensure accuracy, especially when dealing with multiple iterations. This often results in long budget cycles—averaging between 3 to 6 months from initiation to approval in many multinational organizations (Balamurugan et al., 2022). These long timelines are incompatible with the speed of modern business, where real-time responsiveness is increasingly crucial.

In addition, manual processes are non-scalable. As organizations grow or expand across borders, the complexity and volume of data increase exponentially, making it difficult for finance teams to keep up using legacy methods. Every new cost center or regional office adds more spreadsheets and approvals, which slows down the cycle further and increases the potential for human error (Zhang et al., 2023).

Time Consumption, Data Inconsistencies, Human Error, and Lack of Agility

Critical issues emerge from the traditional budgeting approach

Traditional budgeting is extremely time-intensive, consuming up to 25% of senior executives' and financial managers' annual working hours (Perdana et al., 2022). The cycle is burdened by numerous back-and-forth communications, departmental delays, and manual data cleaning processes. With various departments using different formats, assumptions, and timelines, the risk of conflicting figures is high. Disparate spreadsheets may use outdated data, inconsistent KPIs, or unsupported assumptions. This inconsistency reduces the overall reliability of the consolidated budget (Jędrzejka, 2022). Traditional budgeting also leads to human error. Manual data entry is susceptible to errors, such as incorrect formulas, missing decimal points, or copy-paste mistakes. According to a study by Kokina and Blanchette (2020), human error contributes to over 70% of budgeting inaccuracies, leading to flawed forecasts and misguided financial planning. Traditional budgeting lacks the flexibility to respond to rapid internal or external changes. In volatile environments such as during economic downturns, supply chain disruptions, or global pandemics the static nature of annual budgets becomes a liability. Reforecasting takes weeks, which delays corrective actions and hinders strategic responsiveness (Plattfaut & Borghoff, 2022).

Impact on Decision-Making and Responsiveness

The inefficiencies inherent in traditional budgeting can severely hinder organizational decision-making. When finance teams are bogged down by data entry and manual reconciliations, they have limited capacity for analysis and strategic advising. This reduces the effectiveness of the budget as a tool for performance management and future planning.

Slow and inaccurate budgeting cycles impair an organization's ability to forecast accurately and respond proactively to emerging threats and opportunities. Decision-makers may act on outdated or erroneous data, leading to missed investment opportunities, over-allocation or under-allocation of resources, and inability to capitalize on short-term gains (Bellinga, et al., 2022).

Moreover, traditional processes create information silos, where each department works in isolation without visibility into the bigger financial picture. This lack of collaboration often results in misaligned goals and competing resource demands, making it harder for organizations to execute cohesive strategies.

From a leadership perspective, the absence of real-time data undermines performance monitoring and early warning systems. Executives may only discover budget deviations at quarter-end or year-end, at which point mitigation becomes reactive rather than preventive (Oyeniyi et al., 2024).

The Strategic Limitations of Traditional Budgeting

Beyond operational issues, traditional budgeting suffers from strategic limitations. Budgets tend to be incremental, focusing on last year's performance as the baseline. This discourages innovation and critical evaluation of resource allocation. Departments often adopt a "use-it-or-lose-it" mentality to protect future budget allocations, leading to inefficiencies and overspending.

In addition, fixed annual budgets are misaligned with today's dynamic business environment. The pace of change in markets, technology, and regulation necessitates continuous planning and rolling forecasts, which are difficult to implement under traditional budgeting structures (Balamurugan et al., 2022).

The rigidity of traditional budgeting is also a barrier to cross-functional collaboration. Modern business problems require agile, cross-disciplinary solutions, but siloed budgeting limits the organization's ability to reallocate funds or respond dynamically to new priorities.

Need for Modernization

Given these constraints, there is growing recognition that budgeting must evolve. Modern enterprises are exploring more agile budgeting methods—such as zero-based budgeting (ZBB), driver-based planning, and rolling forecasts alongside automation tools like Robotic Process Automation (RPA) and artificial intelligence (AI).

These technologies aim to reduce cycle times, increase data accuracy, and enable real-time decision-making. By eliminating routine manual tasks, finance teams can focus on higher-value activities such as strategic planning, scenario analysis, and stakeholder engagement (Zhang et al., 2023).

Automation tools, particularly RPA, offer a non-invasive way to modernize budgeting processes without overhauling existing systems. They integrate with legacy tools, automate data entry, validate inputs, and generate reports transforming the budgeting process into a more agile, efficient, and strategic function.

MAIN FOCUS OF THE CHAPTER

Implementing RPA Across the Budgeting Lifecycle

The budgeting process remains a critical component of financial planning, performance evaluation, and strategic decision-making within organizations. Yet, it is often plagued by inefficiencies, inconsistencies, and human error. Robotic Process Automation (RPA), as a form of intelligent automation, offers a transformative solution by streamlining the budgeting lifecycle including data collection, aggregation, versioning, and reporting. This section introduces a conceptual framework and methodology grounded in case study analysis and process mapping and also highlights key implementation issues.

Data Collection

Data collection in budgeting typically involves the gathering of historical financial data, cost projections, and operational inputs from multiple departments. Traditionally, this task is manual, repetitive, and time-intensive, often involving email exchanges, spreadsheet updates, and disparate enterprise systems. RPA enhances this stage by deploying bots that automatically extract data from various sources, including enterprise resource planning (ERP) systems, customer relationship management (CRM) platforms, and cloud storage systems (Balamurugan et al., 2022).

For instance, RPA bots can be configured to log into systems such as Oracle or SAP, retrieve key budget data (e.g., sales forecasts, inventory costs), and export them into structured templates in Excel or Google Sheets. Additionally, bots using optical character recognition (OCR) can collect data from scanned documents or PDFs, significantly expanding the range of sources from which data can be drawn (Zhang et al., 2023).

In one multinational telecommunications company, RPA reduced the time taken for initial budget data collection by 70%, while simultaneously reducing human error in data entry (Perdana et al., 2022). This shift freed up finance professionals to focus more on value-added tasks, such as scenario planning and stakeholder consultations.

Data Aggregation and Validation

Following data collection, the next phase involves consolidating and validating inputs to ensure budget accuracy. RPA enables rapid and standardized data aggregation by integrating disparate data sources and applying logical rules. Bots can validate financial inputs based on historical trends, flag unusual variances, and prompt users for justification or clarification (Plattfaut & Borghoff, 2022).

For example, if a business unit projects a 50% increase in operational expenditure without corresponding revenue growth, an RPA bot can automatically detect this discrepancy and send an alert for review. This automated control mechanism enhances data integrity while reducing the reliance on manual reconciliations (Ayinla et al., 2024).

Moreover, bots are capable of standardizing formatting across data sets from different departments, ensuring seamless integration into centralized budget templates. Organizations leveraging this approach have reported up to 90% improvement in budget data consistency and timeliness (Jędrzejka, 2022).

Version Control and Budget Iteration

Budgeting is inherently iterative, involving multiple cycles of review, revision, and approval. Without proper tools, version control becomes a significant challenge, leading to duplication, data misalignment, and communication breakdowns. RPA addresses this issue through automated version tracking and document management.

Bots can archive each iteration, label them with timestamps, notify users of updates, and merge changes from various stakeholders into a single version of truth. Additionally, workflow orchestration tools integrated within modern RPA platforms can automatically route budget drafts to designated approvers based on organizational hierarchies (Balamurugan et al., 2022).

In one use case from a logistics company in Germany, RPA reduced the average budget approval cycle by 35% by eliminating redundant manual coordination and minimizing approval bottlenecks. All changes made by business units were instantly visible to finance controllers, allowing faster review and less back-and-forth communication (Zhang et al., 2023).

Reporting and Dissemination

The final stage of the budgeting lifecycle involves reporting and communicating the final budget to relevant stakeholders. This includes variance analyses, budget summaries, and strategic scenario presentations. Traditionally, finance teams generate these reports manually using Excel and PowerPoint, which is not only time-consuming but also error-prone.

With RPA, bots can automatically generate reports using real-time data, populate dashboards through integration with platforms like Power BI and Tableau, and disseminate reports via email or cloud platforms (Perdana et al., 2022). Furthermore, bots can be programmed to generate reports at scheduled intervals, such as daily, weekly, or monthly updates.

An Australian university deployed RPA bots to prepare faculty-level budget reports and saw a 50% increase in report delivery speed, along with enhanced formatting standardization (Ayinla et al., 2024). Bots also ensured that all submissions adhered to regulatory guidelines, reducing audit discrepancies and manual corrections.

Conceptual Framework for RPA Integration in Budgeting

The successful integration of Robotic Process Automation (RPA) into the budgeting process hinges on a structured and holistic approach that addresses both technical and organizational variables. The proposed conceptual framework consists of four interrelated dimensions: process suitability, technology infrastructure, organizational readiness, and governance and compliance.

Process suitability refers to the identification and prioritization of budgeting tasks that are most amenable to automation. These typically include repetitive, rule-based, and high-volume activities such as data extraction, consolidation, validation, and report generation. Tasks that require minimal human judgment but demand high accuracy and speed are ideal candidates for RPA. Conducting a detailed process audit helps determine automation potential and expected return on investment. For instance, reconciling inter-departmental budget entries or auto-filling budget templates from ERP data are highly suitable for RPA deployment (Axmann, et al., 2021).

Technology infrastructure focuses on evaluating whether the organization's current IT ecosystem can support RPA implementation. This includes assessing the compatibility of legacy systems, the presence of APIs for integration, cloud-readiness for scalability, and data security protocols to protect sensitive financial information. Organizations with fragmented or outdated systems may require middleware solutions or system upgrades to ensure seamless RPA functionality. Additionally, the scalability of the selected RPA software must be considered to accommodate future process automation across broader finance functions (Plattfaut & Borghoff, 2022).

Organizational readiness encompasses the cultural and capability dimensions of automation adoption. It involves gauging the digital literacy of finance and IT teams, identifying training needs, and fostering a ready-to-ready culture that embraces innovation rather than fears disruption. RPA often shifts roles from transactional to analytical, which necessitates upskilling, role redesign, and clear communication about the strategic intent of automation. Without adequate readiness, even technically sound RPA initiatives may face resistance or underutilization (Balamurugan et al., 2022).

Governance and compliance is the fourth and arguably most critical dimension, ensuring that RPA implementations operate within a framework of accountability, transparency, and regulatory conformity. Establishing formal governance protocols including bot lifecycle management, access control, error handling, and audit trail documentation is essential to maintain trust in automated outputs. This is particularly relevant in budgeting processes that influence financial disclosures, investment decisions, and regulatory filings. Proper governance also supports compliance with internal controls, such as SOX or ISO standards, reducing the risk of automation-related breaches or misstatements (Zhang et al., 2023).

Together, these four dimensions provide a comprehensive lens through which organizations can evaluate their readiness, design a phased implementation plan, and align RPA deployment with broader digital transformation objectives. This framework not only guides the technical rollout but also ensures strategic coherence and long-term sustainability of automation in the budgeting function.

Issues, Controversies, and Problems of RPA in Budgeting

Despite the benefits, the deployment of RPA in budgeting is not without its challenges. These include financial, technical, and ethical concerns.

High Implementation and Maintenance Costs

While RPA offers cost savings in the long run, the initial investment can be substantial ranging from USD 100,000 to USD 500,000 depending on the scale and complexity. This includes software licensing, bot development, staff training, and infrastructure upgrades (Zhang et al., 2023). Additionally, bots require ongoing maintenance to adapt to changes in budgeting logic, interfaces, or reporting formats, which can strain IT teams if not adequately resourced. For small and medium enterprises (SMEs), these costs can be a major barrier to entry, limiting access to advanced automation tools unless affordable cloud-based alternatives are considered (Ayinla et al., 2024).

Legacy System Compatibility

Many organizations still operate on outdated or fragmented IT systems. While RPA can mimic human interaction with these interfaces, incompatibilities can arise, especially with systems lacking modern APIs or having unstable UI elements (Jędrzejka, 2022). Bots may break if system upgrades are introduced without updating the underlying scripts. Organizations must evaluate system landscapes before deploying bots and consider hybrid integration approaches that combine RPA with middleware or data warehouses for better reliability (Plattfaut & Borghoff, 2022).

Organizational Resistance and Job Displacement Fears

One of the most cited barriers to RPA adoption is employee resistance, particularly among staff whose tasks are being automated. Studies indicate that 60% of finance professionals express concern over job displacement when automation tools are introduced (Ayinla et al., 2024). Without clear communication, change management, and retraining opportunities, morale may decline and sabotage the implementation effort. Forward-looking organizations address this by upskilling their teams. For example, a Japanese bank implemented a retraining program that converted clerical budgeting staff into data visualization analysts, resulting in higher job satisfaction and reduced turnover (Oyeniyi et al., 2024).

Inconsistent Governance and Compliance Structures

In the absence of standardized governance, RPA deployments can become fragmented and difficult to manage. Poor governance leads to bot duplication, unclear roles and responsibilities, and non-compliance with regulatory standards. Effective RPA governance includes establishing a centralized command center to oversee bot operations, audit trails, error resolution protocols, and compliance alignment. In highly regulated industries such as banking, failure to implement governance controls has led to audit failures and regulatory penalties (Balamurugan et al., 2022).

Risk of Over-Reliance on Automation

While RPA excels at rules-based processes, it lacks the contextual awareness, critical thinking, and ethical reasoning necessary for complex budgeting decisions. Over-relying on bots may lead to overlooked anomalies or poor strategic judgments. In one U.S. retail case, an RPA bot erroneously flagged marketing expenditures for cuts due to a misconfigured logic rule. Without human oversight, the bot would have triggered an inappropriate budget reallocation. Therefore, organizations must adopt a hybrid approach using RPA for operational efficiency and humans for decision-making (Plattfaut & Borghoff, 2022).

The integration of Robotic Process Automation into the budgeting lifecycle presents a significant opportunity for organizations to improve operational efficiency, accuracy, and strategic alignment. From data collection to report dissemination, RPA provides a consistent and scalable solution for managing complex financial planning activities. To fully realize the benefits of RPA, organizations must treat it as a strategic initiative, guided by thoughtful planning, human oversight, and continuous learning. When executed effectively, RPA has the potential to redefine the budgeting process by transforming it from a static, labor-intensive exercise into a dynamic, intelligence-driven function.

CASE STUDY: ROBOTIC PROCESS AUTOMATION IN BUDGETING AT RG GOLD LLP

This chapter uses a case study by Nurpeissova, Dyussembekova, and Issayeva (2024) on the design and implementation of an automated budgeting system at the mining enterprise RG Gold LLP.

Introduction

The Republic of Kazakhstan's mining sector remains a cornerstone of its economic foundation, with gold mining ranking 11th globally in production. In 2021 alone, mining contributed 17% to the nation's GDP, employing over 273,000 workers across 230 enterprises. Yet, this vital industry has faced mounting operational inefficiencies due to outdated technologies and manual management practices. Recognizing the strategic importance of digital transformation in budgeting, RG Gold LLP a leading gold mining company, embarked on a pioneering project to automate its budgeting system with the goal of enhancing financial manageability, forecasting accuracy, and strategic agility.

While earlier automation efforts relied heavily on traditional enterprise resource planning (ERP) systems, RG Gold LLP transitioned toward more intelligent, adaptive solutions. The implementation of Anaplan, a cloud-based financial planning platform, marked the company's move beyond simple RPA (Robotic Process Automation) toward cognitive automation leveraging machine learning, real-time analytics, and AI-assisted decision-making.

Diagnosing the Traditional Budgeting System

Prior to transformation, RG Gold LLP's budgeting process suffered from extensive manual data handling, overreliance on MS Excel, and fragmented workflows. Although the enterprise had a structured and phased budgeting procedure with clear timelines, the lack of digital integration resulted in human errors, data distortions, and limited real-time insight. Furthermore, the absence of advanced business intelligence tools hindered analytical capabilities and delayed strategic responses.

The company undertook a detailed diagnostic using SWOT analysis and in-depth stakeholder interviews to evaluate the current system. Core weaknesses included prevalence of manual labor in data entry and form consolidation, lack of integration between departments, absence of dynamic forecasting and analytics and inability to support rolling or scenario-based budgets. These limitations highlighted the urgent need for a cognitive automation solution capable of integrating real-time data, adapting to complex workflows, and enabling predictive financial modeling.

From Automation to Intelligence: The Anaplan Rollout

RG Gold LLP's transition toward automation began with the evaluation of digital solutions across global providers such as SAP, Oracle, and Microsoft. Ultimately, Anaplan was selected for its modular design, high adaptability, and capacity to combine data, people, and plans into a unified interface in real time.

Unlike traditional budgeting modules embedded within ERP systems, Anaplan supports AI-driven analytics, rolling forecasts, and cognitive modeling capabilities increasingly demanded in volatile, resource-driven sectors like mining. The implementation unfolded in multiple phases as shown in Table 1 below.

Table 1. Stages of design and development (Source: Nurpeissova et al., 2024)

Design phases	Types of work
Phase I – Budgeting	Integration of the 1C accounting system with Anaplan enabled automatic collection of budgeting requests, drastically reducing manual input.
Phase II – Financial statements	works related to revenue modeling (production plan, volume of finished products and revenue; GOGS, taking into account the production and technological specifics of mining enterprises, taxes, CAPEX, etc.)
Phase III: Consolidation and Reporting	Design of a single aggregating form for the consolidation of all budget applications for the analysis and formation of the enterprise budge

Cognitive and AI Features Embedded

RG Gold LLP's budgeting transformation extended beyond basic automation into the realm of cognitive automation and artificial intelligence by embedding several advanced features. One of the key innovations was anomaly detection, where the AI engine automatically flags inconsistencies or outliers in budget submissions, alerting users to potential errors or unusual spending patterns that may require further investigation. The system also introduced rolling forecasts, allowing the company to update its budgets on a monthly or quarterly basis with minimal manual effort. These forecasts adapt dynamically to external factors such as fluctuations in commodity prices, changes in regulatory frameworks, and shifts in geopolitical conditions. Another notable feature was scenario analysis, which enables users to model and compare various financial outcomes such as best-case, worst-case, and most-likely scenarios using historical data and predictive algorithms. Looking ahead, RG Gold LLP plans to incorporate natural language processing (NLP) capabilities into the system to automatically generate budget narratives and commentary. This future improvement is expected to support clearer, AI-assisted communication at the executive and board levels.

Impact and Early Results

The early impact of RG Gold LLP's cognitive automation initiative, while still unfolding across successive budget cycles, has already demonstrated significant qualitative benefits. One of the most notable outcomes was a reduction in manual processing, with form filling and data consolidation efforts cut by over 50%. This efficiency gain allowed finance teams to redirect their focus toward higher-order analytical tasks. Additionally, data accuracy and validation improved markedly, as Anaplan's built-in validation mechanisms minimized both technical and logical errors during data entry, thereby reducing human-induced discrepancies. The system also facilitated real-time collaboration, enabling multiple users to work concurrently on budget inputs across departments. This capability not only ensured consistency but also significantly reduced the turnaround time for budget preparation. Furthermore, advanced analytics dashboards provided senior management with timely access to key performance indicators and visual insights, enhancing the speed and quality of strategic decision-making. Lastly, the adoption of rolling budgets and predictive analytics contributed to more accurate forecasting and agile financial planning, allowing the company to better align resources with dynamic operational conditions.

Organizational and Technological Considerations

The successful implementation of RG Gold LLP's automated budgeting system was underpinned by a comprehensive change management strategy. A key component of this approach was stakeholder engagement, where in-depth interviews with top management and finance teams ensured the system was aligned with the company's practical budgeting needs. This was complemented by phased training programs that empowered users to effectively utilize the platform's advanced capabilities, including analytics dashboards and simulation tools. To ensure a smooth rollout, User Acceptance Testing (UAT) was conducted iteratively, focusing on interface usability, system performance, and adherence to data security protocols. From a technological perspective, the integration of Anaplan with existing platforms such as Oracle, 1C, and various business intelligence (BI) tools was seamless. The implemented solution complies with international standards for data security, user control, and regulatory compliance, all of which are critical for mining enterprises managing sensitive financial and environmental information.

Future Outlook

As RG Gold LLP advances further along its digital transformation journey, the company envisions deeper integration of artificial intelligence (AI), machine learning, and natural language processing (NLP) to further enhance its budgeting system. Planned enhancements include voice-assisted budget inquiries, allowing executives to interact with the system through natural language commands, and automated variance commentary, where AI-generated narratives provide real-time explanations for budget deviations. The company also aims to explore blockchain integration to improve transparency and traceability in capital allocations, as well as digital twin models to simulate the financial impact of operational changes in real-time. To support these initiatives, RG Gold LLP intends to establish an internal Center of Excellence (CoE) focused on budgeting and analytics. This CoE will play a critical role in continuously optimizing system performance, fostering innovation through experimentation, and mentoring cross-functional teams in the effective use of digital financial tools.

The case of RG Gold LLP demonstrates how cognitive automation and AI can revolutionize budgeting practices in capital-intensive industries. The shift from static, manual systems to dynamic, intelligent platforms like Anaplan signifies more than just technological adoption it represents a cultural transformation toward agility, foresight, and strategic resilience. In an era where volatility is the norm, mining enterprises that integrate cognitive tools into financial planning will be better positioned to adapt, compete, and thrive.

SOLUTIONS AND RECOMMENDATIONS

The successful implementation of Robotic Process Automation (RPA) in budgeting and management accounting processes requires more than just technology deployment it demands a holistic, well-coordinated strategy. As organizations navigate the complexities of digital transformation, several key recommendations can guide a smoother and more impactful adoption of RPA. These include strategic planning for phased RPA adoption, integration with change management and employee retraining, leveraging cloud-based solutions for scalability and cost efficiency, establishing strong data governance and audit frameworks, and fostering a hybrid model that balances automation with human expertise.

Strategic Planning for Phased RPA Adoption

Implementing RPA should begin with a clear, phased strategic plan that aligns with the organization's broader goals and operational priorities. Rather than attempting to automate all budgeting or financial reporting processes simultaneously, organizations are encouraged to start with a pilot phase targeting well-defined, repetitive, and rule-based tasks. These might include the automation of data extraction from spreadsheets, generating recurring budget reports, or reconciling monthly financials. A phased approach enables organizations to identify quick wins, gather valuable insights from early implementations, and refine the automation strategy based on actual performance and feedback.

A structured rollout plan also allows for risk mitigation. By focusing first on low-complexity, high-volume processes, organizations can measure the return on investment (ROI) of RPA more effectively and build stakeholder confidence. As internal capabilities grow, the scope of automation can gradually expand to include more complex workflows, such as budget variance analysis or multi-departmental data consolidation. This staged progression reduces disruption, promotes organizational learning, and ensures sustainable transformation over time.

Combining RPA with Change Management and Employee Retraining

One of the most overlooked aspects of RPA deployment is the human factor. Automation inevitably impacts roles and responsibilities, particularly in accounting and finance functions where traditional processes are deeply embedded. Therefore, any RPA strategy must be accompanied by a robust change management plan that addresses employee concerns, encourages adoption, and prepares staff for new roles.

Communication is essential in managing expectations and reducing resistance. Employees must understand that RPA is not about replacing jobs but about augmenting human capabilities and allowing professionals to focus on higher-value tasks. For example, with routine data entry or report generation automated, finance professionals can devote more time to strategic analysis, risk assessment, or scenario planning.

Alongside communication, employee retraining is critical. Organizations should invest in upskilling programs that equip staff with the knowledge and competencies needed to work alongside automation tools. This may include training on RPA platforms, data analytics, process redesign, and digital literacy. By empowering employees with new skills, organizations can create a more agile workforce that embraces innovation and contributes meaningfully to the success of RPA initiatives.

Leveraging Cloud-Based RPA for Scalability and Cost Efficiency

Cloud-based RPA solutions offer significant advantages in terms of scalability, flexibility, and cost management. Unlike on-premises automation tools that often require substantial upfront investments in infrastructure, cloud-based RPA is typically delivered through a subscription model. This reduces the barrier to entry, particularly for small and medium-sized enterprises (SMEs), and allows for rapid deployment without the need for extensive IT support.

Moreover, cloud-based platforms offer scalability on demand. As the organization's needs grow or fluctuate for instance, during peak budgeting seasons or year-end reporting periods—cloud-based RPA can easily accommodate additional workloads without compromising performance. This elasticity supports agile financial operations and ensures that automation efforts remain responsive to changing business conditions.

Security and compliance are also key considerations. Reputable cloud RPA providers offer robust security measures, including data encryption, access controls, and compliance with standards such as ISO 27001 or GDPR. These features help protect sensitive financial data while supporting governance and risk management objectives.

Establishing Strong Data Governance and Audit Trails

Data is the backbone of any budgeting or financial decision-making process. Therefore, the integration of RPA must be underpinned by strong data governance practices. This involves defining clear policies and procedures for data ownership, accuracy, access, and lifecycle management. As RPA bots operate on structured data and perform tasks without human intervention, ensuring data quality and consistency is paramount to avoid automation errors and maintain trust in financial outputs.

In addition, auditability is a critical requirement, particularly in regulated environments. RPA tools must be designed to produce detailed logs and audit trails that document every action performed by the bots. These records enable traceability, facilitate compliance with internal controls, and support external audits. For instance, in the budgeting process, maintaining a complete record of automated data entries, approval workflows, and report generations helps ensure transparency and accountability.

Organizations should also consider integrating RPA tools with existing enterprise resource planning (ERP) systems and accounting software to streamline data flows and reduce silos. Seamless integration enhances the integrity and reliability of financial data across the organization.

Promoting a Hybrid Model of Automation and Human Expertise

While RPA excels at executing rule-based tasks quickly and accurately, it cannot replace the nuanced judgment, contextual understanding, and strategic thinking that human professionals bring to the budgeting process. Therefore, the ideal model is not one of full automation but a hybrid approach that combines the strengths of both machines and people.

In this model, RPA handles repetitive, time-consuming tasks such as data aggregation, formatting, and report generation, freeing up human resources to focus on interpretation, analysis, and strategic planning. For example, once RPA compiles the data for a departmental budget report, a finance manager can analyze trends, identify anomalies, and recommend adjustments based on business objectives or external market conditions.

Furthermore, human oversight remains essential in exception handling and process optimization. When automation encounters errors, ambiguous data, or unstructured inputs, human intervention is needed to resolve issues and update process logic. Continuous collaboration between RPA developers, finance professionals, and IT teams ensures that automation remains aligned with evolving business needs and continues to deliver value over time.

Thus, integrating RPA into budgeting and financial management is a transformative endeavor that requires thoughtful planning and execution. A phased implementation strategy allows for gradual adoption and minimizes disruption, while strong change management and retraining initiatives ensure workforce alignment. Cloud-based RPA platforms enhance scalability and accessibility, and sound data governance practices maintain integrity and transparency. Ultimately, promoting a balanced model of automation and human expertise enables organizations to optimize efficiency without sacrificing strategic insight. By following these recommendations, organizations can unlock the full potential of RPA and position themselves for sustained success in an increasingly digital financial landscape.

Figure 1. Solutions and Recommendations

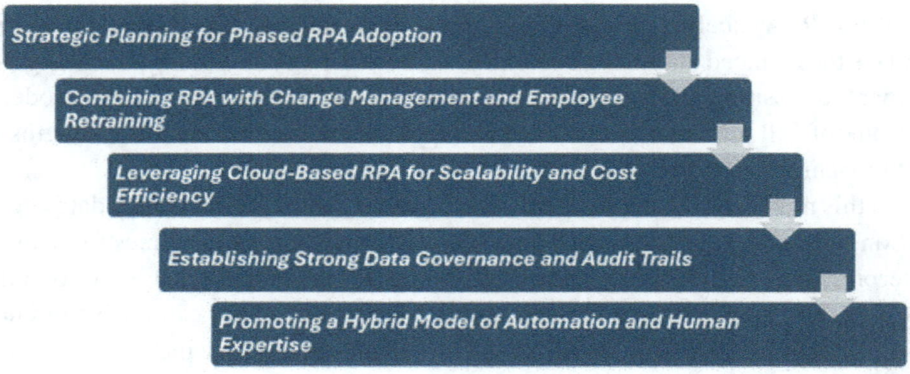

FUTURE RESEARCH DIRECTIONS AND IMPLICATIONS

As Robotic Process Automation (RPA) continues to reshape budgeting and financial operations, future research must extend beyond current capabilities and explore emerging technologies and their convergence with automation. Several promising avenues warrant deeper academic and practical investigation. These include the integration of cognitive RPA and artificial intelligence (AI), the application of natural language processing (NLP) for automating budget narratives, the fusion of blockchain technology for enhanced transparency, the transformation of accounting education to reflect digital competencies, and the development of ethical frameworks to govern responsible automation.

One significant area of future research is the exploration of cognitive RPA and AI-enhanced budgeting processes. While current RPA solutions excel at rule-based, structured tasks, the next frontier involves integrating cognitive capabilities such as machine learning, predictive analytics, and pattern recognition. These technologies enable bots not only to execute predefined tasks but also to learn from data, identify trends, and make recommendations. For example, AI-powered RPA could forecast future budget needs based on historical spending patterns, macroeconomic indicators, and business cycles. Such intelligent automation could transform traditional budgeting from a static, backward-looking exercise into a dynamic, data-driven process that supports strategic agility. Research is needed to explore the feasibility, accuracy, and implementation challenges of cognitive automation in real-world budgeting environments, particularly in complex organizations with decentralized financial operations.

Another area with considerable potential is the use of natural language processing (NLP) to automate budget narratives and financial commentary. Budgeting is not only about numbers; it also involves interpreting variances, explaining projections, and communicating justifications to stakeholders. Currently, much of this narrative generation is manual and time-consuming. With advances in NLP and generative AI, future systems could automatically draft budget reports, executive summaries, or variance analyses by interpreting numerical data and contextual information. For instance, an RPA-enabled system could generate a monthly commentary stating, "Marketing expenses exceeded the budget by 12% due to a mid-quarter campaign launch, offset by savings in administrative costs." This capability could significantly reduce the workload of finance teams while enhancing the clarity and consistency of budget communications. Research can examine the accuracy, linguistic coherence, and organizational acceptance of automated narratives, as well as how NLP tools can be tailored to specific sectors and audiences.

Equally promising is the integration of blockchain technology with budgeting and financial reporting. Blockchain, with its decentralized and immutable ledger, offers transparency, traceability, and security attributes that align well with public budgeting and inter-organizational financial management. A blockchain-based budgeting system could record every transaction or budget amendment in real time, creating an auditable trail accessible to all authorized stakeholders. This could be especially beneficial in public sector budgeting, where accountability and transparency are critical. For example, development funds allocated by a government agency could be tracked through every disbursement stage using smart contracts, ensuring funds are used as intended. Future research should examine the practicalities of integrating RPA with blockchain for secure automation of budget approvals, fund disbursements, and audit documentation. Questions regarding interoperability, data privacy, and regulatory compliance in blockchain-enabled budgeting systems also require scholarly attention.

In parallel with technological advancement, there is a growing need for curriculum transformation in accounting and finance education. As automation becomes integral to financial operations, future accountants must possess a hybrid skillset that blends technical proficiency with strategic thinking. Traditional accounting curricula, which emphasize manual bookkeeping and standard budgeting techniques, must evolve to include topics such as RPA tools, data analytics, process automation, and digital ethics. Research should explore how higher education institutions can integrate RPA into coursework, case studies, and simulations. What pedagogical methods are most effective for teaching automation tools to accounting students? How can academic programs balance technical training with the development of critical thinking and ethical reasoning? Addressing these questions is essential to prepare future finance professionals who can thrive in a digitally transformed workplace.

Lastly, the rapid adoption of automation raises important questions about ethics and responsible use of technology in budgeting and financial management. As bots begin to handle tasks that involve sensitive financial data, approvals, and even compliance checks, the risk of unintended consequences such as bias in data interpretation or unauthorized access—must be addressed. Furthermore, the delegation of decision-making processes to machines raises issues of accountability who is responsible when an automated system makes an error that leads to budget misallocation or financial loss. Developing ethical frameworks and governance structures to guide the deployment of RPA and AI is therefore crucial. Future research should focus on defining principles of transparency, fairness, explainability, and human oversight in automated budgeting environments. Additionally, sector-specific guidelines may be needed to account for varying levels of risk and complexity in government, corporate, and nonprofit financial management.

The implications of these future research directions are profound. By expanding the capabilities of RPA through AI and NLP, budgeting processes can become more intelligent, responsive, and aligned with strategic goals. Integrating blockchain offers new models of trust and accountability in financial governance. Educational reform ensures that the workforce is equipped to design, manage, and ethically evaluate automated systems. And ethical frameworks safeguard against misuse, ensuring that automation enhances rather than undermines public trust and professional standards.

In conclusion, the trajectory of RPA in budgeting is far from static. The next phase of development lies in intelligent, explainable, and integrated systems that can think, communicate, and adapt. Researchers, educators, policymakers, and practitioners must work collaboratively to explore these emerging areas, address the associated challenges, and harness the full potential of automation in financial decision-making. As organizations seek to future-proof their budgeting functions, forward-looking research will be instrumental in shaping the technologies, competencies, and values that define the next generation of financial management.

Implications for Educators, Practitioners, and Policymakers

The rise of Robotic Process Automation (RPA) and other intelligent automation tools is reshaping the accounting profession at an unprecedented pace. As RPA transitions from a back-office enhancement tool to a core component of financial operations, educators, industry practitioners, and policymakers must adapt accordingly. The implications span across curriculum transformation, professional development, ethical considerations, workforce planning, and the establishment of forward-looking policy frameworks. A proactive response to these changes is crucial to ensure that accounting professionals remain relevant, ethical, and capable in a rapidly evolving digital economy.

Curriculum Redesign for Future Accountants

Educational institutions are at the forefront of shaping the next generation of accountants. Traditional accounting curricula, which have long emphasized manual bookkeeping, compliance, and financial reporting, must be reimagined to reflect the realities of a technology-driven profession. The integration of RPA into the accounting syllabus is no longer optional; it is essential.

Firstly, accounting students must be introduced to the principles and mechanics of RPA early in their academic journey. This includes understanding how bots are developed, deployed, and managed across different financial functions such as accounts payable, receivable, general ledger reconciliation, payroll processing, and tax compliance. The curriculum should also encompass key concepts from related domains including business process management, enterprise resource planning (ERP) systems, and data analytics.

Secondly, curriculum redesign must incorporate hands-on experience. Partnerships with RPA software providers such as UiPath, Blue Prism, or Automation Anywhere can enable students to gain practical exposure to designing and testing bots. Project-based learning, capstone projects, and industry collaborations can further bridge the gap between theoretical knowledge and workplace expectations. Case studies involving RPA implementation challenges and ROI analysis will develop critical thinking and decision-making skills.

Moreover, there is a growing need for interdisciplinary education. Future accountants will be required to work in multidisciplinary teams involving IT, cybersecurity, and data governance experts. Hence, accounting programs should integrate relevant modules on programming logic, information systems, and digital transformation strategies. Soft skills such as communication, collaboration, and digital adaptability should also be embedded within technical coursework.

Ultimately, accounting educators must shift from producing technical accountants to developing digitally fluent business advisors. This shift ensures that graduates are not only employable but are also positioned to lead technological innovation in their organizations.

Upskilling and Continuous Professional Development

For current practitioners, RPA presents both a challenge and an opportunity. As automation displaces manual, repetitive tasks, accountants must reposition themselves to focus on strategic, interpretive, and advisory roles. Continuous professional development (CPD) is therefore critical to maintaining relevance in a transforming profession.

Professional bodies such as ACCA, CIMA, CPA Australia, and IFAC must play an active role in facilitating this transition. CPD programs need to offer training on emerging technologies, including RPA, AI, blockchain, and cloud-based accounting solutions. Courses should be modular, flexible, and accessible through digital platforms to accommodate working professionals.

In addition to technical upskilling, accountants must develop higher-order cognitive skills such as analytical thinking, business acumen, ethical reasoning, and change management. These competencies are essential as accountants take on more forward-looking roles such as business analysts, digital transformation consultants, and risk managers.

Workplaces too must create a culture of learning. Employers can support upskilling by offering in-house training, job rotations, mentorship programs, and collaborative projects with IT departments. Investing in the continuous growth of accounting teams will not only improve operational efficiency but also boost employee morale and retention.

Ethical Considerations and Workforce Planning

The integration of RPA into accounting processes introduces complex ethical and workforce-related implications. One of the primary ethical concerns is accountability. When bots perform tasks such as invoice approvals, data entry, or tax computations, who is ultimately responsible for errors or fraudulent outcomes? Clear governance structures must be in place to ensure human oversight and ethical accountability.

Moreover, RPA systems are only as good as the processes and data they are built upon. Biases, data inconsistencies, or flawed logic in bot programming can lead to unintended consequences. Hence, accountants must be trained to critically assess RPA implementation, perform regular audits of bot performance, and escalate anomalies.

Data privacy and cybersecurity are additional areas of concern. Automated processes often involve access to sensitive financial and personal data. As custodians of this information, accountants must ensure that automation tools comply with data protection regulations such as GDPR or Malaysia's PDPA. They must also collaborate with cybersecurity experts to secure automated workflows from breaches and unauthorized access.

From a workforce planning perspective, the fear of job displacement is often cited as a barrier to RPA adoption. While RPA does automate certain roles, it also creates opportunities for reskilling and role redesign. Organizations must adopt a responsible automation approach, one that includes workforce transition strategies, transparent communication, and inclusive planning.

For instance, rather than laying off employees, companies can redeploy them to higher-value roles in analytics, internal controls, and strategic planning. Human capital planning should be informed by a comprehensive understanding of how automation will impact job roles over time, enabling smoother transitions and sustained organizational resilience.

Policy Frameworks for Responsible Automation

Policymakers have a pivotal role to play in setting the regulatory tone for the ethical and effective adoption of RPA. With automation increasingly influencing labor markets, business operations, and compliance frameworks, governments must formulate policies that promote innovation while safeguarding workers' rights and public interest.

Firstly, national and regional policy frameworks should incentivize businesses to invest in upskilling and retraining programs. This can be done through tax credits, training subsidies, or public-private partnerships. Such policies can help mitigate the risk of technological unemployment and create a more inclusive digital economy.

Secondly, regulatory bodies should establish clear guidelines for the governance of RPA systems. This includes standards for transparency, auditability, cybersecurity, and risk management. Accounting standards boards, for example, may need to consider how automation affects internal control systems under frameworks like COSO or COBIT.

Thirdly, education ministries and accreditation councils must collaborate with academic institutions to update learning standards for accounting and business programs. Funding can also be directed toward research into the impacts of automation, particularly in developing economies, small businesses, and public sector accounting.

Finally, policymakers must engage in ongoing dialogue with industry and academia to ensure regulatory relevance. As automation technologies evolve, static regulations will quickly become outdated. A flexible, principles-based regulatory approach combined with stakeholder consultation can enable timely responses to emerging challenges.

CONCLUSION

Robotic Process Automation (RPA) is transforming the budgeting landscape by streamlining repetitive tasks, enhancing data accuracy, and accelerating cycle times. Its integration into budgeting processes marks a paradigm shift from labor-intensive practices to data-driven, agile planning systems. However, the successful adoption of RPA is contingent upon several critical factors, including robust process mapping, stakeholder engagement, cross-functional collaboration, and strong change management strategies. Organizations must ensure that RPA initiatives are not merely seen as IT projects, but as strategic tools for improving financial stewardship and decision-making.

While the benefits of automation are significant, they must be balanced with the irreplaceable value of human judgment. Budgeting involves not just numerical precision but also strategic foresight, ethical considerations, and contextual understanding areas where human insight remains essential. Accountants and finance professionals must therefore reposition themselves as interpreters and advisors, using automation as a tool rather than a replacement.

Furthermore, the rise of RPA underscores the need for a culture of continuous learning. Upskilling, ethical awareness, and adaptive thinking are now fundamental competencies. Educational institutions, professional bodies, and policymakers must collaborate to foster these capabilities and ensure responsible deployment of automation tools. In essence, RPA is not the end of human involvement in budgeting it is the beginning of a new partnership. One is where automation enhances efficiency, and humans provide the wisdom to guide it.

REFERENCES

Axmann, B., Harmoko, H., Herm, L. V., & Janiesch, C. (2021). A framework of cost drivers for robotic process automation projects. In *International Conference on Business Process Management* (pp. 7-22). Springer, Cham.

Ayinla, A., Fadairo, M., & Idris, B. (2024). Workforce reskilling for intelligent automation. *Journal of Emerging Technologies in Accounting*, *21*(1), 45–62.

Balamurugan, A., Krishna, M. V., Bhattacharya, R., Mohammed, S., Haralayya, B., & Kaushik, P. (2022). Robotic process automation (RPA) in accounting and auditing of business and financial information. *British Journal of Administrative Management*, *58*(157), 127–142.

Bellinga, J., Bosman, T., Höcük, S., Janssen, W. H., & Khzam, A. (2022). Robotic process automation for the extraction of audit information: A use case. *Current Issues in Auditing*, *16*(1), A1–A8.

Doğuç, M. (2021). Employee attitudes toward automation and job security. *Journal of Work Psychology*, *12*(2), 33–48.

Hsiung, H. H., & Wang, J. L. (2022). Research on the introduction of a robotic process automation (RPA) system in small accounting firms in Taiwan. *Economies*, *10*(8), 200.

Huang, W. (2019). Developing a better planning, budgeting, and accounting system. In *Built on Value: The Huawei Philosophy of Finance Management* (pp. 303–328). Springer Singapore.

Jędrzejka, D. (2022). Robotic process automation in the finance department: Benefits and implementation challenges. *Journal of Business Transformation*, *8*(1), 59–71.

Kokina, J., & Blanchette, S. (2020). Early evidence on the use of emerging technologies in accounting education. *Journal of Emerging Technologies in Accounting*, *17*(1), 23–42.

Lindgren, I., Johansson, B., Söderström, F., & Toll, D. (2022, August). Why is it difficult to implement robotic process automation? Empirical cases from Swedish municipalities. In *International Conference on Electronic Government* (pp. 353-368). Cham: Springer International Publishing.

Luo, H., & Luo, H. (2021). RPA and artificial intelligence in budget management based on multiperspective recognition based on network communication integration. *Wireless Communications and Mobile Computing*, *2021*(1), 9723379.

Nurpeissova, A. A., Dyussembekova, Zh. M., & Issayeva, A. D. (2024). The design and implementation of an automated budgeting system at the mining enterprise: The case of RG Gold LLP. State and Business: Theory and Practice of Management, *Central. The Asian Economic Review*, *5*(152), 6–21.

Oyeniyi, O., Raman, K., & Malik, A. (2024). Strategic value creation through robotic process automation in finance. *Asian Journal of Accounting Research*, *9*(1), 88–104.

Perdana, A., Yusuf, M., & Lestari, D. (2022). Cognitive automation in financial planning and analysis. *Asian Journal of Accounting Research*, *7*(1), 1–17.

Plattfaut, R., & Borghoff, P. (2022). Intelligent automation in management accounting: Trends and implications. *Journal of Management Control*, *33*(3), 289–307.

Ray, J. K., Sultana, R., Bera, R., Sil, S., & Alfred, Q. M. (2023). A Comprehensive review on artificial intelligence (AI) and robotic process automation (RPA) for the development of smart cities. *Confluence of Artificial Intelligence and Robotic Process Automation*, 289-311.

Tiron-Tudor, A., Lacurezeanu, R., Bresfelean, V. P., & Dontu, A. N. (2024). Perspectives on how robotic process automation is transforming accounting and auditing services. *Accounting Perspectives*, *23*(1), 7–38.

Zhang, Y., Lee, K., & Ng, C. (2023). RPA in budgeting and forecasting: A case-based study. *Global Finance Journal*, *54*(2), 101–123.

Zhou, R., Li, X., Gao, J., & Li, H. (2024, July). Research on Budget Management Digitization of Expressway Company Based on Robotic Process Automation. *In Proceedings of the 2nd International Conference on Educational Knowledge and Informatization* (pp. 427-433).

KEY TERMS AND DEFINITIONS

Artificial Intelligence (AI): The simulation of human intelligence in machines, enabling systems to analyze data, recognize patterns, and support advanced decision-making such as in predictive budgeting, variance analysis, and performance forecasting.

Automation: The application of technology to perform repetitive or structured tasks with minimal human intervention, improving efficiency, accuracy, and consistency in budgeting and accounting operations.

Budgeting: A structured process of planning and controlling financial resources by forecasting future revenues, expenses, and performance targets, supporting effective decision-making and organizational strategy execution.

Cloud-Based Accounting: An accounting system hosted on cloud infrastructure that allows real-time access to financial data, collaboration across departments, and integration with automation tools such as RPA and AI. In budgeting, it facilitates centralized data management, scalability, and remote accessibility.

Digital Transformation: The holistic integration of digital technologies including RPA, AI, cloud computing, and analytics into business operations to fundamentally improve budgeting, reporting, and performance management practices. It promotes agility, real-time insights, and data-driven culture.

Management Accounting: A discipline that provides relevant financial and non-financial information to internal stakeholders to aid in planning, performance evaluation, and decision-making. Budgeting represents a key tool within management accounting for resource allocation and control.

Robotic Process Automation (RPA): A technology that employs software "bots" to replicate human interactions with digital systems, enabling the automation of rule-based processes such as data extraction, consolidation, reconciliation, and report generation in budgeting and financial management.

Chapter 8

Revolutionizing Performance Measurement in Islamic Social Finance Using RPA and AI: A Balanced Scorecard Approach

Azliza Azrah Mohd Zakaria
https://orcid.org/0009-0002-3626-2187
INTI International University, Malaysia

Hamsatulazura Hamzah
Multimedia University, Malaysia

ABSTRACT

This paper explores how Robotic Process Automation (RPA) and Artificial Intelligence (AI) can revolutionize performance measurement in Islamic Social Finance Institutions (ISFIs) through a Balanced Scorecard (BSC) approach. Integrating RPA and AI enables ISFIs to enhance transparency, efficiency, and impact assessment while upholding core values rooted in maqasid al-shariah, such as justice, trust, and public good. The study highlights the need for ISF-specific performance frameworks that combine ethical imperatives with digital innovation, offering practical insights for aligning Islamic social finance with global sustainable development goals in the digital era.

DOI: 10.4018/979-8-3373-3571-1.ch008

INTRODUCTION

Islamic social finance (ISF) is a form of corporate social responsibility that offers non-profit products and services targeting to have social, environmental and economic impact. It does not only benefit individuals but also the society at large. (Rosman et al., 2022). Islamic Social Finance Institutions (ISFIs), encompassing entities such as zakat boards, waqf institutions, and Islamic NGOs, operate with the dual objective of financial accountability and spiritual responsibility. Their governance structures are influenced by maqasid al-shariah, aiming not just at redistribution of wealth but at uplifting communities holistically (Laldin & Furqani, 2013). Laldin and Furqani, in their article titled, The Foundations of Islamic Finance and The Maqasid al-Shariah Requirements, explicitly argued that Shariah is essentially a wider concept than that of Islamic law and regulations. Shariah is a system of life that encompasses all aspects of the creed and belief system (*aqidah*), ethics and moral (*akhlaq*), economic, political, cultural and civilizational matters that concern not only the Muslim community but the entire humanity.

The modern world's emphasis on Sustainable Development Goals (SDGs) aligns with maqasid al-Shariah concept of protecting fundamental human interests such as religion, life, intellect, lineage and wealth. Among the SDG goals are to eradicate poverty, reduce inequality and protect the environment. Awang, Mat Nong, and Wan Chik (2025) presented a conceptual framework that bridges ISF with SDGs offering a value-based model for sustainable development. Hence, ISF becomes an accelerator towards SDG achievement. The introduction of Value-Based Intermediation (VBI) by Bank Negara Malaysia (BNM), a strategic framework that encourages Islamic financial institutions to operate beyond profit maximization will become a push factor for islamic finance (IF) and islamic social finance institutions (ISFIs) to rethink their role—not just as intermediaries of capital but as agents of positive change and social benefit.

Over the past decade, the role of performance management in ISFIs has become increasingly significant as public scrutiny and donor expectations rise. Performance management involves the process of planning, monitoring, evaluating; and rectifying errors and making improvements. However, organisations need to be clear of the objectives to be measured (Hanif and Farooqi, 2023). The purpose of the establishment of profit and non-profit organisations differ, hence require different sets of performance criteria. It is even more vital for ISFIs which considers the concept of maqasid al-Shariah which require broader perspectives beyond financial soundness. The concept of performance management in non-profit and public institutions has evolved from simple financial audits to strategic frameworks that evaluate both efficiency and impact.

In the past, performance measurement in both profit and non-profit institutions primarily focused on financial indicators. While these measures offer valuable insights into institutional performance, they also have notable limitations. Financial performance often represents only the tip of the iceberg—visible and measurable, yet insufficient in explaining the underlying factors that drive those results. Kaplan and Norton (1996) pioneered the Balanced Scorecard (BSC), a tool that has since been adapted by both corporate and non-profit sectors. BSC has been the tool that recognizes non-financial factors that contribute to the success and failures of an institution, However, if BSC is incorporated into the ISFIs performance measurement model without any modifications, the elements upholding the maqasid al-shariah concept may be ignored. Therefore, it is essential to develop a set of performance measurement tools that integrates values like amanah (trust), ihsan (benevolence), 'adl (justice), and maslahah (public good)— into modern performance metrics.

Chatterjee et al. (2020) highlighted that the assimilation of AI-enabled services in government agencies enhances operational efficiency and citizen satisfaction by leveraging tools such as predictive analytics and natural language processing (NLP) to analyze large datasets and support decision-making. While explicit mention of Robotic Process Automation (RPA) is absent, the automation of routine, data-driven tasks described in their research aligns with RPA's role in streamlining public service delivery. ISIFs, like zakat and waqf agencies, can leverage these AI capabilities (predictive analytics, NLP) and automation (RPA) to enhance fund management, beneficiary targeting, and compliance monitoring. The authors also caution about privacy risks that may affect AI adoption, underscoring the importance of ethical considerations in deploying such technologies. The emphasis on privacy and ethical risks in AI adoption is particularly relevant for Islamic finance, which must align with Shariah principles emphasizing justice and protection of individuals' rights.

AI applications in non-profit and philanthropic sectors include donor profiling, fraud detection, and real-time beneficiary needs analysis (Wirtz et al., 2019). Organizations such as UNICEF and Islamic Relief Worldwide have implemented AI-driven impact tracking and digital beneficiary feedback systems (Islamic Relief Worldwide, 2023). Nevertheless, academic research on AI's role in Islamic philanthropy remains limited, despite operational experimentation in countries like Malaysia, Indonesia, and Turkey (Yildiz & Yasar, 2022; Setiawan & Wibowo, 2021).

LITERATURE REVIEW

Islamic Social Finance

Islamic Social Finance (ISF) represents a cornerstone in the Islamic economic system, dedicated to achieving social justice, equitable wealth distribution, and inclusive development. Rooted in Islamic jurisprudential principles and inspired by the maqasid al-Shariah (objectives of Islamic law), ISF encompasses a range of philanthropic tools such as zakat, waqf, sadaqah, microfinance. Qard al-hasan and sukuk (Mustafa et al., 2022). Innovative investments in humanitarian initiatives via ISF contribute to stronger recovery and help at-risk communities adjust and cope with challenges (Mustafa et al., 2022).

Unlike conventional finance, ISF aims to uplift vulnerable communities through mechanisms that reinforce both ethical accountability and spiritual consciousness. Economic exclusion—manifested through poverty, limited financial access, and systemic discrimination—undermines inclusive recovery by marginalizing vulnerable groups. ISF, grounded in Shariah principles, upholds ethical accountability by prohibiting any form of discrimination in the distribution of zakat, infaq, sadaqah, and waqf. This ethical stance reflects a deep spiritual consciousness, emphasizing justice, compassion, and social equity. ISF prioritizes the needs of disadvantaged groups such as women, ethnic minorities, and rural communities. The Proceedings of the 8th International Conference of Zakat (ICONZ) 2024 by Darihan Mubarak and Sri Wahyuni discuss the role of Islamic Social Finance (ISF) in promoting inclusive recovery amid crises.

"Tools like zakat and qard al-hasan are mobilized to support those most in need, reinforcing ISF's role in promoting fairness, dignity, and inclusive development in line with maqasid al-shariah" (Mubarak and Wahyuni, 2024).

Institutionalized Islamic Social Finance

Purpose, practices, and contemporary application of ISF has transitioned from a community-based charitable model to a more institutionalized and digitized ecosystem. Scholars like Obaidullah (2015) discussed the institutionalization of ISF, emphasizing how zakat, waqf, and sadaqah have been integrated into formal financial products and services. This shift is driven by the need for greater transparency, efficiency, and scalability in social welfare efforts.

Furthermore, International institutions such as the Islamic Development Bank (IsDB) and the United Nations Development Programme (UNDP) have established strategic partnerships to harness the potential of ISF in advancing the SDGs. These collaborations exemplify how faith-based financial mechanisms can effectively complement global development frameworks by addressing poverty, enhancing resilience, and promoting inclusive growth (UNDP, 2021; IsDB, 2018). Notably, case studies from countries like Sudan, Nigeria, and Pakistan demonstrate the critical role of ISF instruments—including zakat, waqf, and qard al-hasan—as community safety nets during crises such as floods, pandemics, and internal displacement. These interventions underscore ISF's capacity to provide timely social protection and support vulnerable populations, thereby reinforcing its alignment with the ethical and social objectives embedded within the SDGs (UNDP, 2021).

Theological and Philosophical Foundations of ISF Wealth in Islam is a trust (amanah), and its utilization must reflect fairness, justice, and public interest. Core theological principles such as 'adl (justice), ihsan (benevolence), and maslahah (public good) underpin ISF instruments. These principles are operationalized through mechanisms like zakat—an obligation with detailed guidelines for collection and distribution—and waqf, which facilitates perpetual charitable endowment. The Quran (9:60) and Hadith literature provide clear guidance on these responsibilities, promoting a value system that places social welfare at its core.

Figure 1. The Islamic Social Finance Ecosystem

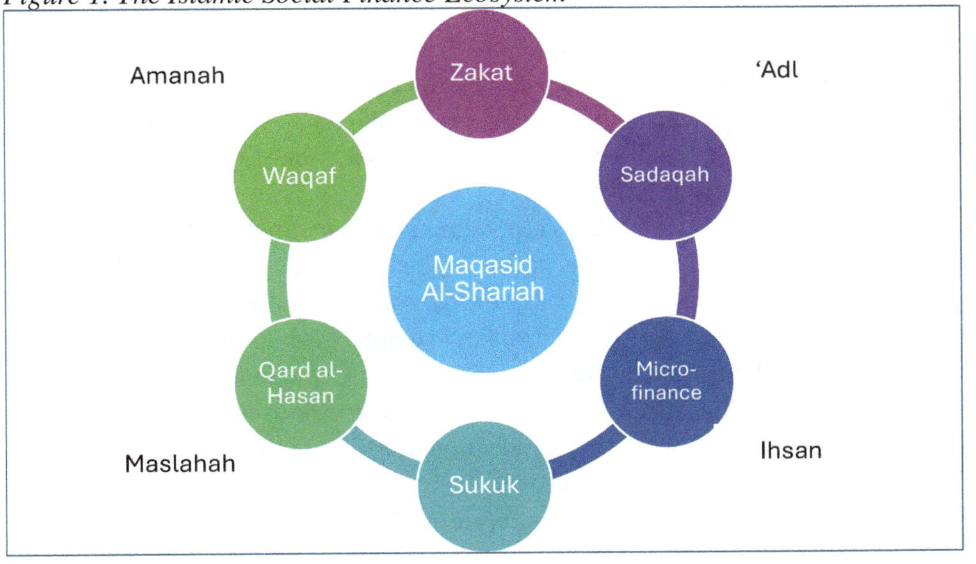

Digitized Islamic Social Finance

Contemporary Islamic legal reasoning (ijtihad) increasingly supports the permissibility of digital giving platforms and AI-assisted donor matching systems. These innovations are generally justified under the principle of *maslahah mursalah* (public interest), provided they comply with Shariah principles and uphold data ethics. This reflects a broader acceptance within Islamic finance of leveraging technology to enhance transparency, efficiency, and inclusivity in charitable giving.

Digitization has significantly transformed the landscape of Islamic Social Finance (ISF) instruments, enhancing accessibility, transparency, and efficiency. In the realm of zakat, institutions like Pusat Pungutan Zakat (PPZ) in Malaysia have implemented online payment platforms, automated eligibility assessments, and real-time donor dashboards, which have notably increased collection and distribution efficiency (International Journal of Economics, Management and Accounting, 2023; Islamic Economist, 2023). Some zakat bodies are also exploring blockchain technology to improve traceability and accountability. Regarding waqf, innovations such as corporate waqf models and blockchain-based asset management are emerging, with platforms like MyWakaf in Malaysia and cash waqf schemes in Turkey exemplifying these trends. Financial institutions, including Turkey's Vakıf Katılım Bank, have integrated waqf into their services, demonstrating successful institutionalization (Razak & Amin, 2024). In terms of sadaqah and infaq, digital crowdfunding platforms and mobile charity applications have expanded donor engagement and increased micro-contributions, especially among younger Muslims (Ismail & Tohirin, 2020). Lastly, qard al-hasan continues to be a vital tool within Islamic microfinance institutions such as Amanah Ikhtiar Malaysia (AIM), which focus on sustainable livelihood programs for women and youth. Emerging technologies like AI credit scoring are being piloted in some countries to ethically assess borrower credibility, although documented cases remain limited.

Governments and Islamic NGOs now administer structured ISF frameworks that are supported by legal, financial, and technological infrastructures. Studies show that zakat institutions in Malaysia have transitioned from traditional management to digital platforms, supported by legal mandates. For example, institutions like the Selangor Zakat Board and Federal Territory Zakat Collection Center have pioneered digital payment systems, online applications, and interactive media to enhance transparency, efficiency, and public trust in zakat collection and distribution (Meerangani et al., 2022). Digital technologies can significantly improve zakat operations when aligned with best practices in governance and service delivery (Farahdina et al., 2025).

Table 1. Comparison of Traditional vs. Digitized ISF Instruments

Instrument	Traditional Practice	Digitized Practice	Example Platforms
Zakat	Manual payment at mosque; paper receipts	Online portals, automated calculators, dashboards	PPZ, BAZNAS
Waqf	Land/property endowment via deed	Corporate waqf, blockchain asset records	MyWakaf, Vakıf Katılım
Sadaqah	In-person cash donation	Crowdfunding, mobile apps	LaunchGood, Kitabisa
Qard Hasan	Local community lending	Microfinance with AI scoring	AIM, fintech pilots
Sukuk	Institutional bonds	E-sukuk platforms	Various banks

Innovating Islamic Social Finance

Robotic Process Automation (RPA) is playing a transformative role in ISF operations, particularly in Malaysia where zakat and waqf institutions have embraced FinTech innovations to improve efficiency and transparency. These institutions utilize robotics and automation to streamline fund collection, distribution, and beneficiary monitoring, which helps reduce manual errors and enhances decision-making. The adoption of these technologies has been especially impactful during the COVID-19 pandemic, leading to increased online zakat collections and more effective fund management. As Kamaruddin et al. (2023) explained, "

Robotics and automation have emerged as viable tools… FinTech assists in the management of zakat funds… FinTech can be used as a tool to monitor the beneficiaries' spending pattern…"

Furthermore, the integration of artificial intelligence (AI) into Islamic finance is transforming the industry by streamlining Shariah compliance, automating risk assessment, and enhancing customer experiences through AI-powered tools. As recent research highlights, AI-driven solutions are enabling Islamic financial institutions to offer more efficient, transparent, and tailored services while upholding ethical and regulatory standards (Iqbal et al., 2025).

Figure 2. Key Applications of AI and RPA in ISF

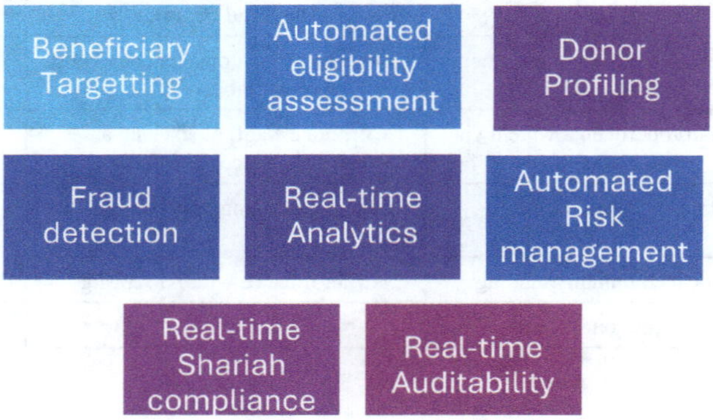

One of the leading examples of innovation in Waqf-based Islamic social finance in Malaysia is the myWakaf initiative, a collaboration between Islamic banks and State Islamic Religious Councils (SIRCs), coordinated by the Association of Islamic Banking and Financial Institutions Malaysia (AIBIM). Launched under Bank Negara Malaysia's Financial Sector Blueprint 2011–2020, myWakaf seeks to optimize Waqf assets for community development through structured governance and financial innovation. A notable aspect of the initiative is the use of digital tools, particularly its online donor portal that allows contributors to track project progress in real time, enhancing transparency and building donor confidence. This reflects an important step toward embedding digital technology within Islamic social finance institutions (Mahmud & Noordin, 2024).

In addition to standard digital tracking, myWakaf is also exploring the integration of blockchain technology to further strengthen transparency and accountability in Waqf fund governance. Blockchain offers a secure and immutable ledger system, which could enable tamper-proof recording and traceability of all financial transactions within Waqf projects. Mahmud and Noordin (2024) note that although blockchain remains in the early experimental phase, its potential for transforming ISF is considerable. It could address long-standing concerns over fund misuse and reporting inconsistencies while supporting real-time compliance and auditability—especially relevant as institutions begin to consider more advanced solutions like Robotic Process Automation (RPA) and Artificial Intelligence (AI).

Performance management in Islamic Social Finance Institutions

Performance management in Islamic Social Finance Institution (ISFI) has evolved beyond traditional financial indicators to include ethical, social, and Shariah-compliant dimensions. Several studies emphasize the importance of integrating Maqasid al-Shariah—the objectives of Islamic law—as a foundational framework for evaluating performance in Islamic financial institutions and social finance entities. For example, Triyuwono (2011) introduces the ANGELS framework, which incorporates Islamic values such as *amanah* (trustworthiness), non-economic wealth, and social responsibility alongside financial metrics, emphasizing process and stakeholder perspectives in performance evaluation. However, these models remain largely conceptual and require empirical validation.

In recent years, there are several research suggesting the aspects to be measured in institutions which upholds the concept of maqasid al-shariah. Qathrunnada et al. (2023) developed a Maqasid Shariah Index alongside the RGEC approach (Risk profile, Good governance, Earnings, Capital) to measure Islamic banking performance in Malaysia and Indonesia, highlighting the need for a multidimensional performance picture that captures both financial and Shariah compliance aspects. Similarly, Maharani and Rahmawati (2021) propose a performance measurement model grounded in Maqasid Shariah principles, critiquing conventional models like CAMELS and BSC for their inability to fully capture Islamic ethical values in banking.

Beyond Islamic banking, performance management in ISFI also considers the interests of multiple stakeholders—managers, donors, and beneficiaries—necessitating comprehensive frameworks that balance financial sustainability with social impact and compliance with Islamic ethics (Hassan et al., 2023). The Balanced Scorecard (BSC) has been adapted in some cases to include Shariah-specific dimensions, but further integration with Maqasid al-Shariah and digital technologies like AI and RPA is needed to enhance measurement accuracy and operational efficiency.

In Malaysia, myWakaf, is an inclusive economic instrument based on philanthropic value under the purview of a State Islamic Religious Council (SIRC). It incorporates outcome-based monitoring, particularly through its alignment with the Sustainable Development Goals (SDGs). The initiative evaluates its socio-economic contributions in areas such as poverty alleviation, education, and environmental sustainability, which reflects a shift from purely financial accountability to impact-based performance measurements. According to Mahmud and Noordin (2024), this approach not only supports long-term community resilience but also increases credibility among stakeholders. These performance areas align closely with the Balanced Scorecard (BSC) framework: financial sustainability through Waqf-featured investments, stakeholder engagement through donor visibility, efficient internal processes via governance reforms, and institutional learning through capacity-building for Waqf managers.

MAIN FOCUS

Understanding Maqasid al-Shariah

Maqasid al-Shariah is an Arabic term that literally means *"the objectives"* or *"the purposes"* of the Shariah (Islamic law). It refers to the deeper wisdom and intended outcomes behind the rules and rulings set by Islamic law. Rather than focusing only on the literal commands, Maqasid al-Shariah looks at the *spirit* and *higher goals* that the law aims to achieve for the well-being of individuals and society as a whole.

The concept of Maqasid al-Shariah is rooted in the belief that Islamic law is designed to promote human welfare and prevent harm. It ensures that the laws not only regulate religious rituals but also guide social, economic, and personal conduct to create a just, balanced, and harmonious society. Islamic scholars, including Imam al-Ghazali and Ibn 'Asyur, have identified five fundamental objectives or essentials that the Shariah seeks to protect and preserve for every human being. These are often called the Five Necessities *(Al-Dharuriyyat Al-Khamsah),* and they form the core framework of Maqasid al-Shariah:

The Five Essential Objectives of Maqasid al-Shariah

1. **Protection of Religion** ***(Hifz al-Din)***

Religion is considered the foundation of human purpose and morality. The Shariah aims to safeguard the freedom to practice faith and maintain religious beliefs without coercion or harm. This protection ensures that individuals and communities can live in accordance with their spiritual values and have access to religious knowledge and worship.

2. **Protection of Life** ***(Hifz al-Nafs)***

The preservation of human life is paramount. Islamic law prohibits harm and violence and promotes the protection of health, safety, and well-being. This objective includes ensuring access to basic necessities such as food, shelter, and healthcare, and protecting individuals from threats that endanger their lives.

3. **Protection of Intellect** ***(Hifz al-'Aql)***

The intellect is essential for reasoning, learning, and making ethical decisions. The Shariah seeks to protect the mind from harm, such as intoxication or ignorance, and encourages education, critical thinking, and the pursuit of knowledge. This objective supports intellectual development and the ability to distinguish right from wrong.

4. **Protection of Progeny *(Hifz al-Nasl)***

This objective focuses on safeguarding family and lineage. It includes protecting the institution of marriage, ensuring the rights and welfare of children, and preserving moral values related to family life. The Shariah promotes stable family structures as the building blocks of society.

5. **Protection of Wealth *(Hifz al-Mal)***

Wealth and property rights are protected to ensure economic justice and prevent exploitation. The Shariah encourages lawful earning, fair trade, and the equitable distribution of wealth through mechanisms like zakat (obligatory almsgiving) and prohibits unjust practices such as usury (riba) and fraud. This objective aims to reduce poverty and promote social welfare.

The five objectives are deeply interconnected, encompassing both individual rights and societal well-being. The ultimate goal of Maqasid al-Shariah is to *promote benefit (maslahah) and prevent harm (mafsadah)* in all aspects of life. This means that every activity, policy, or financial transaction is evaluated not only for its legality but for its alignment with these higher purposes.

Table 2. The Five Essentials of Maqasid al-Shariah

Objective	Meaning	Practical Implications
Protection of Religion (Din)	Safeguarding faith and religious freedom	Ensuring Islamic principles are upheld in all activities
Protection of Life (Nafs)	Preserving human life and health	Providing basic needs, healthcare, and safety
Protection of Intellect ('Aql)	Protecting the mind and encouraging knowledge	Promoting education, preventing harm to intellect
Protection of Progeny (Nasl)	Safeguarding family and lineage	Supporting marriage, child welfare, and moral values
Protection of Wealth (Mal)	Protecting property and economic justice	Encouraging lawful earning, fair distribution of wealth

Maqasid al-Shariah in ISF

In the context of ISF, Maqasid al-Shariah serves as a holistic ethical lens through which institutions can assess their performance. Rather than focusing narrowly on financial returns or operational efficiency, Islamic Social Finance Institutions (ISFIs) must evaluate how well they fulfill these essential objectives—ensuring their activities contribute to religious adherence, protect and enhance human life and intellect, strengthen families, and promote economic justice.

For example, an ISFI's success is not measured solely by the amount of funds collected or disbursed but also by how effectively those funds alleviate poverty, empower beneficiaries, and uphold ethical standards consistent with Shariah. This comprehensive approach ensures that ISFIs remain true to their mission of fostering social justice and sustainable development in line with Islamic values.

Practical Implications for Performance Measurement

By embedding Maqasid al-Shariah into performance measurement frameworks, Islamic Social Finance institutions can develop indicators that reflect both quantitative outcomes (such as poverty reduction or zakat distribution rates) and qualitative values (such as trustworthiness, fairness, and social empowerment). This alignment strengthens accountability to donors, beneficiaries, and regulators, fostering greater trust and legitimacy.

Moreover, Maqasid al-Shariah encourages flexibility and adaptability, allowing ISFs to respond to emerging social challenges while maintaining fidelity to Islamic principles. It supports a dynamic performance measurement system that balances financial sustainability with ethical and social impact, ultimately guiding ISFs toward holistic success.

Introduction to the Balanced Scorecard

The Balanced Scorecard (BSC) is a strategic performance management framework that has revolutionized how organizations measure success beyond traditional financial metrics. Developed by Robert S. Kaplan and David P. Norton in 1992, the BSC emerged as a response to the growing realization that financial indicators alone were insufficient to capture the complexities of organizational performance in modern economies. Prior to the BSC, performance measurement systems predominantly focused on financial outcomes such as profitability, return on investment, and cost control. While these remain critical, they provide a retrospective and narrow view of organizational health, neglecting other vital dimensions that drive long-term success.

Kaplan and Norton's BSC introduced a more holistic approach by integrating four complementary perspectives:

1. **Financial Perspective** – Measures reflecting profitability, cost management, and financial sustainability.
2. **Customer Perspective** – Indicators related to customer satisfaction, retention, and market share.
3. **Internal Business Processes Perspective** – Metrics assessing the efficiency and quality of internal operations.
4. **Learning and Growth Perspective** – Measures of organizational capacity, innovation, employee skills, and knowledge management.

This framework enables organizations to translate their vision and strategy into a coherent set of performance measures that balance short-term financial results with long-term value creation. The BSC's power lies in its ability to link these perspectives through cause-and-effect relationships, often visualized in a strategy map. For example, investing in employee training (Learning & Growth) enhances process efficiency (Internal Processes), which improves customer satisfaction (Customer Perspective), ultimately driving financial performance (Financial Perspective).

The Strategy Map: Visualizing Cause and Effect

A key innovation of the BSC is the strategy map, a visual representation that articulates how strategic objectives interrelate across the four perspectives. The map clarifies how intangible assets such as employee skills and organizational culture contribute to operational excellence and customer loyalty, which in turn generate financial success. This visualization fosters strategic alignment by communicating priorities clearly across all organizational levels, ensuring that day-to-day activities support overarching goals.

For instance, a typical strategy map might show:

- **Learning & Growth:** Enhance employee expertise through targeted training programs.
- **Internal Processes:** Streamline service delivery and reduce errors.
- **Customer:** Increase customer satisfaction and retention rates.
- **Financial:** Achieve revenue growth and cost efficiency.

This cause-effect chain underscores the BSC's role not just as a measurement tool but as a strategic management system that integrates planning, communication, and feedback.

Limitations of the Traditional Balanced Scorecard in Islamic Social Finance

While the BSC has been widely adopted across industries, its traditional four-perspective model has limitations when applied to ISFIs. ISFIs—such as zakat agencies, waqf foundations, Islamic microfinance institutions, and charitable organizations—operate within a unique ethical and religious framework grounded in Shariah principles and the Maqasid al-Shariah. Their mission extends beyond financial sustainability to include social justice, equitable wealth distribution, community welfare, and spiritual accountability.

ISFIs operate under Shariah law, which mandates that financial activities must be free from *riba* (interest*), gharar* (excessive uncertainty*),* and must promote social justice *and equitable wealth* distribution. The traditional financial metrics do not assess compliance with these ethical mandates or the social impact of financial activities.

Issues of that 'mismatch' BSC with ISFIs

Narrow Financial Focus and Inadequate Ethical Dimensions

The financial perspective in the traditional BSC is primarily concerned with profitability, cost control, and financial sustainability. While financial health is undeniably important for ISFs, this perspective alone is insufficient to capture the ethical and social objectives that underpin Islamic social finance. ISFs operate under Shariah law, which mandates that financial activities must be free from riba (interest), gharar (excessive uncertainty), and must promote social justice and equitable wealth distribution. The traditional financial metrics do not assess compliance with these ethical mandates or the social impact of financial activities.

For example, an ISFI may show strong financial returns but fail to distribute zakat effectively or support underserved communities, which contradicts the core objectives of Islamic philanthropy. The traditional BSC's financial perspective lacks the tools to measure such qualitative, value-driven outcomes.

Customer Perspective: Limited to Satisfaction, Overlooking Beneficiary Welfare

The customer perspective in the traditional BSC focuses on customer satisfaction, retention, and market share. In commercial organizations, this often translates to measuring service quality and customer loyalty. However, ISFIs serve a dual role: they are accountable not only to donors but also to beneficiaries who are often vulnerable populations. The traditional customer perspective does not adequately differentiate between these two stakeholder groups or capture the social welfare and empowerment outcomes critical to ISFIs.

Beneficiaries' needs extend beyond transactional satisfaction to include improved quality of life, dignity, and sustainable development. The conventional BSC framework does not measure these dimensions, thus failing to reflect the full scope of ISFI's social mission.

Internal Processes Perspective: Lack of Shariah Compliance and Ethical Governance Measures

The traditional internal business processes perspective evaluates operational efficiency, quality control, and innovation in service delivery. While these are relevant, ISFs require additional focus on Shariah compliance, ethical governance, and transparency. Islamic social finance institutions are bound by religious rulings and ethical standards that govern all aspects of their operations—from fund collection and investment to distribution and reporting.

The traditional BSC does not incorporate metrics that assess adherence to Shariah principles or evaluate the effectiveness of Shariah supervisory boards, compliance audits, or ethical risk management. This omission poses a significant gap, as failure to monitor these dimensions can undermine the legitimacy and trustworthiness of ISFIs.

Learning & Growth Perspective: Overlooking Spiritual and Ethical Development

The learning and growth perspective traditionally focuses on employee training, knowledge management, and innovation capacity. While these are essential, ISFs also emphasize spiritual development, ethical awareness, and community engagement as critical components of organizational growth. The traditional BSC does not explicitly measure these qualitative aspects, which are vital for sustaining the moral and religious ethos of ISFs.

For example, employee understanding of Maqasid al-Shariah and ethical finance principles is crucial for effective service delivery and compliance. The lack of such measures in the conventional BSC limits its applicability in ISFs.

Absence of Explicit Ethical and Social Justice Dimensions

Perhaps the most critical limitation is the absence of explicit ethical and social justice perspectives in the traditional BSC. Islamic social finance is deeply rooted in the principles of *amanah (trustworthiness), adl (justice),* and *ihsan (excellence),* which guide all activities and relationships. The four-perspective BSC does not provide a framework to measure these values, nor does it capture the broader social objectives enshrined in Maqasid al-Shariah—the higher objectives of Islamic law aimed at preserving religion, life, intellect, lineage, and wealth.

Without incorporating these dimensions, performance measurement risks becoming a mechanical exercise that overlooks the spiritual and social imperatives that define ISFs.

Table 3. Limitation of BSC Perspective in ISF Context

Traditional BSC Perspective	Limitation in ISF Context	Maqasid-Driven Enhancement
Financial	Focus on profit and cost control, ignoring Shariah compliance and social redistribution	Include KPIs on Shariah-compliant financing, zakat distribution impact, and equitable wealth protection (Mal)
Customer	Measures customer satisfaction only, overlooking beneficiary empowerment and welfare	Expand to measure social justice, poverty alleviation, and beneficiary dignity (Nafs, Adl)
Internal Process	Focus on operational efficiency, ignoring Shariah governance and ethical compliance	Add metrics for Shariah audit results, transparency, and ethical risk management (Din, Amanah)
Learning and Growth	Emphasizes skills and innovation, neglecting spiritual and ethical development	Include training on Maqasid principles, ethical conduct, and community engagement ('Aql, Ihsan)

Linking Limitations to Maqasid al-Shariah: The Need for a Customized Framework

The Maqasid al-Shariah framework offers a comprehensive ethical and social lens through which ISFIs performance should be evaluated. It emphasizes the preservation and promotion of five essential values:

- *Protection of Religion (Din)*
- *Protection of Life (Nafs)*
- *Protection of Intellect ('Aql)*
- *Protection of Progeny (Nasl)*
- *Protection of Wealth (Mal)*

Aligning the BSC with Maqasid al-Shariah addresses the limitations discussed above by embedding these values into the performance measurement system.

Maqasid as the Outcome and Guiding Principle

Many scholars argue that Maqasid al-Shariah should be the ultimate outcome of ISF performance measurement, guiding the design of KPIs and strategic objectives. For example, effective zakat distribution (*Protection of Wealth and Life*) should be measured not just by amounts disbursed but by the extent to which it alleviates poverty and promotes social welfare. Similarly, investments must be Shariah-compliant and socially responsible, reflecting *Protection of Religion and Intellect.*

This approach shifts the focus from purely financial or operational metrics to value-driven, impact-oriented measures that align with Islamic ethics.

Expanding the BSC Perspectives to Reflect Maqasid

To operationalize Maqasid principles, the BSC must be expanded beyond its traditional four perspectives. Proposed additional perspectives include:

- *Shariah & Ethical Governance Perspective*: Measuring compliance, transparency, and ethical conduct.
- *Social Impact*: Capturing the social outcomes and empowerment of beneficiaries.
- *Stakeholder Trust & Engagement Perspective*: Reflecting donor confidence, community participation, and accountability.

These additions ensure that the BSC comprehensively covers the multidimensional objectives of ISFs.

Figure 3. Proposed Maqasid-aligned Balanced Scorecard

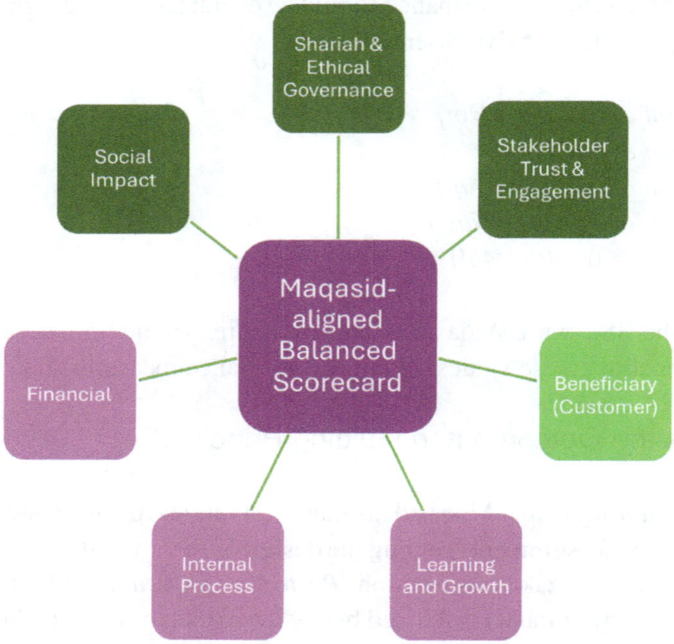

Prioritizing Maqasid Objectives Using Weighted KPIs

Frameworks such as the Maqasid Shariah Index and methods like the Analytic Hierarchy Process (AHP) enable ISFIs to assign weights to Maqasid objectives based on their relative importance. This prioritization helps focus resources and measurement efforts on the most impactful areas, ensuring strategic alignment with Islamic values.

Stakeholders in ISFIs and Their Role in Performance Measurement

Understanding the key stakeholders of ISFIs is essential to designing an effective, Maqasid-aligned Balanced Scorecard. The primary stakeholders include:

Donors

Donors provide the financial resources that enable ISFIs to operate. Their expectations extend beyond financial stewardship to include:

- Assurance of Shariah compliance in fund management.
- Transparency and accountability in fund utilization.
- Evidence of social impact and alignment with Islamic ethical values.

Performance measures must therefore include donor satisfaction, trust indices, and reporting transparency to maintain and grow donor engagement.

Beneficiaries

Beneficiaries are the recipients of zakat, waqf, and other social finance programs. Their welfare and empowerment are central to ISF missions. Performance measurement should capture:

- Improvement in quality of life indicators.
- Access to education, healthcare, and livelihood opportunities.
- Beneficiary feedback and participation in program design.

Traditional customer satisfaction metrics are insufficient; instead, measures must reflect social justice and human development outcomes consistent with Maqasid al-Shariah.

Management and Staff

Management is responsible for strategic planning, compliance, and operational efficiency. Their performance is critical in balancing financial sustainability with ethical and social objectives. Measures should include:

- Effectiveness in Shariah governance and risk management.
- Capacity building in Islamic finance and ethical standards.
- Innovation in service delivery, including digital transformation.

Maqasid-aligned Balanced Scorecard

The traditional Balanced Scorecard, while valuable, falls short in capturing the full spectrum of performance dimensions critical to Islamic Social Finance.

Its narrow focus on financial and operational metrics overlooks the ethical, social, and spiritual imperatives rooted in Maqasid al-Shariah. To effectively measure and manage ISF performance, the BSC must be expanded and customized to include additional perspectives that reflect Shariah compliance, social justice, beneficiary welfare, and stakeholder trust.

Such a Maqasid-aligned Balanced Scorecard not only ensures comprehensive performance measurement but also strengthens the legitimacy, transparency, and impact of ISFs. This expanded framework provides a strategic management tool that balances financial sustainability with the higher objectives of Islamic law, fostering socio-economic development grounded in ethical finance.

In the context of ISF, Maqasid al-Shariah provides the ethical foundation for evaluating the performance of institutions managing zakat, waqf, and charitable funds. Instead of focusing solely on financial returns, ISFs are assessed on how well they preserve and promote these five essential objectives—ensuring that their operations contribute to religious adherence, protect beneficiaries' lives and intellect, strengthen families, and promote economic justice.

To clarify this expanded framework, a strategy map diagram can illustrate how these perspectives interrelate, showing cause-effect linkages similar to the traditional BSC but enriched with Maqasid values. For example, enhanced ethical governance (Shariah & Ethical Perspective) improves beneficiary trust and social impact (Beneficiary Perspective), which supports sustainable financial health (Financial Perspective).

Challenges in Managing and Measuring Performance

ISFIs play a vital role in addressing socio-economic challenges by mobilizing resources such as zakat, waqf, and sadaqah to support vulnerable populations. However, despite their growing importance, many ISFIs face significant challenges in effectively managing and measuring their performance. These limitations hinder their ability to demonstrate impact, optimize operations, and maintain stakeholder trust.

Performance Management System (PMS) provide a structured framework for ISFIs to define strategic goals, monitor implementation, measure results, and adjust operations to optimize social impact. A robust PMS also allows institutions to balance financial efficiency with spiritual mandates, aligning organizational performance with the maqasid al-shariah. Several previous studies have highlighted that Islamic Social Finance programs have not fully succeeded in creating widespread prosperity, and there remains a lack of definitive measurements to demonstrate the sustainability impact of these initiatives. This gap underscores the need for more robust and multidimensional evaluation tools that can capture both the successes and limitations of Islamic Social Finance in achieving long-term socioeconomic upliftment.

Recent research efforts have responded to this challenge by developing comprehensive indices and frameworks for evaluating the impact of Islamic Social Finance programs. These frameworks move beyond traditional financial or quantitative metrics and emphasize broader social and qualitative outcomes (Widiastuti, Robani and Kamal, 2022), such as:

- Beneficiary empowerment
- Improvements in quality of life
- Donor involvement
- Transformation of welfare levels

However, the challenges faced by ISFIs is not only about developing a PMS framework tailored to a particular ISFI but also in execution of PMS. Common shortcomings in existing PMS practices among ISFIs include:

1. *Disjointed Data Across Departments or Branches*

One of the fundamental challenges in ISFIs is the fragmentation of data. Many institutions operate with siloed information systems where data related to beneficiaries, donors, financial transactions, and operational activities are stored separately across different departments or branches. This disjointed data landscape creates several problems:

- *Lack of a unified view:* Decision-makers struggle to obtain a comprehensive picture of institutional performance, as data integration is poor or non-existent.
- *Inefficient reporting:* Generating consolidated reports requires manual effort, increasing the risk of errors and delays.
- *Hindered impact assessment:* Without integrated data, it becomes difficult to link inputs (e.g., funds collected) to outputs (e.g., beneficiaries served) and outcomes (e.g., poverty alleviation), limiting the ability to measure social impact effectively.

Research on Islamic microfinance institutions highlights the infancy of social performance management frameworks tailored to the sector, noting the need for integrated systems that incorporate financial, social, and spiritual dimensions (Khalid & Kamaruddin, 2019). Similarly, inefficiencies in data management have been cited as barriers to effective performance measurement and strategic decision-making in ISFIs (INTI University and Colleges Symposium, 2022).

2. Over-Reliance on Lagging Indicators

Many ISFIs tend to focus heavily on lagging indicators, which measure past activities or outputs rather than forward-looking or outcome-oriented metrics. Common lagging indicators include:

- Number of beneficiaries served
- Total funds disbursed
- Volume of zakat collected

While these metrics provide useful snapshots, they do not capture the quality, sustainability, or long-term impact of ISF interventions. For example, simply counting beneficiaries does not reveal whether their socio-economic conditions have improved or if they have achieved financial independence.

The over-reliance on lagging indicators limits the institution's ability to:

- Anticipate future challenges or opportunities
- Adjust strategies proactively
- Demonstrate meaningful social change aligned with Maqasid al-Shariah objectives

Studies emphasize the need for ISFIs to incorporate leading indicators and impact metrics that reflect beneficiary empowerment, poverty reduction, and ethical compliance (Rosman et.al., 2023). Without this shift, performance measurement risks becoming a compliance exercise rather than a tool for continuous improvement.

3. Limited Use of Analytics and Forecasting Tools

Despite the increasing availability of data, many ISFIs underutilize advanced analytics, predictive modeling, and forecasting tools that can transform raw data into actionable insights. The limited adoption of these technologies results in:

- Reactive rather than proactive management
- Missed opportunities for optimizing resource allocation
- Inability to identify emerging risks or beneficiary needs in real time

4. Difficulty Linking Operational Metrics to Strategic Outcomes

A critical shortcoming in many ISFIs is the disconnect between operational-level metrics and strategic objectives. Performance data often remains fragmented and focused on isolated activities without clear alignment to the institution's mission or Maqasid al-Shariah principles.

This misalignment manifests in several ways:

- Operational metrics such as processing times or fund disbursement volumes are tracked without assessing their contribution to social justice, empowerment, or Shariah compliance.
- Strategic outcomes like poverty alleviation, community development, or spiritual upliftment are difficult to quantify and often lack corresponding measurable indicators.
- Lack of cause-effect understanding limits the institution's ability to prioritize initiatives that drive meaningful impact.

The need for integrated performance management frameworks that connect operational data to strategic goals has been highlighted in recent research, advocating for tools like the Balanced Scorecard customized with Maqasid al-Shariah perspectives (Khalid & Kamaruddin, 2019), (INTI Symposium, 2022).

Beyond these core shortcomings, other issues affecting ISFI performance management include:

- *Inconsistent definitions and understanding of performance metrics*: Different departments or branches may use varying criteria, leading to incomparable or unreliable data (Rosman et.al, 2023).
- *Inadequate capacity and skills*: Limited expertise in social performance measurement and data analytics restricts effective implementation of PMS (INTI Symposium, 2022).
- *Governance and coordination gaps*: Weak coordination among Islamic social finance stakeholders (zakat agencies, waqf institutions, regulators) leads to inefficiencies and duplication of efforts (Rosman et.al, 2023).
- *Regulatory and external constraints*: Compliance requirements and lack of standardized reporting frameworks complicate performance measurement (Meskovic, Kozarevic and Avdukic, 2021).

DISCUSSIONS AND RECOMMENDATIONS

This section explores the rationale, benefits, and implementation strategies for embedding Robotic Process Automation (RPA) and Artificial Intelligence (AI) technologies into PMS of ISFIs', particularly the Balanced Scorecard.

To overcome the common challenges faced by Islamic Social Finance Institutions (ISFIs) in their PMS—such as fragmented data, reliance on lagging indicators, and limited analytics capabilities—emerging literature strongly advocates for the integration of digital technologies. Lestari (2023) emphasize that digital transformation (DT) plays a pivotal role in enhancing data quality, enabling seamless cross-functional reporting, and supporting evidence-based decision-making within ISFIs. Their study highlights how the adoption of advanced digital tools, including online platforms for zakat and infaq payments and digital investment channels, has significantly improved operational efficiency and customer experiences.

Importantly, these technologies facilitate real-time data integration across departments, allowing ISFIs to move beyond basic output metrics towards more sophisticated impact measurement aligned with Shariah principles. Furthermore, Lestari (2023) underscore the necessity of collaborative partnerships between ISF institutions and FinTech firms to drive innovation and sustainable growth in the evolving "New Normal" financial landscape. This synergy between digital transformation and Islamic Social Finance not only enhances financial inclusion but also ensures that ethical and Shariah compliance requirements remain central to institutional strategies.

Besides that, performance management models like Results-Based Management (RBM) and Outcome Mapping are gaining traction in the broader non-profit and humanitarian sectors globally, aiming for more dynamic and responsive structures. The UNFPA (2019) developmental evaluation of RBM underscores the importance of integrating data-driven feedback loops and innovative tools—often enabled by digital solutions—to enhance organizational learning, accountability, and adaptive decision-making in complex environments. Incorporating such models, particularly when combined with digital technologies, holds significant potential for Islamic Social Finance Institutions (ISFIs) to improve their transparency, responsiveness, and overall impact (OECD, 2021).

As ISFIs strive to improve transparency, efficiency, and impact, the integration of RPA and AI into their PMS emerges as a critical innovation. Not only do these technologies modernize operations, they also offer new pathways to overcome the operational limitations, manual processes, and data fragmentation that have long hindered ISFIs effectiveness. They also help align institutional performance with Maqasid al-Shariah such as justice, social welfare, and the preservation of dignity.

RPA in Streamlining ISFI Operations

ISIFs such as zakat boards and waqf organizations, manage complex operations involving large volumes of data, strict Shariah compliance requirements, and a diverse set of stakeholders. To address the challenges of operational inefficiency, data inaccuracies, and compliance risks, many ISFIs are increasingly adopting RPA that automates repetitive, rule-based tasks traditionally performed by humans. Key applications of RPA in ISIFs include:

1. Automated Data Entry and Report Generation

Zakat institutions, including Lembaga Zakat Selangor (LZS), have successfully implemented RPA solutions to automate the extraction of donor and beneficiary information from paper-based forms into cloud-based management systems. This automation has significantly reduced the time required for data entry, minimized human errors, and improved data accuracy. By streamlining these administrative processes, ISFIs can allocate more resources toward strategic activities such as beneficiary engagement and impact assessment. (Zulkifley and Muneeza, 2024).

2. Disbursement Tracking and Reconciliation

RPA tools enable ISFIs to monitor the entire disbursement process meticulously. Automated bots can match transactions against donor instructions and beneficiary records, quickly identifying discrepancies or anomalies. This capability enhances transparency and ensures that funds are disbursed in accordance with Shariah principles, thereby strengthening audit readiness and stakeholder confidence. (Purpose Forward, 2025; Third Sector, 2025)

3. *Regulatory Reporting*

Compliance with regulatory and Shariah supervisory requirements is critical for ISFIs. RPA bots programmed with standard reporting templates can generate timely and consistent compliance reports for Islamic authorities and government agencies. This automation reduces the administrative burden on staff and mitigates risks associated with manual report preparation.

These outcomes demonstrate that RPA not only improves operational efficiency but also contributes to the integrity and trustworthiness of ISFIs—qualities essential for sustaining donor confidence and fulfilling social missions. (Purpose Forward, 2025; Third Sector, 2025)

Artificial Intelligence-Driven Strategic Insights and Forecasting

Artificial Intelligence (AI) is increasingly recognized for its transformative potential in Islamic finance, including ISF. AI technologies offer powerful tools for planning, decision-making, and predictive analytics, which are essential for optimizing resource allocation and enhancing social impact aligned with Maqasid al-Shariah. ISIFs may utilize AI for the following purposes:

1. Beneficiary Segmentation

AI algorithms can analyze large datasets to categorize beneficiaries based on factors such as socio-economic needs, geographic location, and vulnerability (Dataro, 2024). This segmentation enables ISFIs to design targeted and equitable aid distribution strategies, ensuring that resources reach those most in need efficiently. This approach supports the Islamic principle of justice (adl) and welfare (maslahah).

2. Demand Forecasting

Machine learning models can predict fluctuations in funding requirements or beneficiary demand (Dataro, 2024), especially during critical periods such as Ramadan, Qurban, or in response to disasters. Accurate forecasting allows ISFIs to prepare and allocate resources proactively, minimizing delays and maximizing impact.

3. Sentiment Analysis and Feedback Loops

Natural Language Processing (NLP) tools analyze beneficiary feedback, social media sentiment, and donor surveys to gauge satisfaction levels and identify service gaps. This continuous feedback mechanism supports responsive program adjustments and enhances stakeholder engagement. (Helpyousponsor, 2025)

4. Impact Measurement

AI facilitates sophisticated correlation and causal analysis between aid interventions and socio-economic outcomes (Raisely, 2025), such as school attendance and health access. These insights help ISFIs align their programs with Maqasid al-Shariah objectives, ensuring that aid contributes to holistic human development.

Table 4. AI Applications in Islamic Social Finance

AI Application Area	Description	Benefits
Beneficiary Segmentation	Categorizing beneficiaries by needs and vulnerability	Targeted aid, equitable resource distribution
Demand Forecasting	Predicting funding needs during peak periods	Proactive resource allocation, reduced delays
Sentiment Analysis	Analyzing feedback from beneficiaries and donors	Improved service quality, stakeholder engagement
Impact Measurement	Correlating aid with socio-economic outcomes	Evidence-based program evaluation, alignment with Maqasid

Technology Adoption in ISFIs

Malaysian zakat institutions have increasingly embraced technology to enhance their operational efficiency, transparency, and service delivery. While the adoption of digital tools is more advanced in zakat collection, the distribution process and related administrative functions are still evolving, with significant potential for further technological integration.

Research by Salleh and Chowdhury (2020) observes that most Malaysian zakat institutions have implemented technology systems primarily for zakat payment and collection, including online payment platforms and mobile applications. However, the use of technology in zakat distribution, awareness campaigns, and fund reporting remains limited. The authors argue that the management of millions of zakat records and distribution activities can benefit greatly from technology to improve efficiency and public trust.

Similarly, Wahab and Rahman (2013) analyze the efficiency of Malaysian zakat institutions using productivity indices and find that technical progress, including the adoption of computerized zakat systems, has contributed to an average annual productivity increase of 2.4%. Their study highlights that technological advancements, alongside governance factors such as audit committees and decentralization, significantly influence zakat institutions' operational efficiency.

Recent advancements in *Robo-Advisory* systems are transforming Islamic finance by providing automated, data-driven Shariah-compliant investment guidance accessible to retail investors. Katterbauer and Moschetta (2022) developed a robo-advisory framework that integrates Islamic and ESG compliance using machine learning to classify stocks in the S&P 500, enabling real-time, objective Shariah screening that empowers investors to build compliant portfolios without relying solely on costly human advisory services. Complementing this, research from Universiti Sains Islam Malaysia highlights the growing role of robo advisors in promoting financial literacy and education, offering users convenient, AI-powered tools for basic financial planning, including zakat and tax planning, thus broadening access to Islamic financial knowledge (Hishamudin et al., 2024).

Parallel to investment advisory innovations, technology-driven improvements in zakat administration are gaining traction. Muneeza and Nadwi (2019) emphasize the potential of AI-based innovations such as chatbots and automated eligibility screening to enhance zakat distribution efficiency and transparency. These tools facilitate faster data processing, improve beneficiary engagement, and support ethical compliance aligned with Maqasid al-Shariah, marking a significant step towards modernizing Islamic social finance operations through intelligent automation.

Leveraging RPA and AI to Implement the Balanced Scorecard in ISFIs

The Balanced Scorecard (BSC), developed by Kaplan and Norton, is a proven PMS framework that maps performance across four dimensions: financial, customer (or stakeholder), internal processes; and learning and growth. When integrated with RPA and AI, the BSC becomes a living strategy map that allows ISFIs to:

- Align short-term activities with long-term vision
- Incorporate real-time data into scorecard metrics
- Automate performance reporting across departments
- Link key results to Shariah and SDG outcomes

The integration of RPA and AI offers powerful enablers for operationalizing this expanded BSC in ISFs. These technologies can:

- Automate data collection and reporting across multiple perspectives, reducing manual errors and improving timeliness.
- Use AI analytics to assess social impact, detect compliance risks, and predict beneficiary needs.
- Facilitate real-time dashboards that visualize performance aligned with Maqasid objectives, enhancing transparency and decision-making.
- Support stakeholder engagement through AI-powered chatbots and feedback systems that capture donor and beneficiary sentiments.
- Enhance Shariah compliance monitoring by automatically scanning transactions and documents for non-compliant elements.

Together, RPA and AI transform the BSC from a static reporting tool into a dynamic, intelligent system that supports strategic management and continuous improvement in Islamic Social Finance.

To fully operationalize the proposed Maqasid-aligned Balanced Scorecard, it is essential to integrate enabling technologies—particularly AI and RPA—into each perspective. These technologies not only enhance operational efficiency but also strengthen impact measurement, transparency, and alignment with Islamic ethical imperatives.

Table 4 outlines how each perspective in the proposed Maqasid-aligned BSC can be enhanced using AI and RPA, ensuring strategic alignment between institutional performance and Islamic ethical mandates. This technological integration transforms the BSC into a dynamic, intelligent system for continuous spiritual, social, and financial accountability

Table 5. Enhancing Each Maqasid-Aligned BSC Perspective with AI and RPA

Perspective	Role of AI and RPA	Maqasid Dimension
Shariah & Ethical Governance	RPA automates compliance reporting; AI performs document analysis to detect non-Shariah elements; ensures fatwa adherence	Din (Religion), Amanah
Stakeholder Trust & Engagement	NLP-based sentiment analysis gathers real-time donor and beneficiary feedback; AI tailors communications	Amanah, Ihsan
Beneficiary (Customer)	AI segments and targets beneficiaries more equitably; RPA streamlines aid distribution and eligibility verification	Nafs, 'Adl, Maslahah
Learning and Growth	AI-powered e-learning platforms for Maqasid ethics, Shariah finance; tracks learning outcomes via analytics	'Aql, Ihsan
Internal Process	RPA automates repetitive tasks (e.g., data entry, reconciliation); AI optimizes process workflows	Din, Amanah
Financial	AI improves fund forecasting and optimization; RPA enables real-time dashboards for disbursement and collection tracking	Mal, Amanah
Social Impact	AI assesses causal relationships between interventions and outcomes (e.g., poverty alleviation, education upliftment)	Maslahah, Nafs, Nasl

From Efficiency to Impact: RPA and AI for Strategic Value Creation in ISFIs

This section presents a comprehensive discussion of how RPA and AI offer transformative potential for ISFIs. Rather than viewing these tools solely as efficiency drivers, we frame them as instruments of strategic and spiritual renewal — capable of advancing social impact, institutional maturity, and alignment with the Maqasid al-Shariah. The discussion is divided into two parts: the societal and strategic implications of RPA and AI in ISFIs, and the enabling strategies and governance considerations necessary for effective and ethical implementation.

Advancing the Objectives of Maqasid al-Shariah

The central mission of ISFIs is to serve the public good (*maslahah*) in accordance with the five principles of Maqasid al-Shariah: the protection of religion (*din*), life (*nafs*), intellect (*'aql*), progeny (*nasl*), and wealth (*mal*). Technologies such as RPA and AI can become strategic enablers (*wasilah*) for achieving these objectives — provided they are implemented with Shariah-compliant governance and human oversight.

Examples include:

- AI-based predictive targeting of vulnerable groups to optimize zakat distribution.
- RPA-enabled real-time tracking of disbursements during emergencies, enhancing the preservation of life and dignity.
- NLP tools used to gather feedback from beneficiaries, ensuring service remains human-centric even as scale increases.

Such technologies, when governed ethically, serve not only efficiency but also justice (*'adl*) and human dignity (*karamah insaniyyah*), core to the Islamic economic ethos (Laldin & Furqani, 2016).

Enhancing Transparency and Restoring Trust

Donor trust is vital for sustainability. Delays in disbursement, opaque reporting, and lack of impact data erode credibility. RPA and AI can counter this by:

- Generating real-time dashboards for donors, tracking fund usage and social returns.
- Providing automated audit trails and anomaly detection for integrity assurance.
- Facilitating blockchain-integrated ledgers that promote immutable, transparent records.

Multiple studies confirm that digital tools—such as online platforms and blockchain—can improve transparency, accountability, and donor trust in zakat institutions, including the works of Salleh et al. (2022), Abdul-Rahman et al. (2023),, and Nazeri et al. (2023). These improvements are widely recognized as drivers for increased zakat collection and more effective management.

Improving Beneficiary Outcomes

Automation and AI tools can enhance the precision, timeliness, and personalization of aid. Specific innovations include:

- RPA-driven eligibility screening using government databases or biometric validation.
- Machine learning models that recommend customized aid packages based on socio-economic profiling.
- AI-based feedback systems that adapt programmes dynamically in response to beneficiary needs.

BAZNAS, Indonesia's National Zakat Agency, has embraced digital technologies like Robotic Process Automation (RPA) and Artificial Intelligence (AI) to enhance zakat management. These technologies streamline data processing, automate routine tasks, and improve accuracy in beneficiary identification and fund allocation. AI-driven analytics enable better forecasting and targeted aid distribution, helping BAZNAS shift from reactive charity to proactive empowerment. While specific pilot results remain unpublished, these innovations demonstrate BAZNAS's commitment to leveraging technology for greater transparency, efficiency, and positive social impact (Google Cloud, 2023; ICONZ BAZNAS, 2021; Puskas BAZNAS, 2021).

Institutional Transformation and Strategic Alignment via BSC

Digital tools enable ISFIs to evolve into agile, data-informed institutions. The Balanced Scorecard (BSC), when integrated with AI analytics, allows leadership to align actions with long-term goals across financial, stakeholder, internal process, and learning perspectives — as well as a fifth Islamic dimension (ethics/spirituality).

AI-enhanced BSC dashboards offer real-time strategy maps, flagging gaps in stakeholder trust, process inefficiencies, or value misalignment. This promotes forward-looking governance rooted in Islamic values (Ahmed & Suleman, 2023).

From Compliance to Value Creation

Traditionally, ISFIs emphasize compliance — ensuring disbursements follow religious guidelines. However, RPA and AI create space for a value creation paradigm:

- Tracking outcomes (e.g., school graduation) vs. outputs (e.g., scholarship numbers).
- Mapping multiplier effects of aid on family and community economies.
- Data-sharing to form strategic alliances with NGOs and local governments.

By transforming operations into impact-driven ecosystems, ISFIs shift from being guardians of compliance to architects of empowerment.

Challenges in Implementing RPA and AI in ISFIs

Islamic Social Finance Institutions (ISFIs) face multiple challenges in adopting Robotic Process Automation (RPA) and Artificial Intelligence (AI). First, many institutions, especially at the community or state level, suffer from outdated infrastructure and fragmented data systems, which impede automation efforts (Chatterjee et al., 2020; World Bank, 2020). Second, ethical concerns arise from AI's opaque decision-making processes, raising issues of fairness, bias, and Shariah compliance, with no established guidelines to address these concerns (Mittelstadt, 2019). Third, limited digital literacy and technical skills among staff and Shariah scholars create resistance and underutilization of these technologies (Avasant, 2023). Finally, cultural and organizational resistance, rooted in conservative values and prior failed technology initiatives, slows digital transformation despite leadership support.

Strategic Responses and Governance Solutions

To overcome these barriers, ISFIs should pursue a phased digitalization strategy, starting with digitizing records and automating simple, rule-based tasks before scaling AI applications. Infrastructure investments, supported by donors and central authorities, are essential to build foundational capacity (Chatterjee et al., 2020). Improving data governance through standardized formats and secure, interoperable data exchanges will enhance AI effectiveness while respecting privacy and Islamic ethical principles (World Bank, 2020). Developing Shariah-tech ethical frameworks through collaboration among scholars, technologists, and social scientists will ensure fairness, transparency, and accountability in AI use (Dwivedi et al., 2021). Capacity building via cross-disciplinary training that combines technical skills with Shariah knowledge will bridge human capital gaps (Ahmed & Suleman, 2023). Finally, inclusive change management involving all organizational levels and community leaders can foster trust and align technology adoption with ISFIs' spiritual missions.

FUTURE RESEARCH DIRECTIONS

As Islamic Social Finance Institutions (ISFIs) integrate Robotic Process Automation (RPA) and Artificial Intelligence (AI), several critical research areas emerge that span theological, technological, and organizational dimensions. This section highlights ten key directions for future inquiry.

Shariah-Oriented AI Governance Frameworks

There is an urgent need to develop AI governance models grounded explicitly in Islamic jurisprudence. Research should explore how maqasid al-shariah principles can shape AI ethics, ensuring these technologies serve the public good within a faith-based framework. Comparative studies of fatwa interpretations on AI across different Islamic jurisdictions can reveal diverse scholarly perspectives and support harmonized or context-specific guidelines. Collaborative efforts between Shariah scholars, ethicists, and AI developers could lead to Islamic Ethical AI Toolkits that embed values like amanah (trust), maslahah (public interest), and ihsan (benevolence) into AI design.

Performance Measurement with Integrated Ethics and Impact Metrics

While the Balanced Scorecard (BSC) offers a multidimensional performance framework, it requires adaptation to reflect Islamic philanthropy's spiritual and social priorities. Introducing a fifth dimension—"Spiritual and Ethical Impact"—could align operational metrics with institutional values. Developing key performance indicators (KPIs) that measure inclusivity, dignity, and ethical transparency alongside efficiency is crucial and can be difficult. Pilot studies can test these KPIs to evaluate whether AI and RPA improve beneficiary outcomes and donor confidence beyond cost and speed gains.

Comparative Case Studies Across ISFIs

Empirical research comparing digital maturity among national and state-level ISFIs is needed to identify structural, regulatory, and leadership factors influencing success or stagnation. Case studies of AI and RPA implementation can uncover patterns of readiness, resistance, and unintended consequences, helping create practical guidelines for institutions at various digital transformation stages. Examining stakeholder influences—including donors, staff, regulators, and scholars—can inform better alignment and communication strategies.

Digital Literacy and Human-Centric Transformation

Technological adoption depends on human readiness. Research should identify effective training methods for Shariah scholars, operational staff, and leaders, including blended learning and peer mentoring. Investigating how AI adoption reshapes organizational culture—especially in conservative, mission-driven ISFIs—can reveal barriers and enablers. Participatory design approaches involving frontline staff and beneficiaries may produce more trusted, ethical, and effective AI tools.

Cross-Institutional and Cross-Sector Collaboration Models

Siloed digital initiatives risk inefficiency and lost opportunities. Research should focus on governance frameworks enabling secure, privacy-respecting data sharing among zakat, waqf, and Islamic NGOs, guided by Islamic principles like shura (consultation) and hifz al-sirr (confidentiality). Evaluations of shared technology platforms can demonstrate cost and efficiency benefits, especially for smaller ISFIs. Studies on partnerships with fintech startups can explore models that balance innovation access with preservation of Islamic identity.

AI Bias in Beneficiary Selection and Resource Allocation

Algorithmic bias is a critical concern. Future research should conduct audits of AI systems used by ISFIs to detect demographic exclusions or unfair profiling. Islamic ethical analysis can address dilemmas related to justice (adl) and God-consciousness (taqwa) in automated decisions. Developing validation protocols—including fairness testing and human override mechanisms—can establish standards for ethical AI innovation in Islamic social finance.

Impact of AI-Driven Transparency on Donor Behavior

While AI can enhance transparency, it may also complicate donor engagement. Research should examine how real-time dashboards affect donation size, frequency, and retention, and test AI-based personalization tools for donor communication. Understanding donor trust in algorithmic processes within faith contexts will provide nuanced insights into the balance between efficiency and emotional connection.

Policy Innovation and Regulatory Responses

Regulatory frameworks often lag behind technology. Studies should assess national readiness for AI in Islamic finance, focusing on data regulation, AI certification, and religious oversight, particularly within OIC countries. Regulatory sandboxes for digital waqf or AI-assisted zakat can offer controlled environments for innovation. Research on balancing innovation with Shariah compliance will help resolve tensions between agility and accountability.

Cybersecurity, Data Privacy, and Islamic Digital Ethics

As ISFIs digitize, safeguarding data becomes both a moral and technical imperative. Research should identify common cyber vulnerabilities and develop Islamic perspectives on data privacy, consent, and ethical use, drawing on jurisprudence related to amanah (trust) and ghibah (privacy). Establishing Shariah-compliant cybersecurity standards—including encryption and auditability—will build trust among donors and beneficiaries.

Conclusion

The integration of Robotic Process Automation (RPA) and Artificial Intelligence (AI) into Islamic Social Finance Institutions (ISFIs) represents a transformative opportunity — one that extends far beyond digital efficiency. As this chapter has demonstrated, these technologies, when thoughtfully implemented, can enhance transparency, optimize resource allocation, and align institutional practices with the ethical imperatives of Islam. By operationalizing concepts like maqasid al-shariah through intelligent systems, ISFIs can elevate their mission from reactive charity to proactive empowerment.

Section by section, we have explored how AI and RPA can reshape zakat, waqf, and other philanthropic operations; how performance measurement frameworks such as the Balanced Scorecard (BSC) can be adapted to incorporate spiritual and ethical impact; and how institutions must evolve in structure, talent, and governance to support this shift. Moreover, we have highlighted the potential of these technologies to build donor trust, bridge service gaps between urban and rural communities, and transform ISFIs from compliance-driven entities into strategic actors in the socio-economic development of Muslim societies.

However, we have also emphasized that digital transformation is not a technical exercise alone. Critical challenges—ranging from low digital literacy, ethical risks in algorithmic decision-making, to the lack of clear governance models—require careful navigation. These risks are magnified when technologies are adopted without cultural contextualization, Shariah alignment, or adequate stakeholder consultation. Therefore, future adoption must be accompanied by a parallel investment in human capital, policy development, and cross-sector dialogue.

In charting future research, this chapter proposes ten key directions — from Shariah-informed AI ethics, Islamic KPIs, to long-term impact studies — all aimed at advancing scholarly inquiry and guiding practical implementation. As more ISFIs embark on digital transformation journeys, researchers and practitioners must work collaboratively to ensure that RPA and AI do not merely replicate efficiencies from other sectors but instead serve as ethical instruments for systemic good.

Ultimately, the digital transformation of Islamic social finance must not be judged solely by metrics of speed or automation, but by its capacity to enhance dignity, justice, and wellbeing in line with Islamic values. With the right ethical framework, strategic planning, and scholarly support, ISFIs can become global pioneers in faith-based innovation — proving that digital advancement and divine purpose need not be at odds, but can instead walk hand in hand toward a more equitable future.

AI Assistance Declaration

This chapter was developed with the support of AI tools, including OpenAI's ChatGPT, for the purposes of drafting, organizing content, language refinement, and expanding selected sections. All ideas, interpretations, and final edits were reviewed and validated by the author to ensure academic integrity and alignment with the objectives of this work.

REFERENCES

Abdul-Rahman, A., Mohd Nor, S., & Yaacob, S. E. (2023). Technological integration within zakat institutions: A comprehensive review and prospective research directions. *International Journal of Islamic Thought*, *24*(December), 31–43.

Ahmed, S., & Suleman, M. (2023). Capacity building for digital transformation in Islamic finance: Bridging technical and Shariah knowledge. *Journal of Islamic Finance and Technology*, *5*(1), 45–60.

Avasant. (2023). Digital transformation challenges in Islamic social finance institutions. Avasant Research Report.

Awang, M. Z., Mat Nong, N. F., & Wan Chik, W. M. Y. (2025). Integrating Islamic social finance with the United Nations Sustainable Development Goals through Maqasid al-Shariah principles. *International Journal of Academic Research in Business & Social Sciences*, *15*(7).

Chatterjee, S., Rana, N. P., Tamilmani, K., & Sharma, S. K. (2020). Digital transformation and Islamic finance: Challenges and opportunities. *International Journal of Information Management*, *50*, 45–56. DOI: 10.1016/j.ijinfomgt.2019.10.006

Che Mohd Salleh, M., & Chowdhury, M. A. M. (2020). Technological transformation in Malaysian zakat institutions: A qualitative analysis. *International Journal of Zakat*, *5*(3), 44–56.

Dataro. (2024, May 13). $831k in extra funding secured in one year with AI donor predictions [LinkedIn post]. LinkedIn. https://www.linkedin.com/posts/dataro_dataro-ai-artificialintelligence-activity-7195592444657242113-5fra

Dwivedi, Y. K., Hughes, L., Ismagilova, E., Aarts, G., Coombs, C., Crick, T., & Williams, M. D. (2021). Artificial intelligence (AI): Multidisciplinary perspectives on emerging challenges, opportunities, and agenda for research, practice and policy. *International Journal of Information Management*, *57*, 101994. DOI: 10.1016/j.ijinfomgt.2020.101994

Farahdina, F., Md Ishak, M. M., Mohd Taher, M. S. I., & Ghazali, N. (2025). Enhancing zakat management through digitalization: A literature review on emerging technologies and best practices. [JISTM]. *Journal of Information Technology Management*, *10*(38). Advance online publication. DOI: 10.35631/JISTM.1038006

Google Cloud. (2023). BAZNAS case study: AI-driven business transformation in zakat management. https://cloud.google.com/customers/baznas

Hassan, M. K.. (2023). Development and validation of a performance measurement system with ethical emphasis based on Islamic principles. *Journal of Islamic Finance*, *12*(1), 45–60.

Helpyousponsor. (2025, January 18). AI-powered fundraising: 2025 tools that transform donor management [Blog post]. https://blog.helpyousponsor.com/ai-powered-fundraising-2025-tools-that-transform-donor-management/

Hishamudin, M. Z., Kamarudin, N. S., Abdul Hadi, N., & Ahmad, A. (2024). Intention to use digital platforms for Islamic financial education in Malaysia: Structural equation model. Journal of Advanced Research in Applied Sciences and Engineering Technology. https://www.mywakaf.com.my/about-us/

ICONZ BAZNAS. (2021). Development of zakat management digitalization in Indonesia. https://www.iconzbaznas.com/submission/index.php/proceedings/article/download/227/137

International Journal of Economics, Management and Accounting. (2023). [Article on digital zakat platforms]. International Journal of Economics, Management and Accounting.

INTI International University & Colleges. (2022, March 22). Performance & information management: Have Islamic Social Finance institutions done enough? INTI International University & Colleges. Retrieved July 8, 2025, from https://newinti.edu.my/events/e-symposium-performance-information-management-have-islamic-social-finance-institutions-done-enough/

Iqbal, M. S., Sukamto, F. A. M., Norizan, S. N., Mahmood, S., Fatima, A., & Hashmi, F. (2025). AI in Islamic finance: Global trends, ethical implications, and bibliometric insights. *Review of Islamic Social Finance and Entrepreneurship*, *4*(1), 71–85. DOI: 10.20885/RISFE.vol4.iss1.art6

Islamic Development Bank. (2018). SDGs Digest: Islamic Development Bank. https://www.isdb.org/sites/default/files/media/documents/2020-12/SDGs%20Digest%20Issue%20%235%20%20Sep%202018.pdf

Islamic Economist. (2023). *Digital transformation in zakat collection*. Islamic Economist.

Islamic Relief Worldwide. (2023). Annual report and financial statements 2022. https://islamic-relief.org.my/wp-content/uploads/2023/06/IRW-Annual-Report.pdf

Ismail, A., & Tohirin, A. (2020). Digital crowdfunding and mobile charity applications in Islamic finance. *Journal of Islamic Finance*, *9*(1), 45–60.

Kamaruddin, M. I. H., Hanefah, M. M., Shaharuddin, A., Ayedh, A. M. A., & Othman, N. A. (2023). Development of FinTech in Islāmīc social finance in Malaysia. *International Journal of Economics, Management and Accounting*, *31*(1), 177–204.

Khalid, S. N. A., & Kamaruddin, S. S. (2019). A social performance management framework for Islamic microfinance institutions. *The Journal of Muamalat and Islamic Finance Research*, *16*(1), 53–69.

Laldin, M. A., & Furqani, H. (2013). The foundations of Islamic finance and the maqasid al-Shari'ah requirements. *Journal of Islamic Finance*, *2*(1).

Lestari, E. P. (2023). Navigating the New Normal: Digital Transformation and Islamic Social Finance Strategies. In *Indonesian Conference of Zakat-Proceedings* (pp. 187-196).

Maharani, S. N., & Rahmawati, S. A. (2021). *Measuring Islamic banking performance using Islamic ethics perspective*. Atlantis Press.

Mahmud, S., & Noordin, N. H. (2024). Constructive collaboration in ensuring sustainability of Waqf fund in Malaysia: A case study of myWakaf initiative. [IIUM Press.]. *Journal of Islamic Social Finance*, *2*(2), 1–17.

Meerangani, K. A., Mad Sharipp, M. T., Abdul Hamid, M. F., Hashim, S. N. I., Rameli, M. F. P., & Rosele, M. I. (2022). Digitalisation of zakat management system in Malaysia: Potential and challenges. *Journal of Contemporary Islamic Studies*, *11*(2).

Mittelstadt, B. D. (2019). Principles alone cannot guarantee ethical AI. *Nature Machine Intelligence*, *1*(11), 501–507. DOI: 10.1038/s42256-019-0114-4

Mubarak, D., & Wahyuni, S. (2024). The role of Islamic social finance on promoting inclusive recovery amid crises. Paper presented at the 8th International Conference of Zakat (ICONZ), Bandung, Indonesia.

Muneeza, A., & Nadwi, S. (2019, October 3–4). The potential of application of technology-based innovations for zakat administration in India. Paper presented at the International Conference of Zakat 2019, Universitas Padjadjaran, Bandung, Indonesia.

Mustafa, Z., Dzalihu'zzabir, M. F. H., Abdul Moin, N. A., Shamsuddin, A. E., Mohd Fauzi, N. A., & Sabri, N. H. S. (2022). Islamic social finance and its roles in developing Islamic economy in addressing various issues during the COVID-19 pandemic. *International Journal of Academic Research in Business & Social Sciences*, *12*(10), 1–15.

Nazeri, A. N. N., & Mohd Nor, S. (2023). The implementation of blockchain technology in the zakat management system: A conceptual research. *Journal of Information Systems and Digital Technologies*, *5*(2).

Obaidullah, M. (2015). *Islamic social finance: Nature, scope and global landscape. Islamic Research and Training Institute (IRTI).* Islamic Development Bank.

Organisation for Economic Co-operation and Development. (2021). *Global outlook on financing for sustainable development 2021.* OECD Publishing.

Purpose Forward. (2025). How robotic process automation (RPA) can transform the charitable sector. https://www.purposeforward.org/post/how-robotic-process-automation-rpa-can-transform-the-charitable-sector-1

Puskas, B. A. Z. N. A. S. (2021). The National Board of Zakat Republic of Indonesia reaffirms commitment to managing ZIS for poverty alleviation and national development. https://www.puskasbaznas.com/home/activities/2021-the-national-board-of-zakat-republic-of-indonesia-baznas-ri-reaffirms-commitment-to-managing-zis-for-poverty-alleviation-and-national-development

Qathrunnada, E., Samsuri, A., & Hidayat, A. A. (2023). Measuring Islamic banking performance using Maqashid Shariah Index and RGEC: The Malaysian and Indonesian Islamic banking. *Dialektika*, *11*(2), 283–302.

Raisely. (2025). AI for nonprofits: How AI is transforming fundraising in 2025 [Blog post]. https://www.raisely.com/blog/ai-for-nonprofits/

Razak, D. A., & Amin, H. (2024). Institutionalization of waqf: Case studies from Malaysia and Turkey. *Journal of Islamic Social Finance*, *3*(2), 112–130.

Rosman, R., Redzuan, N. H., Mokhtar, N. Z. N., Engku Ali, E. R. A., & Mohammed, M. O. (2022). Islamic social finance and Sustainable Development Goals: Issues and challenges. *Journal of Islamic Finance*, *11*(2), 56–67.

Salaam Gateway. (2023). How AI is powering the future of the Islamic economy. Salaam Gateway. https://salaamgateway.com/story/how-ai-is-powering-the-future-of-the-islamic-economy

Salleh, W. N. A. W. M., Rasid, S. Z. A., & Basiruddin, R. (2022). Optimising Digital Technology in Managing Zakat. *International Journal of Academic Research in Business & Social Sciences*, *12*(8).

Sremack, J. (2018). *An introduction to robotic process automation for nonprofits*. BDO USA, LLP.

United Nations Development Programme. (2021). UNDP and IsDB strengthen partnership to support Sustainable Development Goals. https://www.undp.org/press-releases/undp-and-isdb-strengthen-partnership-support-sustainable-development-goals

United Nations Population Fund Evaluation Office. (2019). Developmental evaluation of results-based management at UNFPA. https://www.unfpa.org/sites/default/files/admin-resource/RBM_Evaluation_final_report.pdf

Virtuous. (2024, October 30). Virtuous + Dataro: How to make AI work for your nonprofit [Blog post]. https://virtuous.org/blog/virtuous-dataro-integration/

Widiastuti, T., Robani, A., & Kamal, M. (2022). Toward developing a sustainability index for the Islamic Social Finance program: An empirical investigation. *PLoS One*. PMID: 36413519

World Bank. (2020). Data governance and interoperability in social finance: Challenges and frameworks. World Bank Group Report.

Zulkifley, M., & Muneeza, A. (2024). Optimizing zakat distribution in Selangor using technology. *Journal of Advanced Research in Applied Sciences and Engineering Technology*, *7*(2).

KEY TERMS AND DEFINITIONS

Artificial Intelligence (AI): The simulation of human intelligence by machines, enabling capabilities such as learning, reasoning, and decision-making. AI is increasingly adopted in Islamic social finance to support data-driven insights, risk analysis, beneficiary targeting, and ethical investment decisions aligned with Islamic principles.

Balanced Scorecard (BSC): A strategic performance management tool that translates an organization's vision and strategy into a coherent set of performance measures across four key perspectives: financial, customer, internal processes, and learning and growth. In Islamic social finance, the BSC can be adapted to integrate Maqasid al-Shariah, enabling both financial and spiritual accountability.

Islamic Social Finance (ISF): A faith-based financial system that encompasses charitable and ethical financial instruments such as zakat (almsgiving), waqf (endowment), sadaqah (voluntary charity), and qard hasan (benevolent loan). ISF aims to address socio-economic imbalances by redistributing wealth and supporting the welfare of the ummah in accordance with Shariah principles.

Maqasid al-Shariah: The objectives of Islamic law, aimed at preserving and promoting five essential values: religion (din), life (nafs), intellect ('aql), progeny (nasl), and wealth (mal). In the context of Islamic social finance, Maqasid al-Shariah serves as a foundational framework to ensure that financial activities lead to social justice, ethical conduct, and communal well-being.

Robotic Process Automation (RPA): A technology that utilizes software robots (bots) to automate routine and rule-based business processes. In Islamic social finance institutions, RPA enhances operational efficiency, improves compliance, and allows human resources to focus on strategic and community-focused functions.

Sustainable Development Goals (SDGs): A universal set of 17 goals established by the United Nations to end poverty, protect the planet, and promote peace and prosperity by 2030. In Islamic finance, SDGs are increasingly aligned with the values of Maqasid al-Shariah, particularly in promoting social justice, responsible resource management, and inclusive economic development.

Chapter 9
Robotic Process Automation in Finance and Accounting:
A Comprehensive Bibliometric Analysis

Mushtaq Ahmad Shah
https://orcid.org/0000-0002-3177-9622
Lovely Professional University, India

ABSTRACT

This chapter presents a comprehensive bibliometric analysis of Robotic Process Automation (RPA) within the finance and accounting sectors, drawing upon a robust body of academic literature from 2010 to 2025 from Scopus database. With the use of RStudio and a specialized bibliometric application called Biblioshiny, data was analyzed and visualized. The review shows a field experiencing a rapid growth with a sharp increase in publications since 2017 culminating in 2023. RPA implementation although has a lot of merits still has a lot of issues notably lack of organizational insight and human centered issues such as job security and data privacy. The next steps of RPA should be intelligent, integrated, and sustainable automation, and it requires closer interactions with such technologies as Natural Language Processing (NLP), Deep Learning, and Blockchain. The findings of this research are useful to financial institutions and accounting firms strategically, as well as researchers, practitioners and policymakers.

DOI: 10.4018/979-8-3373-3571-1.ch009

1. INTRODUCTION

The global landscape of IT robotic automation is experiencing robust expansion. Market projections indicate that the industry will grow from US$ 5.92 billion in 2025 to US$ 17.25 billion by 2032, reflecting a compound annual growth rate (CAGR) of 16.5% during this forecast period. This significant growth is primarily propelled by the escalating adoption of automation technologies across diverse industries, driven by the imperative to enhance operational efficiency, reduce costs, and improve overall productivity (Ghobakhloo et al., 2023; Kedziora et al., 2021; Perdana et al., 2023; Pramod, 2022). Notably, the banking, financial services, and insurance (BFSI) sector stands as a pivotal force in this market (Shah, 2025), commanding a substantial 37.2% of the market share in 2025, closely followed by the healthcare and utilities sectors. Geographically, North America currently leads the global market, holding 40% of the share in 2025, a dominance attributed to considerable investments in Robotic Process Automation (RPA) by major technology corporations and widespread enterprise-level adoption. Concurrently, the Asia-Pacific region is poised for the fastest growth, stimulated by comprehensive digital transformation initiatives in key economies such as China, India, and Japan, where automation is increasingly leveraged to mitigate labor shortages and optimize Business Process Outsourcing (BPO) operations.

This market expansion is further accelerated by continuous technological advancements, particularly the seamless integration of artificial intelligence (AI) and machine learning (ML) capabilities with RPA(Bhoyar, Bhoyar, & Shah, 2023). AI-powered chatbots and cognitive automation tools, for instance, are enabling enterprises to process unstructured data and navigate complex decision-making scenarios, thereby extending the scope and sophistication of automation. Moreover, the discernible shift towards cloud-based RPA solutions, which constituted 62.9% of deployments in 2025, is significantly enhancing scalability and reducing infrastructure overheads for both small and medium-sized enterprises (SMEs) and large corporations alike. Despite these promising opportunities, the broader adoption of RPA is not without its impediments. Challenges such as high implementation costs, the complexities of infrastructure customization, and persistent skill gaps may impede wider penetration, especially among smaller organizations. Nevertheless, the emergence of hyper automation a synergistic approach combining RPA with AI, the Internet of Things (IoT), and process mining is anticipated to unlock novel growth avenues, with Gartner reporting that hyperautomation has already optimized 20% of business processes globally. In essence, the IT robotic automation market is on a strong upward trajectory, fundamentally driven by digital transformation imperatives, advanced AI integration, and dynamic regional economic shifts, firmly positioning it as a cornerstone for future enterprise efficiency and competitiveness.

Against this backdrop of rapid market evolution, this chapter provides a comprehensive bibliometric analysis of Robotic Process Automation (RPA) within the dynamic landscape of finance and accounting. RPA, as a sophisticated software application designed to automate repetitive Graphical User Interface (GUI) tasks, has emerged as a critical phenomenon within Financial Technologies (FinTech), driven by the pervasive digital transformation across sectors. Its relevance to the financial services sector is paramount due to the industry's high transaction volumes, repetitive tasks, and stringent demands for accuracy and compliance.

This chapter systematically maps the academic discourse surrounding RPA in these domains, leveraging rigorous methodologies to identify key trends, influential contributors, evolving research themes, and existing gaps. India has emerged as a prominent contributor, challenging traditional research hubs and signifying a global decentralization of RPA research leadership. Despite its advantages, RPA implementation faces considerable challenges, including a lack of organizational understanding, methodological disconnects, and crucial human-centric concerns regarding job security and data privacy. The future trajectory of RPA points towards intelligent, integrated, and sustainable automation, necessitating deeper integration with advanced technologies like Artificial Intelligence, Natural Language Processing, Deep Learning, and Blockchain to move beyond basic rule-based automation towards more cognitive capabilities. However, significant research gaps remain, particularly in specific functional areas within banking (e.g., sales, ERP), in the technical integration of advanced AI capabilities due to "technical capability gaps," and in exploring the long-term sustainability of RPA through deep learning. This chapter will delve deeper into these aspects, presenting the evolution of RPA research, identifying key stakeholders and their contributions, analyzing the thematic landscape and specific applications in finance and accounting, discussing common implementation challenges, and proposing solutions, future trends, and critical research directions to guide practitioners, researchers, and policymakers in navigating this evolving domain.

2. REVIEW OF LITERATURE

The burgeoning literature on Robotic Process Automation (RPA) within finance and accounting consistently highlights its potential for significant operational improvements however, barriers to widespread and effective adoption. Studies by Perdana et al. (2023) and Eulerich et al. (2022), demonstrate RPA's capacity to enhance efficiency and accuracy in audit tasks and internal accounting processes through structured frameworks and practical scenarios. This is corroborated by Pramod (2022), who notes RPA's extensive use in banking for operational efficiency and productivity. However, a critical challenge lies in a pervasive undervaluing and inadequate leveraging of technologies like RPA, AI, and Blockchain by the accounting profession, as identified by Jackson and Allen (2024). Their research, applying the Technology-Organisation-Environment (TOE) model, indicates that while technological factors are important, paramount concerns around security and privacy significantly influence decision-making, often overshadowing environmental considerations. Similarly, Zhang et al. (2023) identify key themes in RPA adoption including workforce, IT governance, privacy, security, and system sustainability, confirming that human and organizational factors are as critical as technical capabilities. Furthermore, Ghobakhloo et al. (2023) emphasize that Intelligent Automation (IA) implementation, an advanced form of RPA, is primarily determined by internal firm characteristics such as absorptive capacity and employee socio-behavioral concerns, reinforcing the notion that successful adoption transcends mere technological deployment.

The literature reveals a discernible evolution in the conceptualization and application of RPA, moving beyond simple task automation towards more intelligent and integrated solutions within the broader digital transformation agenda. Kommunuri (2022) and Chyzhevska et al. (2021) broadly discuss the increasing influence of AI, Machine Learning (ML), Big Data, and Blockchain in accounting, emphasizing the growing need for their applications. Polak et al. (2020) further elaborate on AI's role in finance/treasury, highlighting its complexity beyond task automation to encompass data management, classification, and integration for financial crisis early warning and risk diagnosis. Gotthardt et al. (2020) explicitly discuss the "greenfield project" of augmenting RPA with AI, noting the lack of sufficient theoretical frameworks to guide such deployments. This shift towards intelligent automation underscores a strategic imperative for organizations to integrate RPA with these advanced technologies, enhancing capabilities for complex, cognitive tasks and leveraging big data for more informed decision-making. However, as Kedziora et al. (2021) point out, while operational benefits like cost savings and efficiency are key drivers for RPA decisions, strategic and external impacts, including on customers and service quality, are often not at the core of decision-making, suggesting a gap between perceived operational gains and broader strategic integration.

Despite the rapid advancements and recognized potential, several critical gaps and future research directions emerge from this body of literature. Plattfaut and Borghoff (2022) notably highlight that the literature has particularly "failed to discuss the quantifiable benefits of RPA," calling for more rigorous empirical studies in this area. Perdana et al. (2023) and Zhang et al. (2023) provide practical insights and case studies, yet the broader implementation challenges and opportunities, especially concerning the socio-technical aspects, remain a focal point for continued investigation. Cooper et al. (2022) reveal a discrepancy in perceptions, where firm leaders believe RPA improves work satisfaction, but lower-level employees do not report similar improvements, indicating a need for deeper understanding of employee experience and proactive strategies to manage workforce impact. Furthermore, while the general embrace of AI is evident (Bakarich & O'Brien, 2021), the extent of its actual extensive utilization in public accounting remains limited, with substantial changes perceived "on the horizon," rather than as a current reality. This collectively calls for future research to not only focus on developing advanced RPA capabilities but also on robust methodologies for measuring their quantifiable benefits, addressing the intricate human and organizational challenges, and developing comprehensive frameworks that guide ethical and sustainable integration of RPA with other emerging technologies in the evolving financial and accounting landscape

A significant gap exists in exploring how RPA can be used to automate a wider range of tasks beyond simple data entry, such as complex reconciliations across multiple modules, automated reporting, and proactive alerts for system inconsistencies. Research could investigate the unique challenges of integrating RPA with these systems, particularly regarding data security, system integrity, and the technical governance required to manage automated processes that touch core business data.

Similarly, the application of RPA in sales processes presents a largely unexplored area. While the literature alludes to its use for efficiency, there's a gap in understanding how RPA can automate and optimize the entire sales cycle, from lead qualification and quote generation to order processing and customer relationship management (CRM) updates. Future research should examine the quantifiable benefits of RPA in sales beyond simple efficiency gains—such as its impact on sales cycle time, order accuracy, and overall customer satisfaction. This would provide a more complete picture of RPA's strategic value and move beyond its current perception as a purely back-office tool.

3. MATERIAL AND METHODS

The bibliometric analysis presented in this chapter was conducted using rigorous methodologies, leveraging the capabilities of RStudio for data processing and visualization. Data for this analysis was primarily drawn from Scopus databases, ensuring a comprehensive collection of relevant scholarly literature. Specialized bibliometric software tools, including Biblioshiny was utilized within the RStudio environment to ensure methodological rigor. These tools facilitated the systematic processing of the collected data, enabling the identification of influential sources, emerging trends, and intricate relationships within the Robotic Process Automation (RPA) research landscape in finance and accounting.

4. DATA ANALYSIS AND DISCUSSION

The analysis of Robotic Process Automation (RPA) in finance and accounting reveals a dynamic and rapidly evolving research landscape, characterized by significant growth, shifting thematic emphases, and persistent implementation challenges. The "Annual scientific production" (Figure 1) provides compelling evidence of this expansion, demonstrating a dramatic surge in publications from a nascent state prior to 2017 to over 100 articles in 2023. This exponential growth underscores a burgeoning academic interest and a strong acknowledgment of RPA's strategic importance within the financial and accounting sectors, mirroring its real-world adoption in industry. While initial research may have been foundational, this sustained increase in scholarly output indicates a maturing field actively exploring diverse facets of RPA.

Figure 1. Annual scientific production

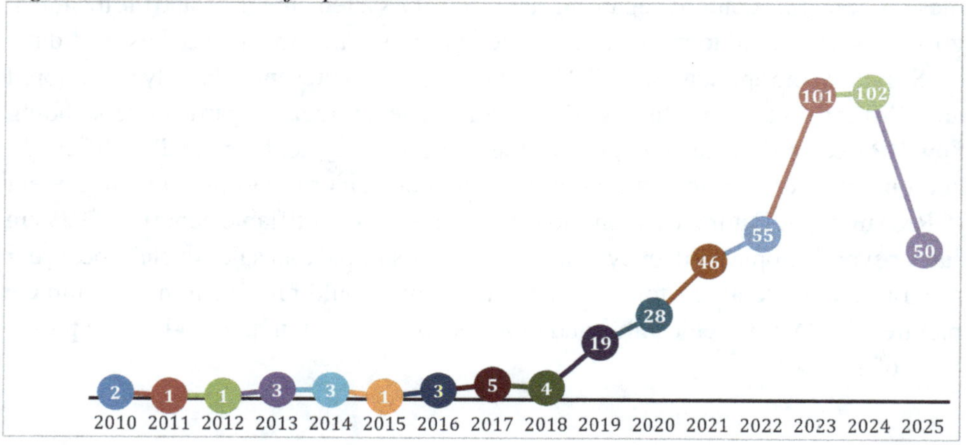

However, the "Mean citations per year " (Figure 2) presents a more nuanced picture of research impact. Despite the sharp increase in publications, the average citations per paper peaked around 2019 and has shown a general decline thereafter. This trend is likely attributable to a combination of factors.

Figure 2. Mean citations per year

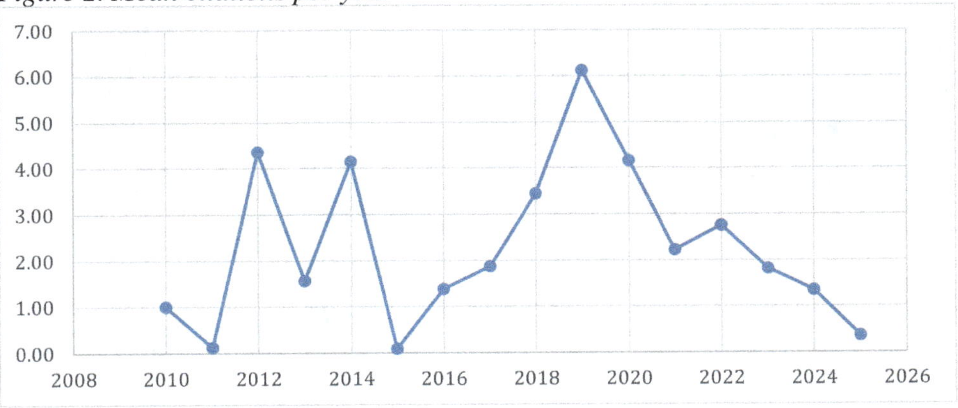

Firstly, the inherent citation lag in academic publishing means that the large volume of recent publications has had less time to accrue citations, thus statistically lowering the mean. Secondly, a "dilution effect" may be at play; as the number of publications increases rapidly, the average citation count per paper can decrease if not all new articles achieve high citation visibility. This suggests that while the quantity of research is growing, the field might be diversifying into more niche or specialized areas that initially generate fewer broad citations, or it could indicate a shift from highly cited foundational works to more incremental or application-specific studies.

The thematic evolution of RPA research, as elucidated by the "Tree Map" (Figure 3) and "Trend Topics" (Figure 4), illustrates a clear progression. Initially, the focus was on core "robotic process automation," "process control," and "robotics." However, the "Trend Topics" vividly show a shift towards broader application and strategic integration, with terms like "financial service," "Information systems," and "enterprise resource management" gaining prominence around 2020-2022

Figure 3. Tree Map Thematic evolution of RPA

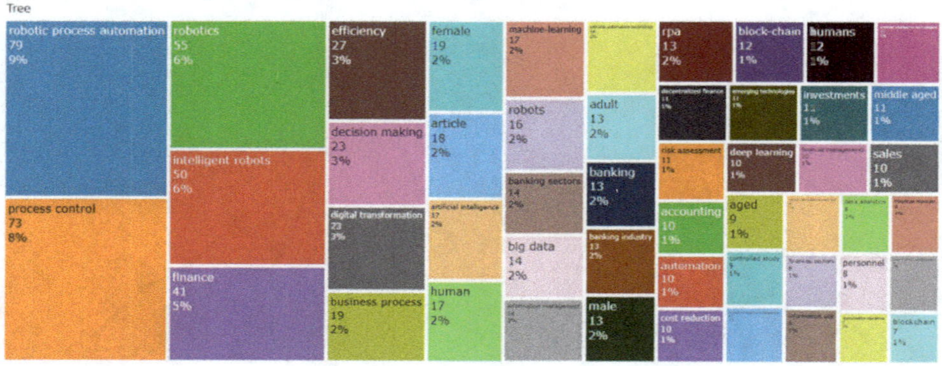

More recently, the discourse has pivoted towards integrating RPA with advanced technologies, as evidenced by the prominence of "machine-learning," "artificial intelligence," "big data," "deep learning," and "block-chain" (Figure 3, Figure 4). This signifies a move beyond basic rule-based automation to more intelligent, data-intensive, and cognitive capabilities, positioning RPA as an indispensable component of comprehensive digital transformation strategies within finance and accounting.

Figure 4. Trend Topics

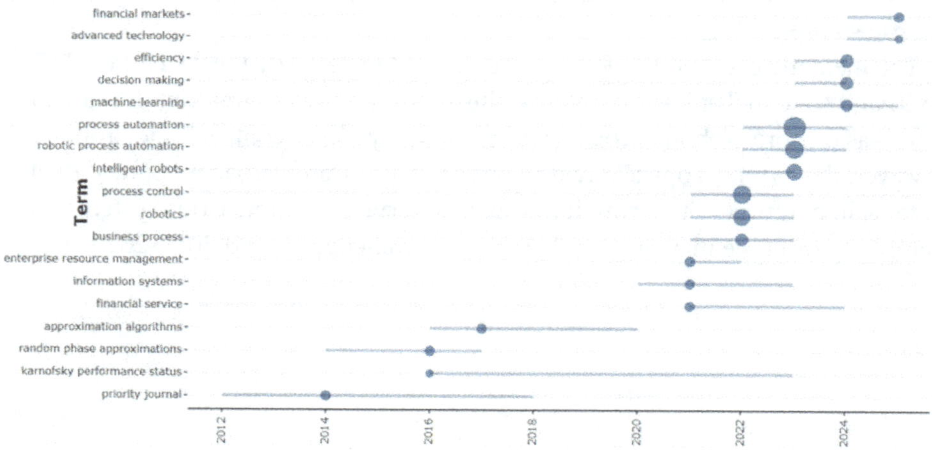

The "Co-occurrence Network" (Figure 5) provides a robust visualization of the interconnectedness of these themes, revealing three distinct clusters. The largest, blue cluster, is centered around "process control," "process automation," and their application in "banking sectors," "accounting financial sectors," and related areas like “risk assessment” and “cost reduction.” This cluster unequivocally highlights the core practical applications and strategic drivers of RPA in financial services. The red cluster signifies the field's embrace of intelligent automation, grouping “intelligent robots,” “artificial intelligence,” “automation,” “data analytics,” “big data,” and “financial markets.” This cluster indicates a strong and growing research front focused on leveraging advanced technologies to enhance RPA capabilities for more complex decision-making.

Figure 5. Co -occurrence network on RPA

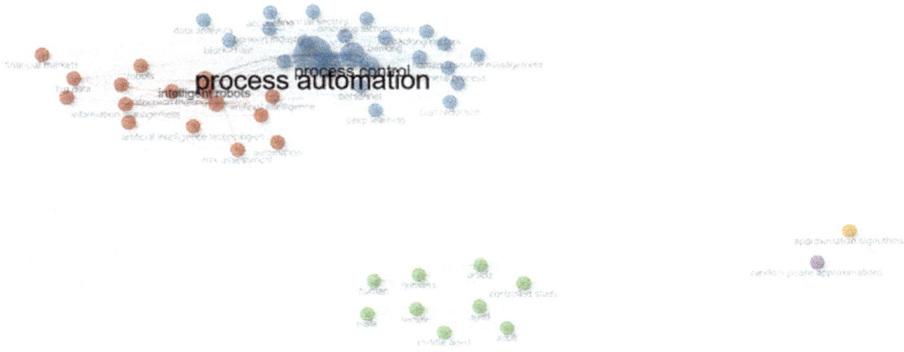

Critically, the distinct green cluster, comprising terms like “human,” “female,” and “male,” points to research explicitly addressing the human dimension of RPA. Its relative isolation from the primary technical and application clusters suggests that while human factors are researched, they may sometimes be treated as a separate line of inquiry rather than being fully integrated into core technical or business process research. This aligns with the “Thematic Map” (Figure 6), which classifies human-centric research as a “niche theme,” indicating a well-developed but specialized area, while core RPA concepts are “motor themes” driving the field.

Figure 6. Thematic Map

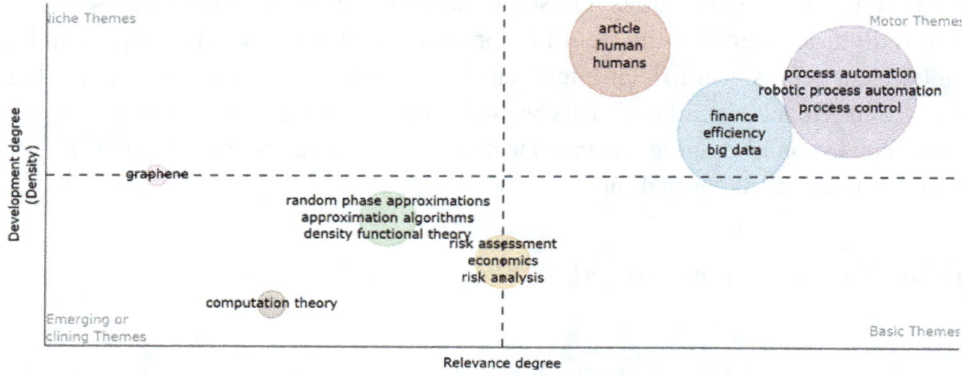

The "Historiograph" (Figure 7) further illuminates the intellectual evolution of the field, showcasing influential authors and their interconnections over time. Authors like "cooper la, 2019" and "kokina j, 2021" emerge as significant contributors, whose works have formed critical nodes in the citation network, shaping subsequent research directions. This visual representation of intellectual lineage highlights the cumulative nature of academic inquiry, where current advancements are built upon foundational works and the collaborative efforts of a growing community of scholars.

Figure 7. Historiograph

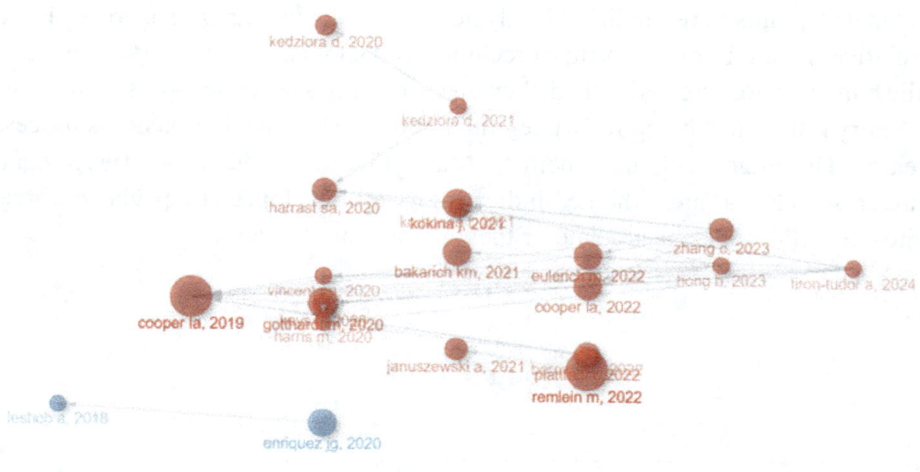

Despite the evident progress, RPA implementation faces significant challenges. A major impediment is a widespread "lack of organizational understanding" regarding

RPA's capabilities, methodological requirements, and true business impact. This often leads to a "methodological disconnect" between human and RPA policies, contributing to implementation failures and data loss. Crucially, human-centric concerns, particularly "job security" for banking employees and customer anxieties about "data privacy" and the lack of human interaction, pose substantial hurdles to successful adoption and scaling. These challenges underscore that RPA success is not solely a technological achievement but a complex socio-technical endeavor requiring robust change management, employee engagement, and clear communication strategies.

The success of any technological implementation hinges on a seamless transition for the people it affects. When it comes to RPA in the banking sector, the apprehension among employees about their roles being made redundant is a significant barrier. This fear can lead to resistance, non-compliance, and even sabotage, ultimately undermining the entire automation effort. Simultaneously, customers are wary of entrusting their sensitive financial data to automated systems and losing the personalized touch of human banking professionals. This highlights a critical need to reframe the narrative around RPA—not as a replacement for human employees but as a tool that augments their capabilities, freeing them from repetitive tasks to focus on more complex, value-added activities like customer relationship management and strategic analysis.

Looking forward, the research points towards a future where RPA transcends basic rule-based automation. The emphasis will be on "intelligent, integrated, and sustainable automation," demanding deeper convergence with technologies like Natural Language Processing (NLP), Deep Learning, and Blockchain to handle unstructured data, perform cognitive tasks, and enhance trust and transparency. However, significant research gaps persist. Functional areas within banking, such as sales and ERP, remain underexplored. Furthermore, there are persistent "technical capability gaps" in effectively integrating Intelligent Process Automation (IPA) with advanced AI techniques like NLP and cognitive automation. A scarcity of research on the long-term sustainability and ethical implications of RPA, particularly with deep learning, also presents a critical void. Addressing these gaps will require interdisciplinary research, moving beyond isolated studies to comprehensive frameworks that integrate technical solutions with human and organizational considerations, ultimately fostering more impactful and responsible RPA deployments

4.1 Bibliometric analysis insights

The bibliometric search that was performed in this work offers an effective empirical prism in which the intellectual framework, development, and thematic focus of the Robotic Process Automation research in the field of finance and accounting could be viewed. The fact that publications have been growing since 2017 is indeed a sign of academic excitement but also a revolution within the industry with automation becoming not an experimental operation but a pillar of strategic competitiveness. The dramatic increase of scientific production, as shown in Figure 1, reflects the pace of fostering the digital transformation initiatives in the global financial services industry over the same time. Since an initial small infiltration into early scholarly databases, the trend indicates a growing field with more than a hundred publications a year, which is becoming more interdisciplinary and is diversifying in methods. This tendency indicates the dispersion of RPA as a body of research, cutting across computer science, information systems, and management accounting, as well as behavioral studies, which makes its socio-technical complexity.

While the quantity of publications demonstrates momentum, the "Mean Citations per Year" (Figure 2) presents a subtler pattern: the citation intensity peaked around 2019 before gradually declining. This phenomenon suggests that foundational conceptual works often those introducing frameworks, typologies, or definitions of RPA have already attained their citation zenith, whereas newer empirical studies, though abundant, have not yet accumulated sufficient academic resonance due to citation lag. Another plausible interpretation is the field's thematic fragmentation. As scholars increasingly specialize in narrower subdomains such as intelligent automation, process mining, and RPA governance the dispersion of attention across multiple micro-fields dilutes overall citation density. Thus, the bibliometric evidence reveals a dynamic yet decentralized intellectual landscape that reflects the real-world complexity of RPA implementation many actors, technologies, and disciplines converge, but with varying degrees of integration and maturity.

4.2 Thematic Shifts and Research Evolution

The "Tree Map" and "Trend Topics" analyses (Figures 3 and 4) further illuminate this intellectual transition. In the early phase, dominant keywords "automation," "workflow," "robotic process," and "efficiency" centered around operational improvement and cost reduction. These themes correspond to the first generation of RPA adoption in banking and accounting, which largely focused on automating repetitive, rule-based tasks such as data entry, invoice processing, and reconciliation. However, the trend shifts markedly post-2020, with the emergence of terms like "machine learning," "artificial intelligence," "blockchain," "digital transformation," and "intelligent automation." This evolution reflects a conceptual leap: RPA is increasingly conceived not merely as a mechanistic efficiency tool but as a component of a broader cognitive and strategic digital infrastructure. The confluence of RPA with AI, ML, and natural language processing (NLP) extends its functional scope from deterministic automation to adaptive decision-making, allowing organizations to handle semi-structured and unstructured data.

The "Co-occurrence Network" (Figure 5) substantiates these thematic shifts by mapping interrelations among keywords. Three major clusters emerge: the largest, focused on "process control," "cost reduction," and "banking sector," represents the field's pragmatic application core. The second, more recent cluster connects "machine learning," "artificial intelligence," and "intelligent systems," denoting a shift toward high-order automation integrating cognitive capabilities. The third cluster, comparatively smaller yet symbolically significant, centers on human-centric themes "employee," "gender," "acceptance," and "behavior." The peripheral position of this cluster suggests that while human factors are acknowledged, they remain somewhat isolated from mainstream research on process efficiency and technological integration.

The "Historiograph" (Figure 7) charts the intellectual lineage of RPA research, linking early conceptual pioneers such as Kokina (2021) and Cooper (2019) with contemporary scholars expanding the field's boundaries into areas such as intelligent auditing, ethical automation, and sustainable digitalization. The networked citation structure underscores a steady migration from descriptive to analytical research, from exploring "what RPA is" to examining "how RPA transforms organizations." This evolution parallels RPA's industry lifecycle from exploratory pilots to enterprise-scale deployments and signals a maturing field that is beginning to grapple with questions of integration, governance, and sustainability.

When these bibliometric insights are combined with industry trends, a very sharp symmetry between trends in research and practical reality is observed. The academic discussion and also the patterns of adoption within corporations focus on optimizing the processes, but at the same time, both neglect human and organizational preparedness. This alignment of priorities and omissions points to the fact that academia and industry are developing in parallel, somewhat convergent ways, each of which strengthens the blind spots of the other one. Academics are interested in algorithms, scalability, and efficiency whereas practitioners are interested in cost and speed of implementation, often neglecting socio-organizational complexities. As a result, the implementation failure, inconsistencies in data and employee resistance are still observed even with the proven potential of the technology. In a more general analytical perspective, the predicaments of RPA adoption can be theorized as a network of mutually reinforcing variables such as organizational, technical, human, regulatory and ethical which all determine the lines of successfulness or failure.

In the organizational setting, the lack of strategic coherence is the most common challenge to Robotic Process Automation (RPA). Many banks and accounting companies are implementing RPA projects on an ad hoc basis, without considering the integration of these projects into an overall enterprise digital transformation strategy. This kind of fragmentation is bound to create redundant work, inadequate governance, and bot sprawl-like behavior, where additional automation initiatives are being started with no central control. According to empirical data provided by Deloitte (2023) almost half of the RPA projects do not go past the pilot stage most of the time simply because of the absence of an automation vision and a sound governance architecture.

These organizational shortcomings are worsened by technical barriers. Although RPA platforms are quite effective in a steady, rule-based setting, they suffer extreme drawbacks when operated in a dynamic, data-intensive system. The weakness of bots, which is vulnerable to interface changes, latency jitter, or changes in data fields, is transformed into considerable maintenance costs, which may include up to thirty-five percent of the overall RPA lifecycle costs (Soni (2024). With core systems that are frequently updated in banking environments and that regulatory changes drive the need to recalibrate core systems, these vulnerabilities cause frequent process failures. In addition, as organizations look to integrate RPA with advanced analytics and artificial intelligence, they face the issues that surround data governance, model interpretability, and Application Programming Interface (API) interoperability.

4.3 Human and Culture RPA Implementation.

The social aspect of human beings is equally challenging. Often, employees tend to view RPA as a tool of redundancy as opposed to empowerment, especially in clerical and administrative sectors. According to surveys of the Indian and European banking industry, sixty to seventy percent of workers first identify automation with job displacement, a notion that creates resistance in the process of discovery and testing of processes. This concern is not baseless, automation when applied without other reskilling and redeployment initiatives may genuinely eliminate human labor.

However, the successful organizations show that the displacement of the workforce does not always happen but depends on the will of the managers and their implementation strategies. As an example, ING bank paired its automation implementation with a massive Human in the Loop reskilling program, retraining employees to handle exceptions, do data analysis and handle customer relationship management, tasks which require empathy and cognitive judgment that is out of scope to automation at present. This strategic reframing shifted the employee moods towards fear to collaboration thereby highlighting the importance of change management as a predeterminer of the RPA success.

4.4 Governance, Compliance and Moral Integration.

Another significant threat in the BFSI sector is regulatory and compliance complexities since automation is subject to strict oversight regimes, including Basel III, the Sarbanes-Oxley Act, GDPR and national data-protection laws. Although RPA has the potential to improve compliance by reducing manual error and creating a consistency, it also presents new risks in the form of auditability and explainability. Regulators are growing more demanding in terms of transparency on automated operations which means that institutions must prove that decisions made by bots are transparent, traceable and just. The inability to offer such transparency can put firms in compliance fines and reputational harm. As such, ethical automation comes out as a moral and operative necessity. The adoption of the structures of responsible automation, with a focus on fairness, accountability, and transparency, is becoming a key source of strategic differentiation. All the algorithmic systems should also have human checkpoints to high-risk operations, as the European Banking Authority (2024) points out that automation should be used to enhance human control, but not to substitute it.

4.5 Sustainable RPA Integration Critical Success Factors.

This chapter determines a strong group of Critical Success Factors (CSFs) that in combination can calculate RPA maturity. The alignment of strategies is a prerequisite; automation projects should be directed to the purpose of supporting organizational goals and not as a cost-cutting initiative with an independent existence. Leadership sponsorship guarantees distribution of the necessary resources and implementation of governance discipline. Another pillar of success is the creation of a Center of Excellence (CoE). CoEs formalize the best practices, implement security and compliance requirements, and promote cooperation between IT, business, and compliance teams. The standardization of processes is supported by tools like process mining to decrease variability and increase the reliability of automation.

Long-term resilience is also based on technology architecture and integration capabilities. Companies that are implementing API-first, modular architectures show an improvement in scalability and downtime. The human factor is also chief. Organizations investing in reskilling of workforce (training employees to work in an analytical and supervisory position) achieve greater levels of acceptance and reduced turnover. The case examples of JPMorgan COIN platform and ICICI Bank automation partner indicate that automation and workforce empowerment make their combination a synergistic benefit in terms of productivity and morale. Mechanisms of continuous improvement maintain the effect of RPA; the accountability and flexibility of the operation of bots are monitored with regular audits of their performance and the outcomes of their cost-benefit analysis. Incorporating compliance and ethical codes directly into the development procedures of the bot, which is known as ethical-by-design automation, is another way to improve institutional integrity.

4.6 Academic Insight Convergence vs. Industry Practice.

These critical success factors are in line with the bibliometric patterns that are established in this research study. The trends of process-centric to intelligence-driven themes are in line with the actual development of RPA in industry. Earlier works were focused on cost and efficiency, whereas the most recent works combine human factors, ethics, and sustainability. This trend is an indication of a narrowing of the gap between theoretical and managerial practice into a holistic, socio-technical view of automation. The study of RPA is widely spread geographically, which supports this trend. The increasing presence of India in academic landscape of RPA is consistent with its industrial dominance in IT enabled services, which makes the academic, technology and financial institutions mutually dependent in terms of exchange of knowledge. Such a co-evolution of research and industry empowers the both the theory and the practice, allowing the emerging economies to be actively involved in the development of global automation standards. In the future, it is possible to predict that the development of RPA will be more enhanced with cognitive intelligence, Explainable AI (XAI), and sustainable automation. The systems in the future will be able to use contextual reasoning, natural-language understanding and autonomous exception management thus the line between robotic performing and cognitive analysis will be blurred. But these innovations also increase the risks in the areas of bias, transparency, and compliance. The application of XAI principles to the governance models of RPA will therefore be an indispensable policy to the financial institutions.

Since the digital transformation is at the cross of ESG goals, RPA should play a role in sustainable work by being energy-efficient in computing, establishing more efficient resource scheduling, and ensuring workforce transition that is socially inclusive. The green automation and ethical AI constructs are moving toward fringe debates to performance metrics used in determining corporate reputation and investor trust. Basically, the next stage of RPA will be intelligent, explainable, and sustainable automation, the continuum of technological strength and human accountability that balances efficiency with ethics.

As the finer analysis reveals, the disruptive capability of RPA is premised on the thin layer of technology and social organization maturity. The bibliometric analysis provides a fast-growing but fragmented discipline, which reflects the complexities of automation adoption in the real world. Multidisciplinary teamwork, moral thinking, and organizational learning are some of the factors that can bridge these gaps.

Fragmentation, the technical fragility, human resistance and regulatory scrutiny are major challenges that can be overcome by strategic alignment, effective governance, effective architecture and human empowerment. The future of RPA in finance and accounting is not this increase in the use of bots but the development of a responsible automation ecosystem that will enhance the power of governance, protect human dignity, and promote the organizational resilience of the digital age.

4.7 RPA, Sustainability, and Inclusion

The conversation of Robotic Process Automation (RPA) is also broadening its scope of discussion on the topics of operational and financial performance to other wider concepts of sustainability and inclusion. With the shift of organizations toward digitally enhanced business models, the need to have responsible, fair, and sustainable automation has taken over the center stage of long-term competitiveness and legitimacy. This change is more pronounced within the finance and accounting sectors where RPA has had significant penetration. The bibliometric analysis performed in this paper indicates that there is an emergent but expanding interest in the concepts of sustainability, ESG, responsible automation, and human-centric innovation that may indicate that both researchers and practitioners are starting to look at automation not in terms of an isolated efficiency project but as an essential contributor to sustainable digital transformation.

It is possible to conceptualize sustainability in RPA in three complementary dimensions, including environmental, social (inclusion and equity), and governance, all of which are sources of resilience and ethical soundness of automation ecosystems. These aspects conform to the concept of the triple bottom line (Elkington, 1997) where economic worth should be as well as social justice and environmental custodianship. The overlap of RPA and Environmental, Social and Governance (ESG) principles marks the paradigmatic shift: it is not only about doing things quicker and cheaper, but also about doing them in a responsible manner, in a way that does not deplete the resources, empower individuals, and enhance institutional accountability.

4.8 Environmental Sustainability of RPA.

Environmentally, RPA sustainability is based on the ability to minimize waste, optimize available resources and minimise carbon emissions relating to repetitive manual processes. Already, digitization based on RPA has shown quantifiable environmental positive impacts in financial institutions. An automated process of documents ensures the reduction of dependence on paper, whereas digital processes do not depend on the energy expenses of physical storage, printing, and transportation. In a McKinsey report on financial digitization published in 2024, banks that used

RPA in back-office operations realized a 35 percent paper savings and 25 percent saving in office power use on average. These cost reductions can be translated into reduced operational carbon footprint, making RPA a provider of green finance and corporate sustainability reporting.

Nevertheless, the issue of sustainability is also based on the management of the technology itself. The growing use of cloud-based RPAs is an issue regarding the digital carbon footprint, as data facilities use a lot of energy. According to the calculations given by the International Energy Agency (IEA, 2023), it is estimated that data centers globally already consume close to 2% of the total electricity demand, and the amount is likely to increase to 2030. Automation workloads are the contributors towards this surge especially where bots have been used continuously or inefficiently. To counter this, organizations are undertaking the so-called green RPA capabilities, which include dynamically scheduled operations, energy-aware scheduling, and server optimization, so the bots are activated during less busy time or consolidated so that idle power is minimized. Even the vendors of technology have started to incorporate the concept of sustainability into their systems. The eco-mode Advanced Scheduling feature of the UiPath and Carbon Tracker plug-in of Blue Prism enable companies to track the energy consumption of their bots and optimize the process towards a net-zero goal. These inventions show how the sustainability of the environment can be implemented, through the smart design of systems. However, as more studies are required to stipulate standardized indicators like the Automation Energy Efficiency Ratio (AEER) to determine the actual net ecological impact of RPA and enable its inclusion in the ESG performance models.

4.9 Digital Inclusion and Social Sustainability.

In addition to the environmental issues, inclusion is the core of social sustainability in RPA. Automation has long been viewed as the job-displacing Weakness but there is growing evidence that when used responsibly, RPA can be an inclusion force, an empowerment force and a digital equity force. This potential in the finance and accounting industries comes to bear in three principal areas namely (1) workforce transformation and upskilling, (2) inclusive access to financial services, and (3) the democratization of digital participation.

To begin with, RPA can facilitate the inclusion of workers by redirecting labor to decision-oriented and creative tasks as well as analytical and low-value tasks. Automation will lead to human capital growth instead of displacement when staff become upskilled with digital training and reskilled into managerial or strategic positions. Examples of programs that work with inclusive workforce transitions include the Human in the Loop Initiative of ING, the Digital Citizens Program of Barclays and the Automation Academy of ICICI Bank. In these programs, employees who were trained to occupy operational positions are retrained to become RPA developers, data analysts, and process controllers-repositioning the source of job loss to skill acquisition. This kind of inclusionary is not only safeguarding employment, but also developing in-house capacity, which makes digital transformation benefits fairly shared among organizational tiers.

Second, financial inclusion can be realized through RPA so that banks and other financial service providers are able to reach out to the underserved populations. When using automated processes the cost of onboarding and servicing of small or rural clients is drastically reduced and the institutions can offer low cost financial products effectively. State Bank of India (SBI) as an example used RPA to automate the process of verifying customers and opening branches in rural locations, which halved the processing time and increased the service opportunities to groups of the population that were previously inaccessible. Likewise, microfinance institutions in Africa and southeast Asia have turned to lightweight RPA to democratize credit assessment of small businesses. RPA is directly connected with the Sustainable Development Goal (SDG) 8 of the United Nations sustainable development initiative (SDG 8) of ensuring inclusive and sustainable economic growth and decent work for everyone by reducing administrative obstacles.

Third, the diversity of participation in the design and governance of RPA systems is also incorporated in automation. The recent studies indicate that women and underrepresented minorities continue to be severely underrepresented in the automation workforce (World Economic Forum, 2024). Inclusive automation programs, which are aimed at hiring, training, and empowering diverse talent, improve creativity, ethical recognition, and system capacity. FinTech gender gap HSBC and Mastercard have created so-called Automation Diversity Labs to teach women engineers and analysts to develop RPA and bridge the gender gap in FinTech innovation. These efforts highlight the idea that an inclusive process of digital transformation is not only a moral necessity but a competitive advantage creating wider horizons and a moral sense in designing automation.

4.10 Governance, Accountability, and Responsible Automation

The structures, policies and ethical standards that guarantee that RPA is transparent, safe, and fair also range on sustainability and inclusion. In the financial sector, where data sensitivity and regulatory oversight are the essential factors, governance is what defines technological plausibility and social trust. According to the bibliometric evidence, there has been a recent explosion of research into the topic of AI governance, explainability, and responsible automation, especially in European settings. This is in line with emerging regulatory frameworks including the EU Artificial Intelligence Act (2024) and OECD AI Principles that require automated decision systems to be classified based on risk, achieve algorithmic transparency, and human oversight.

In the case of RPA, accountability, explainability, and fairness are to be considered as part of the system design. Bots require the production of audit logs, data lineage, and assist humans in the key decision-making. In addition, ethical governance also stretches to the acquisition and lifecycle management processes of the RPA systems making sure the vendors have good practices, safe storage of data, and that the systems comply with sustainability procedures. More and more companies are implementing Responsible Automation Charters, which hold automation programs accountable towards environmental sustainability, human rights and labor protection. To take a specific example, Standard Chartered Bank has centralized the automation management process into its corporate responsibility strategy, whereby each new bot implementation is directly correlated with at least one ESG key performance indicator

One such relevant trend in responsible governance is the emergence of Explainable RPA (XRPA), a new paradigm that integrates conventional automation with machine learning explainability systems. XRPA systems are meant to explain the rationale of automated actions, allowing the human operators and auditors and regulators to comprehend and check decision logic. This method helps to avoid the black-box risk and strengthen the trust of stakeholders- which is essential to the ethical and sustainable financial automation.

4.11 Closure of the Sustainability, Inclusion, and Competitive Advantage Gap.

The inclusion of sustainability and inclusion in RPA is not an issue of ethical ideals but a strategic reality. More data shows that organizations that engage in inclusive automation are more adaptable, innovative, and better at managing their reputations than their peers. In a 2023 study of 250 financial institutions worldwide, Deloitte discovered that companies that had clear automation programs with a defined sustainability orientation reported an increase in employee retention by 22 percent, customer satisfaction by 18 percent, and a 14 percent rise in ESG score in three years. This empirical connection and relationship between responsible automation and performance highlights the fact that ethical change and competitive superiority are not mutually exclusive but rather go hand in hand.

Resilient innovation ecosystems are also sustainable, inclusive, and automated through the synergy of sustainability, inclusion, and automation. With automation objectives aligned to the social and environmental objectives, companies are improving their ability to absorb external shocks, be it during regulatory change, technological disruption, or social examination. In addition, universally applicable automation promotes organizational legitimacy, which strengthens confidence in stakeholders in the age of algorithmic responsibility. In the case of financial institutions that rely essentially on credibility and trust by the people then such legitimacy is a very important intangible asset.

In the future, the intersection of RPA with Artificial Intelligence (AI) and Machine Learning (ML) and Blockchain technologies will create more opportunities and demands. The next generation RPA systems will have to reflect the concepts of the so-called Inclusive AI, which includes the aspect of fairness and accessibility of algorithms to a wide variety of users, as well as the openness of decision-making. Academic studies need to shift to the development of multidimensional assessment paradigms that are able to quantify automation in terms of efficiency indicators, but also assess the impact of automation on sustainability and inclusiveness. An example is a suggested Automation Inclusion Index (AII) that may assess how an organization is committed to diversity in its workforce, investing in reskilling and providing benefits fairly in automation efforts.

Moreover, academic-industry-policymaker partnerships will be crucial to the expansion of inclusive and sustainable automation across the world. All these efforts have already been initiated, with organizations like the Global Partnership on AI and the World Economic Forum Centre on the Fourth Industrial Revolution focusing on how the industry can work together to achieve responsible automation. By integrating RPA into such structures, it is possible to develop global standards of sustainability, inclusion and ethical compliance- harmonizing the standards between jurisdictions.

In short, sustainable and inclusive RPA will be the new frontier of digital evolution (where technological excellence meets with humanistic values to shape the future of finance that is not only automated but also ethical, resilient, and just). Including and embracing sustainability into the genetic makeup of RPA systems, organizations will be able to gain a truly transformative synergy between human and machine, a collaboration that drives corporate performance as well as societal development.

4.12 Regulatory and Ethical Challenges for SMEs Adopting RPA

Small and medium-sized enterprises (SMEs) face significant regulatory and ethical hurdles when adopting Robotic Process Automation (RPA), which are often amplified by limited resources and expertise. A key issue is the management of data privacy and security, as RPA bots frequently handle sensitive information without the robust IT governance and security personnel found in larger corporations (Ghobakhloo et al., 2023). This can lead to a "methodological disconnect" between human and RPA policies, increasing the risk of data loss and regulatory non-compliance (Cooper et al., 2022). Furthermore, ethical concerns around job security are particularly acute in the more tightly-knit environments of SMEs, where a lack of clear communication and a plan for employee upskilling can lead to resistance and undermine implementation success. As research indicates, human factors and organizational understanding are as critical as technical capabilities for successful adoption, making these socio-technical challenges a central impediment for SMEs (Jackson and Allen, 2024; Zhang et al., 2023)

One of the foremost barriers for SMEs adopting RPA is navigating the regulatory landscape that governs data privacy, financial transactions, and labor practices. In many jurisdictions, regulations such as the General Data Protection Regulation in the European Union, the Personal Data Protection Bill in India, and equivalent frameworks across Asia and North America impose stringent requirements on how data must be collected, processed, and stored. For SMEs, which typically lack dedicated compliance departments, ensuring that automated bots adhere to these standards poses a formidable challenge. RPA systems, by design, interact with sensitive data financial records, customer identifiers, and employee information raising significant questions about accountability and data stewardship.

Large corporations often address these challenges by embedding compliance modules and audit trails into their automation platforms. SMEs, however, struggle to invest in such complex infrastructures. A 2023 KPMG survey revealed that 62% of SMEs implementing automation lacked formal data governance protocols, relying instead on ad hoc controls or vendor guidance. This absence of structured governance exposes them to legal liabilities, particularly when automation tasks span cross-border data transfers or financial transactions. Moreover, the traceability requirement the ability to reconstruct the decision-making path of bots is often overlooked in SME environments, where cost considerations outweigh compliance priorities. The resulting lack of transparency can contravene data protection laws, jeopardizing trust among clients, regulators, and partners.

Another layer of regulatory complexity arises from industry-specific compliance mandates. SMEs operating in finance, insurance, healthcare, or logistics are subject to sectoral regulations such as the Payment Card Industry Data Security Standard, Basel III, and (Sarbanes–Oxley Act). These frameworks demand high levels of data integrity, system reliability, and auditability attributes that RPA can enhance, but only if implemented responsibly. Without appropriate oversight, automation may inadvertently propagate non-compliance by replicating human errors at scale or mishandling sensitive records. Therefore, SMEs face a paradox: automation can reduce human error and compliance costs, yet without robust governance, it can amplify regulatory exposure.

4.13 Data Privacy, Security, and Ethical Risk

Data privacy represents one of the most pressing ethical and regulatory challenges in SME automation. Since bots often access multiple systems using elevated privileges, they can become targets for cyberattacks or data breaches if not properly secured. Unlike large institutions with layered cybersecurity architectures, SMEs frequently lack the capacity to implement advanced access controls, encryption, and continuous monitoring. The Ponemon Institute (2023) reports that SMEs are 350% more likely to experience data breaches than large enterprises due to weaker security postures.

Moreover, ethical concerns arise from the opacity of automated decision-making. In financial or accounting contexts, bots may execute transactions, approve invoices, or classify expenses based on pre-set rules. If those rules are inadequately documented or biased, errors can propagate without human awareness. For instance, an SME's automated credit approval system might inadvertently discriminate against certain applicants if historical data reflects systemic biases. Such ethical lapses not only undermine fairness but can expose firms to legal challenges under anti-discrimination and consumer protection laws.

Transparency and accountability are therefore critical. SMEs must ensure that all automation decisions are explainable and auditable a challenge given limited technical resources. The concept of Explainable RPA, which integrates transparency into automation design, holds promise but remains largely inaccessible to smaller firms due to cost and complexity. Industry associations and technology providers could play a pivotal role by developing affordable governance frameworks and compliance templates tailored to SMEs, enabling responsible automation without prohibitive investment.

Workforce Ethics and Social Responsibility

Beyond regulatory compliance, SMEs face ethical dilemmas related to employment, workforce well-being, and organizational culture. Unlike large corporations that can absorb labor displacement through retraining programs, SMEs often operate with small, tightly knit teams where job loss has immediate social and psychological consequences. Employees may perceive automation as a threat to job security, leading to resistance, mistrust, or even active sabotage. Ethical management of workforce transitions thus becomes central to sustainable RPA adoption.

Responsible SMEs can mitigate these challenges by embracing a "human-centric automation" approach that prioritizes augmentation over replacement. Research by Deloitte (2024) shows that SMEs that invest at least 10% of their automation budget in workforce reskilling report 20% higher employee satisfaction and smoother implementation outcomes. Ethical communication—transparently articulating how automation will change roles and introducing continuous learning opportunities—fosters trust and inclusion. Case examples such as Omega Healthcare demonstrate that automation can coexist with human development: by redeploying employees from repetitive documentation to analytical oversight roles, the firm not only retained its workforce but also enhanced service quality and compliance accuracy.

Moreover, ethical automation in SMEs must account for psychological well-being and organizational justice. When bots assume human tasks, questions of fairness—who benefits, who bears the burden, and how accountability is distributed—become salient. Establishing internal ethics committees or automation oversight groups, even informally, can help SMEs evaluate automation decisions through moral as well as operational lenses. Such practices align with the United Nations Guiding Principles on Business and Human Rights, ensuring that technological innovation upholds dignity, equity, and inclusion.

Addressing these challenges requires a principled procurement approach rooted in transparency and shared accountability. SMEs should demand that vendors provide clear documentation on data handling, audit trails, and ethical safeguards embedded in their platforms. Industry associations and government agencies can facilitate this by developing RPA certification programs that evaluate vendors on compliance, security, and sustainability criteria. Public–private partnerships, such as India's Digital MSME initiative or the EU's Digital Europe Programme, can further support SMEs by offering subsidized access to compliant RPA solutions and advisory services.

Building sustainable and inclusive RPA ecosystems for SMEs will require a multi-stakeholder effort linking policymakers, vendors, financial institutions, and academic researchers. Only through collaborative frameworks that balance efficiency with ethics, innovation with inclusion, and technology with trust can SMEs fully realize the transformative promise of automation without compromising their social or moral integrity.

4. 14 RPA Implementation at Omega Health care

Omega Healthcare Management Services, a leading provider supporting over 350 healthcare organizations, implemented Robotic Process Automation (RPA) with the help of UiPath to streamline administrative tasks such as medical billing, insurance claims processing, and documentation. Since 2020, the company has automated processes for 60–70% of its clients, handling over 100 million transactions. As a result, employees are saving more than 15,000 hours per month, documentation time has been reduced by 40%, and document processing is now 50% faster with an accuracy rate of 99.5%. These improvements have led to a 30% return on investment (ROI) for clients. By shifting employees from repetitive tasks to decision-based roles, Omega Healthcare has significantly enhanced operational efficiency and client satisfaction.

5. MANAGERIAL THEORETICAL IMPLICATIONS

For financial institutions and accounting firms, the findings of this analysis carry several critical practical implications. Organizations should strategically view RPA not merely as a standalone efficiency tool but as a foundational and evolving element of their broader digital transformation and AI strategy. This necessitates a long-term vision that aligns RPA initiatives with overall business objectives, fostering a culture of continuous automation and innovation. Successful RPA adoption hinges on addressing both the technical and human-centric aspects; this includes investing in comprehensive methodology understanding, robust change management programs, and proactive communication strategies to alleviate employee anxieties and customer concerns regarding data security and human interaction. To unlock the full potential of RPA, firms must proactively explore and invest in integrating RPA with emerging technologies such as Artificial Intelligence, Machine Learning, Natural Language Processing, and Blockchain. This integration will enable the automation of more complex, cognitive tasks and enhance data security and transparency, moving beyond simple rule-based automation. Finally, organizations should continue to rigorously measure and communicate the clear operational and financial benefits derived from RPA, such as significant reductions in processing time, substantial cost savings, and improved accuracy. This data-driven approach builds a compelling business case and justifies further investment and widespread adoption.

From a theoretical standpoint, this analysis underscores the importance of a socio-technical systems perspective in understanding RPA implementation. The persistent challenges related to organizational understanding and human-centric concerns highlight that technological adoption is not purely a technical endeavor but is deeply intertwined with social structures, human perceptions, and organizational culture. Future theoretical frameworks should explicitly integrate these human factors with technological capabilities to provide a more holistic understanding of successful digital transformation initiatives. Furthermore, the observed evolution of RPA research themes, from basic automation to AI-driven cognitive capabilities, contributes to the theory of technological diffusion and maturation, illustrating how a technology's perceived value and research focus evolve as it integrates with broader technological ecosystems. The emergence of new research hubs also provides empirical evidence for theories of global knowledge transfer and the decentralization of innovation, challenging traditional notions of research dominance. The identified "technical capability gaps" in integrating advanced AI with RPA also point to a need for theoretical development in understanding barriers to complex technological convergence and the role of interdisciplinary knowledge transfer in overcoming these barriers.

6. LIMITATIONS OF THE STUDY

This bibliometric analysis, while comprehensive, is subject to certain limitations inherent to its methodology. Firstly, the data collection was primarily confined to Scopus academic database. While these are highly reputable sources, they may not capture the entirety of relevant literature, particularly industry reports, white papers, or publications in less indexed regional journals, which could offer additional insights into practical applications and challenges. Secondly, the analysis relies on keyword co-occurrence and citation patterns, which reflect published academic discourse but may not fully capture the shades of ongoing industry practices or unpublished innovations. The interpretation of trends and thematic clusters is based on the frequency and relationships of terms, which might not always perfectly align with the qualitative depth of the research content. Lastly, bibliometric analysis, by its nature, provides a quantitative overview of the research landscape and does not delve into the qualitative findings or specific methodologies employed within individual studies, thus offering a broad picture rather than granular detail.

7 FUTURE RESEARCH

Based on this comprehensive analysis and identified limitations, several promising avenues for future research emerge. Firstly, there is a clear need to address the functional gaps identified within the banking sector, specifically in underexplored areas such as sales and Enterprise Resource Planning (ERP), as well as other specific bank functions, to broaden the understanding of RPA's applicability. Secondly, more rigorous studies are required on the underlying techniques and sustainability aspects of advanced RPA, with a particular focus on the practical application of Deep Learning, Natural Language Processing (NLP), and Cognitive Automation to bridge existing “technical capability gaps” and ensure long-term viability. This includes investigating how to effectively integrate these technologies to move beyond rule-based automation. Thirdly, future research should undertake more comprehensive studies on the socio-technical dimensions of RPA implementation, including organizational change management, human-robot interaction, ethical implications of automation, and its impact on workforce dynamics, moving beyond mere identification of these issues to proposing and evaluating solutions. Finally, from a methodological perspective, future bibliometric work could benefit from utilizing advanced machine learning techniques, such as deep learning for pattern discovery, to further refine the understanding of complex research landscapes and uncover hidden trends and relationships that traditional methods might miss. This would allow for a more in-depth and nuanced study of the research domain.

8. CONCLUSION

This comprehensive bibliometric analysis unequivocally demonstrates that Robotic Process Automation (RPA) in finance and accounting is a rapidly expanding and increasingly mature field of academic inquiry, marked by a significant surge in publications since 2017. The thematic landscape has evolved from an initial focus on basic automation to a more sophisticated emphasis on AI-driven digital transformation, integrating concepts such as big data and advanced analytics. Research leadership is becoming more geographically diverse, with emerging economies like India making substantial contributions and challenging traditional research hubs. While RPA offers clear, quantifiable benefits in terms of operational efficiency, cost reduction, and even employee satisfaction, its successful implementation is frequently challenged by socio-technical factors, including organizational readiness, a lack of methodological understanding, and human-centric concerns regarding job security and data privacy. The future trajectory of RPA is firmly set towards intelligent, integrated, and sustainable automation, requiring deeper integration with advanced AI, Natural Language Processing (NLP), and Blockchain technologies to unlock higher-value, cognitive automation capabilities.

9. REFERENCES

Bakarich, K. M., & O'Brien, P. E. (2021). The robots are coming… but aren't here yet: The use of artificial intelligence technologies in the public accounting profession. *Journal of Emerging Technologies in Accounting*, *18*(1), 27–43.

Bhoyar, S., Bhoyar, P., & Shah, M. (2023). A machine learning-based predictive approach in evaluating consumer behavior. *Journal of Statistics and Management Systems*, *26*, 1955–1963. DOI: 10.47974/JSMS-1131

Chyzhevska, L., Voloschuk, L., Shatskova, L., & Sokolenko, L. (2021). Digitalization as a vector of information systems development and accounting system modernization. *Studia Universitatis Vasile Goldi Arad. Seria tiin e Economice*, *31*(4), 18–39.

Cooper, L. A., Holderness, D. K.Jr, Sorensen, T. L., & Wood, D. A. (2019). Robotic process automation in public accounting. *Accounting Horizons*, *33*(4), 15–35.

Cooper, L. A., Holderness, D. K.Jr, Sorensen, T. L., & Wood, D. A. (2022). Perceptions of robotic process automation in Big 4 public accounting firms: Do firm leaders and lower-level employees agree? *Journal of Emerging Technologies in Accounting*, *19*(1), 33–51.

Deloitte. (2023, June 11). *Automation with intelligence*. Deloitte Insights. https://www.deloitte.com/us/en/insights/topics/talent/intelligent-automation-2022-survey-results.html?utm_source=chatgpt.com

Enríquez, J. G., Jiménez-Ramírez, A., Domínguez-Mayo, F. J., & Garcia-Garcia, J. A. (2020). Robotic process automation: A scientific and industrial systematic mapping study. *IEEE Access : Practical Innovations, Open Solutions*, *8*, 39113–39129.

Eulerich, M., Pawlowski, J., Waddoups, N. J., & Wood, D. A. (2022). A framework for using robotic process automation for audit tasks. *Contemporary Accounting Research*, *39*(1), 691–720.

Ghobakhloo, M., Asadi, S., Iranmanesh, M., Foroughi, B., Mubarak, M. F., & Yadegaridehkordi, E. (2023). Intelligent automation implementation and corporate sustainability performance: The enabling role of corporate social responsibility strategy. *Technology in Society*, *74*, 102301.

Gotthardt, M., Koivulaakso, D., Paksoy, O., Saramo, C., Martikainen, M., & Lehner, O. (2020). *Current state and challenges in the implementation of smart robotic process automation in accounting and auditing*. ACRN Journal of Finance and Risk Perspectives.

Huang, F., & Vasarhelyi, M. A. (2019). Applying robotic process automation (RPA) in auditing: A framework. *International Journal of Accounting Information Systems*, *35*, 100433.

Ionescu, L. (2020). Robotic process automation, deep learning, and natural language processing in algorithmic data-driven accounting information systems. *Analysis and Metaphysics*, (19), 59–65.

Jackson, D., & Allen, C. (2024). Enablers, barriers and strategies for adopting new technology in accounting. *International Journal of Accounting Information Systems*, *52*, 100666.

Kedziora, D., Leivonen, A., Piotrowicz, W., & Öörni, A. (2021). Robotic process automation (RPA) implementation drivers: Evidence of selected Nordic companies. *Issues in Information Systems*, *22*(2), 21–40.

Kedziora, D., & Penttinen, E. (2021). Governance models for robotic process automation: The case of Nordea Bank. *Journal of Information Technology Teaching Cases*, *11*(1), 20–29.

Kokina, J., & Blanchette, S. (2019). Early evidence of digital labor in accounting: Innovation with Robotic Process Automation. *International Journal of Accounting Information Systems*, *35*, 100431.

Kokina, J., Gilleran, R., Blanchette, S., & Stoddard, D. (2021). Accountant as digital innovator: Roles and competencies in the age of automation. *Accounting Horizons*, *35*(1), 153–184.

Kommunuri, J. (2022). Artificial intelligence and the changing landscape of accounting: A viewpoint. *Pacific Accounting Review*, *34*(4).

Korhonen, T., Selos, E., Laine, T., & Suomala, P. (2021). Exploring the programmability of management accounting work for increasing automation: An interventionist case study. *Accounting, Auditing & Accountability Journal*, *34*(2), 253–280.

Perdana, A., Lee, W. E., & Mui Kim, C. (2023). Prototyping and implementing Robotic Process Automation in accounting firms: Benefits, challenges and opportunities to audit automation. *International Journal of Accounting Information Systems*, , 51.

Plattfaut, R., & Borghoff, V. (2022). Robotic process automation: A literature-based research agenda. *Journal of Information Systems*, *36*(2), 173–191.

Polak, P., Nelischer, C., Guo, H., & Robertson, D. C. (2020). "Intelligent" finance and treasury management: What we can expect. *AI & Society*, *35*(3), 715–726.

Pramod, D. (2022). Robotic process automation for industry: Adoption status, benefits, challenges and research agenda. *Benchmarking*, *29*(5), 1562–1586.

Risk assessment report,july 2024. (n.d.). https://www.eba.europa.eu/sites/default/files/2024-07/9604ba14-0ec4-4236-94e9-b07cb79db918/Risk%20assessment%20report%20%20July%202024.pdf

Shah, M. A. (2025). The Impact of FinTech on Entrepreneurial Intentions Among Young Indian Entrepreneurs Using Crowdfunding, Blockchain, and Mobile Payments. In *Models, Strategies, and Tools for Competitive SMEs* (pp. 325–340). IGI Global Scientific Publishing., DOI: 10.4018/979-8-3693-4046-2.ch015

Soni, T. (2024). *How much does RPA implementation cost - and how it pays off?* How Much Does RPA Implementation Cost - and How It Pays Off? https://www.prioxis.com/blog/rpa-implementation-cost

Yoon, S. (2020). A study on the transformation of accounting based on new technologies: Evidence from Korea. *Sustainability*, *12*(20), 8669.

Zhang, C., Issa, H., Rozario, A., & Soegaard, J. S. (2023). Robotic process automation (RPA) implementation case studies in accounting: A beginning to end perspective. *Accounting Horizons*, *37*(1), 193–217.

KEY TERMS AND DEFINITIONS

Artificial Intelligence (AI): A broad field of computer science that enables machines to perform tasks that typically require human intelligence, such as processing unstructured data, complex decision-making, and learning from experience. In the context of RPA, AI integration extends the scope and sophistication of automation.

Bibliometric Analysis: A systematic quantitative method used to analyze academic literature, identify key trends, influential contributors, evolving research themes, and existing gaps within a specific field, often utilizing tools like RStudio and Biblioshiny.

Cloud-based RPA Solutions: Robotic Process Automation deployments hosted and delivered over the internet, enhancing scalability and reducing infrastructure overheads for businesses of all sizes, and constituting a significant portion of RPA deployments.

Compound Annual Growth Rate (CAGR): A financial metric that represents the average annual growth rate of an investment over a specified period longer than one year, assuming the profits are reinvested at the end of each year. In the chapter, it's used to project the growth of the IT robotic automation market.

FinTech (Financial Technologies): The technological innovations that aim to improve and automate the delivery and use of financial services. RPA is a critical phenomenon within FinTech due to the financial industry's high transaction volumes, repetitive tasks, and demands for accuracy and compliance.

Hyperautomation: A synergistic approach combining Robotic Process Automation (RPA) with Artificial Intelligence (AI), the Internet of Things (IoT), and process mining, anticipated to unlock novel growth avenues and significantly optimize business processes.

Robotic Process Automation (RPA): A sophisticated software application designed to automate repetitive Graphical User Interface (GUI) tasks, often used to enhance operational efficiency, reduce costs, and improve productivity. It moves beyond basic rule-based automation towards more cognitive capabilities when integrated with advanced technologies.

Chapter 10
Challenges and Ethical Considerations of Robotic Process Automation in Accounting and Finance

Sweta Goel
Jaypee Institute of Information Technology, India

Rahul Sharma
https://orcid.org/0000-0003-2733-8573
Horizon University College, UAE

ABSTRACT

Robotic Process Automation is playing a vital role in rebuilding the perspective of Accounting and Finance. The chapter first, discusses the exhaustive literature on the RPA followed by critically analyzed technological hurdles and operational challenges faced by organizations that includes Processes to be adopted, and its complexity and issues related to the system integration using exploratory research design based on the analysis of secondary data sources. Further the ethical dimensions have been studied in detail. The chapter suggests and discusses mitigation strategies and best practices such as framework of ethical governance, re-skilling of workforce, firm automation protocols and constant monitoring of performance to address these technological, operational and ethical issues. Finally, authors present the future outlook on ethical grounds for AI-Enhanced RPA that includes anticipated regulatory shifts and technological advancements and the Role of human oversight in an increasingly automated environment.

DOI: 10.4018/979-8-3373-3571-1.ch010

1. INTRODUCTION

Robotic Process Automation (RPA) is one of the most important technological advancements of the current era. Fundamentally, RPA involves the use of software bots to automate repetitive, rule-based jobs that were traditionally performed by individuals. These bots respond in the same way as a human would, that is, typing, clicking, extracting data, and navigating systems, making RPA an ideal solution for environments where structured data and repetitive workflows are common (Aguirre & Rodriguez, 2017; van der Aalst, 2018).

With reference to accounting and finance, RPA is changing the way professionals handle day-to-day tasks such as processing invoices, updating ledger, reconciling ledger, financial reporting, and tax filings. Though these activities are very essential for every organization, they often take too much time and are also prone to human error. However, using RPA these tasks can be executed with increased speed, accuracy, and consistency, while also reducing labour costs (Lacity & Willcocks, 2016; Soni et al., 2019). Most importantly, it allows finance teams to shift their focus from routine manual data handling to higher-value strategic functions such as data analysis, financial forecasting, and decision-making support (Fernández & Aman, 2018).

By automating repetitive tasks, RPA not only enhances operational efficiency and internal reporting requirements but also helps organizations respond more quickly to market demands, and regulatory changes. It plays a very significant role in allowing finance departments to meet the fast demands of the digital economy, offering both immediate benefits and long-term strategic value (Syed et al., 2020; Willcocks et al., 2015).

1.1. Rationale for Exploring Challenges and Ethical Implications

There are many advantages that can be seen by organizations that are using RPA in accounting and finance, but these advantages do not stop the possibility of multifaceted sets of technical, operational, and ethical challenges that warrant careful examination. In financial operations regulatory standards, sensitive data management protocols play very important roles. In addition to these high expectations for accuracy and speed, accountability is also expected in financial operations. As such, the use of automation in this context requires more than just a focus on efficiency as it demands a strong ethical framework and robust governance mechanisms (Rozario & Vasarhelyi, 2018).

Organizations face problems on both technical as well as ethical fronts. On technical front, they face various hurdles such as issues involved in consolidating with the legacy system, handling the variability in data, and fulfilling the requirement of sustaining the existing bot (Asatiani & Penttinen, 2016; Stople et al., 2017). From the ethical point of view, concerns for the job replacement, transparency in decision making, threat for privacy and the answerability of the misuse of data exists. For example, if the financial statements or the customer data of a company gets affected by any action of the bots then, who would be accountable for possible outcomes that might contain any blunder or infringement. In addition, organizations have also started involving bots substantially in their decision-making process. This process needs a surety that the decisions taken by bots' system are just and impartial (Davenport & Ronanki, 2018; Gkritzali et al., 2020).

RPA is being spread in the organizations at a very fast pace. If its execution is done with ambiguous ethical guidelines, then this might lead to inadvertent consequences such as defiance of the regulatory system, discontentment in workforce, and the defamation or loss of goodwill for the organization (Cath, 2018). Therefore, it is of paramount importance to decide on the working of RPA and outline its responsibilities. Also, the Governance and management of RPA systems should also be well-defined and aligned within the financial ecosystem (Moffitt et al., 2018; Binns et al., 2018).

This chapter presents an exhaustive and steady investigation of the challenges and ethical considerations of Robotic Process Automation (RPA) in the fields of accounting and finance. Automation has become indispensable in accounting and finance operations. Therefore, on the one hand, it is essential to realize its competence while on the other hand we need to critically examine the limitations and ethical implications associated with it.

This chapter bridges the gap that exists between technical implementation and ethical foresight of RPA. Moreover, it aims to present valuable insights for researchers, professionals, and decision-makers involved in accounting and finance automation.

The chapter is organized into ten interconnected sections. First introduction has been given. Second section presents the literature review followed by methodology in section three. Technological and operational challenges have been discussed in the fourth section. Next section, section five focuses on ethical considerations in RPA implementation. Section six talks about the regulatory and compliance landscape while the seventh section is about mitigation strategies and best practices. Some case studies of RPA adoption have been discussed in section eight. In section nine, future outlook is presented, and section ten concludes.

2. LITERATURE REVIEW ON THE CONCEPT AND CORE COMPONENTS OF RPA

Robotic Process Automation (RPA) has gained recognition in the professional and academic field for its profound role in revamping the proficiency of business processes. RPA enables businesses to automate simple rule-based tasks by using software bots for the work that was earlier performed by human workforce (van der Aalst, 2018). According to Lacity & Willcocks (2016), RPA interconnects with ongoing IT systems via the user interface. Therefore, in comparison to traditional automation tools, it is more efficient in eradicating the requirements of backend programming changes.

RPA acts as a central link between business process management and artificial intelligence although it focuses precisely on deterministic and rule-based actions (Syed et al., 2020). RPA replicates human interactions with digital systems. It can accurately automate the structured tasks such as filling the forms, extracting the data, or navigating the system in a cost-efficient manner with high speed. RPA provides repeatability and scalability in settings where high submissiveness is required along with little process variability (Aguirre and Rodriguez, 2017).

2.1. Key Elements of RPA Systems

A component based, service-oriented and layered architecture is used to build the systems for RPA, that ensures administration, management, flexibility and growth potential. Each element, used in the construction, execution and supervision of the bot system such as bot designer, bot runners, control rooms and audit logs, carries utmost importance to the whole automation lifecycle. These elements are discussed further.

2.1.1. Bot Designer/Developer Studio

This is the environment where automation workflows are created. Through Bot Designer, often presented as a Graphical User Interface (GUI) even developers and business users having minimal programming skills are able to design automation processes using intuitive, drag-and-drop functionalities. GUI includes features such as conditional logic, loops, screen scraping, and pre-configured connectors to enterprise applications. Organizations can speed up its development while maintaining accuracy and consistency by enabling users to define step-by-step bot actions through visual mapping and logic configuration (Syed et al., 2020; Lacity & Willcocks, 2016).

2.1.2. Executor/Runner for Bots

The Bot Runner deploys and runs the automation scripts after they are created. Both attended and unattended modes of operation are available for these bots. While unattended bots operate autonomously on virtual machines or servers, carrying out tasks on a scheduled basis or in response to system events, attended bots collaborate with human users and are frequently activated by user actions on desktop interfaces. RPA bot runners ensure consistency and remove human error by carrying out the automated steps precisely as specified in the workflow. These Bot Runners can manage several processes at once depending upon system configuration and capacity (Aguirre & Rodriguez, 2017).

2.1.3. Control Room/Orchestrator

The primary location for supervising the RPA activities throughout the company is the control room, also known as Orchestrator. It is widely used by the managers for assigning the tasks, deploying the bots, managing the deviations and performance evaluation of the bot system. Orchestrator ensures a secure and productive automation management as it applies chain management, credential vaults, role-based access control, and system health observation. Also, crucial activities needed for large scale deployment such as dynamic resource allotment, load balancing, and scheduling are enabled effectively by Orchestrator (Willcocks, Lacity, & Craig, 2015; Rozario & Vasarhelyi, 2018).

2.1.4. Audit Logs and Analytics

Transparent, auditable logs of bot activity rigorously document every action such as time stamps, system responses, outliers, and user interlinkages and therefore these are a dispensable part of any RPA system. Highly regulated industries like healthcare and finance follow the compliance of this monitoring. As per the study done by Syed et al. (2020), logging systems are attached with the analytics modules that provide dashboards and key performance indicators (KPIs). It facilitates organizations in tracking the bot efficiency, identifying the bottlenecks, and enhancing the effectiveness of the process over time. It further supports in making the system audit ready and increases the confidence in the results obtained by automation.

Jointly, these elements generate an integrated architecture system that supports the implementation of RPA for better monitoring, compliance, transparency and scalability that is very crucial for the organisations especially for finance and accounting sections (Willcocks, Lacity, and Craig, 2015). Apart from technical expertise, governance structures and change management techniques are of utmost importance for the RPA's deployment (Rozario and Vasarhelyi, 2018).

With the development of RPA, professionals, academicians and researchers started to study its interaction with cognitive technologies such as machine learning and language processing. This is a step forward towards Intelligent Process Automation (IPA) and it unifies the flexibility of AI and the speed and accuracy of RPA (Syed et al., 2020).

2.2. Common Applications in Financial Operations

The use of RPA has increased tremendously in the finance sector as it can automate high-volume, rule-based tasks, necessary for daily financial operations. Business organisations have improved the compliance, accuracy, and scalability in financial tasks such as invoicing, reporting and reconciliation by implementing the RPA. Some of these financial applications are discussed further.

2.2.1. Invoice Processing

Invoice management has been one of the most time taking tasks in financial management. Organizations had been performing this task manually which involved extracting the data followed by validating and then entering it into the Enterprise Resource Planning, ERP system. This process can be done with RPA bots which use optical character recognition (OCR) for extracting the invoice data. The data is then compared with the purchase orders and its respective receipt. RPA bot then automatically enters this final data into the accounting systems. RPA bot systems accomplish the task of processing the data in a much lesser time. It also magnifies the processing volume and reduces human inaccuracy (Aguirre & Rodriguez, 2017). RPA bots also enhance the control and precision as it highlights the mismatched invoices or variability in invoice data for further review.

2.2.2. Bank Reconciliation

As per the requirements of bank reconciliation, the finance team must discover and address the irregularities, for which internal financial records are compared with the bank statements. The rules are set for this process, and it is repetitive in nature. Therefore, this process is automated with the help of RPA in which first, the bank statements are collected via automation followed by preparing the comprehensive reconciliation reports. After this, RPA bot highlights the anomalies and reconciles the data with internal records. This process cuts down the time consumed on month-end and year-end closing tasks while ensuring greater consistency and compliance (Rozario & Vasarhelyi, 2018).

2.2.3. Accounts Payable and Receivable

In the area of accounts payables and receivables, onboarding of vendors, approval of invoices and payment processing for accounts payables is automated with the help of RPA. Rules are set for these payment processes and accordingly, RPA bot initiates the pending payments and reminds the users for unsettled approvals. RPA bots trace the customers' accounts for clearing the accounts receivables, remind them for payments, register these incoming payments to pending invoices, and escalate the past-due payments. This process results in an efficient working capital, lower down the outstanding sales for the day and boosts the cash flow (Syed et al., 2020).

2.2.4. Financial Reporting and Close

Financial reporting requires the consolidation and validation of data from different sources. This process may have some inaccuracies if it is done manually and be time consuming. RPA bots extract, accumulate and clean the data from numerous systems and generate the reports. RPA creates both standard and customized reports and sends them to the respective stakeholders within the allotted time frame. In this manner compliance is followed as all the reports are framed as per the internal and legal standards and audit logs as also maintained.

2.2.5. Payroll Processing

Payroll processes include generation of paystubs, application of statutory deductions and deciding on the appropriate employee wages. It is a recurring procedure and is automated with RPA that calculates the net pay, deducts the applicable taxes, compiles the data from HR and timekeeping systems and ensures the workers get payment accurately on time. In addition to this, RPA bots generate and submit the necessary payroll related reports to the Government that enhances the compliance and reduces the possibilities of fine (Fernández & Aman, 2018).

2.2.6. Tax Compliance and Regulatory Reporting

One of the important parts of tax preparation and filing is the compliance of its rules and regulations. RPA bots compile the essential financial data, apply the significant tax rates, produce the reports and submit the tax returns to the concerned authorities. They also keep on updating the moderations in ongoing tax laws and maintain the process compliance in the organization. Automating this process in audits lower down the risk possibility and enhance the transparency (Willcocks et al., 2015).

2.2.7. Budgeting and Forecasting Support

Budgeting and forecasting is an area that requires regular human intervention in the form of judgment and strategic thinking. RPA bots assist in this field as they collect the past data, compile the information obtained from numerous departments and update the forecasting models. This enables the financial experts to afford more time on the data analysis and lower down the time devoted on preparing the spreadsheets and managing the reports.

The acquisition of RPA in the finance industry has significantly improved the quality of workforce, compliance and performance on the operational level. Although RPA's initial appeal frequently stems from its potential to reduce costs. Its wider value proposition also includes process accuracy, transparency, agility, and human capital optimization.

2.3. Benefits Realized in the Finance Domain

2.3.1. Speed and Operational Efficiency

RPA's capacity to significantly boost process efficiency is one of its most obvious advantages in the financial industry. With little latency, tasks like data entry, invoice matching, reconciliations, and financial consolidations can be completed at any time of the day. Software bots consistently achieve high throughput and shorter turnaround times because they are not distracted or tired like human workers are (Lacity & Willcocks, 2016). This has been especially helpful during the time when there is high financial activity, like month-end closings or audit seasons, when accuracy and speed are essential.

2.3.2. Precision and Error Mitigation

Because financial processes affect reporting, compliance, and strategic decision-making, they require a high level of accuracy. Even inadvertent human error can result in expensive misrepresentations or legal infractions. RPA reduces this risk by carrying out tasks with little variation according to deterministic rules. Bots designed to validate entries against compliance parameters or reconcile ledgers, for example, will employ the same logic consistently, minimizing inconsistencies (Rozario & Vasarhelyi, 2018).

2.3.3. Return on Investment (ROI) and Cost Savings

After using RPA, organizations frequently report significant cost savings. Some of the factors contributing towards the savings are labour cost displacement, process efficiency and error mitigation improvements. Software licensing, bot development, and employee training may be a part of the initial investment, but many businesses start getting the return on investment in as little as six to twelve months (Aguirre & Rodriguez, 2017). Furthermore, once created, bots can be replicated across departments without incurring corresponding cost increases, making them scalable assets.

2.3.4. Improved Readiness for Compliance and Audits

Compliance is a must in a time of regulatory scrutiny, particularly in the financial sector. By establishing consistent audit trails, keeping time-stamped logs of all actions, and guaranteeing procedural adherence to financial standards like Sarbanes-Oxley (SOX), International Financial Reporting Standards (IFRS), and local tax laws, RPA aids compliance efforts (Syed et al., 2020). Proactive risk management is made possible by the ability to program bots to instantly identify anomalies or transactions that are not in compliance.

2.3.5. Strategic Redeployment and Enhanced Employee Experience

Although there are worries that automation will replace jobs, RPA actually frequently changes the nature of financial roles rather than doing away with them. Finance professionals can concentrate on higher-value tasks like forecasting, business partnering, strategy formulation, and budgeting by automating repetitive and tedious tasks. According to Fernández and Aman (2018), this change not only improves job satisfaction but also help in employee engagement and retention. Upskilling and role redefinition initiatives have been implemented alongside RPA in many organizations, highlighting the human aspect of digital transformation.

2.3.6. Business Agility and Scalability

The quality of agility and scalability is very important for modern business organizations. The business must be able to adjust to the ongoing internal changes, different market conditions and new rules and regulations. RPA offers workable solutions according to the workload demands in the businesses. RPA bots are rapidly positioned to handle the projections in the transactions. During the steady time periods, they can be ceased or redirected to other processes. According to

Willcocks, Lacity, and Craig (2015), this elasticity makes the finance function more responsive and nimbler.

2.3.7. Real-Time Insights and Data Integrity

The standard and accuracy of the financial data is improved by RPA as it standardizes the handling of data and removes the points of manual contacts. This provides a powerful base to the dashboards, analytics, and decision support systems. Businesses have improved their strategic decision-making capability by combining the RPA with data analytics tools.

3. METHODOLOGY

This chapter examines the role of robotic process automation (RPA) in accounting and finance using exploratory research design based on the analysis of secondary data sources. In this study authors offer a detailed analysis without requiring direct empirical observation, which is advantageous given the speed at which technology is evolving and enhancing the integration of RPA into financial workflows (Snyder, 2019).

Research Approach

The study analyzes literature review and documentary analysis. This approach is suitable for exploratory research projects where the goal is to synthesize and evaluate a wide range of previously published works to produce conceptual clarity and useful insight (Boell & Cecez-Kecmanovic, 2015).

Authors have considered peer-reviewed academic journals, industry reports and professional white papers from top companies like Accenture, Ernst and Young (EY), KPMG and Gartner etc. Case studies and book chapters on digital finance and automation have also been considered.

4. TECHNOLOGICAL AND OPERATIONAL CHALLENGES

Though robotic process automation (RPA) promises efficiency and cost effectiveness, a few operational and technological challenges intercepts usage of practical application. In this section, we will be discussing the technological and operational challenges faced in application of RPA along with the significance of each of the challenges in relation to financial operations.

4.1. Process Suitability

There are various financial functions/ processes but identifying the appropriate process for automation is the biggest challenge being faced by finance leaders. The selection of the appropriate process is a very important factor in the success of RPA. Syed et al. 2020 mentioned that the settings which have structured inputs to be entered or a specific rule to follow are the ideal one for RPA. Data entry, invoice processing and bank reconciliations are the perfect examples of processes to be automated.

This clearly indicates that every financial function is not appropriate. The automation is affected by several factors and one of such factors is subjective judgement like assessment of risk, interpretation of audit reports and fraud investigation. Van der Aalst (2018) claimed that too much variability or load can result in inconsistent outputs.

This conversation is further enhanced by a human-cantered viewpoint. When RPA is improperly used for jobs that call for discretion and insight, workers doing complex tasks could feel threatened or dislocated. Therefore, it is not only a technical but also moral choice to match the scope of automation with the nature of the process.

4.2. Process Complexity

Complexity can appear even in financial procedures that appear to be routine. Human intervention is frequently necessary due to interdependencies, exceptions, and non-standard inputs. For instance, a bot can typically match payments to invoices, but it might have trouble when vendor remittances are unclear or contain foreign exchange discrepancies (Lacity & Willcocks, 2016).

Workflows involving conditional logic or multiple systems add to the complexity of the process. Reconciliations in many finance departments involve more than just math; they may also entail cross-border adjustments, time-sensitive approvals, or judgment on materiality thresholds (Aguirre & Rodriguez, 2017). More human oversight is required while working in dynamic environments, as this may lead bots to commit more frequent errors. Ironically, if complicated process logic is not well understood or documented, the intention of minimizing manual labour may backfire.

4.3. System Integration Issues

RPA simulates user actions on interfaces, but when underlying systems change, this surface-level interaction results in brittle automations. According to Rozario and Vasarhelyi (2018), legacy platforms, which are frequently used in finance functions, frequently have unstable APIs and necessitate custom scripting or interface hacks to allow bot access.

Bot workflows can be disrupted by system upgrades, even if they are small ones like changing the name of a field or its layout. According to Willcocks, Lacity, and Craig (2015), businesses with diverse tech environments encounter significant expenses and coordination difficulties when implementing bots.

Furthermore, stringent access and security controls are frequently found in finance systems, such as payroll, tax, treasury, and ERP. More integration complexity results from granting bots the appropriate permissions while preserving audit trails and user accountability. Data silos, workflow bottlenecks, and even compliance infractions can result from inadequate planning in this area.

4.4. Data Dependency

The consistency, structure, and quality of data are critical components of RPA. Because they lack the contextual intelligence to decipher unclear, unstructured, or incomplete data, bots process inputs exactly as they are. This becomes a significant constraint in the finance industry, where data comes from a variety of internal and external sources, including banks, vendors, and customers.

For example, if fields are mislabelled or vendor formats differ, an RPA bot that extracts invoice details may not work. The nuanced understanding that humans bring to incomplete or messy data cannot be replaced by data validation procedures, although they can help to some extent (Syed et al., 2020). Inaccurate data can spread errors into important financial reports or tax filings in addition to stopping bot execution. Penalties and compliance are jeopardized. Fernández and Aman (2018) contended that for optimizing RPA reliability, businesses need to make investments in upstream data governance, cleansing, and standardization.

4.5. Scalability and Flexibility

Within a single team or department, initial RPA pilots are frequently successful. But there are major obstacles when implementing automation throughout the company. Wider adoption is slowed down by inconsistent documentation, inconsistent process logic, and a lack of unified governance.

Standardization is challenging in the finance industry because different business units may adhere to different accounting rules or practices. Furthermore, unless specifically reprogrammed, bots are deterministic and hard-coded; they do not adapt. Their ability to adapt to shifting laws, corporate policies, or geopolitical events is limited by this rigidity (Lacity & Willcocks, 2016). Fernández and Aman

(2018) contend that in order to optimize RPA reliability, businesses need to make investments in upstream data governance, cleansing, and standardization.

Scaling bots without adequate communication or change management can, in human opinion, lead to employee resentment. Teams may resist because they feel excluded or fear losing their jobs. Therefore, organizational agility and cultural sensitivity must go hand in hand with RPA scaling.

4.6. Maintenance

Implementing RPA is not a "set-it-and-forget-it" approach. A routine maintenance is required for keeping the bots in line with changing regulations and systems. This covers modifications to process logic, security protocols, application interfaces, and compliance standards (Rozario & Vasarhelyi, 2018).

Error correction, log analysis, performance optimization, and documentation updates are additional components of maintenance. Bot dependability can rapidly decline in the absence of a committed support staff. When neglected maintenance results in frequent process failures, the appearance of "hands-free automation" frequently crumbles.

Therefore, in their RPA business case, finance departments need to account for long-term maintenance costs as well as initial development. Businesses that don't set up a support system for bot lifecycle management frequently see a decrease in Return on Investment (ROI) and a rise in user discontent, claim Syed et al. (2020).

4.7. Cost-Effectiveness and ROI Uncertainty

Cost reductions are promised by RPA, but actual outcomes are different. Unexpected costs like licensing fees, training, consulting, rework, and system upgrades could offset the initial benefits. Organizations must create realistic ROI models based on full lifecycle costs, not just development or deployment metrics, according to Willcocks et al. (2015).

Finance executives are also frequently under pressure to defend automation initiatives in terms of quarterly results. The most significant benefits of RPA, however, are long-term and strategic in nature and include workforce enablement, error reduction, and compliance enhancements.

The "false economy," which arises when poorly thought-out automation fails to reduce manual workload, is another frequently disregarded factor. Employees are forced to double the effort in these situations by stepping in to confirm, fix, or reprocess transactions. Disillusionment and low adoption rates may result from this mismatch between automation goals and reality (Fernández & Aman, 2018).

5. ETHICAL CONSIDERATIONS IN RPA IMPLEMENTATION

Speed, consistency and lower error rates are some of the important advantages of robotic process automation (RPA) which have completely changed the way administrative work is done in the organizations. However, this does not mean that there is no need to address ethical conflicts that come with such efficiency gains. The use of technology to perform human tasks raises concerns about employees' lives, rights, fairness, governance and trust. In this section, such ethical considerations are discussed. The objective of this section is to integrate human impact awareness with scholarly insights.

5.1. Workforce Displacement and Employment Ethics

5.1.1. The Human Cost of Scale

Exceptional productivity, one of RPA's strongest points, may also result in widespread workforce displacement. Automation runs the risk of mass employee replacement, particularly for routine tasks. According to a March 2025 review of hub.robomotion.pro, when mitigation measures are not implemented, automation's knock-on effects touch on "job displacement, privacy concerns, and social inequality".

5.1.2. Reskilling as a Simple Rule

Reskilling, redeployment, and upskilling are three strategies that organizations must use to demonstrate responsibility. According to a Thomson Reuters Foundation survey conducted in November 2024, only 25% of companies reported engaging in any kind of AI-related reskilling, and none of them gave any significant information (reuters.com). This opacity exacerbates employees' fears, erodes confidence, and sparks criticism. Companies should instead take proactive measures, such as putting in place training programs that direct displaced employees into RPA tool oversight positions or retraining them for open analytics and decision-support duties.

5.1.3. Adoption of Ethical AI Deployment

Humanizing automation with compassion and intention is beneficial. Organizations should emphasize opportunities over obsolescence and convey the message, "We're not eliminating people—we're realigning them." Examining trailblazers such as Amazon, where male-biased AI CV screening surfaced after deployment, demonstrates ethical oversight when designing AI systems especially in sensitive areas. Needless to mention that recruitment is just an example in this case.

5.2. Transparency and Trust in Automated Decisions

5.2.1. The Issue of The "Black Box"

AI-enabled RPA systems frequently function in an opaque manner as they offer little to no insight into their internal workings. This raises questions on trust, fairness, regulatory issues and debugging the problem. Particularly in fields like finance where choices have real repercussions, trust erodes. The significance of switching from black-box to glass-box systems not only function but can also provide an explanation for their actions.

5.2.2. Documentation and Explainability

Local Interpretable Model-agnostic Explanations (LIME) and SHapley Additive exPlanations (SHAP), two model-agnostic techniques, provide transparency at scale.

It becomes crucial to record procedures using thorough model logs that include training data, versioning, and decision pathways. For instance, a finance company that uses RPA to determine eligibility should have enough audit records to demonstrate how a "declined" outcome was generated. According to PwC (2024), in addition to outcomes, investors are becoming more concerned with the disclosed decision criteria in a clear and understandable manner.

5.2.3. Increasing User Confidence

The trust increases when clients and staff are aware of the rationale behind decisions. Confidence is increased when one explains: "We used X data, baseline Y, and here's the logic." To increase user confidence, managers should receive alerts even from automated backgrounds, such as corn jobs. This will empower stakeholders rather than confusing.

5.3. Accountability and Governance

5.3.1. Shared Accountability

In case of a mistake in a tax file or misplaces a payment, the question arises who is responsible? The literature cautions against "imperfect delegation," in which human actors reject the results of the machine. It is necessary to set up a clear chain of accountability that includes developers, operational managers, audit teams, and compliance officers. If the chain of accountability is missing, it may lead to developing moral hazard.

5.3.2. Frameworks for Audits

Structure is provided by evolving governance tools. Annual audits of AEDTs for employment decisions were required in New York City by Local Law 144 (2023). Although it is still in its infancy, it provides an RPA prototype that includes transparent reporting, independent audits, and precise definitions. Similar to this, ethics-based auditing frameworks (EBA) offer strong standards as a methodical approach to accountability, encompassing everything from consistency to stakeholder rights.

5.3.3. Architectures of Organizations

The best way to embed responsibility is through multi-layered governance, which includes cross-functional AI governance teams, ethics committees, and board-level oversight. Companies such as SAP have incorporated "Responsible AI" into their yearly ESG reports (pwc.com). When RPA implementations go through ethical review boards that examine risk, equity, transparency, and appeal procedures, exposure is reduced.

5.4. Data Privacy and Confidentiality

5.4.1. Minimization of Data

RPA frequently manages sensitive and personal data, including financial records, employee information, and customer data. Purpose limitation and data minimization are required by ethical design; only the information required for each task should be processed (resources.workable.com). Respect for individual privacy is communicated from the outset by this principle.

5.4.2. Mindfulness and Consent

Customers and staff need to be aware of when their data is processed. They also need to know whether this complies with local data protection laws, the EU AI Act, and the GDPR. For example, in HR settings, organizations should grant employees control over records and secure informed consent for RPA-driven data collection. In October 2023, Italy's privacy authority placed a strong emphasis on decision transparency, algorithmic non-discrimination, and human oversight.

5.4.3. Design Security

Data security cannot be compromised. Strict access controls, anonymization, and encryption form the technical foundation of trust. RPA surveillance, including keystroke analysis, walks a tightrope; ethical behaviour necessitates a clear purpose, opt-in procedures, and regular evaluation.

5.5. Ethical Design and Bias

5.5.1. Bias Sources

Patterns in data, such as skewed training sets or past hiring biases, allow bias to infiltrate. Examples include HR bots prioritizing based on education or gender proxies on optezo.com and male-favouring recruitment AI at Amazon, according to reporters.com. There are ample evidence of algorithmic bias in many fields.

5.5.2. Identification and Prevention

Multi-channel tactics are necessary to mitigate the biases such as a variety of training data, systematic internal and external audits and code that considers fairness. Also, ongoing future performance monitoring is important. Techniques such as SHAP and LIME reveal demographic correlations that indicate bias in addition to explaining decisions. Data anonymization has increased fairness in HR with IBM and BBC's blind-recruitment systems.

5.5.3. Frameworks of Ethics

The first step in creating ethical RPA is to incorporate inclusion and fairness into design objectives. Early bias remediation is supported by the UK and US's shift to mandatory auditing in conjunction with oversight bodies. Diverse viewpoints are recorded early on when diverse teams are integrated into development stages.

5.6. Regulatory Compliance

5.6.1. Important Factors Influencing Regulations

The regulatory environment is expanding globally. The EU AI Act, which was approved in 2024, requires human oversight, bias checks, documentation, and risk-based classification. General Data Protection Regulation (GDPR) upholds a “right to explanation,” purpose limitation, and data rights. AI risk disclosures are now mandatory in several US states, including California and Colorado, according to PwC.com. Defective credit-affordability algorithms are penalized by the UK Financial Conduct Authority (FCA).

5.6.2. Integrating Compliance and Ethics

Ethics and compliance are complementary to each other, so don't think of them as twin burdens. Documented oversight, meticulous bias tracking, and transparent procedures all demonstrate corporate responsibility in addition to legal compliance. Investors prefer transparent, well-organized AI governance frameworks, according to PwC (pwc.com). Ethical RPA is not only moral, but also necessary for operating licenses in regulated sectors like accounting and finance.

6. REGULATORY AND COMPLIANCE LANDSCAPE

In today's ever changing global regulatory and compliance environment the implementation of Robotic Process Automation (RPA) has complex interactions in the automated financial environment. RPA is becoming an important component in the banking, finance and other related industries. This is emerging as a key factor of digital transformation strategies as the finance industry is focusing to improve operational efficiency and lowering costs. This adoption process also has some challenges especially when considering compliance with dynamic regulatory frameworks. In this section we will examine the main components of the regulatory and compliance environment of RPA functions.

6.1. Overview Of Relevant Financial Regulations

Globally, various laws have been created to ensure different stakeholders' protection in the financial system. These laws ensure and provide a mandate for using RPA in financial operations of any organization. Some of these regulations are discussed here:

6.1.1. The Sarbanes-Oxley Act (SOX Act)

In reaction to well-known corporate scandals like Enron and WorldCom, the United States passed the Sarbanes-Oxley Act of 2002 (SOX). By requiring stringent reforms, it seeks to rebuild public confidence in corporate governance and financial reporting (Coates, 2007). One of its most notable clauses, Section 404, mandates that management and outside auditors provide a report on how well a company's internal controls over financial reporting are working.

SOX is especially important from an RPA standpoint. Controls must be appropriately designed, documented, and auditable for any RPA implementation that affects financial reporting systems. A Deloitte report from 2019 states that RPA systems have the potential to either improve SOX compliance by decreasing manual errors and increasing process consistency, or they can compromise it if bots are implemented without proper control or change management procedures.

6.1.2. The General Data Protection RegulatioN (GDPR)

Due to its extraterritorial reach, the General Data Protection Regulation (GDPR), which was implemented in the European Union in 2018, has had a significant worldwide impact. The GDPR places stringent restrictions on how personal data is gathered, stored, processed, and transferred (Voigt & Von dem Bussche, 2017).

GDPR requirements, such as data minimization, access control, and the right to erasure, must be met by RPA platforms that interact with Personally Identifiable Information (PII), such as in supplier management, employee payroll, or customer onboarding. The problem here is that bots can quickly and repeatedly access large amounts of data because that is their design. They might handle or store data in ways that are against privacy laws if they are not monitored (Kumar & Tripathi, 2020).

6.1.3. Reporting Requirements and Tax Laws

These days tax authorities in the majority of countries are trying to integrate national and international tax regimes digitally. This is required as a compliance for better accuracy, consistency and timely reporting of business transactions in the electronic form. For example, Indian tax authorities require businesses to e-invoice and file real-time returns as per the Goods and Services Tax (GST). As per the Common Reporting Standard (CRS) and the OECD's Base Erosion and Profit Shifting (BEPS) initiative multinational corporations are also subject to strict data reporting requirements (OECD, 2021).

Finance professionals are using RPA to automate data validation, tax computations, and return filings across borders. This also increases the risk of regulatory violations if the automating is out-of-date or incorrectly interpreted tax logic is used.

6.1.4. Regulations Particular to Industry

A number of industry-specific rules are pertinent in addition to general financial and data privacy laws. For example, Know Your Customer (KYC) and Anti-Money

Laundering (AML) regulations are crucial in the financial industry, and the Payment Card Industry Data Security Standard (PCI DSS) is applicable to any organization that processes card payments. Document verification, transaction monitoring, and exception handling are frequently required to comply with these mandates; these are tasks that are increasingly being assigned to RPA solutions (Accenture, 2021).

6.2. RPA Interaction With Compliance Obligations

There are many chances to improve compliance through the strategic application of RPA. However, this potential can only be achieved if the implementation is carefully incorporated into the organization's larger governance, risk, and compliance (GRC) framework.

6.2.1. Facilitating Ongoing Adherence

RPA's capacity to encourage ongoing compliance is among its most frequently mentioned advantages. Bots follow defined workflows with high repeatability, in contrast to human workers who may vary in accuracy or efficiency. When properly constructed, RPA bots are able to record each action, transaction, and change to data, generating a digital audit trail that facilitates both external regulatory inspections and internal audits (EY, 2023).

Additionally, bots can be set up to carry out real-time compliance checks. They can, for instance, check the accuracy of financial disclosures, compare vendor data to government blacklists, or notify compliance officers of irregularities—tasks that would otherwise involve a great deal of manual labour.

6.2.2. Control Activity Automation

Additionally, RPA facilitates the automation of internal control tasks. These consist of workflows for approval, documentation archiving, reconciliation procedures, and job segregation. A bot, for instance, can guarantee that no one person has access to both the payment authorization and invoice creation systems, which is a crucial SOX control (PwC, 2021).

Additionally, by automating regulatory reporting, RPA can lessen the workload for compliance departments. In countries like India and the United Arab Emirates (UAE) the compliance deadlines are getting shorter whereas the penalties for non fulfillment of compliance requirements are getting higher and higher. In this scenario RPA can be used to extract data from enterprise systems and submit it electronically, which reduces the possibility of errors or omissions.

6.2.3. RPA System Governance

On the one hand RPA improves efficiency and saves time whereas on the other hand it creates a burden of new compliance when processes are automated in the organization. Some of the examples of such burdens may be management of bots' lifecycle, access control, version control and change approval procedures that are important for effective RPA governance. Avoidance of these things may question the reliability and auditability of bots.

Information Systems Audit and Control Association (ISACA) (2020) states that there should be a framework of IT general controls (ITGCs) which should govern the RPA. This governance will ensure that bots are being used in a secure environment, any modification in bots is being recorded and tested, and if there is any unusual or unauthorised identity then it is being supervised.

6.3. Risks of non-Compliance due to Faulty Automation

Like everything has both positive and negative aspects, RPA is also not unaffected by negatives. One of the negative points is mistakes or defects in automation. Automation costs heavily when automation goes wrong in a regulatory setting. Some of the reasons for non-compliance due to faulty automation can be poor process design, lack of proper testing, lack of proper monitoring or poor change management.

6.3.1. Automating Inaccurate Procedures

The basic mistake in robotic process automation (RPA) is due to the flaws in the process. This is possible if the automation is set up using the inaccurate rules or wrong data. This can be understood with an example where input itself is not accurate, how can one expect the output to be correct. For example, if the RPA is fetching wrong data from invoices, obviously it will lead to wrong calculation of Tax and consequently wrong filing of return which will be subject to penalties, interest as well as damage to the goodwill of the organization.

This focuses on the importance of automation processes. Understanding the right process, data as well as expectations from such data and processes play very important roles in RPA. This understanding may avoid various losses as discussed above like penalties, interest as well as damage to the goodwill of the organization.

6.3.2. Inadequate Escalation and Error Management

Generally, RPA systems are based on some cause-and-effect relationship or in simple words we can say that they are based on meeting specific conditions unless mapped with artificial intelligence (AI). If the organization is using strong error-handling procedures for RPA, errors can be identified and corrected. But in the absence of such error-handling procedures it will fail.

According to a 2020 research study by Forrester, not handling exceptions properly was the cause of RPA failures in finance functions. The study revealed that it caused 40% of RPA failures. This failure not only creates operational hurdles but also puts regulators at risk for noncompliance if the data is not accurate.

6.3.3. Unauthorized Bots and Shadow IT

The spread of "shadow bots," or automated systems created by business users without IT or compliance oversight, is another new danger. Models of citizen development can circumvent official change management and security procedures, even though they might spur innovation. Unauthorized bots may operate in non-compliant ways, circumvent audit trails, or access sensitive data (Gartner, 2022).

To reduce this risk, a centre of excellence (CoE) for RPA governance must be established, with clear responsibilities for IT departments, business process owners, and compliance officers.

6.3.4. Privacy Violations and Data Leakage

The possibility of data leakage becomes a major worry as bots handle vast amounts of financial and personal data. Bots with incorrect configurations may store private information in unsafe places, send reports to the incorrect people, or disregard GDPR or other privacy laws' requirements for data deletion and retention.

In addition to monetary penalties (such as the GDPR's maximum fine of 4% of worldwide turnover), a breach damages stakeholder trust, which is more difficult to measure but more difficult to regain.

6.3.5. Lag and Regulatory Ambiguity

And lastly, the problem of regulatory lag. Regulations don't always reflect the latest advancements in technology. For instance, even though RPA might technically do the same task as a human clerk, it's not always obvious if the current regulations governing access, liability, and reporting apply in the same way. Especially in the fields which are highly regulated like insurance underwriting. To minimize such uncertainties, organizations need to actively interact with regulators and industry forums in addition to legal counsel.

7. MITIGATION STRATEGIES AND BEST PRACTICES

Worldwide the businesses have adopted a cautious and responsible implementation of Robotic Process Automation (RPA). It has entirely reshaped the business processes by cost cutting, enhancing productivity and mitigating the repetitive activities for the workforce. However, these benefits are associated with some risks such as operational nonuniformity, security faults, job displacement, and ethical concerns that need to be controlled carefully. For these, businesses must adopt a holistic reduction strategy and adhere to the best practices. In this manner RPA can ensure its optimum performance in an ethical and sustainable manner and thus clear the hindrances. Some strategies that may play crucial role in mitigating risks in RPA are:

7.1. Implementing Ethical Guidelines and Automation Governance

Fusion of automation and ethics seems like an odd combination, but they are highly interrelated. A number of tasks in the businesses are well defined and therefore these are being automated effectively by the bots which are gaining the ability to make decisions. The possibility of any ethical hazard such as biases in data processing, lack of accountability, and unfair results for staff and customers can result from improperly managed assignments.

Explicit rules for the suitable use of automation can be achieved by preparing and practicing the proper ethical guidelines. These rules give priority to the transparency, accuracy, equity and accountability in the work. Therefore, for the tasks that directly impact human rights, business should not adopt a complete automation process unless a perfect mechanism of human insight is achieved.

It is very important to frame a Governance framework that requires fixing the review policies, determining the roles and responsibilities, and setting up the governance committees. Many business organizations have set up a Centre of Excellence (CoE) for the RPA system. This unit administers the selection, growth and management of bots that in turn ensures the adherence of ethical standards across the organization (Aguirre & Rodriguez, 2017).

Algorithmic Impact Assessment (AIA) is an ethical framework that evaluates the operational and societal effects of automation (Reisman et al., 2018). These assessments analyse the possible risk and present an organised method to settle down these risks in advance.

7.2. Workforce Reskilling and Change Management

It is a common belief that RPA bots are replacing human jobs. However, instead of reducing the jobs, RPA is reducing the time taken in accomplishing any task by removing the repetitive work. To adopt automation and create its benefits, organizations must focus on enhancing the skills of the workforce through training and development so that they acquire the capability of performing more strategic work and analytical tasks.

It has been predicted by the World Economic Forum (2020) that the loss of up to 85 million jobs may be the consequences of automation but at the same time it may also lead to the creation of about 97 million new ones. In order to keep up with this change, businesses must invest in up-skilling their workforce. The areas where humans still perform better than machines—digital literacy, data analysis, creative problem-solving, and interpersonal skills—should be the main emphasis of training programs.

In addition to training, businesses need to manage change effectively. Change can be uncomfortable, particularly for workers who are afraid of being replaced. This shift can be facilitated by open communication, leadership backing, and staff participation in the automation process. Long-term success depends on forming a coalition, empowering action, and acknowledging short-term victories, as John Kotter (2012) argues in his well-known change management model. Better RPA adoption, fewer disruptions, and even higher employee morale are frequently observed in organizations that proactively manage change and assist their employees through it (Deloitte, 2021).

7.3. Ensuring Auditability and Traceability in RPA Processes

RPA must be auditable and traceable, just like any other business process. This implies that companies ought to be able to see what bots did, when they did it, and why. Not only is this a good practice, but it's frequently required by law, particularly in sectors like public services, banking, and healthcare. RPA tools ought to offer comprehensive logs that document each action a bot takes, including inputs, outputs, time stamps, and criteria for making decisions. These logs are necessary for pinpointing the problems, conducting a root cause analysis and ensuring that the RPA bots are working as per the requirements.

Some regulations such as General Data Protection Regulation (GDPR), Health Insurance Portability and Accountability Act (HIPAA) and Sarbanes-Oxley Act (SOX) need rigorous controls on data processing and its access. Traceability supports these regulations and ensures that bots are accessing the data that is actually required. It also guarantees that all the interlinkages, a crucial component, are securely recorded (Willcocks et al., 2015).

For enhancing the accountability, business firms connect the RPA tools with their centralized monitoring platforms. It sends out the alert signals whenever something wrong happens. In addition to this, version control is used to inspect the updation on the script and modifications over the time period. Therefore, bots can function only under vigilance as these are not humans. A strict logging and traceability is required for creating trust and ensuring compliance.

7.4. Regular Monitoring, Review, and Stakeholder Communication

RPA is a technology that after setting up requires further attention and management. Even after the bots go live, ongoing supervision and analysis are needed to ensure that they are functioning properly and providing values. Some of the common causes of errors are unanticipated boundary cases in data, upgrade in the regulations, and modification in the source systems. Therefore, a sound system is necessary in real time to monitor the RPA bots on a regular basis.

Monitoring is done on the various performance indicators such as completion of task on time, minimum execution time and reduction in error rate. Team insight is provided to the bot operations through various tools such as altering systems and dashboards. In case of the occurrence of some problem, these tools enable the bot system to take quick actions (UiPath, 2020).

It is also necessary to examine the RPA system from time to time on whether the automation should be continued or not. It needs to be tested on the parameters if some better effective methods exist or the task has become obsolete. For keeping the alignment between the company objectives and automation strategy, these reviews are to be conducted regularly on a quarterly or semi-annually basis.

Alongwith the technical aspects, communication with stakeholders is also an important aspect. As the employees, management and IT teams are kept informed, there is trust and transparency in the organization. A feeling of common aim and equity is created by automation, staff and management meetings, regular updates and imparting a feedback system in the organization. As per the study conducted by Davenport & Ronanki (2018), a culture of unbarred communication and constant improvement enhances the ability of business houses to adjust to the changing technology and concur the potential risks and threats.

7.5. Security Protocols

One of the most vital ingredients of RPA that cannot be compromised is security. RPA bots have accessibility to the most crucial components of the business that is its critical system and sensitive data. If these are not properly secured then it may result in the security violation in the form of cyberattacks, fraud and data breach.

One of the preliminary steps is the use of Role Based Access Control (RBAC) that ensures that RPA bots can access only the required information (Kroll et al., 2020). It suggests that the developers should manage and keep the passwords using the secure credential vaults. This method is more promising than hardcoding the credentials into various scripts. In addition to this, encrypting the sensitive data both during the storage and transit is very important. This avoids the interception or leakage of data especially when personally identifiable information (PII) or financial figures are being processed by the RPA bots.

It is highly recommended that the RPA bots should be completely isolated from the production systems. Sandboxing Technique is very successful in minimizing the damages in cases where bot exhibits unanticipated behaviour. Integration of Security Information and Event Management (SIEM) tools with the bot system provide the advantages of anomaly recognition and real-time alerts.

Lastly it is recommended that the business houses must conduct penetration tests and vulnerability assessments on a regular basis. Adherence to the established frameworks such as ISO/IEC 27001 and NIST standards ensures an additional shielding and assurance (World Economic Forum, 2020).

8. CASE STUDIES OF RPA ADOPTION AND LESSONS LEARNED

8.1. Case Studies

8.1.1. American Express Global Business Travel (Amex GBT)

Amex GBT is the world's leading B2B travel platform. It provides software and services to manage travel, expenses, and meetings & events for companies of all sizes. The company has travel professionals in more than 140 countries. The company adopted RPA to automate invoice processing and expense management. RPA was used by Amex GBT to automatically update accounting systems, match purchase orders, and extract and validate invoice data.

After incorporation of RPA, it was found that 75% less time was being spent on processing invoices. This RPA also helped in attaining Over 99% data accuracy. In addition to enhancing contentment among employees by enabling them to concentrate on tasks that add value, strong executive support, well-integrated IT-business collaboration, and transparent RPA governance also contributed to the success of Amex GBT. This case highlights the significance of identifying rule-based, high-volume processes early in the automation process, according to Deloitte (2021).

8.1.2. ANZ Bank

Australia and New Zealand Banking Group Limited (ANZ) used RPA to enhance operational efficiency, customer service, and decision-making processes. ANZ used RPA to revolutionize customer interactions by understanding customer queries, providing personalized responses, and assisting in routine transactions. This helped ANZ in increasing customer satisfaction in addition to reducing service cost.

ANZ also employed RPA to analyse vast amounts of transactional data in real-time, identifying unusual patterns indicative of fraud to enhance accuracy in detecting fraudulent activities, safeguarding both the bank and its customers. Further, AI-driven predictive analytics was also used by ANZ in wealth management, retirement planning, and investment strategies for individual customers by analysing historical data.

The impact of adopting RPA is tremendous not only in terms of improved quality but also in terms of processes. Today, staff are able to focus more on analysing the data, rather than try to make the repetitive process work. And that is the kind of value-addition that drives success for ANZ Bank.

8.1.3. KPMG India

RPA was implemented by KPMG India to optimize risk assurance and internal auditing processes. To manage repetitive data collection, ledger analysis, and compliance reporting, KPMG deployed bots. This RPA adoption improved precision, audit productivity, internal controls and service turnaround time in addition to decreasing operational risk. This proved that RPA works best when it is viewed as a strategic endeavour rather than just a cost-cutting measure by incorporating it into its larger digital transformation agenda.

8.2. Lessons Learned From Real Organizations

8.2.1. Strategic Alignment and Leadership buy-in

The organizations that considered RPA with business strategic alignment rather than a cost cutting measure were more successful. One example of such an organization is ANZ Bank. The leadership of ANZ Bank took RPA as a broader transformation vision rather than a cost-cutting measure. This move helped the bank in attaining internal support as well as resources.

On the other hand, companies with disjointed business and IT strategies—like the failed European bank case—had trouble coordinating execution. Digital transformation is not possible without effective leadership and cross-functional alignment (Dery et al., 2020).

8.2.2. Importance of Process Selection

It is essential to choose the appropriate processes for automation. High-volume, rule-based, stable processes with few exceptions were the focus of those cases that adopted RPA successfully. On the other hand, failure was the result of processes that were subject to frequent changes, judgmental decisions, or unstructured inputs. Inappropriate process selection accounts for almost 30% of failed RPA projects, according to the Institute for Robotic Process Automation & AI (IRPAAI) (IRPAAI, 2022). This risk can be reduced with the use of tools like task mining and process mining for a comprehensive process discovery and suitability analysis.

8.2.3. Incremental Implementation and Pilot Program

Starting at a small scale is another important lesson. Before expanding, the majority of prosperous businesses have tested RPA in one or two departments. KPMG India, for example, began with internal auditing before branching out into more general financial operations.

However, extensive rollouts without pilot testing frequently led to low adoption and resource strain. Testing, feedback, and course correction are made possible by incremental approaches.

8.2.4. Change Management and Workforce Engagement

One of the most frequent obstacles to RPA implementation is resistance to change. As demonstrated by ANZ Bank's example, humanizing bots (e.g., naming them or framing them as assistants) and incorporating staff members early in the design process increase acceptance. If the workforce is not involved in this process, then it results in poor adoption, job insecurity, and fear. Successful businesses frequently combine RPA with reskilling initiatives, giving staff members new responsibilities in analytics, exception handling, or automation supervision.

8.2.5. Governance, Monitoring, and Compliance

Automation may be risky without governance. Effective RPA implementers create governance frameworks to keep an eye on bot activity, guarantee adherence, and carry out frequent audits.

Data security breaches and untracked bot failures were some of the problems caused by inadequate or nonexistent governance in unsuccessful implementations. It's very important for an organization to create exception logs and automated alerts.

8.2.6. Integration and Technological Readiness

RPA success is determined by technological preparedness, which includes infrastructure, integration capabilities, and API availability. Poor tech stack compatibility stalls performance, as the insurer who failed to integrate RPA with legacy systems.

Businesses that make sure RPA tools work with both legacy and modern systems and get ready for future scalability with intelligent automation (AI + RPA) do better.

9. FUTURE OUTLOOK

The robotic process automation (RPA) was started as a tool of cost reduction, but it has now made its own standing in the business world. The RPA was started for straight forward rule-based functions using machine learning which is now gradually shifting in different directions after the incorporation of Artificial Intelligence (AI). In this section, the future outlook of RPA considering three important areas: ethical debates with AI-enhanced RPA, anticipated regulatory shifts and technological advancements and the role of human oversight in an increasingly automated environment are discussed further.

9.1. Evolving Ethical Debates With AI-Enhanced RPA

Initially, ethical issues were limited to labour displacement and transparency while using RPA. The discussion over ethical issues has increased further after the incorporation of AI into RPA. After incorporating AI into RPA, it can be used to make decisions in unknown circumstances in addition to drawing conclusions. This has resulted into various ethical concerns regarding autonomy, accountability and bias as a result of this integration.

Another ethical issue is about the bias of algorithms. RPA works on the basis of algorithms and the way historical data has been used to train AI models built into RPA platforms which may be influenced by societal biases. This is very much possible, especially, in the finance sector which can further lead to unethical practices (Mehrabi et al., 2021).

Another issue regarding AI-enhanced RPA is known as the “”black box” problem (Burrell, 2016). Sometimes AI-enhanced RPA provides decisions which are not even understood by their developers. This also means that there is lack of logic in the decision given by the AI-enhanced RPA and becomes more crucial when it comes to set accountability.

Job displacement is still one of the major issues. The supporters of RPA assert that it allows the human workforce to concentrate on higher-valued tasks whereas, the critics argue that RPA may lead to growth in digital divide and unemployment (Bessen, 2019). This discussion turns out to be an ethical discussion demanding to maintain a balance between economic efficiency and social responsibility, especially in the developing countries where job automation may lead to the worst impact on income equality.

Ethical governance frameworks such as “Ethics by Design” have been introduced for inculcating the ethical aspects in the evolution lifecycle of AI-enhanced RPA systems (Floridi & Cowls, 2019). This approach includes existing stakeholder engagement, ongoing ethical auditing, and transparent data usage. Corporations such as IBM and Microsoft have addressed the ethical administration of AI-driven automation by incorporating internal ethics boards and review processes in the organizations.

Privacy and surveillance concerns are high as the RPA tools are capable of tracing and examining human behaviour via biometric data, keystrokes and logs. Although productivity and compliance has enhanced because of these abilities, it bears the risk of extensive monitoring and a loss in workplace autonomy. It also provides a model but legal safeguarding such as the General Data Protection Regulation (GDPR) in the European Union are not always relevant at global level.

Incorporation of AI into RPA has completely changed the ethical prospect. It has aroused a necessity of adopting a much higher proactive and inclusive ethical framework by technologists, regulators and businesses. Ethical automation is not just a compliance concern but is a basic ingredient of long-term digital transformation.

9.2. Anticipated Regulatory Shifts and Technological Advancements

Rapid technological advancement and changing regulatory environments will also influence RPA's future. It is anticipated that regulatory actions will become more organized, complex, and internationally standardized as governments and international organizations have started to understand the profound effects of AI-driven automation.

The formalization of AI governance frameworks is one of the most anticipated regulatory changes. Under negotiation, the European Union's AI Act aims to categorize AI applications according to risk levels, from low to unacceptable, and imposes requirements for compliance (European Commission, 2021). RPA systems with embedded AI are probably going to be classified as high-risk, requiring strict oversight, transparency, and data protection procedures, especially when they are used in decision-making processes that have an impact on individuals.

In the US, states like California and New York have started enacting local laws pertaining to algorithmic transparency and data privacy, even though federal AI regulations are still in their infancy. Globally, nations like the UAE, Canada, and Singapore are creating data governance and AI ethics regulations that will have a direct impact on the use of RPA (Jobin, Ienca, & Vayena, 2019).

Furthermore, sector-specific rules are being framed. Regulators in banking and finance are becoming more and more worried about the application of AI to risk assessment, fraud detection, and credit scoring. Discussion papers stressing the necessity of AI accountability, transparency, and dependability in financial automation have been released by the Basel Committee on Banking Supervision (BCBS, 2021). The Health Insurance Portability and Accountability Act (HIPAA) is undertaking an assessment to solve AI-related processing of health data in the US.

Technological advancement is resulting in RPA being more and more cognitively knowledgeable. Bots can even apprehend the unstructured text. Bots can be engaged in discussion with the users with the help of the integration with Natural Language Processing (NLP). Technologies such as context-aware computing, autonomous agents and computer vision are playing an important role in redefining the capabilities of RPA.

Intelligent Process Automation (IPA), which integrates RPA with AI, that is, process mining and analytics, is evolving as a significant trend. Gartner report forecasted that by the end of the year 2030, IPA is going to contribute significantly in assisting hyper automation with businesses that focus to automate as much as 70% of their operations (Gartner, 2022). To guarantee a proper conduct, these advancements will need embedded controls, dynamic compliance procedures and real-time audit trails.

Because of cloud-based RPA, low-code/no-code platforms, and API-first architecture, the use of Bot design and functioning has become easier for non-technical users. The emergence of this democratic system of automation is gaining strength and if it is not supervised properly, then it will raise the risk of decentralized governance and "automation sprawl".

One more emerging area is the amalgamation of RPA and blockchain. Trust in automated transactions, security, and traceability is enhanced by blockchain's smart contract capabilities and immutable audit logs (Mendling et al., 2020). However, there is a lack of clarity in the regulations regarding the use of blockchain and cryptocurrency that make the integrated solutions much more difficult.

The regulatory landscape of the future must be multi-stakeholder, anticipatory, and flexible. Keeping up with technological advancements will require ongoing regulatory discussions, sandbox environments, and international collaboration rather than static compliance checklists.

9.3. The Role of Human Oversight in an Increasingly Automated Environment

Even with RPA's claims of accuracy, scalability, and efficiency, human oversight is still crucial. Human-in-the-loop mechanisms must develop in tandem with increasingly autonomous and AI-driven systems to guarantee resilience, accountability, and trust.

First and foremost, risk management and governance depend on human oversight. Automated systems are susceptible to error, compromise, and malfunction. An excellent illustration of AI decision-making gone wrong because of a lack of human intervention is the 2020 incident involving a UK government algorithm that degraded student grades during the pandemic (The Guardian, 2020).

In addition to monitoring, human oversight also entails creating, honing, and verifying AI models integrated into RPA. To develop context-aware and organizationally values-aligned systems, domain experts, data scientists, ethicists, and operational personnel must work together. This is especially important in fields like healthcare, where choices have a direct impact on people's lives.

A more inclusive and balanced paradigm is provided by the idea of augmented intelligence, in which human and machine intelligence collaborate. While humans concentrate on interpretation, empathy, and strategic thinking, RPA manages repetitive, high-volume tasks in augmented workflows. Bots can answer questions in customer service, for instance, but human judgment and emotional intelligence are still needed for escalation and resolution.

The workforce must also be continuously trained and upskilled. To enable staff to oversee, manage, and develop RPA systems, organizations must make investments in digital literacy, data ethics, and automation fluency. According to research, businesses that successfully incorporate RPA into their HR strategies have reported increased employee satisfaction and productivity (Willcocks, Lacity, & Craig, 2017).

Organizations should create oversight committees for automation and incorporate "kill switch" mechanisms to step in the event of unanticipated outcomes from a governance perspective. Standard procedures should include independent audits and ethical review boards, especially for high-stakes automation in industries like finance, defence, and the legal system.

Lastly, cultural adjustment is essential. Workers need to view automation as a partner rather than a threat. A narrative of "humans and machines" must be promoted by leadership instead of "humans versus machines." Social cohesiveness and trust are just as important to the adoption of RPA as technological prowess.

10. CONCLUSION

Robotic Process Automation (RPA) in accounting and finance is more than just a new piece of technology. RPA has evolved as a new style of performing, managing, and experiencing financial operations. Organizations have realized that it has operational benefits like speed, accuracy, and cost-effectiveness. But they need to understand that RPA is not the size that fits all solutions. To make RPA implementation successful, organizations need to carefully plan and monitor the RPA along with focus on its people. To deal with issues regarding process suitability, data integrity, systems fragility and moral dilemmas, organizations need to deploy both technical and empathetic leadership.

The leadership of the organizations also need to see that automation should help people in reskilling themselves, not force them out of jobs without support. This is why reskilling the workforce, clear communication, and cultural inclusion are so important. This is the reason that transparency, accountability and fairness are beliefs of ethical RPA implementation. As the RPA evolves further and integrates with AI and machine learning, businesses require a system that can responsibly govern it. Organizations should focus on formalizing governance systems like ethical review boards, bot supervision and audit trails in the organization rather than simply improving the systems. Organisations can stay forward in this journey by putting efforts in spreading digital literacy, enhancing the collaboration of humans and machine systems, and keeping the automation as per the long-term societal and strategic values. Additionally, Organizations should consider compliance as an initial step instead of final destination. Integrating ethics into automation requires giving organizational integrity and social trust equal weight with financial benefits. In accounting and finance, RPA is successful if it carries speed along with intellect, equity and foresightedness.

REFERENCES

Accenture. (2021). Driving Value with Intelligent Automation. Accenture Digital Finance Report.

Aguirre, S., & Rodriguez, A. (2017). Automation of a Business Process Using Robotic Process Automation (RPA): A Case Study. In Figueroa-García, J., López-Santana, E., Villa-Ramírez, J., & Ferro-Escobar, R. (Eds.), *Applied Computer Sciences in Engineering. WEA 2017. Communications in Computer and Information Science* (Vol. 742). Springer., DOI: 10.1007/978-3-319-66963-2_7

Asatiani, A., & Penttinen, E. (2016). Turning robotic process automation into commercial success–Case OpusCapita. *Journal of Information Technology Teaching Cases*, *6*(2), 67–74.

Bessen, J. (2019). *AI and Jobs: The Role of Demand*. MIT Press.

Binns, R., Van Kleek, M., Veale, M., Lyngs, U., Zhao, J., & Shadbolt, N. (2018, April). 'It's Reducing a Human Being to a Percentage' Perceptions of Justice in Algorithmic Decisions. In *Proceedings of the 2018 Chi conference on human factors in computing systems* (pp. 1-14).

Boell, S. K., & Cecez-Kecmanovic, D. (2015). On being 'systematic' in literature reviews. *Formulating Research Methods for Information Systems*, *48*(2), 161–173.

. Burrell, J. (2016). How the machine 'thinks': Understanding opacity in machine learning algorithms. Big Data & Society, 3(1). https://doi.org/ (Original work published 2016)DOI: 10.1177/2053951715622512

Cath, C. (2018). Governing artificial intelligence: Ethical, legal and technical opportunities and challenges. *Philosophical Transactions. Series A, Mathematical, Physical, and Engineering Sciences*, *376*(2133), 20180080. PMID: 30322996

Coates, H. (2007). A Model of Online and General Campus-Based Student Engagement. *Assessment & Evaluation in Higher Education*, *32*, 121–141. DOI: 10.1080/02602930600801878

Davenport, T. H., & Ronanki, R. (2018). Artificial Intelligence for the Real World. Harvard Business Review (HBR). https://www.bizjournals.com/boston/news/2018/01/09/hbr-artificial-intelligence-for-the-real-world.html

Deloitte. (2021). Global RPA Survey. Deloitte Insights. https://www2.deloitte.com

Dery, K., Sebastian, I. M., & Ross, J. W. (2020). The Digital Transformation Journey: Capabilities, Outcomes, and Lessons Learned. *MIS Quarterly Executive, 19*(2), 123–139.

EY. (2023). RPA Governance and Risk Management in Financial Services. Ernst & Young Insights. https://www.ey.com

Fernandez, D., & Aman, A. (2018). Impacts of Robotic Process Automation on Global Accounting Services. *Asian Journal of Accounting and Governance, 9*, 123–132. DOI: 10.17576/ajag-2018-09-11

Fernandez, D., & Aman, A. (2018). Impacts of robotic process automation on global accounting services. Asian Journal of Accounting & Governance, 9.

Floridi, L., & Cowls, J. (2019). A Unified Framework of Five Principles for AI in Society. *Harvard Data Science Review, 1*, 2–15. DOI: 10.1162/99608f92.8cd550d1

. Gartner. (2022). RPA Market Trends and Technology Readiness. Gartner Research.

Gkritzali, A., Mavragani, E., & Gritzalis, D. (2020). Negative MWOM and value co-destruction during destination crises. *Business Process Management Journal, 26*(4), 839–856.

Institute for Robotic Process Automation & AI. (2022). Process Discovery and RPA Success Rates. https://irpaai.com

Jobin, A., Ienca, M., & Vayena, E. (2019). The Global Landscape of AI Ethics Guidelines. *Nature Machine Intelligence, 1*, 389–399. DOI: 10.1038/s42256-019-0088-2

Kotter, J. (2012). *Leading Change*. Harvard Business Review Press.

KPMG. (2020). Automation in audit: A practical guide. KPMG Insights. https://home.kpmg

Kumar, R., & Tripathi, R. (2020). Blockchain-based framework for data storage in peer-to-peer scheme using interplanetary file system. Handbook of research on blockchain technology. Academic Press. 35-59.

Lacity, M., & Willcocks, L. (2016). Robotic Process Automation at Telefonica O2. *MIS Quarterly Executive, 15*(1), 4. https://aisel.aisnet.org/misqe/vol15/iss1/4

Mehrabi, N., Morstatter, F., Saxena, N., Lerman, K., & Galstyan, A. (2021). A Survey on Bias and Fairness in Machine Learning. *ACM Computing Surveys, 54*, 1–35. DOI: 10.1145/3457607

Mendling, J., Pentland, B. T., & Recker, J. (2020). Building a complementary agenda for business process management and digital innovation. *European Journal of Information Systems*, *29*(3), 208–219. DOI: 10.1080/0960085X.2020.1755207

Moffitt, K. C., Rozario, A. M., & Vasarhelyi, M. A. (2018). Robotic process automation for auditing. *Journal of Emerging Technologies in Accounting*, *15*(1), 1–10.

Reisman, D., Schultz, J., Crawford, K., & Whittaker, M. (2018). *Algorithmic Impact Assessments: A Practical Framework for Public Agency Accountability*. AI Now.

Rozario, A. M., & Vasarhelyi, M. A. (2018). Auditing with smart contracts. *The International Journal of Digital Accounting Research*, *18*, 1–27. DOI: 10.4192/1577-8517-v18_1

Snyder, H. (2019). Literature review as a research methodology: An overview and guidelines. *Journal of Business Research*, *104*, 333–339. DOI: 10.1016/j.jbusres.2019.07.039

Stople, A., Steinsund, H., Iden, J., & Bygstad, B. (2017). Lightweight IT and the IT function: Experiences from robotic process automation in a Norwegian bank. *Bibsys Open Journal Systems*, *25*(1), 1–11.

Syed, R., Suriadi, S., Adams, M., Bandara, W., Leemans, S. J. J., Ouyang, C., Hofstede, A. H. M., Weerd, I. V., Wynn, M. T., & Reijers, H. A. (2020). Robotic Process Automation: Contemporary themes and challenges. Computers in Industry, Volume 115, pp. 103-162, ISSN 0166-3615, https://doi.org/DOI: 10.1016/j.compind.2019.103162

Van der Aalst, W. M. P., Bichler, M., & Heinzl, A. (2018). Robotic Process Automation. *Business & Information Systems Engineering*, *60*, 269–272. DOI: 10.1007/s12599-018-0542-4

Voigt, P., & Von dem Bussche, A. (2017). *The EU General Data Protection Regulation (GDPR): A Practical Guide* (1st ed.). Springer International Publishing., DOI: 10.1007/978-3-319-57959-7

Willcocks, L., Lacity, M., & Craig, A. (2017). Robotic Process Automation: Strategic Transformation Lever for Global Business Services? *Journal of Information Technology Teaching Cases*, *7*, 17–28. DOI: 10.1057/s41266-016-0016-9

Willcocks, L. P., Lacity, M., & Craig, A. (2015). The IT function and robotic process automation. The Outsourcing Unit. Working Research Paper Series.

KEY TERMS AND DEFINITIONS

Data Dependency: It represents the reliance of RPA on structured, high-grade, and persistent data inputs to accomplish the tasks effectively when human intervention is not provided.

Ethical Governance: To ensure the responsibility, compliance and fairness, a framework of principles and policies in the form of ethical governance is prepared for conducting an accountable deployment and use of automation technologies.

Process Suitability: It is the degree to which a business process can be valuably automated. Task repetitiveness, low variability, and rule-based structure are some of the factors that determine the process suitability.

Regulatory Compliance: To avoid the fines and maintain the trustworthiness, RPA- driven processes are being adhered to industry-specific financial regulations, such as Sarbanes-Oxley Act (SOX), General Data Protection Regulation (GDPR), and tax laws.

RPA: Robotic Process Automation is a technology that automates rule-based, high-volume, repetitive tasks by using software robots or "bots".

System Integration: It is a process that enables RPA bots to interface with the existing enterprise systems, for example, accounting software and ERP.

Workforce Displacement: It represents the possibility of the transformation or loss of jobs because of the automation of tasks that were traditionally performed by human workers.

Chapter 11
Integrating RPA in Investment and Wealth Management:
Transforming Portfolio Strategy and Efficiency

Chengcheng Wang
Software Engineering Institute of Guangzhou, China

Kelvin Yong Ming Lee
https://orcid.org/0000-0001-6884-1787
Taylor's University, Malaysia

Amirah Fatini Johari
Taylor's University, Malaysia

ABSTRACT

Robotic Process Automation (RPA) is rapidly transforming the landscape of investment and wealth management by enhancing efficiency, accuracy, and decision-making. This chapter explores how RPA is being integrated into portfolio management strategies. It examines the impact of RPA on client onboarding, compliance, portfolio rebalancing, and risk management, highlighting real-world use cases in asset and wealth management firms. The discussion also covers challenges related to implementation, such as regulatory compliance, data security, and integration with legacy systems. Furthermore, the chapter investigates how RPA, when combined with artificial intelligence and analytics, can support personalized investment solutions and improve customer experience. By redefining traditional processes, RPA not

DOI: 10.4018/979-8-3373-3571-1.ch011

only reduces operational costs but also empowers financial advisors to focus more on strategic planning and client engagement. This chapter offers practical insights for institutions looking to adopt RPA-driven solutions in their portfolio strategies.

1.0 INTRODUCTION

Digital transformation refers to the comprehensive integration of digital technologies into all aspects of a business, leading to fundamental changes in how organizations operate, deliver value to customers, and remain competitive in an increasingly digital economy (Kraus et al., 2022). In the financial sector, this transformation has been especially impactful and accelerated rapidly during the COVID-19 pandemic, as institutions were compelled to adopt digital solutions to meet shifting customer expectations and ensure business continuity (Fu & Mishra, 2022). The adoption of disruptive technologies such as Artificial Intelligence (AI), blockchain, and cloud computing has significantly reshaped financial services—transforming business models, enabling innovative product offerings, and enhancing operational efficiency (Naimi-Sadigh et al., 2022).

Digital transformation in finance has manifested in various ways, including the digitization of payment systems, the emergence of new forms of electronic money, and a widespread transition toward online and mobile banking platforms, especially in the post-pandemic landscape (Bueno et al., 2024). However, this transformation is not solely about implementing new technologies. It also involves rethinking organizational culture, structures, and strategic direction to harness the full potential of digital innovation (Ratna et al., 2024). A key technology driving this evolution is Robotic Process Automation (RPA), which has become increasingly popular in the financial industry. RPA involves the deployment of software robots or virtual assistants to automate repetitive, rule-based tasks that were traditionally carried out by humans. These include data entry, transaction processing, form filling, and information retrieval—functions that are central to banking, insurance, and investment operations (Meiryani et al., 2023).

By implementing RPA, financial institutions can streamline workflows, increase speed and accuracy, and reduce operational costs. Unlike traditional automation, RPA tools work across legacy systems without the need for significant infrastructure changes, making it an efficient and scalable solution for digital transformation. Moreover, when RPA is combined with other technologies such as AI and machine learning, it enables smarter automation—capable of learning from data, adapting to changes, and supporting more complex decision-making processes. This positions RPA not only as a cost-saving tool but also as a strategic enabler of digital transformation in the evolving financial ecosystem.

1.1 The Rise of RPA in the Financial Sector

Robotic Process Automation (RPA) has emerged as a key enabler of digital transformation in the financial sector, offering a practical and scalable solution for automating routine and rule-based operations. RPA refers to the use of software robots or "bots" that can mimic human actions to perform tasks such as data entry, information extraction, reconciliation, and document processing. By replacing manual processes with automated workflows, financial institutions can significantly improve operational efficiency, enhance accuracy, and maintain better compliance with regulatory standards (Zareen et al., 2024). This automation helps reduce processing time, eliminate human errors, and lower operational costs, making it especially beneficial for high-volume tasks like transaction processing, regulatory reporting, and customer onboarding (Abdullah & Almaqtari, 2024).

Beyond improving operations, the widespread adoption of RPA is also reshaping workforce dynamics in the financial sector. Rather than replacing jobs outright, RPA is shifting the nature of work, freeing up human employees from mundane tasks. This shift enables them to focus on strategic decision-making, client engagement, and analytical roles. This evolution has led to the emergence of new job roles such as RPA developers, automation analysts, and AI data scientists, reflecting the changing demands of the modern financial workplace (Yang et al., 2024). Furthermore, the integration of RPA with complementary technologies like artificial intelligence (AI), machine learning, and natural language processing is driving the next phase of automation known as hyper-automation. This advanced approach combines cognitive capabilities with process automation to deliver smarter, faster, and more adaptive solutions across various financial services functions. As a result, RPA is no longer viewed solely as a tool for operational efficiency but also as a strategic asset that enables personalized customer service, real-time risk management, and agile business innovation (Haleem et al., 2021).

The existing literature underscores the operational and strategic benefits of RPA adoption in financial institutions; there remains a notable gap in understanding its long-term implications for investment management practices. Most prior studies (e.g., Babina et al., 2024; Prasetyo et al., 2025) emphasize process efficiency and cost reduction but give limited attention to how automation reshapes portfolio decision-making, investor behavior, and risk governance frameworks. Furthermore, the integration of RPA into broader digital ecosystems is still in its nascent stage in financial research. Hyper-automation, which combines RPA with artificial intelligence (AI), machine learning (ML), natural language processing (NLP), and advanced analytics, enables end-to-end automation of cognitive and decision-intensive tasks. However, few empirical studies explore its application in investment management, particularly in areas such as predictive analytics, robo-advisory systems, and automated compliance monitoring. This highlights a significant opportunity for future research to examine how hyper-automation transforms financial decision-making and enhances strategic agility in investment and wealth management.

1.1.1 Statistics of RPA[1]

Robotic Process Automation (RPA) is rapidly transforming the financial sector by streamlining operations, increasing accuracy, and enhancing compliance. According to the findings of Baker Tilly (2024), nearly 75% of financial institutions have partially implemented RPA in their operations, while 25% have fully integrated RPA across multiple functions. This shows a predominant preference for phased or incremental adoption, allowing firms to test automation in low-risk areas before scaling up. Institutions are using this strategy to monitor impact, refine deployment, and tailor RPA according to their specific business processes. Baker Tilly (2024) also reported that over 80% of surveyed institutions are investing either in outsourcing RPA development or in training internal teams. However, a growing number of institutions are prioritizing in-house RPA capabilities. This reflects a broader understanding that automation isn't just about adopting a tool, but also requires internal ownership, skill development, and institutional learning to truly leverage RPA for long-term value.

The move toward internal talent development signals a strategic shift: rather than relying solely on vendors, firms are building dedicated RPA teams. Around 50% of respondents operate with small teams (1–3 members), testing targeted projects. 20% manage mid-sized teams (4–10 members), indicating scaling efforts. The remaining 20% operate large teams (11–25 members), handling advanced and enterprise-wide automation initiatives. This progression shows how organizations evolve through maturity stages. Financial institutions are strategically positioning themselves for a future where automation, data analytics, and intelligent systems are tightly integrated. As RPA evolves, institutions are expected to move toward intelligent automation, capable of adapting, learning, and optimizing in real time.

1.2 Key Benefits of RPA in Finance

In the evolving landscape of the financial sector, Robotic Process Automation (RPA) has become a strategic tool not just for operational efficiency but also for enhancing the overall agility and responsiveness of financial institutions. By automating routine, rule-based tasks, RPA enables institutions to reallocate human resources to higher-value activities, such as client engagement, analytics, and strategic planning. This shift supports better alignment with digital transformation goals, where real-time data access, process transparency, and service personalization are essential. Moreover, RPA can be integrated with technologies like artificial intelligence and machine learning to enable more intelligent automation, allowing financial firms to anticipate trends, mitigate risks, and improve decision-making. As financial institutions face rising customer expectations and complex regulatory environments, RPA serves as a foundational technology that helps build resilient, scalable, and innovative-driven operations.

1.2.1 Increased Efficiency

RPA automates high-volume, rule-based tasks such as data entry, transaction processing, and report generation. This reduces the time and effort required to complete routine activities, allowing human employees to shift their focus to more strategic, value-added functions such as client advisory, complex problem-solving, and innovation. By eliminating manual steps, RPA helps institutions accelerate business cycles and improve overall productivity (Zareen et al., 2024).

1.2.2 Enhanced Accuracy

Manual processes are prone to human error, particularly when dealing with large volumes of financial data. RPA mitigates this risk by executing tasks with consistent precision, leading to fewer discrepancies and enhanced data integrity. In areas such as transaction handling, reconciliation, and compliance reporting, RPA's accuracy significantly reduces the risk of costly mistakes or regulatory penalties (Yadav et al., 2025).

1.2.3 Cost Savings

One of the most attractive benefits of RPA implementation is its potential to reduce labor costs and operational expenditures. By automating routine processes, institutions can lower their reliance on large teams for manual processing. This does not necessarily mean replacing human employees but rather allowing organizations to optimize their workforce allocation and reduce the need for overtime, rework, and error correction (Ma & Jia, 2022).

1.2.4 Scalability

RPA systems are highly scalable and can adapt to changing business needs without requiring a proportional increase in headcount or infrastructure. For instance, during peak periods such as financial year-end or tax season, RPA bots can manage increased workloads efficiently and without interruption. This flexibility is crucial for organizations operating in dynamic financial environments (Flechsig et al., 2022).

1.2.5 Regulatory Compliance

RPA contributes to stronger regulatory compliance by ensuring consistency and transparency in reporting. Automated workflows can be designed to align with regulatory requirements, generate accurate reports, and maintain audit trails. This reduces the risk of non-compliance and enables quicker, more reliable responses to regulatory reviews or audits (Kokina et al., 2025).

1.3 Key drivers of RPA Adoption in Financial Institutions[2]

Robotic Process Automation (RPA) has become an essential component in the digital transformation strategies of financial institutions. The banking and financial services industry, characterized by vast volumes of transactions, regulatory compliance demands, and ever-evolving customer expectations, is an ideal candidate for RPA deployment. As financial institutions navigate a rapidly changing environment influenced by technological innovations, regulatory changes, globalization, and customer-centric demands, the adoption of RPA has proven to be a strategic response. This discussion explores the major drivers propelling the adoption of RPA in financial institutions, linking each factor with the transformative potential of automation technologies.

1.3.1 Evolving Consumer Expectations

One of the primary catalysts for RPA adoption in the financial sector is the shift in consumer expectations. The modern banking customer demands quick, seamless, and personalized services. The rise of digital-native generations and the increasing ubiquity of mobile banking have redefined what customers expect from their financial service providers. Today's consumers expect real-time services and fast responses, whether it's processing a loan application, updating account details, or transferring funds. RPA addresses this demand by automating time-consuming and repetitive tasks, such as data entry, document verification, and customer onboarding. This automation reduces processing time from day to minute, thereby significantly enhancing customer satisfaction.

With its advanced data handling capabilities, RPA enables institutions to offer personalized financial services. For example, bots can collect and analyze customer data to provide customized investment recommendations or loan products. Integration with AI further enhances this personalization by enabling predictive analytics and customer behavior modeling. This shift from standardized to personalized service models is increasingly becoming a differentiator for banks and other financial service providers.

1.3.2 Competitive Landscape

Financial institutions operate in a highly competitive environment where innovation and customer experience are key to gaining market share. The entry of fintech firms, digital banks, and other non-traditional players has intensified this competition. To maintain a competitive edge, traditional banks are leveraging RPA to improve operational efficiency and service delivery. RPA helps them reduce turnaround times for customer requests, improve service accuracy, and deliver consistent experiences across channels. For instance, automating the back-office processing of loan applications not only speeds up service but also minimizes human errors.

RPA adoption also supports innovation by enabling banks to develop new financial products and services more efficiently. By reducing the operational load on human employees, RPA allows more resources to be allocated to innovation-focused initiatives. Additionally, when combined with emerging technologies like AI, blockchain, and cloud computing, RPA can drive the creation of advanced financial services such as robo-advisors, smart contracts, and real-time fraud detection systems.

1.3.3 Regulatory Environment

Regulatory compliance is a top priority for financial institutions. The global financial ecosystem is governed by complex and ever-changing regulations, making compliance a significant challenge. In recent years, regulatory bodies have become more supportive of automation technologies, recognizing their potential to enhance accuracy and transparency. Regulators are increasingly encouraging institutions to adopt RPA for standardized reporting, audit trails, and risk management. RPA systems can generate compliance reports quickly and with high accuracy, ensuring timely and consistent reporting. With the rise in data breaches and heightened concerns over data privacy, RPA offers a secure solution for handling sensitive customer information. RPA bots operate based on pre-defined rules, reducing the risk of unauthorized data access and ensuring adherence to data protection laws such as GDPR, CCPA, and others. Automated processes also ensure data is processed consistently, minimizing the risk of non-compliance.

1.3.4. Globalization and Digital Transformation

The globalization of financial services and the digitalization of banking processes have brought new complexities that RPA is well-suited to address. The shift towards digital banking platforms has accelerated post-COVID-19, driven by consumer preferences and safety concerns. RPA plays a crucial role in supporting digital banking infrastructure by automating background processes such as transaction processing, account reconciliation, and fraud checks. This ensures uninterrupted service delivery and helps maintain customer trust. For banks with international operations, RPA facilitates the automation of cross-border activities like currency conversions, international payment settlements, and global compliance reporting. By streamlining these processes, institutions can enhance efficiency and reduce operational risks in their international dealings. RPA also ensures consistent adherence to local regulations across multiple jurisdictions.

1.3.5. Cost Efficiency

Cost management is a critical concern for all financial institutions, particularly in the face of economic uncertainty and market volatility. RPA offers substantial cost benefits by improving process efficiency and reducing reliance on manual labor. By automating routine tasks, RPA allows financial institutions to reduce their headcount or reallocate human resources to more value-adding activities. This results in lower operational costs and improved profit margins. Moreover, RPA reduces the frequency and cost of errors, rework, and customer complaints. RPA provides a scalable solution that can grow with the institution. As banks expand their services or experience seasonal fluctuations in transaction volumes, RPA bots can be scaled up or down accordingly. This flexibility ensures operational continuity without the need for proportionally increasing human resources, making it ideal for growing organizations.

1.3.6 Technological Advancements

The evolution of technology has further reinforced the value proposition of RPA in financial institutions. Integration with advanced technologies is expanding RPA's capabilities and enabling more intelligent automation solutions. The fusion of RPA with AI and machine learning has given rise to intelligent automation, where bots are capable of learning, reasoning, and making decisions. This is particularly useful in areas like fraud detection, risk assessment, and customer service. For instance, AI-powered bots can analyze historical data to detect unusual transaction patterns, raising alerts for potential fraud. The adoption of cloud computing is making RPA solutions more accessible and cost-effective, especially for small and medium-sized institutions. Cloud-based RPA eliminates the need for extensive on-premises infrastructure, reduces implementation costs, and enables faster deployment. Additionally, it provides the flexibility to access and manage bots from anywhere, supporting remote work and global operations.

1.4 Five Strategic Uses of RPA in Finance[3]

IBM has stated the following five strategic uses of RPA in Finance

1.4.1 Supporting Sustainable Revenue Growth

In an environment marked by tight margins and intense competition, especially under low-interest conditions, financial institutions must find innovative ways to grow. RPA enables institutions to unlock new revenue streams by automating the identification of cross-selling opportunities. With the help of ML, customer behavior can be analyzed and segmented to match clients with suitable products or services. For instance, data can reveal customers who are likely to benefit from a new line of credit, allowing for more targeted and timely engagement.

1.4.2 Enhancing Operational Efficiency

RPA significantly reduces operational costs by automating data-heavy, repetitive tasks that previously required manual intervention. Digital bots can access multiple systems, reconcile discrepancies (such as invoice mismatches), and resolve issues instantly. Natural Language Processing (NLP) further enhances these capabilities by interpreting invoice text and directing unresolved matters to the appropriate departments automatically, saving time and reducing errors.

1.4.3 Improving Customer Experience

With customers demanding faster, more tailored services, RPA offers a means to meet expectations across the customer journey, from onboarding to service updates. New clients can open accounts or apply for financial products within minutes using automated Know Your Customer (KYC) processes. RPA also improves customer responsiveness by alerting institutions to complaints or service issues, while ML analyzes historical patterns to suggest actionable improvements and optimize service delivery.

1.4.4 Enhancing Fraud Detection and Financial Crime Prevention

Financial institutions face increasing pressure to detect and prevent fraudulent activities. RPA strengthens cybersecurity measures through faster and more precise execution of due diligence tasks, including sanctions screening and transaction monitoring. Bots can instantly validate data compliance with anti-money laundering (AML) regulations, while ML identifies patterns or anomalies that may indicate suspicious activity, enabling institutions to take proactive steps.

1.4.5 Strengthening Regulatory Compliance

Maintaining compliance with regulatory standards is critical to avoid penalties and reputational harm. RPA assists in consolidating and extracting required data from various systems or documents, minimizing manual workload and improving the reliability of compliance reporting. AI-powered models can anticipate auditors' information needs, locate relevant records, and store them in accessible formats, thereby expediting audit processes and supporting informed decision-making.

1.5 Challenges and Considerations of RPA[4]

Despite its many benefits, Robotic Process Automation (RPA) implementation is not without challenges. While RPA can streamline operations and reduce costs, its deployment often involves complex considerations that organizations must manage effectively. Successful adoption requires more than just installing software bots, it demands a thorough understanding of existing workflows, integration with legacy systems, and alignment with broader organizational strategies. Additionally, long-term sustainability depends on continuous monitoring, updating of automated processes, and adapting to regulatory changes. Resistance to change, lack of technical expertise, and insufficient employee training can also hinder implementation efforts. Therefore, organizations must adopt a holistic approach that includes strategic planning, stakeholder engagement, change management, and robust governance frameworks to realize the full potential of RPA and ensure its long-term value.

1.5.1 Integration with Legacy Systems

One of the major barriers to RPA adoption in finance is integration with existing legacy systems. Many financial institutions still rely on outdated software platforms that were not designed to interface with modern automation tools. Ensuring seamless integration requires technical expertise and often involves reengineering business processes or upgrading core systems. Without proper integration, the effectiveness of RPA can be significantly limited, leading to operational disruptions and increased IT costs.

1.5.2 Data Security

As RPA interacts with sensitive financial data, such as customer information, payment details, and regulatory documents, ensuring data privacy and security becomes critical. Improperly configured bots, unauthorized access, or cybersecurity vulnerabilities can pose serious risks. Financial institutions must establish strict governance policies, enforce encryption protocols, and conduct regular audits to safeguard data integrity during RPA operations.

1.5.3 Workforce Adaptation

Implementing RPA impacts the workforce, as employees must adapt to new ways of working alongside digital tools. This transition requires effective change management, including clear communication, training, and upskilling programs. Employees need to understand the role of RPA, not as a threat to job security but as a tool to enhance their capabilities and reduce mundane tasks. Organizations that invest in workforce development are more likely to experience smoother RPA adoption and higher employee engagement.

Existing literature has largely focused on the technical or operational barriers to RPA adoption, such as integration challenges and workforce adaptation, with relatively limited attention to strategic gaps such as organizational readiness for hyper-automation. Future research should therefore explore how financial institutions can transition from isolated automation to fully integrated hyper-automation ecosystems, which align automation technologies with business intelligence and data analytics. This transition is crucial to unlocking RPA's full potential for innovation in investment and wealth management.

1.6 Overcoming Barriers to RPA Implementation in Financial Institutions

Despite the proven benefits of Robotic Process Automation (RPA), many financial institutions face critical hurdles in implementation. Successfully overcoming these challenges is essential to realizing the full value of RPA initiatives and achieving long-term strategic transformation. Several key barriers can be addressed with targeted strategies.

A major obstacle in RPA adoption is the lack of skilled personnel who understand both automation tools and financial processes. To overcome this, financial institutions must invest in comprehensive training programs that equip employees with the technical and operational knowledge necessary for RPA deployment. Collaborations with educational institutions and professional development centers can facilitate a steady pipeline of RPA-skilled talent. Moreover, recruiting experienced professionals in RPA technologies can ensure smoother implementation and better alignment with business goals.

Budget limitations often hinder large-scale digital transformation. However, RPA can still be implemented effectively through an incremental approach. Financial institutions are advised to prioritize automating critical or high-volume processes first, thereby demonstrating quick wins and building confidence among stakeholders. Exploring open-source or lower-cost RPA platforms can also help reduce the initial investment burden while delivering measurable value. This strategic phasing allows institutions to maximize return on investment (ROI) with minimal upfront costs.

Uncertainty around the return on investment is another frequent concern, especially in institutions seeking approval from senior leadership. To address this, organizations should conduct detailed assessments of potential gains, including reductions in processing time, error rates, and operational costs. Establishing clear, measurable Key Performance Indicators (KPIs) from the outset can help track the impact of RPA and validate its benefits. Additionally, using case studies and real-life examples to showcase tangible outcomes can instill greater confidence in decision-makers and support broader organizational buy-in.

By proactively addressing these barriers through strategic initiatives, financial institutions can move beyond pilot programs and scale their RPA efforts effectively. This enables them to unlock new levels of efficiency, foster innovation, and stay competitive in a rapidly evolving digital finance ecosystem.

1.7 Ethical, Cybersecurity, and Regulatory Considerations in RPA Adoption

As financial institutions increasingly rely on RPA and hyper-automation, cybersecurity and ethical considerations have become critical dimensions of sustainable automation. RPA systems interact with vast amounts of sensitive client and transaction data, creating new vulnerabilities that extend beyond traditional IT risks. In investment and wealth management, data breaches or bot manipulation could compromise portfolio integrity, market positions, or confidential advisory communications. Recent studies emphasize the need for *secure bot governance*, multi-factor authentication, and real-time monitoring frameworks to prevent unauthorized access or malicious bot activity (Kaur et al., 2023; Nazir et al., 2023).

Effective cybersecurity architecture must also include continuous audit trails and encryption of automated data flows across systems.

Ethical implications are equally important, particularly in client-advisory roles increasingly mediated by AI-driven automation. While RPA enhances efficiency and personalization, overreliance on algorithms risks diminishing human judgment, empathy, and fiduciary responsibility. Financial advisors must ensure transparency in automated recommendations and preserve accountability for final investment decisions. The emerging debate on *algorithmic fairness* in wealth management suggests that AI-enabled RPA tools must avoid bias in portfolio selection and client segmentation, maintaining equity across investor demographics (Ali et al., 2025).

Global regulators are beginning to establish frameworks governing AI-integrated automation. However, regulatory maturity varies widely across jurisdictions, highlighting the need for harmonized standards and international cooperation. For developing markets, particularly in Asia, regulators are still exploring proportionate frameworks that balance innovation with consumer protection. Addressing these ethical and regulatory aspects is crucial to building stakeholder trust and ensuring that automation in investment and wealth management evolves responsibly, aligning technological innovation with professional integrity.

1.8 Strategic Limitations and Long-Term Risks of RPA

RPA delivers substantial short-term efficiency and accuracy gains; its long-term strategic implications warrant critical evaluation. One significant risk involves over-reliance on automation, where excessive dependence on bots may reduce organizational resilience and adaptability. When decision-making becomes fully automated, institutions risk losing tacit knowledge embedded in human judgment, especially under unprecedented market conditions where algorithms lack contextual awareness. This dependency could undermine the firm's ability to respond creatively to crises or anomalies, highlighting the need for hybrid human–automation governance structures (Hao et al., 2025).

Another long-term concern is the erosion of human advisory skills. As RPA assumes more analytical and procedural responsibilities, employees may experience skill stagnation or devaluation of their professional roles. Wealth management firms in particular face the risk of diminishing interpersonal trust between clients and advisors if automation replaces human consultation entirely. Continuous reskilling programs, scenario-based training, and job redesign are essential to ensure that automation complements rather than substitutes human expertise (Zirar et al., 2023).

Additionally, strategic rigidity may emerge when automation systems constrain innovation. RPA tends to codify existing workflows, potentially institutionalizing inefficiencies or outdated processes if not periodically reviewed. Financial institutions must therefore adopt dynamic RPA governance models that periodically reassess automation's relevance, effectiveness, and alignment with evolving strategic goals.

Finally, sustainability and ethical dimensions deserve consideration. Large-scale automation could increase digital energy consumption and operational complexity, introducing environmental and governance risks. Hence, long-term RPA strategy should balance efficiency objectives with social and environmental accountability. Incorporating these dimensions elevates RPA discussions beyond technical execution toward responsible innovation in finance.

2.0 BENEFITS FOR INVESTMENT AND WEALTH MANAGEMENT

Robotic Process Automation (RPA) offers a wide array of benefits for the investment and wealth management sector, playing a critical role in enhancing operational efficiency, precision, and overall service quality. In an industry where time-sensitive decisions and large volumes of repetitive processes are common, RPA proves particularly valuable. By automating routine tasks such as data entry, transaction reconciliation, report generation, and client onboarding, RPA significantly reduces labor costs and operational overheads. It also allows firms to reallocate human resources to more strategic and client-focused activities (Susilo et al., 2021). This automation not only saves time but also minimizes the risk of human error, which is essential in ensuring the accuracy and integrity of financial transactions and data handling. Maintaining such accuracy is vital not only for operational success but also for compliance with stringent regulatory frameworks that govern the financial services industry (Bin-Nashwan et al., 2025).

Moreover, RPA strengthens regulatory adherence by providing clear, traceable, and auditable digital footprints for every automated transaction or decision made. This capability is particularly crucial in wealth and investment management, where firms must consistently meet compliance standards while handling vast amounts of sensitive financial data (Bhambri et al., 2024). Another strategic advantage of RPA lies in its scalability. Unlike human labor, RPA systems can be scaled up quickly to accommodate seasonal surges or market volatility without proportional increases in cost or staffing (Flechsig et al., 2022). This agility helps firms maintain service levels and operational continuity, especially during periods of market stress or client onboarding peaks.

In addition, RPA does not operate in isolation. Its integration with advanced technologies such as Artificial Intelligence (AI) and machine learning enhances its utility, enabling the development of intelligent automation systems that can adapt to changing market conditions, learn from historical data, and respond proactively to client needs (Policepatil et al., 2025). For example, AI-enhanced RPA can help in portfolio rebalancing, predictive analytics, and real-time fraud detection. By automating data gathering and processing across multiple platforms, RPA provides investment professionals with timely, consistent, and reliable data. This real-time access to accurate information enhances analytical capabilities and supports more informed, strategic decision-making (Garad et al., 2024).

3.0 APPLICATION OF RPA IN INVESTMENT MANAGEMENT

Robotic Process Automation (RPA) plays a critical role in enhancing operational efficiency and accuracy within investment management. As the industry handles vast volumes of transactions, regulatory obligations, and market data, RPA offers powerful solutions to automate repetitive and rule-based processes, thereby reducing errors, cutting costs, and freeing up human resources for higher-value tasks. Key areas where RPA is making a significant impact include trade processing and settlement, portfolio monitoring and rebalancing, and compliance reporting with audit trails. These applications not only improve internal workflows but also enhance client service by enabling faster response times and more reliable outputs. As investment firms increasingly embrace digital transformation, RPA serves as a foundational technology that helps bridge operational needs with strategic agility.

3.1 Trade Processing and Settlement

Robotic Process Automation (RPA) can significantly streamline trade processing and settlement functions by automating repetitive, rule-based tasks that are traditionally performed manually. These tasks often include trade matching, confirmation, allocation, and settlement instruction processing. By automating such processes, RPA reduces the need for manual intervention, thereby increasing operational efficiency and minimizing the risk of errors that could lead to failed trades or regulatory breaches. This is particularly important in high-volume trading environments, where even small mistakes can result in substantial financial and reputational costs. RPA systems can process large volumes of transactions accurately and within tight timeframes, ensuring both speed and reliability in trade execution and settlement (E-Fatima et al., 2023).

Furthermore, one of RPA's key strengths lies in its ability to integrate with legacy systems and modern financial platforms alike. This interoperability enables financial institutions to implement RPA without the need for complete infrastructure overhauls, making it a cost-effective solution for optimizing trade processing workflows. In the context of trade settlements, the RPA ensures smoother and more transparent operations. It facilitates seamless data exchange across systems, enhances compliance by generating consistent audit trails, and reduces the chances of reconciliation issues or mismatches (Waduge et al., 2024). As financial institutions strive for greater operational resilience and efficiency, RPA serves as a powerful tool for modernizing trade processing while maintaining accuracy, compliance, and scalability.

3.2 Portfolio Monitoring and Rebalancing

Robotic Process Automation (RPA) can significantly enhance portfolio monitoring and rebalancing by automating the end-to-end processes of data collection, analysis, and execution. In traditional portfolio management, monitoring asset performance and maintaining target allocations often require continuous tracking of market data and manual recalculations. By automating these functions, RPA enables real-time portfolio performance tracking, allowing investment managers to respond promptly to market fluctuations and deviations from the desired asset allocation (Abdullah & Almaqtari, 2024). Through pre-defined rules and thresholds, RPA can trigger alerts or even initiate rebalancing actions automatically when specific conditions are met.

Moreover, RPA-powered virtual robots can simulate various trading strategies using historical and real-time data, generating performance histories that provide valuable insights for optimizing portfolio composition. These simulations help in evaluating risk-return trade-offs and improving the decision-making process for rebalancing. With automation, rebalancing can be done more frequently and precisely. The rebalancing is also an advantage in volatile markets where timely adjustments are critical to maintaining portfolio objectives and risk levels. Ultimately, RPA contributes to improved accuracy, speed, and efficiency in portfolio management, enabling investment firms to offer more responsive and data-driven services to their clients (E-Fatima et al., 2023).

3.3 Compliance Reporting and Audit Trails

Compliance reporting and the maintenance of audit trails are essential components of investment management, playing a critical role in ensuring regulatory adherence, transparency, and investor trust. With increasing regulatory scrutiny and complex reporting standards, financial institutions must consistently deliver accurate, timely, and well-documented compliance reports. Robotic Process Automation (RPA) offers a powerful solution by automating the generation of these reports and maintaining comprehensive audit trails. Through programmed workflows, RPA can systematically capture, log, and store every transaction and decision made during the investment process, ensuring that all actions are traceable and verifiable. This not only reduces the risk of human error but also ensures consistency and reliability in meeting compliance obligations (Kokina et al., 2025).

Moreover, RPA supports secure and reliable data management by interfacing with core systems to retrieve and validate information, minimizing the chance of discrepancies or data manipulation. As RPA bots operate with predefined rules, they eliminate subjectivity in compliance processes and maintain objectivity in report generation. This automated oversight provides regulators and auditors with clear, real-time access to activity logs, significantly improving the audit readiness of the organization. In an environment where non-compliance can lead to substantial financial penalties and reputational damage, RPA ensures institutions maintain a high standard of regulatory integrity while improving operational efficiency (Kokina et al., 2025).

3.4 Challenges in Applying RPA to Investment Management

RPA enhances efficiency in investment management, several domain-specific challenges persist. One major issue involves algorithmic bias in automated decision-making. RPA and AI-driven trading or portfolio rebalancing systems often depend on historical market data that may embed biases or reflect limited market conditions. Consequently, automated recommendations might systematically favor certain asset classes, regions, or investor profiles, thereby undermining diversification and fairness (Kattnig et al., 2024). To mitigate these biases, investment firms must adopt transparent data-governance protocols, conduct periodic model audits, and integrate human oversight to validate automated outputs.

Another critical challenge concerns market-volatility handling. RPA bots execute pre-programmed rules without contextual judgment; during periods of extreme volatility or unforeseen macroeconomic shocks, these bots may overreact to short-term signals, amplifying trading risks. Integrating adaptive AI modules and real-time monitoring dashboards can help ensure that automation responds dynamically to market anomalies (Gao et al., 2025).

Furthermore, model explainability is an emerging concern. As hyper-automation fuses RPA with machine learning, investment decisions become increasingly opaque to both managers and regulators. Maintaining transparency and auditability is vital to preserving client trust and meeting regulatory expectations. Addressing these challenges requires an integrated approach that combines robust technical design, ethical governance, and continuous performance evaluation of automated investment systems.

4.0 APPLICATION OF RPA IN WEALTH MANAGEMENT

Robotic Process Automation (RPA) is increasingly being adopted in wealth management to streamline operations, improve customer experience, and enhance data accuracy. As wealth management firms deal with high volumes of client data, regulatory requirements, and the demand for personalized services, RPA provides an efficient and scalable solution. Key applications include automating client onboarding and Know Your Customer (KYC) processes, integrating and managing data across multiple platforms, and delivering tailored investment solutions through real-time analysis and insights. By minimizing manual intervention and standardizing workflows, RPA not only reduces processing time and operational costs but also improves compliance and decision-making. The integration of RPA with advanced technologies such as artificial intelligence (AI) further enhances its ability to support dynamic, client-centric wealth management practices in a competitive and rapidly evolving financial landscape.

4.1 Client Onboarding and KYC Processes

RPA can significantly streamline client onboarding and Know Your Customer (KYC) processes by automating repetitive tasks such as data entry, document verification, and compliance checks. This reduces the time required for onboarding and minimizes human errors. By automating these processes, financial institutions can reduce the need for a large workforce, thereby cutting costs and reallocating human resources to more complex tasks (Shidaganti et al., 2023).

Robotic Process Automation (RPA) can play a transformative role in streamlining client onboarding and Know Your Customer (KYC) processes, which are often resource-intensive and time-consuming in the financial services sector. These processes involve numerous repetitive tasks such as collecting customer information, entering data across multiple systems, verifying identification documents, and conducting compliance and risk checks. By automating these tasks, RPA greatly reduces the manual workload, shortens processing time, and minimizes the potential for human error. This leads to a faster, more seamless onboarding experience for clients, which is especially critical in a competitive financial landscape where customer expectations for speed and convenience are continually rising (Shidaganti et al., 2023).

Beyond improving efficiency, RPA also contributes to cost reduction. With fewer staff required for routine onboarding tasks, financial institutions can optimize workforce allocation and focus human resources on more value-added or judgment-based functions, such as complex risk assessments or personalized client interactions. Additionally, the automation of KYC processes ensures consistent application of regulatory standards, supports the maintenance of accurate records, and reduces the likelihood of compliance breaches. As regulations continue to evolve and the demand for digital-first services increases, RPA provides a scalable and reliable solution that supports both regulatory adherence and operational excellence in client onboarding (Huang & Vasarhelyi, 2019).

4.2 Data Integration and Reporting

Robotic Process Automation (RPA) is particularly effective in managing the vast volumes of structured and unstructured data generated and used in wealth and investment management. In an industry where data accuracy, consistency, and timeliness are vital for sound decision-making, RPA offers a robust solution by automating key data processes. These include data cleansing, normalization, validation, and integration. By automating these processes, RPA ensures that the data used for portfolio analysis, risk assessment, and client reporting is both accurate and up to date (Flechsig et al., 2022).

One of RPA's key strengths is its ability to extract data from multiple disparate sources, such as spreadsheets, databases, emails, and legacy systems, and compile it into a coherent, standardized format for real-time reporting and analytics. This capability significantly enhances the speed and quality of data-driven insights, allowing wealth managers to make timely, well-informed decisions. Moreover, RPA bots can be configured to work across multiple systems and platforms without requiring significant changes to the existing IT infrastructure. This interoperability makes RPA a cost-effective and scalable solution for financial institutions looking to improve data integration and operational efficiency without overhauling their entire technology stack. Ultimately, RPA enables better data governance and empowers investment professionals with reliable information to support strategic planning and client service (Sharma & Cupek, 2023).

4.3 Personalized Investment Solutions

Robotic Process Automation (RPA), particularly when integrated with Artificial Intelligence (AI), is increasingly being used to deliver personalized investment solutions by analyzing client behavior, enabling real-time decision-making, and enhancing risk management strategies. These capabilities collectively allow financial institutions to deliver more tailored and responsive services that align closely with individual client goals and risk appetites (Mariani & Dwivedi, 2024).

One of the core features of personalized investment solutions is customer behavior analysis. Advanced RPA systems, when combined with AI-driven analytics, can process large volumes of customer data from diverse sources, such as transaction histories, online interactions, and financial goals. The system can also detect patterns, preferences, and behavioral trends. This deep understanding allows investment firms to craft customized portfolio strategies and service offerings that are specifically tailored to each client's unique financial profile. By automating this analysis, firms can deliver personalization on a scale, which would be difficult and time-consuming to achieve manually (Abdullah & Almaqtari, 2024).

RPA also enables real-time recommendations, which are becoming a critical expectation among modern investors. By continuously pulling and analyzing data from market feeds, client portfolios, and economic indicators, RPA systems can assist in delivering up-to-date investment advice. These timely insights enhance the client's experience and potentially lead to improved investment outcomes. For instance, if a market fluctuation impacts a particular asset class in a client's portfolio, the system can instantly suggest rebalancing options or alternative investment strategies based on the client's predefined goals and risk profile (Alfzari et al., 2025).

Furthermore, RPA contributes to risk management by constantly monitoring both market conditions and individual portfolios. Automated systems can detect deviations from expected performance or emerging risk indicators, generating alerts and proactive recommendations to mitigate losses. These features are particularly valuable in volatile markets, where swift action is essential to protect portfolio value. RPA allows wealth managers to stay ahead of risk by embedding intelligence into portfolio oversight, enhancing both compliance and client trust (Prasetyo et al., 2025).

Beyond personalization, RPA fundamentally transforms the decision-making architecture in wealth management. Traditionally, portfolio decisions relied heavily on human advisors interpreting financial statements, market trends, and client preferences. With RPA, decision-making becomes increasingly *data-driven, consistent*, and *scalable*. Automation facilitates continuous aggregation and analysis of diverse data streams, such as market prices, client risk profiles, macroeconomic indicators, and ESG factors, allowing wealth managers to make timely, evidence-based decisions.

Moreover, RPA's integration with AI and machine learning supports predictive analytics, enabling systems to simulate market scenarios and recommend optimal asset allocations under varying risk conditions (Policepatil et al., 2025). This predictive capability enhances proactive decision-making and helps advisors identify portfolio vulnerabilities before adverse market movements occur.

Importantly, RPA also contributes to augmented decision-making, where human expertise and automation interact symbiotically. Rather than replacing advisors, RPA serves as a *decision-support system* that filters information, identifies anomalies, and provides data-driven recommendations. This allows human advisors to focus on qualitative judgments, such as behavioral cues, ethical considerations, and long-term client goals, while automation handles quantitative precision and process consistency.

Furthermore, RPA enhances transparency and auditability in decisions. Each automated recommendation can be traced, reviewed, and validated through digital logs, ensuring accountability in advisory processes. This transparency not only improves regulatory compliance but also strengthens client trust, as investors can understand how decisions are generated. As a result, RPA redefines decision-making from being intuition-based and episodic to being continuous, auditable, and analytically optimized.

5.0 STATISTICS ABOUT RPA[5]

A Deloitte survey indicates that 53% of organizations have initiated their RPA journey, with an anticipated 72% of companies expected to adopt RPA in some capacity by 2025. This signifies a substantial transformation in organizational strategies regarding business processes and human resource distribution, enabling employees to concentrate on value-enhancing activities such as creativity, decision-making, and strategic planning.

RPA is becoming a fundamental element in the larger framework of digital transformation. Especially attractive to organizations seeking immediate benefits, it provides readily quantifiable outcomes and operational enhancements without the intricacies of more sophisticated AI or machine learning technologies. This positions RPA as the preferred solution for organizations seeking to expedite their transformation initiatives while reducing disruptions to essential operations.

According to Precedence Research, the global robotic process automation (RPA) market is projected to grow from USD 28.31 billion in 2025 to approximately USD 211.06 billion by 2034, reflecting a compound annual growth rate (CAGR) of 25.01% over the forecast period. In North America, the market was valued at USD 8.87 billion in 2024 and is expected to grow at a CAGR of 23.74%. Beyond North America, RPA adoption in financial services shows diverse regional trajectories. Europe leads in compliance-driven automation, emphasizing transparency and data protection. Asia-Pacific, driven by fintech ecosystems in Singapore, India, and Japan, focuses on cost-efficiency and customer-centric automation. In contrast, emerging markets in Africa and Latin America prioritize financial inclusion and operational scalability through cloud-based RPA solutions. Cross-country comparisons reveal that institutional readiness, digital infrastructure, and regulatory openness significantly influence the pace and scope of RPA adoption. Understanding these variations provides a global perspective and contextualizes how investment and wealth management firms can benchmark their digital transformation strategies internationally.

6.0 RPA IMPLEMENTATION CASE STUDIES[6]

6.1 Nike

Nike's inventory management system was severely inefficient, with manual procedures causing delays and inaccurate stock level readings. To automate replenishment, inventory tracking, and real-time stock updates, the company deployed RPA bots. With this solution, Nike's inventory discrepancies decreased by 30%, audit frequency increased by 100%, and warehouse space utilisation increased by 20%. More precise forecasting and increased supply chain efficiency were the outcomes of the RPA implementation.

6.2 JP Morgan Chase

RPA was utilised by JP Morgan Chase to optimise their audit and compliance procedures. The bank used RPA to automate the compilation of tax filings and regulatory reports in response to the increasing complexity of regulations. This decreased the likelihood of human error and cut down on the time needed to report compliance by 50%. Additionally, the deployment allowed their audit team to concentrate on more strategic financial analysis and planning by freeing up 25% of their time.

6.3 Credigy

Credigy uses RPA to improve the efficiency and accuracy of many internal processes. One of its key uses is in handling legal documents. RPA tools scan these documents for important keywords and return them with the keywords clearly highlighted, helping staff review content faster and more accurately. The company also automates the processing of account-related documents. Whether these documents arrive by email or other digital channels, RPA downloads and sorts them into the correct accounts within their internal system. This allows users to quickly access the documents they need without manual sorting.

To improve cybersecurity, Credigy uses RPA to identify suspicious emails. These are sent for URL scanning, and users are alerted if any links are found to be harmful. This proactive step helps reduce the risk of phishing attacks. RPA also assists customer service by generating automatic replies to incoming emails. These initial responses help manage response times and keep customers informed while waiting for further action. In terms of compliance, RPA is used to download updates from official websites and load the information into specific internal tables for staff review. This ensures Credigy stays current with regulatory changes. Through these applications, Credigy uses RPA to streamline operations, improve security, support compliance, and enhance overall productivity.

6.4 Cost–Benefit Analysis of RPA Adoption

Evaluating the feasibility of RPA deployment in investment and wealth management requires a structured cost–benefit analysis (CBA). Implementation costs typically include software licensing, infrastructure integration, staff training, and ongoing maintenance. On the benefit side, firms can quantify savings from reduced labor hours, fewer operational errors, improved compliance efficiency, and faster turnaround times. Qualitative gains, such as enhanced client satisfaction and improved decision accuracy, should also be considered.

Simple return-on-investment (ROI) models can be extended with discounted cash flow (DCF) or net present value (NPV) frameworks to capture long-term efficiency gains and risk mitigation benefits. For instance, a three-to-five-year horizon analysis can compare cumulative cost savings against upfront investment to assess payback periods. Incorporating sensitivity analysis further allows decision-makers to evaluate varying scenarios of automation scale, error reduction, and regulatory cost avoidance. Presenting such models equips practitioners and policymakers with tangible tools to determine the economic viability of RPA adoption in financial institutions.

7.0 FUTURE OUTLOOK: THE NEXT DECADE OF RPA IN FINANCE

Looking ahead, RPA will evolve from process-centric automation toward cognitive, predictive, and ethical automation ecosystems. The convergence of RPA with generative AI, blockchain, and quantum computing will enable real-time portfolio optimization, fraud detection, and ESG-aligned investment analytics. These technologies will transform investment management from rule-based execution to intelligent decision support systems capable of learning and adapting continuously.

Moreover, hyper-automation will redefine operational architecture, where RPA, AI, and data analytics operate in an integrated environment, allowing end-to-end automation from client onboarding to risk management. Global regulatory bodies are expected to establish harmonized standards for algorithmic transparency, data privacy, and accountability, shaping the ethical boundaries of financial automation.

From a strategic perspective, future competitiveness will depend on how institutions balance automation efficiency with human expertise. Firms that integrate human judgment, transparent algorithms, and responsible AI governance are likely to lead the next wave of digital transformation in investment and wealth management.

Summary

In summary, digital transformation and the rise of RPA are reshaping the financial sector by enhancing operational efficiency, reducing costs, and enabling financial institutions to offer more innovative and customer-centric services. The ongoing integration of these technologies will continue to drive significant changes in how financial services are delivered and managed. RPA also enables a shift from standardized investment services to personalize, data-driven advisory models. Its ability to automate and scale behavioral analysis, generate real-time advice, and actively manage risk positions financial firms to offer a higher level of client engagement and satisfaction, while simultaneously increasing operational efficiency. This fusion of automation and personalization reflects the future of wealth and investment management.

REFERENCES

Abdullah, A. A. H., & Almaqtari, F. A. (2024). The impact of artificial intelligence and Industry 4.0 on transforming accounting and auditing practices. *Journal of Open Innovation, 10*(1), 100218. DOI: 10.1016/j.joitmc.2024.100218

Alfzari, S., Al-Shboul, M., & Alshurideh, M. (2025). Predictive analytics in portfolio management: A fusion of AI and investment economics for optimal risk-return trade-offs. *International Review of Management and Marketing*, *15*(2), 365–380. DOI: 10.32479/irmm.18594

Ali, O., Murray, P. A., Al-Ahmad, A., Jeon, I., & Dwivedi, Y. K. (2025). Comprehending the theoretical knowledge and practice around AI-enabled innovations in the finance sector. *Journal of Innovation & Knowledge, 10*(5), 100762. DOI: 10.1016/j.jik.2025.100762

Babina, T., Fedyk, A., He, A., & Hodson, J. (2024). Artificial intelligence, firm growth, and product innovation. *Journal of Financial Economics, 151*, 103745. DOI: 10.1016/j.jfineco.2023.103745

Bhambri, P., Rani, S., & Pareek, P. K. (2024). Financial innovations: Intelligent automation in finance and insurance sectors. In Darwish, D. (Ed.), *Hyperautomation in business and society* (pp. 226–243). IGI Global., DOI: 10.4018/979-8-3693-3354-9.ch012

Bin-Nashwan, S. A., Li, J. Z., Jiang, H., Bajary, A. R., & Ma'aji, M. M. (2025). Does AI adoption redefine financial reporting accuracy, auditing efficiency, and information asymmetry? An integrated model of TOE-TAM-RDT and big data governance. *Computers in Human Behavior Reports, 17*, 100572. DOI: 10.1016/j.chbr.2024.100572

Bueno, L. A., Sigahi, T. F. A. C., Rampasso, I. S., Leal Filho, W., & Anholon, R. (2024). Impacts of digitization on operational efficiency in the banking sector: Thematic analysis and research agenda proposal. *International Journal of Information Management Data Insights*, *4*(1), 100230. DOI: 10.1016/j.jjimei.2024.100230

E-Fatima, K., Khandan, R., Hosseinian-Far, A., & Sarwar, D. (2023). The adoption of robotic process automation considering financial aspects in beef supply chains: An approach towards sustainability. *Sustainability*, *15*(9), 7236. DOI: 10.3390/su15097236

Flechsig, C., Anslinger, F., & Lasch, R. (2022). Robotic Process Automation in purchasing and supply management: A multiple case study on potentials, barriers, and implementation. *Journal of Purchasing and Supply Management*, *28*(1), 100718. DOI: 10.1016/j.pursup.2021.100718

Fu, J., & Mishra, M. (2022). Fintech in the time of COVID–19: Technological adoption during crises. *Journal of Financial Intermediation*, *50*, 100945. DOI: 10.1016/j.jfi.2021.100945

Gao, Y., Liu, S., & Yang, L. (2025). Artificial intelligence and innovation capability: A dynamic capabilities perspective. *International Review of Economics & Finance*, *98*, 103923. DOI: 10.1016/j.iref.2025.103923

Garad, A., Riyadh, H. A., Al-Ansi, A. M., & Beshr, B. A. H. (2024). Unlocking financial innovation through strategic investments in information management: A systematic review. *Discover Sustainability*, *5*(381). Advance online publication. DOI: 10.1007/s43621-024-00542-6

Haleem, A., Javaid, M., Singh, R. P., Rab, S., & Suman, R. (2021). Hyperautomation for the enhancement of automation in industries. *Sensors International*, *2*, 100124. DOI: 10.1016/j.sintl.2021.100124

Hao, X., Demir, E., & Eyers, D. (2025). Beyond human-in-the-loop: Sensemaking between artificial intelligence and human intelligence collaboration. *Sustainable Futures : An Applied Journal of Technology, Environment and Society*, *10*, 101152. DOI: 10.1016/j.sftr.2025.101152

Huang, F., & Vasarhelyi, M. A. (2019). Applying robotic process automation (RPA) in auditing: A framework. *International Journal of Accounting Information Systems*, *35*, 100433. DOI: 10.1016/j.accinf.2019.100433

Kattnig, M., Angerschmid, A., Reichel, T., & Kern, R. (2024). Assessing trustworthy AI: Technical and legal perspectives of fairness in AI. *Computer Law & Security Review*, *55*, 106053. DOI: 10.1016/j.clsr.2024.106053

Kaur, R., Gabrijelčič, D., & Klobučar, T. (2023). Artificial intelligence for cybersecurity: Literature review and future research directions. *Information Fusion*, *97*, 101804. DOI: 10.1016/j.inffus.2023.101804

Kokina, J., Blanchette, S., Davenport, T. H., & Pachamanova, D. (2025). Challenges and opportunities for artificial intelligence in auditing: Evidence from the field. *International Journal of Accounting Information Systems*, *56*, 100734. DOI: 10.1016/j.accinf.2025.100734

Kraus, S., Durst, S., Ferreira, J. J., Veiga, P., Kailer, N., & Weinmann, A. (2022). Digital transformation in business and management research: An overview of the current status quo. *International Journal of Information Management, 63*, 102466. DOI: 10.1016/j.ijinfomgt.2021.102466

Ma, J., & Jia, H. (2022). *Application of financial robots based on RPA technology in small and medium-sized enterprises*. In *2022 International Conference on Knowledge Engineering and Communication Systems (ICKECS)*. IEEE. https://doi.org/ DOI: 10.1109/ICKECS56523.2022.10060387

Mariani, M., & Dwivedi, Y. K. (2024). Generative artificial intelligence in innovation management: A preview of future research developments. *Journal of Business Research, 175*, 114542. DOI: 10.1016/j.jbusres.2024.114542

Meiryani, Z. A. D., Chandra, F. C., & Warganegara, D. L. (2023). Analysis role of robotic process automation in accounting and business. *Journal of Theoretical and Applied Information Technology, 101*(11), 4390–4399. https://www.jatit.org/volumes/Vol101No11/22Vol101No11.pdf

Naimi-Sadigh, A., Asgari, T., & Rabiei, M. (2022). Digital transformation in the value chain disruption of banking services. *Journal of the Knowledge Economy, 13*, 1212–1242. DOI: 10.1007/s13132-021-00759-0

Nazir, A., He, J., Zhu, N., Wajahat, A., Ma, X., Ullah, F., Qureshi, S., & Pathan, M. S. (2023). Advancing IoT security: A systematic review of machine learning approaches for the detection of IoT botnets. *Journal of King Saud University. Computer and Information Sciences, 35*(10), 101820. DOI: 10.1016/j.jksuci.2023.101820

Policepatil, S., Sharma, J., Kumar, B., Singh, D., Pramanik, S., Gupta, A., & Mahabub, B. S. (2025). Financial sector hyper-automation: Transforming banking and investing procedures. In Adibi, M. A., Anigol, R., Chandel, K., Kapur, N., & Sayed, K. (Eds.), *Examining global regulations during the rise of fintech* (pp. 1–20). IGI Global., DOI: 10.4018/979-8-3693-3803-2.ch012

Prasetyo, M. L., Peranginangin, R. A., Martinovic, N., Ichsan, M., & Wicaksono, H. (2025). Artificial intelligence in open innovation project management: A systematic literature review on technologies, applications, and integration requirements. *Journal of Open Innovation, 11*(1), 100445. DOI: 10.1016/j.joitmc.2024.100445

Ratna, S., Saide, S., Putri, A. M., Soleha, A., & Andini, P. R. (2024). Financial systems transformation in the digital age: A systematic review and future research directions. *Procedia Computer Science, 234*, 1538–1545. DOI: 10.1016/j.procs.2024.03.155

Sharma, N., & Cupek, R. (2023). Real-time control and optimization of internal logistics systems with collaborative robots. *Procedia Computer Science, 225*, 248–258. DOI: 10.1016/j.procs.2023.09.027

Shidaganti, G., Khandelwal, L., Gupta, K., & Vaswani, H. (2023). Challenges in banking and solving them using RPA. In *Confluence of Artificial Intelligence and Robotic Process Automation* (pp. 133–156). Springer., DOI: 10.1007/978-981-19-8296-5_6

Susilo, A., Prabowo, H., Kosasih, W., Kartono, R., & Utami, V. (2021). The implementation of robotic process automation for banking sector: Case study of a private bank in Indonesia. *In ICIT '21: Proceedings of the 2021 9th International Conference on Information Technology: IoT and Smart City* (pp. 365–371). ACM. https://doi.org/DOI: 10.1145/3512576.3512641

Waduge, S., Sugathadasa, R., Piyatilake, A., & Nanayakkara, S. (2024). A process analysis framework to adopt intelligent robotic process automation (IRPA) in supply chains. *Sustainability*, *16*(22), 9753. DOI: 10.3390/su16229753

Yadav, R., Singh, S. K., Luu, T. M. N., & Najeeb, A. (2025). RPA in banking and finance. In Chaudhary, T., Rajvanshi, J., Mehta, K., Wang, M., & Nohanty, S. (Eds.), *Intelligent robotic process automation: Development, vulnerability and applications* (p. 20). IGI Global., DOI: 10.4018/978-8-3693-4365-4.ch002

Yang, J., Blount, Y., & Amrollahi, A. (2024). Artificial intelligence adoption in a professional service industry: A multiple case study. *Technological Forecasting and Social Change*, *201*, 123251. DOI: 10.1016/j.techfore.2024.123251

Zareen, M., Hamdan, A., & Binsaddig, R. (2024). The impact of Robotic Process Automation (RPA) on accounting profession. In Hamdan, A., & Harraf, A. (Eds.), *Business development via AI and digitalization* (Vol. 538). Springer., DOI: 10.1007/978-3-031-62102-4_16

Zirar, A., Ali, S. I., & Islam, N. (2023). Worker and workplace Artificial Intelligence (AI) coexistence: Emerging themes and research agenda. *Technovation, 124*, 102747. DOI: 10.1016/j.technovation.2023.102747

ENDNOTES

1 The discussion is extracted from the following source: https://www.bakertilly.com/insights/the-state-of-rpa-in-financial-institutions-2024

2 The discussion in this section is extracted from the following source: https://www.bakertilly.com/insights/rpa-in-financial-institutions-applications-challenges-and-strategies-for-success

3 The discussion in this section is extracted from the following source: https://www.ibm.com/think/topics/rpa-for-finance

4 The discussion in this section is extracted from the following source: https://www.novasarc.com/challenges-in-rpa-infrastructure-and-how-to-overcome-them

5 The discussion in this section is extracted from the following source: https://mark-bridges.medium.com/robotic-process-automation-revolutionizing-efficiency-across-value-chains-827521b75778; https://www.precedenceresearch.com/robotic-process-automation-market

6 The discussion in this section is extracted from the following source: https://mark-bridges.medium.com/robotic-process-automation-revolutionizing-efficiency-across-value-chains-827521b75778 and https://www.ibm.com/case-studies/credigy-solution

Chapter 12
Robo-Advisory in Finance:
Mapping the Research Landscape

Babalpreet Kaur
https://orcid.org/0009-0004-0286-8054
Chandigarh University, India

Manjit Kour
https://orcid.org/0000-0003-1043-3187
Chandigarh University, India

ABSTRACT

Robo-advisors refer to a great leap for the financial services sector, offering automated investment advice via cutting-edge algorithms and online interfaces. This review of the literature considers the evolution and current research on robo-advisors, highlighting their development, utilization, and impact on the financial system. The study examines a comprehensive list of academic publications to assess how much robo-advisors are reshaping the traditional advisory structures and influencing investor behavior. It also outlines the wider consequences of automation in finance decision-making and lays out where additional scholarly research needs to occur. The review forms an extensive overview of the scholarly discussion regarding robo-advisors and extracts lessons for the future of finance.

DOI: 10.4018/979-8-3373-3571-1.ch012

1. INTRODUCTION

1.1 Background of Robo-Advisors

The robo-advisors and digital revolution in the financial sector have been significantly improved the financial services. They enhance long-term financial responsibility by employing data-led approaches to curb emotional decision-making (Kaur & Kour, 2025). Companies are able to accommodate more customers with fewer assets because of robo-advisors' scalability. They were designed to overcome most of the limitations of traditional financial advisors, to deliver algorithm-based, affordable, and unbiased investment solutions. They emerged for the first time after the 2008 financial crisis (Fein, 2015; Sironi, 2016). Wealthfront and Betterment were two of the initial companies to implement Modern Portfolio Theory (MPT) in creating diversified portfolios that aligned with their clients' investment goals, time horizon, as well as risk tolerance (D'Acunto et al., 2019). Their popularity has grown with the rise in the usage of mobile technology, the accessibility of the internet, and growing demands from consumers for more openness and control in financial decision-making (Baker &Dellaert, 2019). Their user-friendly interfaces and educational materials enable individuals to grasp financial planning. Moreover, robo-advisors are also emerging as a crucial factor for promoting financial inclusion through offering lower capital requirements for investment opportunities.

1.2 Role of Automation in Financial Advisory Services

The financial sector is experiencing a true revolution that is fundamentally shifting from simple automation. The invention of robo-advisors, used in the quest to provide and receive financial services, represents the peak of the way automation is revolutionizing the financial sector (Kaur & Kour, 2025). Due to the transformation, individuals will be in a better position to make healthier financial decisions. Automation frees rule-based, data-driven investment decisions from human planners' budgetary constraints and emotional biases (Fein, 2015; D'Acunto et al., 2019). Through algorithms that offer personalized personal finance recommendations, the technology increases the number of individuals who can access investment advice (Kaur & Kour, 2025). These websites enhance efficiency in operations and customer satisfaction by the availability of scalable features like tax-loss harvesting, real-time tracking of monetary goals, and rebalancing of portfolios on an automatic basis. Moreover, automation is needed to expand access and inclusion in wealth management by reducing entry barriers to novice as well as low-net-worth investors (Jung et al., 2018). Thus, robo-advisors with automation have the capability of redesigning efficiency through transparency, and trust established in financial decision-making.

Their algorithmic methods provide a new way of delivering financial advice because of a lack of human bias. Digitally born investors and Millennials, are especially attracted to robo-advisors for self-management of finance and online communication. It is confirmed by literature review that such investors are looking for affordability, simplicity and expertise—all of which robo-advisors can offer (Kaur & Kour, 2024; Belanche et al., 2019). Younger generations who might consider traditional advisory services as too expensive, will benefit greatly from this improved accessibility. Moreover, robo-advisors provide financial inclusion and investor friendly platforms to low-capital individuals investor that have been historically excluded by traditional wealth management services by reaching masses. (Jung et al., 2018; Oehler et al., 2021). Robo-advisors enable people to take control over their investing plan and strategy by providing tutorial content and user friendly interfaces (Kaur & Kour, 2024).

However, with the rate of adoption of robo-advisories, there are also impactful concerns on data privacy, secure use of customer data, algorithm transparency, and adequacy of investor protection in an entirely automated position that need to be thoroughly investigated (Kim et al., 2015; Arner et al., 2017). Weakness of current market is Insufficient human oversight on critical decision-making, When critical decisions are taken without human oversight, issues go awry in unforeseen manners. Particularly under unclear or complicated financial situations, Scholars wonder whether it is possible for algorithms to realistically offer the trust, good sense, and emotional consciousness human counselors could. (Lo, 2019). Furthermore, the increased utilization of algorithms in decision-making can lead to systemic risks that are not yet known and being properly addressed. Finally, a policy that harnesses the potential dangers of automation with its advantages is needed to achieve a sustainable and ethical financial future,.

1.3 Research Objectives and Scope of the Literature Review

The emergence of robo-advisor has drawn significant attention and controversy It is important to clarify the implications of the robo-platforms for directing their future development in a responsible manner and rendering their role answerable in the financial system. The literature review, has been critically and extensively examined the academic discourse surrounding robo-advisors. It investigates the effect of robo advisors on investment behaviour, financial inclusion and market structures, as well as their influence on conventional financial advisory frameworks through technological transformation. The review synthesizes insights from a range of disciplines, including behavioral economics, finance, law and information systems, to highlight major academic findings, controversies and outstanding issues. Additionally, it also identifies areas and suggests avenues that require further academic investigation, especially as robo-advisors evolve into more sophisticated hybrid models combining artificial intelligence and human oversight. The objective is to provide an integrated consciousness about robo advisors that how they are transforming the financial services industry and the implications of these shifts for both institutional and retail players. In doing so, the review contributes to the ongoing discussion about digitization, automation, and the direction of financial decision-making.

1.4 Organization of Content

The paper is organized into six parts. After the introduction, the next part presents a thematic literature review, categorizing the main scholarly contributions on robo-advisors. This is followed by an explanation of the methodology used to select and analyze the literature. The following part synthesizes findings on key themes such as user adoption, investment behavior, market disruption, and regulatory implications. The subsequent sections identify existing research gaps and suggest future directions in academic study and policy formation for researchers. The final part offers concluding insights and reflects on the evolving role of automation in financial advisory services.

2. METHODOLOGY

This literature review employs a qualitative, thematic synthesis approach to critically analyze existing research on robo-advisors and their implications for the financial services industry. The methodology was designed to ensure comprehensive coverage of academic, regulatory, and industry perspectives on the subject, capturing both foundational theories and emerging insights.

The review spans the period from 2009 to 2024, coinciding with the emergence and maturation of robo-advisory platforms. The study focuses on scholarly publications and high-quality grey literature addressing key dimensions such as technological architecture, investor psychology, regulatory frameworks, ethical implications, and theevolution of financial advisory services.

Databases including JSTOR, Scopus, SSRN, and Google Scholar were used with search terms such as "robo-advisors," "automated financial advice," "digital investment platforms," and "fintech." Studies were selected based on relevance, methodological robustness, and contribution to key themes such as investor behavior, technological frameworks, cost-effectiveness, trust, and regulation. Approximately

35 scholarly articles and reports were analyzed to identify core thematic patterns in the research.

3. THEMATIC LITERATURE REVIEW

3.1 Origins and Technological Foundations

Robo-advisors had originated in the period following the global financial crisis of 2008, which was characterized by public mistrust of conventional financial institutions and there was a greater need for investing solutions that were transparent, affordable, and easily accessible online. According to Fein (2015),the emergence of algorithm-driven investment platforms was facilitated by the decline in investor trust and the development of financial technologies.

Early Pioneers and Adoption Trajectory

The official introduction of robo-advisors into the financial services industry was marked with the launch of Betterment in 2008 and Wealthfront in 2011. The intention of these systems was created with reducing human biases, automating portfolio allocation, and streamlining the investment process. Retail investors, especially millennials, who were looking for accessible, user-friendly alternatives to traditional wealth management, were the target market for their business models (Sironi, 2016; Lam, 2016).

Despite early criticism about the use of algorithms in sensitive financial decision-making, early adopters quickly gained concentration because of low fees, modest account balance restrictions, and automated rebalancing, . For people who were previously shut out of traditional advising services because of high minimum investment limits, robo-advisors quickly emerged as a competitive alternative (Jung et al., 2018).

Modern Portfolio Theory and Goal-Based Algorithms

A significant number to adjustments have been made to the basic principles of portfolio management by automated systems that use well-known financial theories to make investment options easy to find and effective. Harry Markowitz developed Modern Portfolio Theory (MPT), in 1952 is the foundation of the majority of robo-advisory platforms. For best possible portfolio construction, MPT provides the fundamental framework based on risk-return trade-offs and diversification principles.Robo-advisors implement MPT in order to distribute assets among diverse portfolios of ETFs by utilizing client responds to standardized risk tolerance questionnaires (Exchange-Traded Funds) (Fein, 2015; Lam, 2016).

These platforms attempt to identify and address the unique financial goals of each investor in addition to determining an investor's risk tolerance. Goal-based investing models are also employed by many platforms that systematically convert their customers' regular long term financial objectives, for example, retirement, education, and home ownership - into their respective investment horizons and appetite for risk. This practice enables a more meaningful and personalized investing procedure, as portfolios are created based on some milestones in life and funds objectives.

In this manner, investment proposals remain in conformity with user's goals not only in theory but also in a broader practical scenario (Baker &Dellaert, 2019).

Integration of Artificial Intelligence and Machine Learning

Over the years, robo-advisors have evolved using machine learning (ML) and artificial intelligence (AI). The dynamic technologies enable platforms break free from rule-based paradigms to flexible data-centric offerings. Robo-advisors are able to learn from data and improve by time, to deliver better investing results. AI models are better than default models in that they have the ability to update to investment portfolios using real-time market data, customer behavioral patterns, and economic signals outside of the firm (D'Acunto et al., 2019). Machine learning is a key process of analyzing vast amounts of user data in a bid to ascertain individual needs and tastes. It enables Robotic advisors to improve the accuracy of client profiling and adapt interactions based on past behavior, predictive analytics and real-time feedback. The more information platforms have about their customers, the stronger their predictive functions and the better they are at producing personalized financial guidance (Belanche et al., 2019; Kim et al., 2015).

Behavioral Finance and User Interaction

In robo-advisory system behavioral finance concepts and their design have been integrated in order to enhance user interaction and decision-making quality. For instance, in order to protect the users from common behavioral mistakes like overtrading or selling at a loss during market decline, loss aversion, and platforms are now utilizing visual aids, notifications, and reminders (Lo, 2019; Thaler & Sunstein, 2009).

Certain robo-advisors now combine behavioral signals with algorithmic advice—so-called “hybrid intelligence”—to mimic elements of human advisory connections. This more human-like digital interaction enhances client satisfaction and confidence in robo advisors, specifically for users who may be hesitant to rely completely algorithm-based systems (Oehler et al., 2021).

Scalability and Cost Efficiency

Scalability is one of the most cited technological advantages of robo-advisors. Once the system is developed and compared to traditional advisory models, the marginal cost of serving each additional client is extremely low. Automated backend operations reduce staffing, compliance, and operational costs, anabling firms to offer high-quality services and operate efficiently on these margins (Sironi, 2016; Arner et al., 2017).

FurtherMore, with financial planning tools in mobile applications and banking platforms, APIs and cloud infrastructure allow for seamless integration. These integrations support an ecosystem model, in which robo-advisors serve as foundational layers within broader fintech platforms, offering customers a comprehensive suite of financial tools.

3.2 Impact on Traditional Financial Advisory Models

In the financial advisory ecosystem, the emergence of robo-advisors represents a disruptive innovation initiating the service standards, rearrangement of roles and pricing strategies. As algorithm-driven platforms scale rapidly to direct scholars and practitioners alike have explored their implications for traditional human advisors, raising important questions about the future of financial intermediation.

Cost Efficiency and Market Disruption

As the cost structure is a key area of distinction when, it compared to traditional human financial advisors who charge 1% or more, robo-advisors usually charge yearly advice fees between 0.25% and 0.50% of assets under management (AUM) (Baker &Dellaert, 2019). Retail investors, particularly millennials and Gen Z, who prioritize digital access, affordability, and accessibility, find robo-advisors highly appealing due to their cost advantage and low or no account minimums (PwC, 2022).

As a result, the market is experiencing a democratization of investing services, allowing previously marginalized or underserved groups to now have access to portfolio management tools which were previously only available to wealthy individuals (Sironi, 2016). Traditional advising firms are being forced to reconsider their value propositions and operational efficiencies as a result of this challenge to their established business models.

Service Comparison and Functional Limits

Despite the scalability and cost appeal of robo-advisors, research consistently represents their functional limitations. While algorithms are excel at automatic rebalancing, risk profiling, and tax-loss harvesting, but they are not very good at providing complicated planning, estate administration, or succession planning, or providing emotional support or nuanced financial guidance (Fein, 2015; Lo, 2019).

Relational intelligence is provided by human advisors, that is the capacity to comprehend life events, react to emotional cues, and build enduring relationships based on trust, . Clients frequently seek for human *reassurance,* adaptive and empathetic planning during periods of market volatility or personal financial difficulties, which is something that modern robo-advisors find difficult to match (Kim et al., 2015).

The Rise of Hybrid Advisory Models

The scale and automation of robo-platforms are combined with the individualized attention of human advisers towards hybrid advisory models, which have been adopted by numerous financial institutions in response. in accordance to this convergence paradigm, companies like Vanguard Personal Advisor Services, Charles Schwab Intelligent Portfolios Premium, and Personal Capital is providing algorithmic portfolio design along with human advice for strategic decision-making (Lam, 2016; Jung et al., 2018).

In "cyborg advisor" model, human and machine intelligence integrated, which enables companies to cater their client segments who are both advice-dependent and cost-sensitive. According to Belanche et al. (2019) and D'Acunto et al. (2019), hybrid models might offer the optimal trade-off between efficiency and personalization, resulting in stronger financial performance, greater client satisfaction and improved customer retention.

Evolving Role of Human Advisors

By concentrating on areas where human judgment and empathy are essential, robo-advisors seem to be changing their roles, instead of replacing human advisors. Robo-advisors are currently serving as behavioral coaches, financial educators, and life planners (Lo, 2019). Advisor competencies need to be reassessing and up skilling for this change, ranging from asset allocation and product selection to emotional intelligence and holistic life planning (Oehler et al., 2021).

Moreover, traditional firms are investing in digital transformation by integrating client portals, analytics dashboards, and AI-powered tools, in order to augment human productivity and deliver more insightful, data-driven advice (Arner et al., 2017).

Strategic Complementarity, Not Substitution

According to the majority of academics, robo-advisors and human advisors are acting as more complementary than competitive. Human advisors are still essential for non-standardized, context-sensitive interactions, even though standardized tasks can be effectively handled by robo-platforms. The human touch is still essential for wealthy clients or those with complicated financial requirements (Jung et al., 2018; Baker &Dellaert, 2019).

According to market statistics, high-net-worth individuals still favor in-person interactions while robo-advisors mainly cater to mass-affluent and entry-level investors. This differentiation of every model to meet different investor risk tolerance and preferences has been stated to entail market cohabitation according to PwC (2022) and Kim et al. (2015), where digital and human-assisted services run parallel to one another to meet distinct investor needs.

3.3 Investor Behaviour and Trust

Investor behavior and trust relationships are of essential importance in the current usage and adoption of robo-advisory services. Building trust through computer-mediated interfaces, algorithmic explainability, and multiple performance is essential to Robo-advisors, as opposed to the traditional financial advice framework, which relies predominantly on human interaction. This section outlines previous work into how different demographic groups use robo-advisors as well as behavioral factors that influence their trust and decision-making.

Demographic Variations in Adoption

Studies indicate that the most significant and primary user of robo-advisory websites are young, digitally born investors, namely millennials and Gen Z (Belanche et al., 2019; PwC, 2022). These groups are more laid-back when it comes to algorithmic decision-making over their finances as they are fond of ease of use, lower charges, and convenience. As per Oehler et al., (2021), older clients and high net-worth individuals are conservative, however, as they tend to value people interaction, personalized advice, and the reassurance that comes from personal relationships. In addition to this, research by Kim et al., (2015) illustrates that financial literacy and education also have an impact on robo-advisor adoption intention. Financially more literate people are more critical about the non-transparency of algorithms, whereas people with less investment knowledge are either over-reliant or extremely skeptical based on past experience with digital technology.

Dimensions of Trust in Robo-Advisors

Trust towards robo-advisors is multidimensional, encompassing perceived competence (the correctness and effectiveness of the algorithm),Security and data protection, and transparency in decision-making processes.

Based on Jung et al. (2018), Users judge robo-advisors on the basis of how explicitly they define investment decisions and the overall trustworthiness of the system. The lack of transparency is a major concern, generally referred to as the "black box problem" which continues to be an ongoing issue. Numerous users are reluctant to believe that platforms avoid stating explicitly how asset allocations are calculated or how risk preferences are integrated into portfolio design (Lo, 2019; D'Acunto et al., 2019). Algorithm aversion is another serious hindrance, especially in emotionally charged situations such as market crashes.

Algorithm aversion is also a major challenge, particularly in emotionally charged situations such as a market decline. Dietvorst et al., (2015) establish that individuals tend to be unwilling to use algorithms after observing a single mistake despite the fact that the algorithm performs better than human judgment in the long term.

Behavioral Finance and Bias Reduction

To reduce behavioral biases is one of their most celebrated benefits of the capacity of robo-advisors. As compare to human investors, algorithms are immune to emotions, herd mentality, overconfidence, and loss aversion,. According to empirical data presented by D'Acunto et al., (2019), robo-advisors promote long-term, disciplined investment, particularly by automating procedures like diversification and periodic rebalancing. This automation encourages logical financial conduct In order to reduce rash responses towards market swings.

Furthermore, it has been discovered that the propensity of investors is more to make choices on recent market developments rather than long-term fundamentals as robo-advisors lessen regency bias (Lo, 2019). Robo-advisors increase consistency and impartiality in portfolio management by offering data-driven suggestions.

Skepticism and Emotional Disconnect

In spite of the benefits a significant barrier is emotional disconnection. The emotional comfort that many investors associate with traditional advising system is diminished by the lack of in-person meetings, algorithms' incapacity to express empathy and the absence of personalized life counsel (Kim et al., 2015). When making high-stakes decisions like planning for retirement, organizing inheritances, or managing financial disasters, investors usually choose traditional human advisory system because human advisor are able to understand reveals beyond what data can provide.

Interface design is a critical factor influencing user trust. Human-like interfaces including Natural language processing (NLP) chatbots, have been demonstrating to improve user comfort and the perceived competence of the robo advisory platform (Belanche et al., 2019). To prevent the "uncanny valley" effect, efforts to humanize robo-advisors must carefully balance functionality with consumer expectations, as overly anthropomorphic bots cause discomfort instead of fostering trust. (PwC, 2022).

3.4 Financial Inclusion and Democratization

One of their most ground-breaking claims is that robo-advisors can increase financial inclusion by expanding access to affordable, user-friendly wealth management services. By using automation, these platforms significantly reduce the cost and complexity of investing, opening it up to previously underrepresented groups such as young investors, gig economy workers, and those with low to middle incomes. Reher and Sokolinski (2024) shows how lowering account minimums increased participation among middle-class investors, and Abraham et al., (2019) highlights robo-advisors' lower fees and efficient features like tax-loss harvesting.

Lowering barriers to Entry

Traditional wealth management services sometimes charge exorbitant fees and have a high minimum investment requirement, which effectively shuts out a large number of potential investors. In contrast, robo-advisors often provide flat or low-percentage fees (as low as 0.25% annually), Low or zero account minimums, and automated portfolio management tools that are accessible through websites or mobile apps (Baker &Dellaert, 2019; OECD,2021). Automated saving and Micro-investment are made possible through these platforms such as Betterment and Acorns, which enable users to begin investing with as little as no more than a few dollars. These practices are especially attractive to younger users who have less money to spend (Fein, 2015; Sironi, 2016).

Serving the Underbanked and Gig Economy

Large-scale studies testified to the financial vulnerability of freelancers, part-timers, and gig economy workers, and many of them do not benefit from financial advice or employer-sponsored pension plans (Zelma, 2024). Robo-advisors help to bridge these gaps through automatic rebalancing, goal-oriented investment tools, and self-directed retirement planning. As per a study conducted by Belanche et al.

(2019), robo-advisors enables consumers to manage their finances employ simple interfaces, thus, it encourages financial literacy and independence. Consumers should seek experts advice and have knowledge about limitations of robo-advisors while handling sophisticated financial issues.

Second, evidence indicates that robo-advisors encourage inclusive saving behavior, particularly among digitally native millennials, who might financially undeserved. These platforms integrate investing into everyday financial activities by offering personalized, low-cost investment options aligned with user-defined goals, such as paying off student loans, building an emergency fund, or planning for early retiring (D'Acunto et al., 2019).

Bias Reduction and Equal Treatment

Unlike human financial advisors who might exhibit conscious or unconscious biases related to gender, ethnicity, or income level, robo advisors offer consistent, data-driven recommendations based solely on objective input parameters (Fein, 2015; Lo, 2019). Technical neutrality can enhance fairness in financial ecosystems as it lessens discriminatory treatment in financial advisory services. Researchers were concerned, that if the algorithms are not properly vetted, they can inherit prejudice in training data or design choices (Barocas et al., 2019). Monitoring and auditing must be done carefully in order to prevent unintentional bias and deliver fair results for everyone.

Challenges to Universal Access

There are still major obstacles to overcome, even with these benefits. Digital literacy which is the capacity to use technological platforms and efficiency to analyze financial data, One of the main obstacles is, Especially in developing or rural regions, many potential users lack the financial literacy necessary to navigate investment platforms, in addition to lacking digital connectivity (Sironi, 2016; OECD, 2021). Additionally, inexperienced investors may be discouraged by services of complicated interfaces in form of inadequate language localization, and a lack of individualized help (Jung et al., 2018).

Uneven distribution of robo-advisor benefits is also caused by the "digital divide," which refers to disparities in access to mobile technology, dependable internet, and cyber security. Younger, urban, and educated populations are well-positioned to benefit, but unless design interventions and proactive policy implemented, older, less educated, or marginalized groups may continue to be excluded (Lo, 2019).

3.5 Regulatory and Ethical Considerations

As robo-advisors are continuously reshaping the financial services industry, complex ethical and regulatory issues with this regard are transparency, fiduciary duty, data privacy, and algorithmic fairness are brought up with their transformation. These challenges have intensified as a growing number of investors place their trust in automated systems which, despite being created by human experts to make financial decisions.

Regulatory Uncertainty and Compliance Gaps

The suitability of existing regulatory frameworks for monitoring robo-advisory services is the central concern in the literature. Researchers contend that even though the majority of jurisdictions categorize robo-advisors under conventional financial advisory laws, these regulations may not be adequate to regulate algorithm-driven digital platforms that function more autonomously than human advisors (Arner et al., 2016; Kim et al., 2015). The legal standing of algorithms is still important issue for debate, including whether they are subject to fiduciary standards and concern regarding responsibility when automated advice produces subpar results.

For instance, the Investment Advisers Act of 1940 in the United States generally treats robo-advisors as Registered Investment Advisers; nevertheless, digital platforms were never meant to be governed by this regulation (Baker &Dellaert, 2019). European regulators have attempted to address these concerns through MiFID II and the General Data Protection Regulation (GDPR), which impose disclosure for data protection requirements. However, Enforcing laws is still challenging, especially with international platforms and proprietary algorithms whose inner workings are rarely disclosed (Fein, 2015).

Algorithmic Transparency and Accountability

Transparency of robo-advisors in recommendation-making process is one of the main topics of regulatory discussions. Unlike human advisors who explain their reasoning about robo-advisors rely on black-box algorithms, Investors find it difficult to comprehend how decisions are made by robo-advisers (Jung et al., 2018). This lack of consistency can harmed investor trust, particularly if suggestions seem erratic and portfoliosperform poorly.

Scholars recommended for the implementation of "explainable AI" (XAI) in the financial advisory domain to provide users and regulators with better insight into how algorithms work, (Bussmann, 2020). There are still issues with data manipulation, conflicts of interest, and algorithmic drift, which occur when machine learning models change in ways that their creators did not expect and may violate appropriateness standards (Lo, 2019).

Ethical Concerns: Bias, Privacy, and Discrimination

Robo-advisors must contend with a growing series of ethical concerns as well as compliance with the law. When models are trained on data Algorithmic bias is the most prevalent bias, which can happen with demonstration of past injustices or structure prejudice (Barocas et al., 2019). This may lead to the discriminatory treatment of certain groups in portfolio construction or the assessment of risk like women or minorities.

User data and privacy use is another ethical issue involves in adoption of Robo advisors. They collect vast personal and financial data in the attempt to provide personalized advice. In the absence of strong protections, the client's data may be vulnerable to hacking or misuse, such as intentional or accidental exploitation for purposes such as behavioral profiling, marketing, or resale to third parties (Arner et al., 2016). Though GDPR and similar legislation have enhanced consent and data protection requirements, academics contend that moral regulation of data should extend beyond compliance with law, insisting on active respect for user agency and digital rights protection (Sironi, 2016).

The Role of Regulatory Innovation

Some scholars advocate for regulatory innovation such as the development of RegTech and SupTech solutions to minimize such issues that utilize technology to improve regulation. For instance, many fintech firms can test new products within regulatory sandboxes where innovations are trialed under close observation. This approach refines the services of firms by fostering innovation while keeping consumers safety remains a priority (Zetzsche et al., 2017). Besides, regulators are also challenged to possess adaptive risk-based models that enable them to observe algorithmic activity continuously and update compliance procedures accordingly as technology changes (OECD, 2021).

3.6 Future Developments and Research Opportunities

The robo-advisor evolution is where technology innovation, user innovation, and regulation evolution converge. Future literature shows that the future of robo-advisory services will be shaped by how artificial intelligence (AI) interacts with blockchain technology, open banking, and personalized finance (Jung et al., 2018; Arner et al., 2016).

Technological Innovations and Integration

New robo-advisors are likely to provide full-service financial planning alongside core portfolio management. The services may extend beyond the typical investment to estate planning, insurance optimization, tax-loss harvesting, and AI-driven and NLP-driven real-time financial coaching (Baker &Dellaert, 2019). Blockchain integration will likely improve security in transactions, improve identity verification of the client, and improve data transparency, with more audit capacity and creating greater trust between platforms (Sironi, 2016).

In addition, open banking that calls for banks to share customer information with third-party providers based on consent by users will possibly facilitate extremely personalized financial services. This will allow end-to-end financial ecosystems to be created by facilitating the ability of robo-advisors to develop complete financial profiles of users and deliver personalized recommendations across different areas of finance (OECD, 2021).

Shifts towards Holistic and Predictive Advisory Models

Such a shift in paradigm from reactive to forward-looking robo-advisory systems is also inferred by research that leverage the use of AI to predict client's financial needs, analyze potential risks, and forecast market volatility (Lo, 2019). Such a proactive approach could potentially avoid loss aversion and other behavioral traps during economic downturn through warning mechanisms for customers early on as well as dynamic rebalancing portfolios (D'Acunto et al., 2019). Second, robo-advisors can utilize life-cycle investment models to adapt strategies based on users' evolving risk profiles, income levels, and life stages. Such models could better follow human advisors' approaches, which can be assisted to enhance the range of robo-advisors' functionality beyond wealth accumulation to include intergenerational wealth transfer and decumulation (planning during retirement) (Fein, 2015).

3.7 Research Gaps and Areas for Further Study

There are still a number of important research gaps in the collection of current literature, despite the promising trajectory.

1. ***Cross-Cultural and Global Perspectives***: In developing regions like Asia, Africa, or Latin America, Limited research exists on the use of robo-advisors as the majority of studies have focused on North America and Europe. Comparative cross-jurisdictional studies are required to examine how cultural, legal, and so-

cioeconomic factors influence the adoption and trust of robo advisory services (Jung et al., 2018; Belanche et al., 2019).

2. ***Long-Term Behavioral Outcomes***: The psychological and behavioral effects of using robo-advisor in context of Long-term are still remain underexplored. There is a lack of longitudinal studies that track user satisfaction, portfolio performance, and behavioral consistency over time, despite the fact that short-term adoption trends are extensively documented (D'Acunto et al., 2019).
3. ***Algorithmic Auditability and Explainability***: Academicians urge for the further research on techniques to audit and interpret robo-advisor algorithm. Concerns are increasing over "black box" systems in which decision-making procedures are unclear to developers, regulators and even to clients also (Bussmann, 2020). There is an urgent need to develop standardized audit protocols within AI governance frameworks.
4. ***Ethical and Bias Detection Mechanisms***: There is limited research on reliable methods to identify, prevent and rectify algorithmic biases, particularly when these biases are immediately visible in financial outputs. Addressing this is crucial for safeguarding vulnerable or marginalized user groups (Barocas et al., 2019).
5. ***Hybrid Advisory Models and Human-Robo Interaction***: As robo-advisors develop, hybrid models that combine automation and human oversight are becoming more popular. However little is known about architecture, dynamics of trust, and client outcomes of these models (Kim et al., 2015).
6. ***Impact on Financial Advisors and Labor Markets***: Another disregarded aspect is the impact of robo-advisors on the financial advising industry including job displacement, role transformation, and upskilling requirements (Arner et al., 2016).

Addressing these gaps will be essential not only for academic insight but also for policymakers, developers, and financial institutions aiming to create more inclusive, ethical, and effective robo-advisory ecosystems.

4. CONCLUSION

Robo-advisors emerged as a disruptive force in the financial services sector driven by advances in algorithmic technology, artificial intelligence, and the accessibility of digital platform. The reviewed literature confirms that robo-advisors significantly democratize investment opportunities by reducing costs and streamlining intricate portfolio management procedures that have been historically handled by human advisors,(Sironi, 2016; Baker &Dellaert, 2019). Robo-advisors,, promote more consistent investing practices by automating decision-making and reducing emotional biases, especially for younger, tech-savvy populations(D'Acunto et al., 2019; Belanche et al., 2019).

Despite of evident benefits of Robo-advisor's, there are significant drawbacks and difficulties in this regard. Trust issue, remain a significant hurdle to wider adoption among older and high-net-worth investors, particularly concerning algorithmic transparency and security (Jung et al., 2018). Furthermore, the hidden human touch needed personal advice for complicated financial planning, and study suggests that role of robo advisors in the future will probably be complementary rather than substitute (Kim et al., 2015). A more sustainable and customer-preferred approach is the development of hybrid advisory models merging automated efficiency with human know-how.

From governance and regulation as well as ethical considerations, automation of financial advice is challenging in terms of governance complexity, algorithmic fairness issues, data privacy and fiduciary obligations (Arner et al., 2016; Belanche et al., 2019). These are issues to be resolved through changing frameworks where AI is used in open, accountable and ethical ways, while finding balance between the innovation and protection of the consumer (Kim et al., 2015; Bussmann, 2020).

For robo advisors, several important directions for future research have emerged. To ensure that these services are operating fairly and transparently, it is extremely important to create standard means of algorithm explainability and auditability (Bussmann, 2020). In addition, to unlock the global patterns of adoption and develop tailored solutions, additional cross-country comparative studies in various cultural and regulatory contexts are required (Jung et al., 2018). Lastly, longitudinal studies need to evaluate how robo-advisor usage long-term affects financial literacy, accumulation of wealth, and investment choices (D'Acunto et al., 2019).

Lastly, the manner of delivery and receipt of financial advice has been revolutionized by robo-advisors. The websites have the potential to create more inclusive, efficient, and ethical financial systems by confronting behavioral, technological, and regulatory issues.

REFERENCES

Abraham, F., Schmukler, S. L., & Tessada, J. (2019, February). *Robo-advisors: Investing through machines* (Research & Policy Brief No. 21). World Bank, Chile Center and Malaysia Hub. https://documents.worldbank.org/en/publication/documentsreports/documentdetail/498211549262503395/robo-advisors-investing-through-machines

Arner, D. W., Barberis, J., & Buckley, R. P. (2017). FinTech, RegTech, and the reconceptualization of financial regulation. *Northwestern Journal of International Law & Business*, *37*(3), 371–413.

Arner, D. W., Barberis, J. N., & Buckley, R. P. (2016). The evolution of FinTech: A new post-crisis paradigm? *Georgetown Journal of International Law*, *47*(4), 1271–1319.

Baker, T., & Dellaert, B. G. (2019). Regulating robo advice across the financial services industry. *Iowa Law Review*, *103*(3), 713–750.

Barocas, S., Hardt, M., & Narayanan, A. (2019). *Fairness and machine learning: Limitations and opportunities*. Retrieved from https://fairmlbook.org/ and https://www.myecole.it/biblio/wp-content/uploads/2020/11/2020-Fairness-book.pdf

Belanche, D., Casaló, L. V., Flavián, C., & Schepers, J. (2019). Service robot implementation: A theoretical framework and research agenda. *Service Industries Journal*, *39*(3–4), 203–225.

Bussmann, N. (2020). Explainable AI in robo-advisory services: Towards transparency in financial algorithms. *Journal of Financial Transformation*, *51*, 124–130.

D'Acunto, F., Prabhala, N., & Rossi, A. G. (2019). The promises and pitfalls of robo-advising. *Review of Financial Studies*, *32*(5), 1983–2020. DOI: 10.1093/rfs/hhz014

Dietvorst, B. J., Simmons, J. P., & Massey, C. (2015). Algorithm aversion: People erroneously avoid algorithms after seeing them err. *Journal of Experimental Psychology. General*, *144*(1), 114–126. DOI: 10.1037/xge0000033 PMID: 25401381

Fein, M. L. (2015). Robo-advisors: A closer look. SSRN *Electronic Journal*. https://doi.org/DOI: 10.2139/ssrn.2658701

Jung, D., Dorner, V., Glaser, F., & Morana, S. (2018). Robo-advisors: Digitalization and automation of financial advisory. *Business & Information Systems Engineering*, *60*(1), 81–86.

Kaur, B., & Kour, M. (2024, September). Exploring the Growth and Future research directions on Robo advisors: A Bibliometric Study. In *2024 International Conference on Advances in Computing Research on Science Engineering and Technology (ACROSET)* (pp. 1-6). IEEE.

Kaur, B., & Kour, M. (2025). Impact of Artificial Intelligence on Investment Decision-Making. In AI, Economic Perspectives, and Firm Business Management (pp. 1-12). IGI Global Scientific Publishing

Kim, Y., Park, Y.-J., Choi, J., & Yeon, J. (2015). An empirical study on the adoption of "fintech" service: Focused on mobile payment services. *Advanced Science and Technology Letters, 114*(Business), 136–140. http://dx.doi.org/DOI: 10.14257/astl.2015.114.26

Lam, J. (2016). Robo-advisors: A portfolio management perspective. *Journal of Financial Planning*, *29*(7), 42–48.

Lo, A. W. (2019). *Adaptive markets: Financial evolution at the speed of thought.* Princeton University Press.

Markowitz, H. (1952). Portfolio selection. *The Journal of Finance*, *7*(1), 77–91.

Oehler, A., Horn, M., & Wendt, S. (2021). Investor characteristics and their impact on the decision to use a robo-advisor. *Journal of Financial Services Research*, *62*, 91–125. DOI: 10.1007/s10693-021-00367-8

Organisation for Economic Co-operation and Development. (2021, October 27). *G20/OECD-INFE report on supporting financial resilience and transformation through digital financial literacy*. OECD Publishing. https://doi.org/DOI: 10.1787/0132c06d-en

PwC. (2022). *Asset and wealth management revolution 2022*. Retrieved from https://www.pwc.se/sv/esg/asset-wealth-management-revolution-2022-se.pdf

Reher, M., & Sokolinski, S. (2024). Robo advisors and access to wealth management. *Journal of Financial Economics*, *155*, 103829. DOI: 10.2139/ssrn.3515707

Sironi, P. (2016). *FinTech innovation: From robo-advisors to goal-based investing and gamification*. Wiley.

Thaler, R. H., & Sunstein, C. R. (2009). *Nudge: Improving decisions about health, wealth, and happiness*. Penguin Books.

Zelma, E. (2024). The gig economy from the freelancer's perspective: The risk of precarianization. *International Journal of Contemporary Management*, *60*(1), 211–226. DOI: 10.2478/ijcm-2024-0013

Zetzsche, D. A., Buckley, R. P., Arner, D. W., & Barberis, J. N. (2017). Regulating a revolution: From regulatory sandboxes to smart regulation. *Fordham Journal of Corporate & Financial Law*, *23*(1), 31–103.

Chapter 13
Robotic Process Automation in Banking: Enhancing Risk Management

Mujahid Tabasum
https://orcid.org/0000-0001-7806-0247
South East Technological University, Ireland

Waqar Hassan
https://orcid.org/0009-0002-7641-9820
Leads Group of Colleges, Punjab, Pakistan

Mohammad Asif Nawaz
Independent Researcher, USA

Sadia Bashir
Bloom Field Hall School, D.G. Khan, Pakistan

ABSTRACT

Technology is making important changes to the banking sector, mostly by helping increase productivity, reduce risk and ensure compliance. Using Robotic Process Automation (RPA), businesses now have an innovative way to automate tasks that are both repeated and need to be performed accurately and correctly. This chapter investigates how RPA helps banking systems handle risks. Despite putting more money into digital transformation, banks face problems with risks related to their manual processes. RPA assists by reducing wrongdoing, ensuring live updates with regulations and promoting more automation. Both artificial intelligence and machine learning allow RPA to perform predictive tasks differently from traditional rule-based solutions. RPA solutions for fraud detection, credit risk analysis and regulatory reporting are discussed in the chapter, looking at their benefits and drawbacks.

DOI: 10.4018/979-8-3373-3571-1.ch013

Based on qualitative studies with help from academic papers and industry reports, this work proposes ways for banking professionals and policymakers to address risk and improve the way banks function.

1 INTRODUCTION

1.1 Contextual Background of the Banking Sector

In recent years, global banks have been transformed by globalization, changes in technology, updates in regulations, and differing customer needs. Economic instability, threats from cyber attacks, and fierce competition from fintech companies have pushed banks to introduce new, more flexible, and modern methods (World Bank, 2022). Because of the 2008 crisis, several governments introduced stricter guidelines, including Basel III. Meanwhile, what consumers expect from companies is immediate and easy access to services online. The changes have resulted in every banking department, including customer help, detecting fraud, and assessing credit, now using new technology more often (PwC, 2021). Due to many changes, both quick action and efficient operations are now crucial. Therefore, banks are assessing their old systems, mainly in areas like managing risks, where adopting technology and data-based methods is highly necessary (EY, 2020). As a result of these changing trends, banking institutions are integrating RPA technology to assist in their operations.

1.2 Defining Risk Management in Banking

It is important in banking to use a step-by-step method to find, judge, watch for, and lower potential threats to a bank's financial health, rules and morality or image. It covers many kinds of risks, such as credit, market, operational, liquidity, and compliance. (Hull, 2018) Proper risk management aims to protect a bank from various dangers and maintain its operation in uncertain conditions, the goals set by the strategy, and the rules that govern the company. Models and tools for assessing risks, test cases, and different scenarios should be used as part of the process (Jorion, 2009). Today, because changes in technology and financial regulations happen fast, managing risks in the financial sector must be adaptable and flexible.

1.3 Limitations of Traditional Risk Management Systems

While risk management systems have been vital for financial institutions, it is becoming clear that they are unable to keep pace with the changes and speed in the banking world. The heavy use of processes not connected contributes to delays, inconsistency, and problems caused by human action (Accenture, 2021). They usually lack the ability to include real-time information, stopping them from detecting threats and responding as needed. Any delays in dealing with sudden threats in the financial sector can have significant consequences. The problem of legacy systems is that they are not very scalable.

When banks operate in several nations, they must have dependable and flexible risk management systems. Following separate rules in each region is difficult for traditional systems, especially since the regulations are updated regularly. Besides, it costs a lot to fix and upgrade these old infrastructure systems, and this often reduces the funds available for innovations (Deloitte, 2020). Since advanced analytics and automation cannot be used with legacy systems, they have difficulty recognizing risks that may come from unusual activities. In essence, basic risk management models lack the tools to handle the huge amounts of data required in today's fast-paced banking world.

1.4 Emergence of Robotic Process Automation

Automation of mundane functions has come from Robotic Process Automation (RPA) in the financial sector. RPA relies on bots to carry out digital tasks just as a person does, to handle repetitive work that would take a lot of time otherwise. Work processes must be accurate and consistent when we think about work (Willcocks, Lacity, & Craig, 2017). With RPA, many banks are streamlining customer onboarding, monitoring financial transactions, handling reports, and identifying fraud, all of which are closely linked to risk management. RPA helps banks renovate their work easily because it does not require upgrading existing IT systems. Auditors also find RPA helpful, as it records all activity in logs that can be reviewed for requirements and accuracy. RPA enables institutions to easily react to any significant changes that might happen in the financial market. Chiefly, RPA makes it possible to monitor risks continuously, allowing quick reactions to anything unusual and fewer risks related to failures or regulations. The combination of RPA with AI and ML will likely make RPA an important factor in transforming risk management and the ability to respond to any risks.

Figure 1.

Source: sstech.us

1.5 The Value Proposition of RPA in Banking Risk Management

RPA improves risk management in banking by making work more efficient, reliable, and compliant. RPA is significant because it automates processes for regulatory reporting, customer due diligence, know-your-customer services, and transaction monitoring when these depend on rules. In these fields, any delays or mistakes made by people can lead to serious trouble for the company or for authorities (Deloitte, 2022). Introducing RPA in banking reduces operational risk, as it eliminates errors and extra work from human hands. They could also be developed to review in real-time a high number of transactions, ensuring there is no suspicious activity regarding AML policies. Moreover, the logs kept by RPA help ensure transparency and ease the review process when the company is being audited (Willcocks, Lacity, & Craig, 2017).

Using RPA can lower expenses and facilitate scaling processes. Banks are able to use bots in different areas or departments with modest training, which allows them to expand rapidly with only a minor additional burden related to risks. Furthermore, since RPA can be easily updated, it allows banks to adjust to new rules without replacing the entire system (PwC, 2021). On the whole, RPA is important in today's banks, providing the required adaptability, strength, and obedience to support better risk management methods.

1.6 RPA's Strategic Role in Enterprise Risk Frameworks

About a complete system for managing risks called ERM, RPA greatly improves the way the bank handles risks and operates smoothly. Because traditional ERM focuses on hand-entered information and several different data sets, there may be breaches in risk management. Using RPA, data from several parts of the company can be combined, while testing, analysis, and reporting of risks can be automated (Accenture, 2021). By using RPA, companies now have up-to-date views of risks and can also make better risk-related decisions. Because of this, leaders allocate resources better, evaluate ongoing risks as situations change, and swiftly deal with new risks. RPA allows an organization to adopt a proactive attitude in managing risks, which fits with the main goals of a flexible ERM framework. As a result, risk management systems benefit from AI and machine learning, enabling technology to take over routine tasks in the field.

1.7 RPA vs Other Automation Tools

RPA is not the same as other automation tools, such as Business Process Management (BPM) and Artificial Intelligence (AI). While BPM redesigns every step of a process and AI gives insights into decision making, RPA is good for automating routine repetitive jobs without changing the IT system (Willcocks et al., 2017). Because it is not invasive and can interact with old software, banks using mature and vital software find it especially useful. RPA bots are not like AI; they are fast to set up and immediately improve efficiency and follow rules in the business.

1.8 Problem Statement and Knowledge Gap

Although RPA is being used more in banking, many do not fully understand how it works with regard to risk management. Most research has looked at the generic benefits of RPA, but few have considered how it can be applied to risk management. Moreover, research findings cannot always measure the exact benefits of RPA in terms of risks, costs, and following rules (Deloitte, 2022). The chapter aims to solve this issue by discussing how RPA affects risk management in banks, its benefits over other similar technologies, and its agreement with risk objectives in a strict and risky market.

1.9 Objectives of the Chapter

The primary objective of this chapter is to explore and critically analyze the **application of Robotic Process Automation (RPA) in banking risk management**. Specifically, it aims to:

- Evaluate the **value proposition** of RPA in mitigating operational, compliance, and fraud-related risks.
- Assess how RPA fits into **enterprise risk management (ERM) frameworks**.
- Compare RPA with other automation tools in the context of risk governance.
- Identify the **existing knowledge gap** and outline opportunities for future research in this domain.

By achieving these objectives, the chapter contributes to a deeper understanding of how automation technologies can fortify risk management in banking institutions.

1.10 Scope and Structure of the Chapter

This chapter is arranged to comprehensively study how RPA affects banking risk management. The research starts with setting the background, which is followed by looking at theory and existing evidence. After discussing existing systems, the article considers adding RPA, lists the challenges and suggestions, and summarizes possible directions for further study.

2 BACKGROUND / LITERATURE REVIEW

2.1 Evolution of Risk Management in the Banking Industry

For years, banks have depended on risk management to guard their finances and make sure they are profitable. There are six main varieties of banking risks: credit, operational, market, liquidity, compliance, and cyber risks (BCBS, 2019). For a long time, banks used person-driven checks, paper record-keeping, systems that operated alone, and policy guidelines to supervise risk management. Previously, these processes worked, but now they are not as effective as they once were due to modern financial details and high-speed online transactions.

Normally, risk identification would wait for a loss and happen reactively. There have been many developments in the laws that oversee the industry. The Basel Accords (referred to as Basel I, II, and III) marked important moments since they determined capital adequacy and risk-weighted assets around the world. FATCA, AML, KYC, and GDPR have increased the amount of compliance required. Accordingly, banks are required to oversee customers' actions, report any dubious transactions, and ensure data is private. All these tasks require detailed risk management and thorough reporting (Ghosh, 2020). Due to rising regulations, advances in technology, and increased customer demands, current legacy systems are not able to deal with modern risks. Now, experts in the industry recognize that risk management should be more focused on using data and technology. The implementation of Robotic Process Automation (RPA) is helping companies manage financial risks faster, more consistently, and in real time.

2.2 Rise of Robotic Process Automation (RPA) in Financial Services

RPA (Robotic Process Automation) uses robotics to make it possible for software bots to perform the same tasks over and over like humans. RPA imitates human behaviors such as clicking, typing and using various systems without needing to modify what exists currently (Willcocks, Lacity, & Craig, 2017). Using RPA, you do not need to do major rework on your processes or change your existing programs, which is great for banks structured on legacy technologies. RPA was formed using the help of basic technologies including screen scraping, scripting and BPM. As time passed, they became smarter, resulting in a group of digital workers known as bots which can easily handle tasks all day with excellent control and reliability (Aguirre & Rodriguez, 2017).

Because the industry requires more efficiency and accuracy, the BFSI sector has quickly adopted automated processes. Banks are using RPA to handle things such as signing up new customers, verifying their identities, handling loans, detecting fraud and meeting regulatory requirements (PwC, 2021). Because these activities follow the same rules, they can easily be automated. It took RPA only seconds, thanks to JPMorgan Chase, to do work that once required 360,000 hours to complete. Likewise, ICICI Bank introduced 750 bots in both trade finance and retail services. The bots report completing more than one million tasks each day. HSBC used RPA to make its back-office routine operations easier and increase the quality of its customer service (Deloitte, 2022). What matters about RPA is that it curbs costs, improves on accuracy, and allows expansion while keeping infrastructure costs low. RPA is seen as a key factor that drives change in financial services, supporting both day-to-day operations and risk management, compliance, and important decisions.

2.3 Academic Literature on RPA and Banking Operations (Approx. 300 words)

The majority of studies on Robotic Process Automation (RPA) and banking focus on boosting process speed, eliminating costs, and increasing flexibility. It is widely acknowledged by researchers that RPA improves accuracy in financial processes and helps fast-track responses. In the Journal of Information Technology Teaching Cases, Willcocks et al. (2017) looked at how RPA could make shared services more efficient. It was established by researchers that RPA helped reduce processing costs by 20–60% and improved compliance.

Furthermore, Syed et al. (2020), in the Business Process Management Journal, pointed out that RPA contributes to standardizing business operations, mainly by resolving banking transaction issues caused by a lack of consistency. According to Aguirre and Rodriguez (2017), using RPA in banking customer services has improved response times and ensured its expansion of the service, and ensured that customers are happy. Another investigation by Hofmann, Samp, and Urbach (2020) in IEEE Access outlined how financial companies can achieve their digital transformation goals using RPA. While a significant amount has been written about RPA, little is mentioned about its part in managing risk. The focus of most studies is on how RPA affects the company's operations and not how it relates to managing risks in the organization. Moreover, researchers do not yet focus much on the connections between RPA and compliance with regulations, fraud detection, and cyber protection. This chapter focuses on using RPA both to improve how things are done at a bank and to manage and minimize the risks found in credit, operations, compliance, and cyber categories.

2.4 RPA Applications in Risk Management Domains

Thanks to RPA, risk management is now being improved since tasks are automated, allowing for greater speed, no errors and more flexibility. Bringing RPA together with AI and analytics has made it possible for RPA to further ensure risk management.

2.4.1 Fraud Detection and Prevention

There are many dynamic uses for RPA found in fraud detection. Companies now rely on RPA to track transactions in real time and catch situations that vary from the usual behavior recorded in history (McKinsey & Company, 2021). These robots analyze huge sets of data from various systems to detect instances of suspected fraud. Combined with AI and machine learning, RPA bots help set the risk scores for each transaction in a fraud scoring system. When organizations work together, early discovery and quick steps prevent a lot of damage and losses to the company's reputation. Thanks to RPA, many top institutions can now automatically create alerts and cases, allowing fraud analysts to concentrate on the most critical scenarios.

2.4.2 Regulatory Compliance & Reporting

Banks now face much greater challenges complying with rules set by AML, GDPR and Basel III. RPA helps address the issue by automating all steps, including collecting and submitting regulatory reports such as SARs and STRs. I use bots in order to file documents properly and promptly, and I also keep strict records for regulators. RPA for compliance is reported by the report and KPMG to give better on-time compliance, fewer errors, and increased visibility for banks. Also, KYC/ AML updates are supported by bots that continually check customer details in different databases.

2.4.3 Credit Risk Assessment

Robotic process automation allows the verification of borrower data and checks that all materials needed for an application are correct by matching them with other sources. Because of this, the loan origination and underwriting process takes less time, mainly for high-volume loans to retail or small business clients. It was reported in the IMF's working papers that Nordic banks have experienced error reduction of up to 60% and shorter turnaround times for decisions after automating credit and document checks. Along with better customer service, they also help avoid giving credit to those who do not meet the criteria.

2.5 Comparative Technologies in Risk Management

RPA is crucial, yet it works hand in hand with other emerging technologies to improve risk management in the financial industry. Thanks to AI and ML, it is possible for companies to estimate the risks of credit defaults, market fluctuations, and fraud. These models gather insights from past patterns that still rule but do not see. Big Data Analytics works with it by alerting users to new risks as they appear. Financial institutions gather data in various forms to notice patterns and events that can influence risk management. It ensures that transactions and data cannot be changed and are open to all.

Blockchain helps to make sure transactions are reliable when it comes to import/ export handling, managing your KYC, and cross-border transactions. Nonetheless, there are still reasons why RPA is unique. It is easy to work with, does not require extensive coding, and can quickly be applied to systems that banks built years ago. RPA delivers a rapid return on investment and causes little trouble for any organization, in contrast to AI and blockchain's longer process. It is worth noting that RPA can allow companies to steadily digitize their workflow as they prepare themselves for more advanced developments.

2.6 Benefits of RPA in Risk Management (from Literature & Practice)

Using RPA in managing risks improves the company's operations, strategy, and compliance.

2.6.1 Efficiency Gains

Many people agree that processing speed is among the top advantages. Since bots operate all the time, tasks like watching transactions, filing reports and verifying documents are processed faster. Deloitte (2018) showed that using RPA with repetitive tasks, banks have been able to reduce the time needed by up to 80%. A decrease in mistakes made by workers. By following accurate predefined guidelines, RPA bots avoid the usual errors caused by mistakes in data entry or reports. This helps greatly with follow-up in areas such as regulations, because making mistakes in these areas can result in significant penalties or damage to a company's reputation.

Figure 2.

Source Blueprism

2.6.2 Compliance Accuracy

RPA makes it possible for data to be entered into compliance files on time without any delays. This automation means institutions will be certain any new rules from regulators are correctly added to their procedures.

2.6.3 Improved Auditability

Each RPA task is logged and marked with a date and time, ensuring no one can alter the records. It helps ensure that due diligence is shown in reviews by regulators or auditors.

2.6.4 Quantitative Benefits

EY (2019) studies conclude that RPA helps financial institutions save up to 40% on costs within the first year of its use. Shuffling staff from manual to analytical positions has helped to create better business decisions.

All of this indicates that RPA should be seen as more than an operations tool; it can also help with banking risk management.

2.7 Implementation Challenges Highlighted in the Literature

Although RPA has greatly improved financial services, implementing it still encounters some problems. Studies have pointed out several restraints that prevent it from fulfilling its potential in managing risks.

2.7.1 Legacy system support

Many organizations argue that it is difficult to combine RPA with old IT systems. Since financial institutions have used the same systems for years, it is difficult for RPA to integrate well unless additional layers are used. Deloitte (2018) reports that using bots on old platforms might result in failures in day-to-day tasks, limited success in data retrieval, and costly upkeep.

2.7.2 Managing and Supervising

If there is no proper leadership, RPA bots could increase risks within a company. If robots are not frequently examined or upgraded, they might keep repeating the wrong actions. KPMG (2020) points out that the absence of ownership rules and governance in AI may enable errors, audit failures, and violations against regulations.

2.7.3 Change Management

Obstacles can come from both employees and middle managers refusing to accept RPA. Automation can make people in the organization feel threatened that they could lose their jobs or feel cut off from what they do. To encourage cultural acceptance and engagement, change management suggests focusing on communicating, retraining individuals, and aligning leaders (Capgemini, 2020)

2.7.4 Scalability issue

Sometimes, expanding RPA means altering overall business processes apart from simply programming current tasks. If the workflow is not clearly described, the data varies, or people are making changes in the system, the company's main output could decrease, and the deployment of the software across the organization may slow down.

2.7.5 Security Risks

Having bots configured without proper care may cause them to have extra access, resulting in risks. McKinsey (2021) indicates that if bots that have system-level access are not secured, they could be used to avoid security checks and cause problems for companies' compliance with various rules. To achieve results, risk managers must solve the technical, organizational, and strategic barriers that RPA presents.

2.8 Ethical and Regulatory Considerations of RPA Use

Because many financial institutions rely on RPA, there are ethical and regulatory issues that must be addressed right away.

It is a matter of ethics that the use of automation for risk processes could result in job loss among compliance workers and teams in operations. There's a chance that tasks RPA handles with the help of AI can lead to systemic discrimination. There is also the risk that bots will act without proper oversight.

Auditing of bots and how they are owned is often not very well-established. To make sure all goes according to plan, bot developers must specify who is liable for their results, how they are updated, and what access individuals have.

Rules and standards related to RPA are still being developed. Both the FCA and ECB believe that better governance over RPA should involve more attention to risks and data protection. Following new regulations will ensure trust and reduce ethical challenges in AI when progress is made in automation.

2.9 Synthesis of Literature Gaps

Most studies on RPA deal with streamlining operations rather than exploring how it serves in managing risks. Not many studies look into how RPA enhances controls within an organization, decreases financial crime, or helps with following regulations. Still, there are not many studies that have shown how these theories can actually be put to use in real banking situations. In terms of integration, there is little information on how to bring together RPA with systems for managing enterprise risk. For this reason, this chapter examines how to apply RPA in risk management using theories.

3 MAIN FOCUS OF THE CHAPTER / METHODOLOGY

3.1 Section Overview & Purpose

This section has two functions. The chapter is focused on the role Robotic Process Automation (RPA) plays in strengthening risk management for banks. In addition, it describes the way the study was done by using qualitative and thematic analysis of secondary information. Moreover, the section analyzes the actual challenges, issues, and difficulties that arise when banks start using RPA. Upon summarizing existing research, this section moves on to empirical observations and serves to guide new tactics.

3.2 Main Focus of the Chapter

The chapter explores how using RPA technology is affecting the nature of risk management in banking. Unlike how it is commonly used to optimize operations, applying RPA to risks is both hopeful and challenging. Applications of RPA consist of improving KYC and AML checks, using it to make sure the company follows rules like FATCA and Basel III, and allowing the detection of fraud due to fast anomaly spotting.

By making these processes automated, it is much easier to reduce errors and related risks. Leveraging RPA bots helps companies ensure steps involved in their risk management process are reliable, transparent, and can be reviewed. With AI, ML, and big data analytics, RPA can become intelligent enough to support smart decisions in processes. The application of these technologies ensures RPA is useful for banking risk management and also helps save expenses.

Therefore, this chapter explores both the positives and negatives that RPA brings to risk management in banking. To find out what shows up in different organizations and to study the risks and problems each one encounters.

3.3 Methodology Used

3.3.1 Research Design

I have used a qualitative, in-depth approach to examine the influence of RPA on risk management in banks. Instead of just using statistics, the idea is to study how practices, problems, and patterns appear in various universities.

3.3.2 Data Sources

It only uses information found in peer-reviewed academic journals. I have analyzed whitepapers and reports from global and regional banks using Risk Management, as well as those published by top technology companies and consultancies (like McKinsey & Company, PwC, and Deloitte). I looked into JPMorgan Chase, ING, BBVA, DBS Bank, and Standard Chartered, giving an overview of how different regions are covered.

3.3.3 Analytical Framework

A coding approach using the theme of automation efficiency was applied to the study's findings. They disclose both the advanced technical and management elements concerning how RPA is deployed.

3.3.4 Justification

Since RPA is still new in risk management, the most appropriate approach is through case analysis and interpretation. It helps to discuss the finer points of the situation, company traditions, differing regulations, and challenges specific to vendors.

3.4 Issues in RPA-Based Risk Management

3.4.1 Technical Integration Challenges

One major obstacle to using RPA in banks is that bots cannot work with outdated banking systems. The core systems in most banks are not interlinked, and because standardized APIs are unavailable, deploying bots can be risky and often result in errors. Regular changes to software can sometimes prevent RPA from running smoothly due to changes in the user interface or in some of the data fields. Therefore, companies must significantly enhance their IT systems and redesign them to make RPA deployment stable (Damco Group, 2024).

3.4.2 Data Quality and Structuring Problems

To perform risk tasks such as detecting fraud or handling credit analysis, you need quality and organized data. Many banks face the tough challenge of diverse formats, messy records, and data that is not well organized due to its age. As a result, using bots can create additional risks because incorrect outcomes may occur.

3.4.3 Over-Reliance on Rules

RPA relies on rules and is neither flexible nor capable of interpreting situations. Bots are great at handling routine tasks, but they stumble when meeting unexplained cases, vague data, or jobs requiring careful judgment. As a result, they cannot be used where making decisions relies on human judgment.

3.4.4 Limited Scalability

Extending RPA across a company increases how complicated things can be. With the introduction of each new bot, the business must pause and add more monitoring, maintenance, and integration steps.

If governance is weak, the results from scaling end up being divided automations instead of integrated improvements at the company level (FinRiskAlert, 2024).

3.4.5 Incomplete Risk Visibility

RPA systems are more concerned with carrying out each step than understanding all the parts of the process. Bots cannot detect major risks or vulnerabilities rising across the financial sector unless they are tied into advanced analytics or ERM systems.

3.5 Controversies Surrounding RPA in Banking Risk Functions

3.5.1 Job Displacement

Many people discuss whether RPA will cut jobs, especially for those who monitor compliance, examine documents, and handle internal audits. Even though using automation can make things more efficient, it tends to cause people to lose their jobs or be reassigned, leading to problems within organizations and ethical disputes. The pressure from innovation and employment makes RPA adoption more difficult and reduces teamwork.

Figure 3.

Source talentelgia.com

3.5.2 Accountability and Legal Ambiguities

If a bot doesn't notice or stop a fraudulent transaction, it is not clear who should be held accountable. Either the developers working on the product, the designated risk managers, or third-party vendors may be responsible. It's not clear in law how bots can be used, which makes governments reluctant to use them for key tasks.

3.5.3 Regulatory Uncertainty

Ways of governing RPA depend on different regions. In contrast, European regulators set strict data requirements (GDPR), but other countries are not so strict. Bots that work in various locations often end up unintentionally breaking compliance rules. Because there is no one set of global standards, public policies can be difficult to enforce (Global Banking & Finance Review, 2024).

3.5.4 Black Box Nature of AI-Driven Bots

When RPA works with AI, the way the system operates becomes unclear. Because of how this automation works, it's hard to examine decisions or spot where errors occur, especially in areas such as fraud detection or approval of loans.

3.6 Problems Emerging from RPA Deployment in Risk Areas

3.6.1 Implementation Misalignment

Most banks that use RPA do so mostly to lower costs, rather than to strengthen their risk management. Such a short-term mindset leads developers to waste effort, automating tasks that aren't necessary or repeating actions that could be avoided. If implementation goals become out of sync, it slows the improvement of a company's ability to withstand risks (McKinsey & Company, 2021).

3.6.2 Lack of Change Management

Many organizations are missing a key part of successful RPA: the approval of leaders. Employees can see bots as competitors instead of as something useful. The absence of proper change management, such as training, evaluation, and interaction with machines, leads to employees resisting the changes and automation systems working less well (Upcore Technologies, 2024).

3.6.3 Vendor Dependency

RPA tasks such as design, upkeep, and changes are often left to outside RPA vendors when managing banks. Because of this, projects face risks from intruders, problems with intellectual property, and it is difficult to manage key process knowledge. This results in fewer ways for a state to build up its abilities, which reduces its ability to survive over the long run.

3.6.4 Bot Governance and Monitoring Gaps

It is important to oversee bots well in order to confirm they are functioning the way they should. Unfortunately, many institutions do not have version control systems, real-time dashboards, or procedures for managing exceptions. Because operations cannot be fully monitored, there is a risk that errors will be missed and automated systems may fail.

3.7 Summary and Transition to Solutions

While RPA can greatly improve banking risk management, getting it into use is challenging due to technical, ethical, and operational issues. Unless dealt with smartly, these challenges can stop automation from helping. Next, we will look at RPA-related solutions and frameworks that help maximize benefits while keeping the risks in check.

4 SOLUTIONS AND RECOMMENDATIONS / DISCUSSIONS

4.1 Part 1: Strategic and Enterprise-Level Solutions

RPA is being introduced into banking procedures more and more due to reasons beyond simply cutting costs. Implementing RPA in banks should always support the company's main strategies and help drive transformation in more than just one part of the business. In this stage, RPA is planned alongside the organization's digital transformation, clear links are made between automation targets and business goals and top leaders are fully involved. A good basis begins with including RPA as a core aspect of the bank's digital transformation. You should find repetitive, low-complexity work processes that have fixed rules in each of the departments, including compliance, customer service, finance, and risk management. RPA can be part of the bank's strategy to have automation projects grow and endure (Willcocks, Lacity, & Craig, 2017).

Strategic alignment is greatly helped by having executive sponsorship. With senior executives not involved, RPA is likely to have divided involvement, disorganized routines, and tiny effects on the organization. Top managers should encourage RPA and show others in the company how to keep improving and embracing new developments. Among other things, this means setting clear targets for using automation, deciding on important KPIs, and including the RPA efficiency in the bank's balanced scorecard (Kroll, 2021). Overseeing and monitoring RPA deployment requires a framework that covers the whole business. They need to define rules for how to develop bots, code them, maintain them, and manage exceptions and audits.

Figure 4. RPA

Ensuring one Center of Excellence for governance can simplify work practices, guarantee compliance, and allow different departments to collaborate. The CoE ought to offer staff training on how to make, oversee, and measure the performance of bots. In addition, banks should set up special programs to support employees as they use new automated tools. A lot of times, people's fear of change stands in the way of RPA's progress. So, launching campaigns for communication inside the company, gathering feedback, and rewarding those working on automation can encourage employees and ease worries about losing their jobs (According to Deloitte, 2020). When using RPA strategically, you must also link it to Artificial Intelligence (AI), Machine Learning (ML), and Data Analytics. Thanks to intelligent automation, which uses RPA and AI together, banks can automate activities like fraud identification, customer registration, and loan work (Purdy & Daugherty, 2017). A network of standardized and interchangeable modules is required to automatically handle workflows between different platforms. Banks must also craft a sustainable plan for moving to greater levels of automation. Usually, the process starts with automating tasks, moves on to linking processes, and finishes with cognitive automation. For every level, a company must strategically buy tools, recruit top talent, and enable change.

Tracking their maturity, comparing themselves with others, and constantly updating processes can show banks where they need to improve in automation (ISG, 2021). Furthermore, working with RPA vendors, fintech companies, and consulting firms can help you reach your automation objectives more quickly. It's important for such partnerships to share objectives, maintain transparency, and distribute risks to each business. Connecting with outside partners allows banks to try the latest RPA technologies, benefit from industry experience, and quicken the process of building and putting automation solutions in place. Summing up, banking institutions need to use a complete strategy that involves leaders, smart management, linked technology, and cultural transformation to perform well with RPA. When its uses are connected to the overall business strategy, RPA can boost innovation, increase agility, and give a firm an edge in the fast-changing financial services industry.

4.2 Part 2: Organizational and Human-Centric Recommendations

The advancement of Robotic Process Automation in banks means we cannot ignore the people and teams who are impacted. Although the main aim of RPA is to automate repetitive tasks, the effects it has on employees, the company's framework, and its culture are just as important. Banking risk management needs to find a balance between adopting new technologies and preparing people for those changes.

Figure 5.

Source Blueprism

A main suggestion is to focus on providing workforce training and development for everyone. When RPA is adopted, the roles in compliance, auditing, and risk areas tend to change. Instead of making people redundant, banks should shift employees into roles that need their judgment, along with what automation can do. The curriculum in training programs should concentrate on data analytics, process reengineering, and technology governance skills. Those in manager roles related to RPA must have skills in risk activities and an understanding of digital automation tools. Organizations not paying attention to workforce development may find employees feeling disengaged and their operations becoming less effective because bots have not been managed properly (Asatiani et al., 2021)

This workforce transformation can be backed up by a human-in-the-loop (HITL) process within banks. The model supports human review at times when it is important, mainly in areas such as fraud detection, granting credit, and complying with regulations (Davenport & Ronanki, 2018). The HITL method guarantees that bots are responsible and that automated results are in accordance with the policies and values of the organization. Furthermore, it helps overcome the difficulties that RPA systems have with ambiguous or unusual situations that need recalling context and decision making.

Creating well-defined rules for managing the digital workforce is an important recommendation for organizations. Because hundreds or thousands of bots may be used by banks, a lack of control can cause the organization to break compliance rules or repeat work unnecessarily. Therefore, banks should establish a framework including SOPs that cover running bots through every stage, including planning, trying them out, putting them into action, monitoring their performance, and ending their lifecycle (Lacity & Willcocks, 2021).

To manage bots effectively, organizations should create groups of risk management, IT, HR, and compliance professionals who will oversee bots and ensure they are used to support the organization's main objectives. Banks ought to encourage digital collaboration instead of making it a point of conflict. Fears about automation resulting in job cuts can cause workers to report fewer mistakes and to keep IT teams at arm's length (Vial, 2019). Leaders are responsible for explaining what RPA means for employees and the organization's mission. It's important that change strategies consist of workshops, storytelling of successful examples, and incentives for those who respond quickly to change. Assessing culture can enable organizations to create management strategies that suit specific teams or places (Syed et al., 2020).

An important ethical consideration is the psychological impact of working alongside or managing digital entities. Supervising bots, especially in high-stakes environments like risk management, can lead to new forms of cognitive workload and stress. This is especially true when human employees are accountable for errors triggered by automated processes beyond their immediate control. Therefore, mental health and occupational well-being support mechanisms should be integrated into digital transformation strategies. HR departments must update job descriptions, performance evaluation criteria, and employee support systems to reflect this new reality (Gartner, 2023).

A main ethical issue arises from the mental effects that come with dealing with automated systems. Looking after bots, especially when the stakes are high in risk management, can cause more cognitive workload and stress. This holds when staff do not control all systems, causing errors that can have major consequences for the company. Consequently, organizations need to add mental health and job satisfaction to their plans for digital transformation. HR should change job descriptions, standards for evaluating performance, and workers' support systems in response to the new situation (Gartner, 2023). In addition, organizational success with RPA requires defining who is responsible for automated processes.

Usually in traditional banking, who makes decisions is very clear. RPA causes these areas to overlap, thanks to bots that perform or generate tasks without constant human oversight. The audit trail, developed by banks, should log bot activities with timings, the system's decision process, and past escalation steps. It is very important for anti-money laundering (AML), know-your-customer (KYC), and credit risk assessments since this area requires documented proof of decision making (FSB, 2022). As well as formal rules, banks should motivate cooperation among IT, operations, compliance, and risk management teams. If IT or operations alone are deciding on RPA projects, often there isn't enough awareness of the rules and nuances needed (Siderska & Wach, 2021).

Using agile techniques lets stakeholders communicate frequently, allowing automated workflows to improve step by step and lowering the chances of an issue with alignment. Finally, companies should prepare for RPA systems that can grow and remain strong. Often, banks start their RPA adoption with small, independent pilots that later prove challenging to combine with other parts of the business. Investment in central RPA CoEs will allow banks to set up standard procedures, tools, manage places where knowledge is kept, and manage relationships with vendors (van der Aalst et al., 2018). CoEs provide a significant reference point for learning and implementing best practices in large and scattered banking groups. Overall, carrying out RPA in risk management in banking calls for both expertise in technology and in human organization. With proper reskilling, strong management, and emphasis on ethics and support, banks are better able to embrace automation without losing control over their operations or the faith their employees have in them. Strategies within organizations must change together with technological progress to secure and sustain RPA's use in risk management.

4.3 Part 3: Regulatory, Compliance, and Risk Governance Solutions

With RPA becoming involved in banks' processes, vital components for its success are compliance regulations and ensuring effective risk governance. This means that automated compliance, control, and reporting tools must keep up with updates in regulations and risk-related requirements.

A lack of compliance with multiple regulations can result in fines, a breach of data, and loss of reputation for financial institutions. Previously, banks should have made it mandatory to use compliance-by-design ideas when they begin developing RPA. It requires including regulations, including those for data retention, trails for review, and access controls, in the blueprint of the bots from the start. Automated compliance makes sure each activity stays error-free, which eliminates the need for many post-deployment reviews. Implementing compliance frameworks early in the design process, according to KPMG (2023), allows banks to be consistent with Basel III, MiFID II, and PSD2, which reduces risks from both operations and regulations. In addition, RPA platforms must be regularly checked to ensure their records follow regulatory standards for reporting.

Financial institutions are now expected by many regulators to make their automated processes clear in AML, KYC, and stress testing (European Central Bank [ECB] 2023). When banks bring RPA solutions into these key compliance fields, they must put in place effective validation to prove that the systems are working properly and stay in line with local laws. Submitting incorrectly may result in penalties from regulators or legal issues. In order to accomplish these objectives, setting up risk governance frameworks exclusively for RPA is important. Common enterprise risk systems are usually not able to tackle the specific risks that come with automated solutions, such as bias, errors spreading, or a shift in how processes are carried out.

As noted by the Financial Stability Board (2022), banks should put in place a detailed system of automation risks, make sure bot actions have proper owners and have routines for handling unusual cases. For these control standards to help with monitoring and risk control, their procedures must keep in line with internal audit activities. A further suggestion is the development of single systems to monitor compliance called Compliance Control Centers (CCCs). These units should join forces with RPA Centers of Excellence (CoEs) to make certain that bot compliance is always maintained, especially when regulations are changing. As a result, both groups are responsible for keeping the company's regulatory systems up to date. Based on Deloitte (2023), using this model strengthens a company's resilience, promotes the development of new ideas, and ensures that automation is handled by the main compliance rules.

There is a big challenge in guaranteeing that RPA follows the transparency and explainability standards now required by lawmakers. Bots used in risk modeling, determining credit, and fraud detection rely on explainable AI (XAI), which is increasingly important. Bodies like the European Banking Authority (EBA, 2023) and the UK Financial Conduct Authority (FCA, 2023) now expect financial organizations to explain the process behind their automated decisions when dealing with customers. As a result, risk decisions made with robotic process automation must be logged, provide explanations, and have decision trees that enable auditors and compliance employees to follow and understand each bot's decisions.

Figure 6.

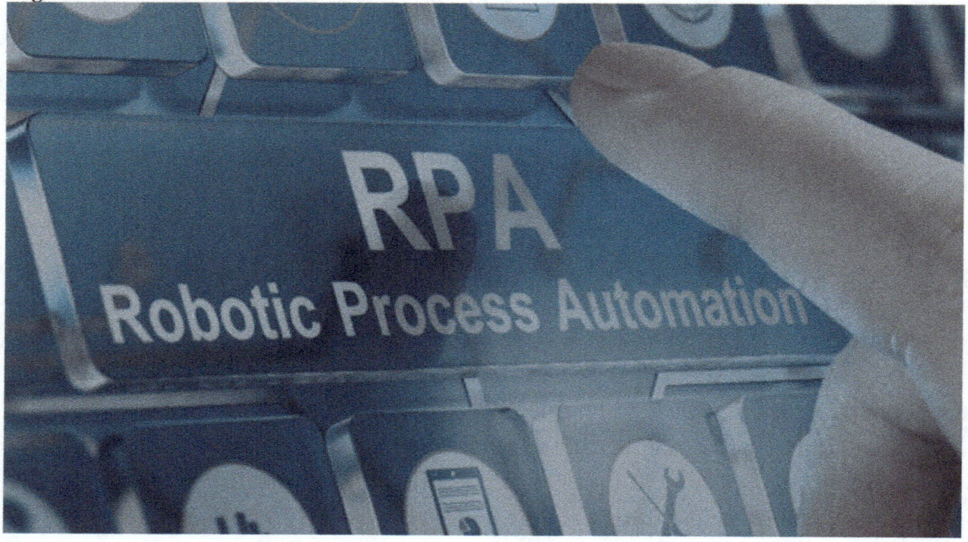

Citizens should be able to monitor activities in real-time and control processes before illegal conduct reaches a serious level. Systems should be in place that give automated alerts as soon as suspicious transactions or unusual behavior is noticed by bots in either AML or loan production applications. Employing AI and machine learning-powered surveillance tools can help companies respond well to regulations while relying less on audits from the past, says PwC (2024). Access to dynamic dashboards, reporting compliance KPIs, should be given to both RPA managers and risk officers to monitor performance in all areas. In addition, banks need to follow the rules in each region when using RPA outside their domestic markets. Because privacy laws can be quite different from place to place, bots need to be tailored in each region to manage private data properly. Without correct geo-tagging, RPA has the possibility of unintentionally moving data across borders. As mentioned by IBM (2023), doing automatic mapping of compliance requirements and routing data according to locations prevents IT systems from triggering regulatory violations and fines in worldwide operations.

Regulators are gradually acknowledging the problem of who is responsible for the outcomes of automated processes. Automation can comply with rules more efficiently, but it must not take away the responsibilities of human decision-makers, the BIS emphasizes in recent discussions. Because of this, banks should ask a person to review decisions such as regulatory disclosures, clients being denied, and the calculation of risk-weighted assets when bots handle these jobs. Including this treatment of people within governance frameworks helps lower the chances of legal fights and criticism about morality. In addition, RPA in risk and compliance requires teamwork and active participation of regulatory bodies. Instead of only waiting for what is required, banks should discuss regulations and test new solutions with industry consultations and innovation tests run by regulators.

It leads to better understanding between financial services and ensures they can direct regulatory expectations related to RPA technology. As examples, the UK FCA Regulatory Sandbox and the Singapore MAS FinTech Regulatory Sandbox support the development of automation solutions in protected and closely regulated areas (FCA, 2023; MAS, 2023). All things considered, strong regulatory, compliance, and risk governance solutions for RPA in banking should be more than just the automation of simple processes. They need to include compliance-by-design, clear audits, sharing of policies across nations, ethical approval, and continuous risk monitoring. Active engagement with regulators and solid governance procedures are needed by banks to make sure RPA helps, rather than hinders, compliance with regulatory requirements and risk matters.

4.4 Part 4: Data Integrity, Cybersecurity, and Operational Resilience

Since banks automate risk management and compliance with RPA, protecting data, computer systems, and ongoing operations becomes crucial. Adding bots to important tasks like credit checking, preventing fraud, and reporting compliance makes it much easier for attacks to arise, speeds up the exposure of private data, and adds new dependencies in the organization's IT environment.

Even if you have the most efficient RPA system, it may still be exposed without reliable cybersecurity and resilience rules, which can be the result of data breaches, cyberattacks, or significant systems failing to keep operations going. Data security must be maintained during all parts of developing RPA systems, but it is most critical in sensitive processes. Since the data supporting bots mostly comes from old bank systems, central platforms, and feeds from outside sources, the possibility of misinterpreting or missing crucial data is great. To overcome these issues, banks must use automated ways to make sure inputs are correct and consistent before bots get involved.

To ensure reliable data, firms depend on PwC (2024), which explains that having data governance, such as tracking, version control, and exception handling, is foundational in complex automation environments. Given the increasing complexity of the threats, protecting cybersecurity matters a lot for financial organizations. Bots are becoming more vulnerable to attacks by those who aim to use unattended procedures and undefended service accounts. If they are compromised, these bots could give an attacker entry to sensitive systems within the company. Hence, banks should rely on zero-trust systems, verifying and granting access to every user or robot they interact with continuously. Both human users and bots must benefit from using multi-factor authentication (MFA), role-based access controls (RBAC), and identity federation protocols (Gartner, 2023).

RPA scripts must be kept encrypted, centrally controlled, and watched for any modifying activity to prevent tampering or malware introduction. To ensure RPA environments are safe, banks need to connect their SIEM with their RPA software. Because of this integration, it is possible to oversee bots in real time, find suspicious patterns, and handle incidents efficiently. IBM reports (2023) that using behavioral analytics on bot logs allows companies to notice odd patterns such as bots working at the wrong times and reaching for unapproved data, helping to find and address dangers faster. Moreover, any audit trails made by bots need to be unchangeable and marked with a time stamp for easier investigation and compliance checks after incidents.

Figure 7.

By being operationally resilient, banks should get ready for interruptions in bot processes because of outages, faulty data, or third parties. Developing special BC/DR strategies for RPA is very important. You should have backup bots, duplicate data streams, and systems that take over automatically without manual input. In an example, having a mirrored version of your critical bots in a secondary environment and redeploying them automatically whenever there is a production failure keeps downtime short.

According to the Bank of England (2022), operational resilience applies to third-party suppliers and vendors offering RPA platforms or services, so clear SLAs and tests of their resilience are necessary. Banks should follow the rules about cyber resilience that stress the importance of safe automation. The Digital Operational Resilience Act (DORA), which starts in 2025, means companies in the financial sector must have formalized ICT risk measures, including control mechanisms used in automated areas such as RPA (European Commission, 2023). Monitoring, performing TLPT, and reporting incidents are included in DORA, which should be part of the way RPA is managed.

Similarly, the UK Financial Conduct Authority (FCA, 2023) says firms should figure out their key business services, identify how they rely on them, and test their effect, including rules that apply to RPA-integrated services as well. Administering the lifecycle of bots makes sure they are not vulnerable, up to date, and still meet all company policies. Without a unified system for keeping all bots updated, older bots may introduce problems that allow vulnerabilities. On this basis, banks should install a BMS that supports frequent updates, vulnerability checks, and the assessment of how well systems operate. These approaches keep bots safe when they are retired and avoid uncontrolled automation scripts being a threat to shadow IT (McKinsey, 2023).

Ensuring data is localized and that data in the cloud is secure is crucial when RPA platforms are being established globally by banks. Using RPA and APIs, together with outside services, introduces more threats from moving data across borders. Meeting GDPR, CCPA, and other regulations calls for geo-fencing, encrypting both data resting and moving, and closely watching cloud access to maintain data sovereignty and avoid breaking these laws (Deloitte, 2023). Altogether, RPA can greatly improve the efficiency of risk and compliance functions only if it is built on top of reliable data, strong security, and reliable operations. Banks should focus on secure creation, constant monitoring, secure chatting, and upkeep of their bots. Taking a proactive and integrated approach to resilience protects customer trust, adheres to regulations, and guarantees both business continuity and digital reliability in a highly automated banking world.

4.5 Part 5: Outlook and Strategic Recommendations

With financial institutions adopting digital change, RPA in risk management and compliance looks set to grow and improve. As regulations worldwide become more complex, data grows faster, and financial crimes evolve, banks are expected to use intelligent automation systems. Thanks to AI, ML, NLP, and predictive analytics, such ecosystems will help RPA take on new duties such as predicting risks and ensuring compliance. The evolution of CRPA will allow machines to detect context, understand data, make decisions based on rules, and operate with very little human help. This will support real-time compliance, better fraud recognition, and fewer mistakes in compliance alerts within a bank. This means that if AI helps analyze a company's transaction patterns, it can detect dishonest behavior earlier and support constantly updating risk models. Applying AI and RPA together in banking can reduce risks and cut down on compliance costs by a third, says McKinsey (2023), and also improve the accuracy and speed of control operations. We also suggest establishing an Automation CoE in every financial institution. With CoE, all of the departments can be guided by a single governance plan, develop employees with similar skillsets, secure all systems, and use RPA consistently in each department. Moreover, it encourages team members to share what they know, speeds up innovation efforts, and aligns RPA projects with the organization's general goals.

Forrester (2022) points out that mature automation governance is connected to having better results from automation, greater agility and increased resilience compared to others. There should be dedicated people in the CoE who lead on risk and compliance to influence how automation is planned. More and more, it will be vital for RPA to be linked with GRC platforms that are used across an entire organization. Connecting the outcomes from RPA to GRC dashboards allows banks to view and manage risks from all major aspects: finance, operations, compliance, and cybersecurity. With this integration, automatic policy enforcement is possible, and both internal audits and risk scenario simulations can be supported.

Gartner (2023) expects half of all banks to apply AI and RPA to GRC and internal audit by 2026, making compliance a useful area for the organization instead of just a cost. Going forward, industry initiatives and shared spaces for development will guide the proper adoption of RPA in the banking industry. All four groups should co-create automation systems, guide AI ethics, and figure out cross-border controls for handling data. Alliances may make it simpler to create standard APIs that link RPA to both older and more recent systems, which helps improve compatibility and lessens the risks in setup. The European Banking Authority has already encouraged by use of collaborative sandboxes; it is possible to test AI-RPA applications safely.

Forward-looking approaches to talent must now focus on creating a union of bots and humans in the workplace. Banks ought to train their risk managers, compliance officers, and IT personnel in areas of automation, governance, machine learning monitoring, and protection of data. The digital operating model will bring new jobs, including Bot Supervisors, Digital Compliance Analysts, and Automation Risk Officers. It is said by Accenture (2023) that humans and AI will join forces in the financial sector, helping human experts deal with complicated and risky work.

Since current regulations are not enough, banks should get ready for stricter rules and must clearly explain and make transparent how they use AI and automation tools.

Figure 8.

The EU AI Act as well as the principles from the Financial Stability Board (FSB, 2023) will affect the way RPA is developed, explained, and examined to ensure its ethical use in financial services. "Explainable RPA" is important for banks, requiring the results from RPA in credit decision-making, AML alerting, and customer profiling to be transparent, traceable, and easy to audit. Non-compliance can damage your organization's reputation and result in fines from authorities. Also, new cloud-native and low-code/no-code RPA tools will make it possible for more people to use process automation, saving time, increasing the scale of use, and empowering users throughout the business. Cloud technology guarantees resilience, the ability to scale up easily, and savings, yet businesses should still have solid policies, encryption, and comply with rules in each region (KPMG, 2024). Hybrid cloud is the future because it allows for higher security of sensitive workloads by running them on local servers, while less sensitive ones are put in the cloud.

Strategically, banks should make sure that technology supports their people, their customers, and also protects the environment. Good ethical practices in automation are found in limiting the bias of algorithms, making services available to all customers, and lessening the environmental effects of data centers and data processing. As Green FinTech is adopted more, RPA can help eliminate paper processing, energy-consuming audits, and tasks that pollute the environment, in line with ESG guidelines and the European Green Deal (OECD, 2023). Overall, the future of RPA in banking risk and compliance will be molded by a link with AI, better regulations, joint collaboration, and a focus on people. Organizations in the industry must change from using automation when possible, to planning a complete enterprise update based on governance, flexibility, and anticipation. Such advancement will decrease risks, promote compliance, and make it possible to gain long-term advantages, build trust, and differentiate in a rapidly changing financial world.

5 FUTURE RESEARCH DIRECTIONS / IMPLICATIONS

5.1 Introduction

The industry relies on Robotic Process Automation (RPA) for managing risks in banking. By using RPA together with AI and ML, modern banking can now achieve better efficiency, accuracy, and compliance with rules. Even so, progress in technology makes it more complex to ensure regulations are met, stay safe from cyber attacks, and consider ethical problems. Consequently, new research must keep up with challenges and help make the most of what RPA can do for the industry.

5.2 Technological Integration and Innovation

5.2.1 AI and ML Synergy with RPA

Intelligent Process Automation (IPA) has come about due to the association of RPA with AI and ML, allowing banks to be adaptive and predict the outcome of operations. Studies in the future should explore how predictive analytics can be used within RPA to boost both fraud detection and credit scoring. It has been discovered that using RPA with AI can lower costs within a company and increase compliance with regulations (FinTech Weekly, n.d.). Also, using unsupervised machine learning methods for anomaly detection is a beneficial idea.

5.2.2 Cybersecurity Considerations

Using RPA in banking means that new cybersecurity threats emerge. Studies show that poorly managed RPA tools may be at high risk for cyber issues such as bot takeovers and losing data, as was described by Security Magazine (n.d.). More research is needed on safe ways to identify and authorize bots, secure methods for bots to interact, and systems to protect against fraud when using bots within a company.

5.2.3 Cloud-Based and API-Driven RPA

Mixing cloud-native and API-driven systems for RPA opens up both new prospects and issues to consider. While their offer of scalability and increased performance is welcome, these technologies also cause alarm about being locked into a single vendor and security risks. There is a need for research to develop guidelines for the right ways to govern RPA and ensure security and compatibility when using numerous cloud platforms (FinTech Weekly, n.d.).

5.3 Risk Management and Compliance

5.3.1 Regulatory Compliance Efficacy

The part that RPA plays in complying with regulations, including Basel III, GDPR, and AML, has often been mentioned. Even so, research that proves the success of RPA in these fields is not broadly available. Future analysis of AML compliance should compare manual processes with those that are automated to find out how efficiency and penalties change, as well as the improvement in real-time alert management (FinTech Futures, n.d.).

5.3.2 Dynamic Risk Modeling

Banks should respond to rising dangers linked to pandemics and climate change. Experts need to see if RPA allows for quick and flexible testing and preparing different scenarios to handle such risks. When RPA is connected to ESG risk monitoring, you can gain a more complete view of managing risks (FinTech Weekly, n.d.).

5.3.3 Ethical Considerations and Explainability

Since RPA systems are becoming more independent, it is especially important that everything remains clear and ethical. Research for the future should study how RPA frameworks can help create plain-language AI models that make loan approvals or transaction blocks more accountable. Having AI ethics boards will help banking institutions use RPA technologies responsibly (Baum et al., 2019).

5.4 Organizational and Human-Centric Aspects

5.4.1 Workforce Transformation

Figure 9.

When RPA is introduced to banking, people's jobs and tasks need to be reconsidered. Studies should focus on how job structures change for RPA professionals, which skills gaps may appear, and what psychological influence supervising machines has on these experts. Ensuring people have reskilling and mixed-model training can help employees move forward at a good pace (FinTech Weekly, n.d.).

5.4.2 Governance and Oversight Frameworks

For RPA to be successful, effective governance is very important. It is important that future research sets up accountability approaches, brings together various groups for oversight, and establishes rules for handling bots from their development to their end of life. The use of these frameworks supports RPA systems to be in line with the targets of the company and all rules set by the government (FinTech Futures).

5.4.3 Change Management and Cultural Adaptation

Using RPA in banking demands that bank staff are prepared to change their daily routines. Longitudinal studies allow us to discover methods and principles that help people and businesses adapt to digital change. To adopt RPA sensitively, you need to understand such dynamics (FinTech Weekly, n.d.).

5.5 Global and Industry-Wide Perspectives

5.5.1 Cross-Country Comparative Studies

Little research is available that looks at how RPA is adopted and how financial regulations and risk models differ in different parts of the world. Studies to come should study how various regions influence the success of RPA in banking, focusing on AML compliance and how cultures view automation (FinTech Futures, n.d.).

5.5.2 RPA in Small and Medium-Sized Enterprises (SMEs)

Large banks can afford to use RPA solutions for their business, but SMEs often have a tougher time doing so. Studies ought to examine how well open-source automation tools work in risk management among smaller institutions and how efficient they are compared to complete FinTech solutions (FinTech Weekly, n.d.).

5.5.3 Collaborative Risk Networks

Connecting major banks' RPA systems for fraud and identity verification creates a chance for banks to manage risks together. Research in the future could explore how blockchain and RPA can support shared compliance reporting and what their joint use could accomplish, supporting better financial security (FinTech Futures, n.d.).

5.5.4 Policy and Regulatory Implications

As RPA grows in popularity, governments must act quickly to create solid policy rules. Experts should concentrate on the regulation of jobs done by AI, bots authorized for certain work, and measures for cross-border exchange of data. As a result, RPA regulatory sandboxes with standardization bodies can make it easier to introduce standards for audits and to carry out regular reviews on the security of bots (FinTech Weekly, n.d.).

5.5.5 Academic and Educational Considerations

Designing finance and IT programs to place RPA and risk management at their heart will train future experts well. It is important to study the creation of ability frameworks for auditors and risk managers, and also to offer courses that combine topics from information technology, finance, ethics, and law. Working together, universities and banks can bring about educational programs focused on automation (FinTech Weekly, n.d.).

5.5.6 Recommendations for Future Research

1. **Adopt Mixed Methods**: Combine case study research with simulation analysis to get a deep understanding of how RPA functions.
2. **Longitudinal Tracking**: Test how RPA affects banking operations and has an impact on handling banking risks both now and in the long run.
3. **Focus on Multi-Agent Systems**: Look into the ways RPA connects with other technologies such as chatbots, nodes on the blockchain, and devices in the Internet of Things when it comes to risk.
4. **Leverage Real-Time Data Analytics**: Combine data from banking bots to watch RPA performance live in real time.
5. **Include Ethical and Legal Lenses**: Focus on data rights, the influence of biases in algorithms, and the way RPA technologies affect workers' roles when planning future studies.

6 CONCLUSION

With RPA, banking risk management and compliance have been transformed in a constantly changing and strict market. This chapter has explored the evolving role of RPA, the challenges and opportunities it presents, and a comprehensive set of strategic, organizational, technical, and governance-related solutions. As digital transformation becomes indispensable for survival and competitive advantage in the financial services sector, RPA emerges as a pivotal enabler for efficiency, accuracy, resilience, and compliance.

We have looked into the development of RPA, the issues it presents, and several effective solutions in strategy, organization, technology, and governance. Since digital transformation is key to success and competitiveness in financial services, RPA helps companies become more efficient, accurate, reliable, and compliant. The beginning of the chapter highlighted that as regulations, costs, and cybercrime have risen, companies in finance have needed to turn to RPA as a solution. Firms in the financial sector are being asked to follow global and national laws such as Basel III, the EU's Anti-Money Laundering Directives, and the U.S. Dodd-Frank Act. RPA has made it possible to automate certain tasks using rules that must be followed, without human errors, at less cost and with more consistency (Deloitte, 2023). More banks are turning to bots to automate their KYC activities, watch for suspicious activities, create reports needed by regulators, and spot fraud. The first part of the report concentrated on using RPA wisely and its initial phases in the fintech industry. Participants were reminded that RPA should be aligned with company goals, all regulations, and the needs of the customers. If RPA is adopted by an organization without a defined digital strategy, its efforts may be inefficient and noncompliant. RPA adoption should be managed through a blueprint that handles risk issues, uses data efficiently, and matches what client stakeholders expect (Gartner, 2023). Using AI and ML, RPA evolves from a traditional, fixed rules system to a dynamic tool that can handle shifts in regulations and respond in real time to changes in risks. In Part Two, suggestions were made to help organizations foster an atmosphere of teamwork and creativity as RPA is implemented. Many people worry that jobs will be lost to automation, but banks can make things easier by hiring a combination of human and automation workers, helping people retrain, and highlighting why bots are needed.

as ways to improve work rather than to take jobs away from people. The suggestion to form an Automation Center of Excellence (CoE) was given as a best approach for handling governance, standardization and the exchange of knowledge among departments (Forrester, 2022). Building strong talents is important for long-term automation results, as bank staff must learn how to look after RPA, analyze data, know about ethical AI and understand cybersecurity. Process automation tools must be accepted by the entire institution so that workers can work together and have a sense of involvement. In Part Three, RPA in banking was looked at from the standpoint of regulatory, compliance and risk governance. As regulators are paying more attention to the use of new technologies, banks have to manage their RPA activities transparently and in a proactive way. Experts said that RPA needs to be included in GRC systems to keep the organization auditable and accountable. Self-regulation software should never replace a bank's review of riskier processes like credit investigations and transactions monitoring. For the process to follow the law, ethical criteria and standards, risk management, escalation procedures and documentation rules must be included in RPA development. The European Banking Authority (EBA) and the Financial Stability Board (FSB) have issued statements encouraging the correct and fair use of automation in financial decisions (EBA, 2023; FSB, 2023). Part Four looked at the technical aspects and operations involved in RPA such as data security, secure systems and being prepared for problems.

When automating, data should always be accurate, of high quality and properly governed from start to finish. RPA works only if the data it uses is secure, so companies should focus on pipe security, use encryption, limit access and ensure regular monitoring. It is also important to prevent Cyber Security risks by only granting access to people with the right roles, by using secure bot accounts and by linking identity access management.

KPMG (2024) points out that effective systems are very important (para. 12). Since banks depend more on RPA for these vital tasks, preparing for unexpected problems and having real-time alerts and extra systems reduces the risk to their operations. However, using cloud-native and low/no-code RPA tools poses new security and compliance risks which can best be managed by strong governance frameworks in the cloud. In Part Five, the topic turned to planning the future of RPA in banking, with a strong focus on future-thinking and new solutions. Automation will reach new limits as cognitive RPA, predictive analytics and AI compliance systems become more common.

Banks need to use intelligent automation that connects to governance, risk and compliance tools as well as their online banking systems. Plans for future talent should add experts like bot supervisors and digital compliance officers to help make sure the work system is both automated and together-driven. Regulators are required to write regulations that ensure AI and automation systems are open and fair in all their decisions. Actions taken together by regulators, banks, fintechs and academics will be important for designing responsible automated systems.

Automation is expected to support societal priorities that include digital inclusion, protecting the environment and making platforms more inclusive. To conclude, the chapter illustrates that using RPA has become a need for financial institutions and is not an optional trial. Implementation of RPA with a focus on strategy and ethics gives businesses the opportunity to improve compliance, reduce risks, smooth their operations and please their customers. But there is more to automation success than just technology itself. It relies on leaders with ideas, solid management, people from different parts of the company joining forces, involved workers and anticipating necessary rules. Banks ought to always develop their knowledge and methods, so that automation remains valuable to both businesses and the entire community. RPA can form the foundation for future-resilient banking operations if it is implemented with suitable technologies.

7 REFERENCES

Accenture. (2021). *The future of risk management in banking: Reimagining risk in the digital era*. https://www.accenture.com

Accenture. (2023). *Automation maturity in financial services: From pilot to scale*. https://www.accenture.com

Aguirre, S., & Rodriguez, A. (2017). Automation in financial services: Robotic Process Automation in banking. *International Journal of Computer Applications, 179*(19), 47–51.

Asatiani, A., Malo, P., Nagbøl, P. R., Penttinen, E., Rinta-Kahila, T., & Salovaara, A. (2021). Sociotechnical challenges of robotic process automation (RPA): A contingency model for RPA implementation. *Journal of Business Research, 136*, 473–484. DOI: 10.1016/j.jbusres.2021.07.039

Bank of England. (2022). *Operational resilience: Impact tolerances for important business services*. https://www.bankofengland.co.uk

Basel Committee on Banking Supervision. (2019). *Principles for operational resilience*. Bank for International Settlements. https://www.bis.org

Baum, K., Hermanns, H., & Speith, T. (2019). Towards a framework combining machine ethics and machine explainability. *arXiv preprint arXiv:1901.00590*. https://arxiv.org/abs/1901.00590

Capgemini. (2020). *World FinTech Report 2020*. https://www.capgemini.com

Damco Group. (2024). *RPA in Banking: Challenges, Benefits, and Use Cases*. Retrieved from https://www.damcogroup.com/blogs/rpa-in-banking-industry

Davenport, T. H., & Ronanki, R. (2018). Artificial intelligence for the real world. *Harvard Business Review, 96*(1), 108–116.

Deloitte. (2018). *Robotic Process Automation: A path to the cognitive enterprise*. https://www2.deloitte.com

Deloitte. (2020). *Automation with intelligence: Pursuing organization-wide reimagination*. Deloitte Insights. https://www2.deloitte.com

Deloitte. (2022). *RPA in financial services: Automating the future*. https://www2.deloitte.com

Deloitte. (2022). *RPA in financial services: Automating the future*. https://www2.deloitte.com

Deloitte. (2023). *Cloud and data governance in RPA environments*. https://www2.deloitte.com

Deloitte. (2023). *Future-ready risk management: How AI and RPA are reshaping compliance*. https://www2.deloitte.com

EBA. (2023). *Report on the use of digital platforms and AI in EU banking*. European Banking Authority. https://www.eba.europa.eu

European Banking Authority (EBA). (2023). *Guidelines on the use of remote customer onboarding solutions*. https://www.eba.europa.eu

European Commission. (2023). *Digital Operational Resilience Act (DORA)*. https://finance.ec.europa.eu/dora_en

EY. (2019). *RPA in risk and compliance: A transformative advantage*. https://www.ey.com

EY. (2020). *Global banking outlook: Banking in the new normal*. https://www.ey.com

FCA. (2023). *Building operational resilience in financial services*. https://www.fca.org.uk

FCA. (2023). *Regulatory sandbox: Promoting responsible innovation*. https://www.fca.org.uk/firms/innovation/regulatory-sandbox

Financial Stability Board. (2022). *Supervisory and regulatory approaches to climate-related risks*. https://www.fsb.org

Financial Stability Board. (2022). *Supervisory and regulatory approaches to climate-related risks*. https://www.fsb.org/2022/10/supervisory-and-regulatory-approaches-to-climate-related-risks/

FinRiskAlert. (2024). *The Current State of the Robotic Process Automation (RPA)*. Retrieved from https://www.finriskalert.it/the-current-state-of-the-robotic-process-automation-rpa-a-cura-di-deloitte/

FinTech Futures. (n.d.). Convergence of RPA and AI: The changing landscape in banking compliance. https://www.fintechfutures.com/regulations-compliance/convergence-of-rpa-and-ai-the-changing-landscape-in-banking-compliance

FinTech Weekly. (n.d.). RPA & AI in the financial sector: Key functions, technologies, future decade trends. https://magazine.fintechweekly.com/magazine/articles/rpa-and-ai-in-the-financial-sector

Forrester. (2022). *The state of automation governance*. https://www.forrester.com

FSB. (2023). *Regulation, supervision and oversight of AI in financial services.* Financial Stability Board. https://www.fsb.org

Gartner. (2023). *How to manage employee experience during digital transformation.* https://www.gartner.com/en/articles/manage-employee-experience-digital-transformation

Ghosh, S. (2020). *Banking regulation and compliance in the digital era.* Routledge.

Global Banking & Finance Review. (2024). *Robotics Process Automation (RPA): A Transformation Lever for Risk & Compliance Management.* Retrieved from https://www.globalbankingandfinance.com/robotics-process-automation-rpa-a-transformation-lever-for-risk-compliance-management/

Hofmann, P., Samp, C., & Urbach, N. (2020). Robotic Process Automation. *IEEE Access : Practical Innovations, Open Solutions, 8*, 205946–205962. DOI: 10.1109/ACCESS.2020.3032964

Hull, J. C. (2018). *Risk management and financial institutions* (5th ed.). Wiley.

IBM. (2023). *Building trust in automation: Compliance and ethical AI strategies.* https://www.ibm.com

IMF. (2021). *Nordic Banking: Efficiency, Competition, and Risk.* https://www.imf.org

ISG. (2021). *ISG Automation Index 2021.* Information Services Group. https://isg-one.com

Jorion, P. (2009). *Financial risk manager handbook* (6th ed.). Wiley.

KPMG. (2023). *Rethinking compliance: From regulatory burden to strategic advantage.* https://advisory.kpmg.us/

KPMG. (2024). *Cloud-enabled intelligent automation in banking: Balancing agility with control.* https://home.kpmg

Kroll, C. (2021). *Leading digital transformation: Strategies for success.* Harvard Business Review Press.

Lacity, M. C., & Willcocks, L. P. (2021). Becoming strategic with intelligent automation: Aligning organizational and technological change. *MIS Quarterly Executive, 20*(2), 103–119.

McKinsey & Company. (2021). *The Transformative Power of Automation in Banking.* Retrieved from https://www.mckinsey.com/industries/financial-services/our-insights/the-transformative-power-of-automation-in-banking

Monetary Authority of Singapore (MAS). (2023). *FinTech Regulatory Sandbox*. https://www.mas.gov.sg

OECD. (2023). *Greening the financial sector: Digital tools and sustainability*. https://www.oecd.org

Purdy, M., & Daugherty, P. (2017). *Why artificial intelligence is the future of growth*. Accenture. https://www.accenture.com

PwC. (2021). *Retail banking 2025 and beyond: The future is open*. https://www.pwc.com

PwC. (2021). *Retail banking 2025 and beyond: The future is open*. https://www.pwc.com

PwC. (2021). *Retail banking 2025 and beyond: The future is open*. https://www.pwc.com

PwC. (2024). *Data governance and automation controls for risk-sensitive industries*. https://www.pwc.com

PwC. (2024). *Emerging technologies in regulatory compliance: The rise of RPA and AI in banking*. https://www.pwc.com

Security Magazine. (n.d.). Reducing RPA security risk. https://www.securitymagazine.com/articles/98953-reducing-rpa-security-risk

Siderska, J., & Wach, K. (2021). Robotic process automation: A sectoral and regional exploration. *Sustainability*, *13*(10), 5436. DOI: 10.3390/su13105436

Syed, R., Bandara, W., French, E., Stewart, G., & Fielt, E. (2020). Robotic process automation: Contemporary themes and challenges. *Computers in Industry*, *115*, 103162. DOI: 10.1016/j.compind.2019.103162

Upcore Technologies. (2024). *RPA in Finance - Exploring Use Cases, Benefits & Challenges*. Retrieved from https://www.upcoretech.com/insights/robotic-process-automation-in-finance/

van der Aalst, W., Bichler, M., & Heinzl, A. (2018). Robotic process automation. *Business & Information Systems Engineering*, *60*(4), 269–272. DOI: 10.1007/s12599-018-0542-4

Vial, G. (2019). Understanding digital transformation: A review and a research agenda. *The Journal of Strategic Information Systems*, *28*(2), 118–144. DOI: 10.1016/j.jsis.2019.01.003

Willcocks, L. P., Lacity, M. C., & Craig, A. (2017). Robotic Process Automation: Strategic transformation lever for global business services. *Journal of Information Technology Teaching Cases*, *7*(1), 17–28.

Willcocks, L. P., Lacity, M. C., & Craig, A. (2017). Robotic process automation: The next transformation lever for shared services. *Journal of Information Technology Teaching Cases*, *7*(1), 17–28. DOI: 10.1057/s41266-016-0002-8

World Bank. (2022). *Global financial development report 2022/2023: Financial inclusion*. https://www.worldbank.org

Chapter 14
The Future of Finance: RPA, Blockchain, and AI in Action

Kelvin Yong Ming Lee
https://orcid.org/0000-0001-6884-1787
Taylor's University, Malaysia

Irfah Najihah Binti Basir Malan
https://orcid.org/0009-0007-5757-4469
Universiti Teknologi MARA, Malaysia

Isma'il Tijjani Idris
Ahmadu Bello University, Zaria, Nigeria

ABSTRACT

This chapter explores the transformative potential of Robotic Process Automation (RPA), Artificial Intelligence (AI), and Blockchain in the financial sector. It begins by defining each technology and outlining their individual contributions to improving financial operations. The chapter highlights the current applications of RPA in areas such as transaction processing, customer onboarding, and compliance reporting, emphasizing benefits like cost savings and productivity gains. It then discusses how AI technologies are reshaping risk management, fraud detection, and personalized financial services. The role of blockchain is also examined, focusing on its ability to enhance transparency, security, and efficiency in processes like cross-border payments, identity verification, and trade finance. While each technology offers significant advantages, the chapter also addresses their limitations and the challenges of integration. Ultimately, it provides a forward-looking perspective on how the combined use of RPA, AI, and blockchain can redefine financial services in a digital era.

DOI: 10.4018/979-8-3373-3571-1.ch014

1.0 INTRODUCTION

Technological advancements have significantly transformed the financial sector, leading to the emergence of Financial Technology (FinTech). AI and ML are pivotal in handling large volumes of data, identifying trends, and making decisions with minimal human intervention. They are used in algorithmic trading, customer service, credit scoring, lending, investment management, and fraud prevention (Pattnaik et al., 2024). These technologies not only streamline operations but also enable financial institutions to offer more personalized and timely services. AI-driven automation and predictive analytics improve efficiency, risk management, and customer satisfaction (Manta et al., 2022), allowing firms to respond quickly to market changes and customer needs. At the same time, blockchain technology ensures secure, transparent, and efficient transactions. It supports cryptocurrencies like Bitcoin and Ethereum, which offer decentralized financial solutions (Tripathi et al., 2023). Beyond cryptocurrencies, blockchain is playing an increasingly important role in digital payments and decentralized finance (DeFi), offering innovative alternatives to traditional financial systems and creating pathways for broader financial inclusion (Javaid et al., 2022). The rise of digital payment systems and mobile banking has revolutionized how financial transactions are conducted, making them faster, more secure, and more accessible (Dinh, 2024). Mobile payments and digital wallets are becoming increasingly popular, driven by consumer demand for convenience and efficiency, and supported by the rapid expansion of internet and smartphone access. As the financial landscape continues to evolve, the integration of these technologies is expected to further disrupt traditional models, pushing institutions to innovate while ensuring regulatory compliance and cybersecurity.

As these technologies become more embedded in financial services, they also raise important questions around regulation, ethics, and long-term sustainability. The speed of innovation often outpaces regulatory frameworks, leading to uncertainties around data privacy, cybersecurity, and accountability (Brodny & Tutak, 2025). For instance, while AI enhances decision-making, it can also introduce algorithmic bias or lack transparency in how decisions are made, particularly in credit approval or fraud detection processes. Similarly, the decentralized nature of blockchain, while beneficial for security and transparency, poses challenges in terms of regulatory oversight and legal jurisdiction (Andrew et al., 2023). Governments and financial regulators are increasingly exploring adaptive regulatory approaches such as sandbox environments to strike a balance between fostering innovation and protecting consumers (Khan et al., 2023). In parallel, financial institutions must invest in responsible technology use. They need to ensure that the adoption of AI, ML, and blockchain is not only efficient but also ethical, inclusive, and aligned with long-term social and economic goals (Vuković et al, 2025).

2.0 ROBOTIC PROCESS AUTOMATION (RPA) IN FINANCE

Robotic Process Automation (RPA) is a technology designed to automate repetitive, rule-based tasks typically performed by human workers. It uses software robots, or "bots," to mimic human interactions with digital systems and applications (Bédard et al., 2024). RPA works on top of existing systems without requiring changes to the underlying infrastructure, making it easy to implement and scale (Costa et al., 2023). By automating tasks, RPA reduces the likelihood of human errors, enhancing accuracy and reliability in business processes (Shamsuzzoha & Pelkonen, 2025). Notably, RPA significantly improves operational efficiency and reduces costs by automating high-volume, repetitive tasks (Bhardwaj & Kumar, 2025). These benefits have made RPA an attractive solution for organizations seeking to improve productivity while maintaining lean operations. In the financial sector, where compliance, transaction processing, and reporting are critical and time-sensitive, RPA provides a practical way to streamline operations without compromising accuracy. In 2020, the global market for robotic process automation (RPA) in financial services was worth $340.95 million. By 2030, it is expected to be worth $4,883.41 million, rising at a rate of 30.9% per year from 2021 to 2030 (Allied Market Research, 2022). This rapid growth reflects not only increasing adoption across various segments of the financial industry but also growing confidence in RPA as a foundational technology for digital transformation. As businesses face pressure to stay competitive and agile, the role of RPA in reshaping operational models is becoming more prominent, paving the way for broader integration with AI and data analytics in the near future.

2.1 Banking Operations

Robotic Process Automation (RPA) is increasingly adopted in the finance sector to enhance efficiency, accuracy, and productivity by automating repetitive and time-consuming tasks. For instance, RPA is used to automate transaction processing, reducing manual effort and errors. This includes tasks such as verifying transactions, transferring data between systems, and processing credit card and mortgage applications (Sonik et al., 2024). These processes, which once required significant human effort and were prone to delays or mistakes, can now be completed rapidly and with greater consistency. In customer service, automating the onboarding process helps in quickly and accurately gathering and verifying customer information, thus improving customer experience and reducing processing time (Choubey & Sharma, 2021). This not only shortens waiting times for customers but also allows staff to focus on more complex, value-adding activities. Additionally, RPA assists in compliance reporting by automating data collection and report generation, ensuring adherence to regulatory requirements and reducing the risk of non-compliance.

As regulatory expectations grow more complex, the ability of RPA to maintain accurate audit trails and ensure timely submissions becomes increasingly valuable. Overall, RPA is not merely a tool for cost-saving. It is becoming a strategic asset that supports financial institutions in maintaining competitiveness and operational resilience in a rapidly evolving industry.

2.2 Financial Management

RPA automates the reconciliation of financial data, ensuring accuracy and consistency across different systems and significantly reducing the manual workload on financial personnel. This not only saves time but also minimizes the risk of errors that could impact financial decision-making. Furthermore, RPA can automate the generation of financial reports, ensuring timely and accurate reporting that supports both internal management and external regulatory requirements (Tang et al., 2024). These reports can be generated at regular intervals without human intervention, allowing finance teams to focus on analysis rather than data compilation. When integrated with artificial intelligence, RPA becomes even more powerful. AI-enhanced RPA can assist in complex tasks such as risk assessment and fraud detection by quickly analyzing large volumes of data to identify unusual patterns or anomalies that may indicate fraudulent activity (Bhambri et al., 2024). This proactive approach helps financial institutions improve their risk management capabilities while maintaining compliance and protecting customer trust.

2.3 Accounting

Automating invoice processing through RPA reduces the time and effort required for manual data entry and validation, leading to faster turnaround times and fewer errors. This is particularly beneficial for organizations that handle a large volume of invoices, as it ensures that payments are processed accurately and on time, improving relationships with suppliers and maintaining healthy cash flow. In addition, RPA can efficiently manage payroll processing by automating tasks such as calculating wages, deductions, and tax withholdings, and ensuring timely and accurate salary disbursements (Ma & Jia, 2022; Kumar & Khanna, 2023). This reduces the administrative burden on HR and finance teams while also minimizing the risk of payroll errors that could lead to employee dissatisfaction or regulatory penalties. Overall, the automation of routine financial operations enhances operational efficiency, improves accuracy, and allows staff to focus on more strategic and value-added activities.

2.4 Customer Service

RPA bots can manage customer inquiries by handling routine questions and directing customers to relevant information or services, providing quick and accurate responses that significantly improve customer satisfaction (Treacy et al., 2023; Salila et al., 2025). This not only reduces response time but also ensures 24/7 support, allowing financial institutions to deliver consistent service even outside business hours. By taking over repetitive customer service tasks, RPA allows human agents to focus on more complex issues that require personalized attention (Coombs et al., 2020). In addition, automating loan processing tasks such as data verification, credit checks, and document validation helps streamline the loan approval process, leading to faster decision-making and disbursement. This improves operational efficiency and enhances customer experience by reducing waiting times. The integration of RPA into customer service and loan operations reflects its growing role in enhancing responsiveness, accuracy, and overall service quality in the financial sector.

2.5 Benefits of RPA

Robotic Process Automation (RPA) offers several benefits to the financial sector. It improves operational efficiency by reducing the time required to complete repetitive, rule-based tasks, allowing employees to shift their focus to more complex, value-added work such as decision-making and strategic planning (Kassa & Worku, 2025). This shift not only enhances overall productivity but also leads to better service delivery and customer satisfaction. Another important advantage is cost reduction. By automating routine processes, financial institutions can significantly lower labor costs, reduce error-related expenses, and boost profitability. Furthermore, RPA is highly scalable, which means it can quickly adapt to changes in workload or business demands without requiring major investments in additional resources. This flexibility makes RPA especially valuable in fast-changing and competitive environments, where speed, accuracy, and cost control are essential (Flechsig et al., 2022). As a result, RPA is becoming an integral part of digital transformation strategies across the financial industry.

2.6 Challenges of RPA

Despite these advantages, there are still several challenges that financial institutions must consider when implementing Robotic Process Automation (RPA). One major challenge is integrating RPA with older legacy systems, which can be complex, time-consuming, and may require additional customization to ensure compatibility (Patrício et al., 2025). These legacy systems were not designed with automation in mind, making seamless integration difficult. Another critical concern is data security. Since financial processes involve highly sensitive customer and transactional data, any weakness in RPA deployment could expose the system to cyber threats or data breaches. Ensuring robust security protocols and compliance with data protection regulations is vital. Additionally, employee adaptation plays a key role in the success of RPA initiatives (Syed et al., 2020). Staff members need proper training to understand and work alongside automation tools, and resistance to change can slow down adoption. Therefore, clear communication, ongoing support, and change management strategies are essential to help employees transition smoothly and fully leverage the benefits of RPA.

3.0 ARTIFICIAL INTELLIGENCE (AI) TRANSFORMING FINANCIAL SERVICES

Artificial Intelligence (AI) encompasses a broad range of technologies designed to mimic human intelligence and cognitive processes. These include machine learning (ML), natural language processing (NLP), and other techniques that enable systems to learn from data, recognize patterns, and make informed decisions. AI technologies such as ML and NLP are increasingly used in the financial sector for tasks such as fraud detection, predictive marketing, customer service automation, and financial decision-making. FinTech companies integrate AI to enhance data analysis capabilities, deliver personalized services, and streamline operations, ultimately improving the quality and speed of financial services (Qatawneh & Lutfi, 2024). However, the deployment of AI technologies also brings significant challenges. Concerns about data privacy, algorithmic bias, and the ethical implications of autonomous decision-making are growing, especially in high-stakes financial environments where fairness and transparency are critical (Mellouli et al., 2024). Addressing these concerns requires more than just technical solutions—it demands careful planning, multidisciplinary collaboration among data scientists, financial experts, and ethicists, and continuous research to overcome both technical and institutional barriers to responsible AI integration (Faiyazuddin et al., 2025).

3.1 Risk Management and Credit Scoring

AI enhances credit scoring by analyzing a wide range of data sources, including non-traditional and unstructured data such as social media activity, utility payments, and online behavior. This broader data input allows for more inclusive credit assessments, particularly benefiting individuals and small businesses with limited credit histories. As a result, AI contributes to improved financial inclusion by enabling lenders to make informed decisions about applicants who may have been excluded by traditional scoring methods (Sadok et al., 2022). Advanced AI techniques such as decision trees, neural networks, and deep learning algorithms are now widely used to assess creditworthiness, offering more accurate, flexible, and transparent credit scoring models (Metawa et al., 2025). These models can adapt and improve over time, allowing for dynamic risk evaluation based on evolving borrower behavior and market conditions.

In addition to credit assessment, AI significantly enhances risk assessment and mitigation across various financial domains. Through data-driven methods, AI enables institutions to identify and manage potential risks before they materialize. For instance, real-time fraud detection systems can monitor transaction patterns and flag suspicious activities, reduce financial losses and enhance customer trust. AI-based stress testing tools simulate adverse economic scenarios, allowing banks and financial institutions to evaluate the resilience of their portfolios and develop proactive risk management strategies (Ma & Wei, 2025). Moreover, AI's strength in analyzing vast amounts of structured and unstructured data enables more effective management of credit, market, operational, and compliance risks.

By identifying hidden trends and potential threats within large datasets, AI helps financial institutions respond more quickly to risk events and regulatory demands. It supports a shift from reactive to predictive risk management, enhancing the stability and performance of financial operations. As regulatory bodies increasingly emphasize the importance of robust risk frameworks, AI stands out as a critical tool for institutions aiming to maintain compliance while staying competitive in a data-driven environment (Tian et al., 2025). Overall, the integration of AI in credit and risk management not only improves decision accuracy but also strengthens the foundation for sustainable and responsible financial practices.

3.2 Algorithmic Trading and Asset Management

AI-driven algorithmic trading is transforming financial markets by leveraging machine learning and deep reinforcement learning to develop and execute trading strategies with high precision and speed (Dakalbab et al., 2024; Bhuiyan et al., 2025). These AI systems continuously learn from market data, adapt to changing conditions, and make split-second decisions that would be impossible for human traders to replicate. By analyzing vast amounts of structured and unstructured data, such as historical price movements, economic indicators, social media sentiment, and real-time news feeds. AI algorithms can identify patterns and signals that inform profitable investment opportunities (Gómez-Martínez & Medrano-Garcia, 2025).

This ability to process diverse data sources enables AI to make more informed and timely investment decisions, providing a competitive edge in fast-paced markets.

In addition to trading, AI plays a growing role in portfolio management by enhancing how assets are allocated and adjusted over time. AI systems can tailor investment strategies to align with an individual's risk tolerance, financial goals, and investment horizons. These systems dynamically rebalance portfolios in response to market changes, optimizing returns while managing risk (Sutiene et al., 2024). This level of personalization and responsiveness was traditionally limited to high-net-worth clients but is now increasingly accessible through robot advisors and other AI-driven platforms. Furthermore, AI tools enhance decision-making accuracy and consistency in asset management, allowing portfolio managers to scale operations and manage larger portfolios without compromising performance (Babina et al., 2024).

Overall, AI's contribution to investment and asset management lies in its ability to improve analytical depth, automate complex processes, and provide more efficient, data-driven strategies. As financial markets grow in complexity and volume, AI-driven tools offer a vital solution for institutions and investors seeking to navigate uncertainty and achieve sustainable financial outcomes.

3.3 Fraud Detection and Cybersecurity

AI significantly strengthens fraud detection by analyzing large and complex financial datasets to identify anomalies and flag potentially fraudulent activities in real time (Odufisan et al., 2025). Traditional rule-based systems are often limited in scope and slow to adapt, whereas AI-driven models continuously learn from new data, allowing them to detect emerging fraud patterns and reduce false positives. Machine learning algorithms, including supervised, unsupervised, and deep learning methods, are widely used to improve both the accuracy and efficiency of fraud detection systems (Afriyie et al., 2023). These algorithms can uncover subtle and hidden relationships in data that human analysts might overlook, making them highly effective for detecting unusual transactions, identity theft, or account takeovers.

Beyond fraud detection, AI also plays a vital role in enhancing cybersecurity across financial services. AI-powered tools such as biometric authentication systems, AI chatbots, and smart virtual assistants help secure customer identities and protect financial transactions (Najem et al., 2025). Biometric features like facial recognition and fingerprint scanning add an extra layer of security, reducing reliance on traditional passwords that are vulnerable to breaches. AI chatbots and virtual assistants also help detect and respond to suspicious user behavior, acting as a first line of defense against potential threats (Zahidi et al., 2024).

Moreover, AI is increasingly applied in fraud risk monitoring and anti-money laundering (AML) processes, further reinforcing financial security. By analyzing transaction patterns and customer behavior, AI systems can identify potential money laundering schemes and automatically flag suspicious activity for further review (Bello & Olufemi, 2024). This reduces the workload for compliance teams and ensures more consistent monitoring in high-risk areas. As financial crimes become more sophisticated, AI offers a scalable and adaptive solution that not only improves threat detection but also strengthens the overall resilience of financial institutions (Bello, 2025).

3.4 Personalization of Financial Advice (Robo-Advisors)

AI-powered robo-advisors offer personalized financial advice by analyzing individual factors such as risk tolerance, financial goals, and investment horizons, allowing for the creation of customized investment strategies that align with clients' unique profiles (Zhu et al., 2023; Zhu et al., 2024). These systems rely on machine learning algorithms and real-time data analysis to continuously adapt to market changes and user preferences, ensuring that financial plans remain relevant and effective. By automating complex advisory processes, robo-advisors make professional financial planning more affordable and accessible, especially for individuals who may not have access to traditional wealth management services. This democratization of financial advice enhances financial inclusion and empowers more people to make informed investment decisions.

In addition to investment guidance, AI significantly improves customer service through the use of intelligent chatbots and virtual assistants. These tools provide real-time support, handle routine inquiries, and offer personalized interactions that were once limited to human advisors (Li et al., 2023). AI chatbots can assist with tasks such as account management, transaction tracking, and answering product-related questions, all while learning from past interactions to improve future responses. This not only reduces response time and operational costs for financial institutions but also ensures consistent, 24/7 service delivery (Li et al., 2023).

Furthermore, by leveraging customer data and behavioral insights, AI enables financial institutions to deliver more tailored financial products and services on a scale (Babina et al., 2024). This includes recommending specific savings plans, insurance packages, or investment portfolios based on customer preferences and financial behavior. As a result, AI plays a central role in reshaping customer experience, making financial services more responsive, personalized, and efficient.

3.5 Challenges of AI

Despite AI offering transformative potential in the financial sector, its adoption comes with a range of critical challenges that must be addressed to ensure responsible and sustainable implementation. Poor data quality is a fundamental issue that can lead to inaccurate outcomes, erode stakeholder confidence, and undermine the credibility of AI systems (Hanna et al., 2025). Moreover, the extensive use of personal and financial data in AI applications raises significant privacy concerns, making robust data protection frameworks essential to safeguard user information and maintain regulatory compliance (Wieringa et al., 2021). AI models are also susceptible to inheriting biases present in the data they are trained on, which can result in unfair or discriminatory decisions. particularly in sensitive areas like credit scoring, fraud detection, or lending. Addressing these biases is vital for promoting fairness and integrity within financial services (Ferrara, 2024). Effective strategies to mitigate such biases include using diverse and representative datasets, conducting regular algorithmic audits, and employing transparent model development processes (González-Sendino et al., 2024).

Another major challenge lies in the complexity of AI algorithms, which often function as "black boxes," making it difficult for stakeholders to understand how decisions are made. This lack of transparency poses issues for accountability, particularly when AI-driven decisions have significant financial implications for individuals or businesses. In response, the development of Explainable AI (XAI) has become increasingly important. XAI provides insights into how models function and why specific decisions are made, helping build trust among users and enabling financial institutions to validate and justify AI-based outcomes (Ali et al., 2023). Furthermore, as regulatory scrutiny intensifies, financial institutions must navigate a complex legal landscape to ensure that their AI systems comply with evolving standards and regulations. The development of standardized frameworks and industry guidelines is essential to harmonize AI implementation across the financial sector and ensure consistency in ethical and legal practices.

In addition to ethical and regulatory issues, technical and operational challenges also hinder AI adoption. AI systems are vulnerable to cyberattacks, which can compromise sensitive data and disrupt critical financial operations (Parambil et al., 2024). As AI becomes more integrated into core financial infrastructure, implementing strong cybersecurity measures is crucial to protect both systems and users. However, high implementation costs and difficulties in integrating AI with outdated legacy systems remain significant barriers, especially for smaller institutions or those with limited digital infrastructure (Robertson et al., 2025). Overcoming these obstacles requires strategic investment in AI talent, tools, and infrastructure. Building internal capabilities and fostering a culture of innovation are key to unlocking the full potential of AI in finance and ensuring long-term success (Haefner et al., 2023).

3.6 Ethical Considerations

Ethical issues in using AI in financial institutions need to be handled carefully to ensure that technology benefits both the organization and its customers (Hanna et al., 2025). First, there must be clear rules and responsibilities to guide the proper use of AI systems, especially when these systems make decisions that affect people's finances and lives. Financial institutions should also have well-defined governance structures to manage any ethical concerns that may arise, such as discrimination, data misuse, or lack of transparency (Papagiannidis et al., 2025). Even with advanced AI, human oversight remains essential. People working alongside AI should use their judgment to monitor outcomes and intervene when needed (Zirar et al., 2023). They also need to ensure that AI tools are supporting, not replacing, responsible decision-making. Another concern is the impact of AI on jobs. As automation reduces the need for some roles, companies have a responsibility to support their employees through retraining and upskilling programs, helping them transition into new positions. Importantly, ethics should be built into the design of AI from the beginning. This means developing systems that are fair, transparent, and accountable, reducing the risk of harm. In short, while AI can bring many benefits, it must be used thoughtfully, with attention to both technical performance and human impact.

4.0 BLOCKCHAIN AS THE FOUNDATION FOR SECURE FINANCIAL TRANSACTIONS

Blockchain technology is a decentralized, distributed ledger system that records transactions across a network of independent nodes. Each transaction is grouped into a block, and these blocks are cryptographically linked to one another, forming a continuous and secure chain of data (Tripathi et al., 2023). Once a transaction is recorded, it cannot be altered or deleted. Consequently, integrity, transparency, and security emerge as the core advantages of blockchain-based systems. Unlike traditional centralized systems, blockchain operates without the need for a central authority, relying instead on a peer-to-peer network to validate and record transactions (Saleh, 2024). Each block includes transaction data, a hash of its own content, and the hash of the previous block, which creates a secure chain that is resistant to tampering. To achieve consensus and validate transactions across the network, blockchain employs mechanisms such as Proof of Work (PoW), Proof of Stake (PoS), and Proof of Elapsed Time (PoET), each offering different approaches to maintaining network integrity and security (Kaur et al., 2021).

Blockchain technology offers several important advantages, particularly in terms of security, transparency, and efficiency. Its decentralized structure and use of cryptographic techniques make it highly resistant to fraud and unauthorized tampering, which is especially valuable in sectors like finance and supply chain management where data accuracy and trust are critical (Saleh, 2024). Moreover, since all transactions are visible to authorized participants in the network, blockchain enhances accountability and fosters trust among users (Zantalis et al., 2023). This level of transparency reduces the risk of information asymmetry and enables real-time verification of records. Additionally, blockchain can streamline processes by eliminating intermediaries, which reduces transaction times and lowers operational costs. For instance, in financial services, blockchain can facilitate faster payments, simplify cross-border transfers, and improve settlement processes. These benefits are driving growing interest in blockchain as a foundational technology for next-generation digital infrastructure across various industries.

4.1 Use Cases of Blockchain in Finance

4.1.1 Cross-Border Payments and Settlements

Blockchain technology offers significant advantages for cross-border payments and settlements, particularly in terms of enhanced security, transparency, and operational efficiency. Traditional cross-border payment systems are often slow, expensive, and complex, involving multiple intermediaries such as correspondent banks, which can introduce delays and increase the risk of fraud or data breaches. Blockchain addresses these inefficiencies by providing a decentralized and immutable ledger that records transactions in real time, enabling faster settlement times, reduced transaction costs, and heightened security (Khan & Ahmad, 2024). One practical application involves the use of blockchain to facilitate the conversion of local currencies into digital tokens. These tokens can be transferred across the blockchain network and then converted back into the recipient's local currency, significantly reducing processing time and bypassing the need for intermediaries (Hariyani et al., 2025).

Moreover, blockchain's built-in transparency and traceability features make it easier for financial institutions to meet regulatory requirements. Each transaction is permanently recorded and time-stamped, allowing for accurate audit trails and real-time monitoring. This is especially useful for compliance with anti-money laundering (AML) and counter-financing of terrorism (CFT) regulations, as authorities and institutions can track and verify transactions more easily (Wang, 2022). The ability to enhance compliance while simultaneously improving speed and cost-efficiency positions blockchain as a transformative solution for international payments. As global trade and cross-border financial activities continue to grow, blockchain technology offers a promising framework for building a more secure, inclusive, and efficient global payment ecosystem.

4.1.2 Identity Verification and KYC Processes

The Know Your Customer (KYC) process is essential for financial institutions to verify the identity of their clients and prevent illicit activities such as money laundering, fraud, and terrorist financing. However, traditional KYC procedures are often manual, paper-based, and labor-intensive, making them time-consuming, costly, and prone to human error. These inefficiencies not only slow down customer onboarding but also burden institutions with high compliance costs. Blockchain technology presents a transformative solution by offering a decentralized, secure, and tamper-proof system for storing and verifying identity documents (Malhotra et al., 2022; Patil & Sangeetha, 2022).

Through the use of smart contracts, blockchain-based KYC systems can automate the verification process, significantly reduce processing time and operational costs while maintain a high level of data security and privacy. Once a customer's identity is verified and recorded on the blockchain, the information can be securely shared with other authorized institutions without requiring the customer to repeat the KYC process multiple times. This not only enhances efficiency but also gives users greater control over their personal data, allowing them to grant access selectively and maintain transparency about how their information is used (Elveny et al., 2025). By streamlining identity verification and improving data security, blockchain technology has the potential to revolutionize KYC processes and set a new standard for regulatory compliance in the financial sector.

4.1.3 Trade Finance and Supply Chain Finance

Blockchain technology has the potential to significantly transform trade finance and supply chain finance by addressing long-standing issues such as information asymmetry, fraud, and inefficiencies that are common in traditional systems. In conventional trade finance processes, the lack of real-time information and reliance on intermediaries can lead to delays, increased costs, and reduced trust between trading partners. Blockchain offers a decentralized and transparent platform that enables all stakeholders to access a single, tamper-proof version of transaction data. This shared visibility enhances the traceability and accountability of every step within the supply chain, reducing disputes and promoting smoother operations (Kumar et al., 2025).

Financial institutions benefit from blockchain by gaining more accurate and timely data to support credit identification, transaction monitoring, and risk management. This improved visibility allows them to assess the creditworthiness of suppliers more effectively and provide financing with greater confidence, ultimately enhancing the efficiency and security of supply chain finance (Zheng et al., 2022). Furthermore, blockchain-based platforms can automate key processes such as invoice validation, payment execution, and goods tracking. These capabilities make it possible to track products from origin to destination in real time and verify compliance with contractual terms, thereby reducing the risk of fraud and building stronger trust among all participants in the supply chain (Duan et al., 2023). As global trade continues to grow in scale and complexity, blockchain offers a promising solution for streamlining operations and reinforcing trust in supply chain finance.

4.1.4 Technical Advantages of Blockchain Adoption

Blockchain offers several technical advantages that make it attractive for financial and business applications. Its decentralized structure provides high security and allows transparent, traceable transactions, which is especially useful in the financial sector (Javaid et al., 2022). Additionally, blockchain improves service capacities by enabling better performance and operational efficiency across industries (Kumar et al., 2025). Another key benefit is consensus-based validation, which helps ensure data accuracy, synchronization, and security across all users in the network (Alzoubi & Mishra, 2024).

From an economic perspective, blockchain adoption brings clear cost and efficiency benefits. It can reduce transactional costs for firms, suppliers, and financial institutions by streamlining processes and eliminating intermediaries (Al-Dmour et al., 2024). This cost reduction may lead to new business models and more efficient operations. Furthermore, blockchain reduces information gaps between consumers and service providers, which helps enhance consumer welfare. Better access to accurate information allows consumers to make more informed decisions and can lead to improved satisfaction and trust (Hina et al., 2025).

Despite these benefits, blockchain adoption also comes with notable security challenges. One concern is the overall security of blockchain systems, which can influence whether organizations decide to adopt the technology (Ostern et al., 2022). Institutions may hesitate due to fears of data breaches or weak security frameworks (Park, 2019). Privacy issues are another major challenge, as users worry about how their data is stored and accessed. Additionally, threats from earlier blockchain platforms have shown the need for stronger and more reliable security measures. Addressing these challenges is essential to encourage wider adoption of blockchain technologies.

4.1.5 Challenges of Blockchain

Despite its many advantages, blockchain technology also faces several limitations that can impact its widespread adoption. One key concern is performance—blockchain's decentralized nature, while enhancing security and trust, can lead to slower transaction speeds and higher energy consumption, particularly in networks that rely on consensus mechanisms like Proof of Work (Zheng et al., 2019). This inefficiency may limit blockchain's scalability, especially in high-volume environments such as global finance or supply chain systems. Despite blockchain promoting transparency by allowing all participants to view transaction data, this feature can raise privacy concerns. Sensitive information being visible to all nodes in the network may not align with data protection standards, posing challenges for financial institutions and industries handling confidential data (Wei et al., 2024).

Beyond technical concerns, there are also regulatory and practical barriers. Legal and regulatory uncertainties remain a major obstacle, as many jurisdictions lack clear guidelines for blockchain implementation. Questions surrounding data ownership, liability, and cross-border transactions are yet to be fully resolved. Moreover, high transaction costs on certain blockchain platforms can undermine its cost-saving promise, especially for smaller organizations. Finally, a lack of awareness and understanding of blockchain's capabilities and limitations among business leaders and policymakers can slow down adoption efforts (Xiao et al., 2024). Addressing these challenges will require continued innovation, regulatory clarity, and broader educational initiatives to ensure blockchain's potential is realized in a secure, efficient, and responsible manner.

5.0 CASE STUDIES AND PRACTICAL INSIGHTS

The convergence of Robotic Process Automation (RPA), Artificial Intelligence (AI), and blockchain technology are reshaping the financial services industry. Institutions are moving beyond experimental phases and increasingly deploying these technologies to streamline operations, enhance customer experience, improve compliance, and enable new business models. This section explores real-world case studies of financial institutions implementing integrated RPA, AI, and blockchain, analyzes their outcomes, and provides actionable insights and best practices drawn from these implementations.

5.1 JPMorgan - COiN[1]

AI in Finance has changed the way JP Morgan works so much that it now takes only seconds instead of 360,000 hours a year to read and understand business credit agreements. The bank has automated the document review process for a certain type of contract using its AI-powered Contract Intelligence (COiN) platform. COiN uses picture recognition to find and compare distinct clauses. It runs on a private cloud network. In its first use, COiN pulled out over 150 relevant attributes from annual business credit agreements in a matter of seconds, saving 360,000 hours of manual review time. The algorithm finds patterns based on phrases or places in the contracts. This saves a lot of time and money while also making the process more efficient and less error prone.

This initiative offers several key lessons:

i. Automating high-volume legal processes leads to significant cost and time savings.
ii. AI improves consistency and reduces human error in contract analysis.
iii. Successful AI implementation depends on clearly defined tasks and robust data infrastructure.

5.2 Santander: Blockchain for Cross-Border Payments[2]

Santander has started a foreign exchange service that lets customers send money to other countries on the same day using blockchain technology. Blockchain company (Ripple) built the technology that powers the service, which is named Santander One Pay FX. Santander claims that it is the first bank to offer a cross-border payments service that uses blockchain. At first, it was just used to support bitcoin, but banks are now more interested in other uses, such as clearing and settling payments. The blockchain-based foreign exchange platform from Santander is now available in Spain, the UK, Brazil, and Poland.

This initiative offers several key lessons:

i. Blockchain can enhance traditional banking services by **reducing reliance on intermediaries** and speeding up settlement times.
ii. Collaboration with fintech firms (like Ripple) accelerates innovation and product deployment.
iii. Early adoption in multiple countries (Spain, UK, Brazil, Poland) showcases the scalability and global potential of blockchain solutions in finance.

5.3 ING and the Launch of Contour – Revolutionizing Trade Finance with Blockchain[3]

The successful launch of Contour, a blockchain-based trade finance platform co-founded by ING Ventures, marks a significant milestone in the transformation of the global trade finance landscape. Entering a market valued at over $18 trillion, Contour addresses long-standing inefficiencies associated with the use of letters of credit. Notably, letter of credit was a 400-year-old process that has remained largely manual, paper-based, and time-consuming.

By leveraging R3's Corda blockchain network, Contour has digitized the issuance, exchange, and tracking of letters of credit, enabling all participants in a trade transaction to share trusted information in real time. The platform has demonstrated remarkable success, reducing processing times from five to ten days to under 24 hours. This timesaving alone presents a competitive advantage for banks and corporates alike, enabling faster settlement, lower administrative burden, and improved liquidity management.

What sets Contour apart is its collaborative and scalable approach. It was co-created by ING and seven other banks, signaling a strong commitment from major financial institutions to drive blockchain adoption in trade finance. The live pilots conducted across 14 countries and involving over 50 banks and corporates, including ING-led trials with HSBC (Cargill and Tricon) and CTBC Bank, provided strong validation of the platform's utility. The fact that 96% of participants acknowledged improvements in speed, efficiency, and cost-effectiveness demonstrates that Contour addresses real operational pain points in trade finance.

The launch of Contour as an independent network headquartered in Singapore also underscores its global ambitions and readiness for commercial scale. By inviting banks and corporates to join its growing ecosystem, Contour is moving beyond a pilot phase into a live, functioning infrastructure that has the potential to become a new standard in digital trade finance.

This initiative offers several key lessons:

i. Co-creation and industry collaboration are critical for the adoption of emerging technologies in conservative, risk-averse sectors like trade finance.
ii. Pilot programs and sandbox environments help build trust among stakeholders and fine-tune technological and operational features before commercial launch.
iii. Blockchain's value lies not just in security, but in real-time visibility, interoperability, and the removal of friction across transactional processes.

From a broader perspective, ING's leadership in developing Contour reflects its strategic commitment to innovation and digital transformation. The project aligns with global trends toward paperless trade, reduced reliance on intermediaries, and real-time transaction processing, all of which are especially crucial in an increasingly digital and globalized economy.

5.4 Goldman Sachs QIS – Using AI to Decode Executive Sentiment in Earnings Calls[4]

Goldman Sachs Asset Management's Quantitative Investment Strategies (QIS) team presents a compelling example of how artificial intelligence (AI) can be harnessed to uncover alpha-generating insights from unstructured data. The teams hypothesize that the sentiment expressed by corporate executives during earnings calls can signal future company performance. At the core of this strategy lies the use of Natural Language Processing (NLP), particularly models like BERT (Bidirectional Encoder Representations from Transformers), to analyze the textual content of earnings call transcripts. These transcripts, rich in qualitative data, provide critical insight into management's outlook, confidence levels, and strategic direction. Executives, given their insider understanding of operations, may unconsciously convey subtle cues that hint at future earnings or operational shifts. The QIS team uses these tools to quantify sentiment and feeds that information into predictive investment models.

However, what makes Goldman Sachs' approach particularly innovative is their expansion into audio analysis. Since 2023, the QIS team has deepened its methodology to capture not just *what* corporate officers say, but *how* they say it. By examining audio recordings, AI models can detect tonal variations, emotional cues, hesitations, and other speech characteristics that do not appear in written transcripts. The Q&A sections of earnings calls, being less scripted and more spontaneous, are especially valuable in this analysis. Emotional signals such as uncertainty, defensiveness, or overconfidence can provide additional depth to the investment signal. To achieve this, the QIS team employs deep learning techniques on more than 400,000 hours of raw audio data, a volume unmanageable for human analysts. This approach allows the team to control potential analyst bias and ensures consistency across evaluations. One particularly noteworthy element is the use of personalized vocal baselines, comparing a speaker's current delivery against their own historical vocal patterns. This controls for natural differences in speech styles and helps identify significant deviations in tone or delivery that may be meaningful.

Despite its promise, audio-based sentiment analysis is not without challenges. Unlike clean, structured text, audio files can be messy filled with silences, background noise, music, varied accents, and inconsistent speaking styles. The QIS team addresses this by cleaning raw data in-house to retain nuanced features and eliminate irrelevant disturbances. They avoid over-sanitizing the data to ensure that potentially meaningful emotional or tonal shifts remain detectable. These carefully processed inputs are then used to feed advanced models that convert speech characteristics into quantitative alpha signals.

This case offers several important insights and lessons:

i. *Integration of Text and Audio Enhances Signal Strength*

By combining textual NLP with audio sentiment analysis, the QIS team creates a multidimensional view of executive sentiment. This holistic approach improves model robustness and helps identify signals that would be missed through text or voice analysis alone.

ii. *AI Enables Scalability and Speed*

While human analysts can interpret tone and language, AI allows this analysis to be performed across thousands of calls, providing scale, speed, and consistency. This is particularly valuable in today's data-rich financial environment.

iii. *Customization and Context Matter*

The use of individualized vocal profiles shows the importance of context when analyzing unstructured data. Understanding a speaker's typical delivery pattern is essential for detecting meaningful deviations.

iv. *Noise Management is Critical*

Effective pre-processing of audio data is essential. Cleaning the data while preserving emotional nuances is a delicate but necessary balance.

v. *AI Reduces Human Bias but Requires Human Oversight*

While AI models offer consistency and avoid subjective interpretation, human involvement remains vital in model validation, error handling, and refining assumptions over time.

vi. *Expanding the Role of AI in Active Investing*

The work of Goldman Sachs QIS demonstrates how AI is no longer confined to back-office tasks or automation but is now at the heart of active portfolio strategy and alpha generation.

This approach can extend to other qualitative financial events such as investor briefings, annual general meetings (AGMs), and regulatory calls, further broadening the scope of sentiment-based forecasting. The ability to process and interpret unstructured data at scale may become a defining capability for asset managers aiming to maintain an edge in increasingly efficient markets.

5.5 Maybank Asset Management – Pioneering AI-Driven Shariah-Compliant Investing with Arabesque AI[5]

The collaboration between Maybank Asset Management Sdn Bhd (MAM Malaysia) and Arabesque AI represents a groundbreaking development in the Malaysian financial sector, particularly within the realm of Islamic wealth management. By launching a Shariah-compliant, AI-powered Hyper-Customized Discretionary Portfolio Mandate (DPM), MAM Malaysia has not only elevated the standards of personalized investing but also aligned technological innovation with ethical finance.

This strategic partnership marks a convergence of three transformative elements in modern investing: artificial intelligence (AI), Shariah compliance, and environmental, social, and governance (ESG) integration. Arabesque AI's platform leverages vast datasets, including both financial and non-financial indicators, to generate AI-driven alpha forecasts across a universe of over 25,000 listed equities. This breadth of analysis, coupled with a deep commitment to Shariah principles, allows for the construction of portfolios that are not only performance-driven but also values-based and ethically guided.

The DPM is particularly significant for High-Net-Worth (HNW) clients, who increasingly demand bespoke investment solutions that reflect both their financial goals and personal beliefs. Through the use of AI, the platform can rapidly process investor preferences, including risk appetite, ethical considerations, and sectoral exclusions. This level of personalization, once reserved for institutional clients, is now accessible to individual investors at scale, thanks to the computational power and real-time capabilities of AI.

From an operational standpoint, the platform also offers robust portfolio monitoring and dynamic rebalancing, ensuring that investments remain aligned with investor profiles and market changes. Such automation reduces human error, enhances efficiency, and provides a seamless user experience.

Beyond the technological achievement, this initiative sends a strong message about the future of Islamic finance. As noted by Omar Selim, CEO of Arabesque Group, the intersection of sustainability, Islamic principles, and AI signals a shift in how wealth is managed. Islamic finance, rooted in ethical and responsible investing, naturally aligns with the growing global demand for sustainable and socially conscious financial solutions. The partnership reflects a broader movement toward investment strategies that are profitable, purposeful, and personalized.

Maybank Islamic's support, through the voice of CEO Dato' Mohamed Rafique Merican, further emphasizes the institution's commitment to being at the forefront of ethical digital innovation. By offering the DPM across Maybank branches, the bank ensures wide accessibility, reinforcing its mission to democratise ethical investment opportunities while remaining grounded in Islamic values.

6.0 SUMMARY

Key Insights and Lessons Learned:

i. *AI Enables Unprecedented Personalization*

DPM demonstrates how AI can process individual preferences at scale, creating highly tailored portfolios that meet both financial and ethical criteria.

ii. *Alignment of Islamic and ESG Principles*

The values underpinning Islamic finance—moderation, fairness, and responsibility—closely align with ESG standards. This natural synergy is effectively captured and operationalized by Arabesque AI's models.

iii. *Shariah Compliance Enhanced by Technology*

Ensuring portfolios meet Islamic requirements traditionally involved manual screening. AI now automates this, increasing efficiency and improving compliance accuracy.

iv. *From Mass Customizations to Hyper-Customizations*

Moving beyond generic risk profiling, the platform enables hyper-customization, offering distinct portfolios for each client based on real-time analysis and input.

v. *Bridging Innovation and Trust*

In Islamic finance, trust and transparency are vital. By combining AI with clear ethical guidelines, the DPM builds investor confidence in the use of advanced technologies.

vi. *Scaling High-Touch Services*

What was once the domain of wealth managers serving a few elite clients can now be scaled to a broader segment through AI-driven automation.

The launch of the Hyper-Customized DPM reflects a broader transformation, where investors no longer need to choose between returns and values. As the financial landscape evolves, such innovations will be key to meeting the expectations of a new generation of investors who are digitally savvy, ethically driven, and increasingly seeking bespoke, intelligent solutions.

REFERENCES

Afriyie, J. K., Tawiah, K., Pels, W. A., Addai-Henne, S., Dwamena, H. A., Owired, U., Ay, E. O., Eh, S. A., & Eshun, J. (2023). A supervised machine learning algorithm for detecting and predicting fraud in credit card transactions. *Decision Analytics Journal, 6*, 100163. DOI: 10.1016/j.dajour.2023.100163

Al-Dmour, A., Al-Dmour, R., Al-Dmour, H., & Al-Adwan, A. (2024). Blockchain applications and commercial bank performance: The mediating role of AIS quality. *Journal of Open Innovation, 10*(2), 100302. DOI: 10.1016/j.joitmc.2024.100302

Ali, S., Abuhmed, T., El-Sappagh, S., Muhammad, K., Alonso-Moral, J. M., Confalonieri, R., Guidotti, R., Del Ser, J., Díaz-Rodríguez, N., & Herrera, F. (2023). Explainable Artificial Intelligence (XAI): What we know and what is left to attain Trustworthy Artificial Intelligence. *Information Fusion, 99*, 101805. DOI: 10.1016/j.inffus.2023.101805

Allied Market Research. (2022). *Robotic process automation (RPA) in financial services market size, share, competitive landscape and trend analysis report, by component, deployment mode, enterprise size, application and end user: Global opportunity analysis and industry forecast, 2021–2030.* https://www.alliedmarketresearch.com/robotic-process-automation-rpa-in-financial-services-market-A06933

Alzoubi, Y. I., & Mishra, A. (2024). Blockchain consensus mechanisms comparison in fog computing: A systematic review. *ICT Express, 10*(2), 342–373. DOI: 10.1016/j.icte.2024.02.008

Andrew, J., Israevel, D. P., Sagayam, K. M., Bhushan, B., Sei, Y., & Eunice, J. (2023). Blockchain for healthcare systems: Architecture, security challenges, trends and future directions. *Journal of Network and Computer Applications, 215*, 103633. DOI: 10.1016/j.jnca.2023.103633

Babina, T., Fedyk, A., He, A., & Hodson, J. (2024). Artificial intelligence, firm growth, and product innovation. *Journal of Financial Economics, 151*, 103745. DOI: 10.1016/j.jfineco.2023.103745

Bédard, M., Leshob, A., Benzarti, I., Mili, H., Rab, R., & Hussain, O. (2024). A rule-based method to effectively adopt robotic process automation. *Software: Evolution and Process.* Advance online publication. https://doi.org/DOI: 10.1002/smr.2709

Bello, H. O. (2025). Developing predictive financial fraud models using AI-driven analytics within cybercrime-resilient security ecosystems. *International Journal of Research Publication and Reviews, 6*(1), 5055–5069. DOI: 10.55284/ijrpr.6.0125.0651

Bello, O. A., & Olufemi, K. (2024). Artificial intelligence in fraud prevention: Exploring techniques and applications challenges and opportunities. *Computer Science & IT Research Journal*, *5*(6), 1505–1520. DOI: 10.55195/csitjv5i6.1252

Bhambri, P., Rani, S., & Pareek, P. K. (2024). Financial innovations: Intelligent automation in finance and insurance sectors. In *Hyperautomation in business and society* (Chapter 12). https://doi.org/DOI: 10.4018/979-8-3693-3354-9.ch012

Bhardwaj, V., & Kumar, M. (2025). Transforming higher education with robotic process automation: Enhancing efficiency, innovation, and student-centered learning. *Discover Sustainability*, *6*, 356. DOI: 10.1007/s43621-025-01198-6

Bhuiyan, M. S. M., Rafi, M. A., Rodrigues, G. N., Mir, M. N. H., Ishraq, A., Mridha, M. F., & Shin, J. (2025). Deep learning for algorithmic trading: A systematic review of predictive models and optimization strategies. *Array (New York, N.Y.)*, *26*, 100390. DOI: 10.1016/j.array.2025.100390

Brodny, J., & Tutak, M. (2025). Stakeholder interactions and ethical imperatives in big data and AI development. *Journal of Open Innovation*, *11*(1), 100491. DOI: 10.1016/j.joitmc.2025.100491

Choubey, A., & Sharma, M. (2021). Implementation of robotics and its impact on sustainable banking: A futuristic study. *Journal of Physics: Conference Series*, *1911*, 012013. DOI: 10.1088/1742-6596/1911/1/012013

Coombs, C., Hislop, D., Taneva, S. K., & Barnard, S. (2020). The strategic impacts of Intelligent Automation for knowledge and service work: An interdisciplinary review. *The Journal of Strategic Information Systems*, *29*(4), 101600. DOI: 10.1016/j.jsis.2020.101600

Costa, D. S., Mamede, H. S., & da Silva, M. M. (2023). A method for selecting processes for automation with AHP and TOPSIS. *Heliyon*, *9*(3), e13683. DOI: 10.1016/j.heliyon.2023.e13683 PMID: 36873512

Dakalbab, F., Abu Talib, M., Nasir, Q., & Saroufi, T. (2024). Artificial intelligence techniques in financial trading: A systematic literature review. *Journal of King Saud University. Computer and Information Sciences*, *36*(3), 102015. DOI: 10.1016/j.jksuci.2024.102015

Dinh, D. V. (2024). Digital economy and the electronic payment behavior: An empirical analysis. *Transnational Corporations Review*, *16*(4), 200078. DOI: 10.1016/j.tncr.2024.200078

Duan, K., Pang, G., & Lin, Y. (2023). Exploring the current status and future opportunities of blockchain technology adoption and application in supply chain management. *Journal of Digital Economy*, *2*, 244–288. DOI: 10.1016/j.jdec.2024.01.005

Elveny, M., Nasution, M. K. M., Purnamasari, F., & Wook, T. S. M. T. (2025). Blockchain-enabled KYC integration for CLV optimization with robust M-Estimation and IRLS method. *ICT Express*, *11*(3), 402–410. DOI: 10.1016/j.icte.2025.03.006

Faiyazuddin, M., Rahman, S. J. Q., Anand, G., Siddiqui, R. K., Mehta, R., Khatib, M. N., Gaidhane, S., Zahiruddin, Q. S., Hussain, A., & Sah, R. (2025). The impact of artificial intelligence on healthcare: A comprehensive review of advancements in diagnostics, treatment, and operational efficiency. *Health Science Reports*, *8*(1), 1–18. DOI: 10.1002/hsr2.70312 PMID: 39763580

Ferrara, E. (2024). Fairness and bias in artificial intelligence: A brief survey of sources, impacts, and mitigation strategies. *Sci*, *6*(1), 3. DOI: 10.3390/sci6010003

Flechsig, C., Anslinger, F., & Lasch, R. (2022). Robotic Process Automation in purchasing and supply management: A multiple case study on potentials, barriers, and implementation. *Journal of Purchasing and Supply Management*, *28*(1), 100718. DOI: 10.1016/j.pursup.2021.100718

Gómez-Martínez, R., & Medrano-Garcia, M. L. (2025). AI technology for developing Bitcoin investment strategies based on altcoin trends. *Sustainable Technology and Entrepreneurship*, *4*(1), 100087. DOI: 10.1016/j.stae.2024.100087

González-Sendino, R., Serrano, E., & Bajo, J. (2024). Mitigating bias in artificial intelligence: Fair data generation via causal models for transparent and explainable decision-making. *Future Generation Computer Systems*, *155*, 384–401. DOI: 10.1016/j.future.2024.02.023

Haefner, N., Parida, V., Gassmann, O., & Wincent, J. (2023). Implementing and scaling artificial intelligence: A review, framework, and research agenda. *Technological Forecasting and Social Change*, *197*, 122878. DOI: 10.1016/j.techfore.2023.122878

Hanna, M. G., Pantanowitz, L., Jackson, B., Palmer, O., Visweswaran, S., Pantanowitz, J., Deebajah, M., & Rashidi, H. H. (2025). Ethical and bias considerations in artificial intelligence/machine learning. *Modern Pathology*, *38*(3), 100686. DOI: 10.1016/j.modpat.2024.100686 PMID: 39694331

Hariyani, D., Hariyani, P., Mishra, S., & Sharma, M. K. (2025). A literature review on transformative impacts of blockchain technology on manufacturing management and industrial engineering practices. *Green Technologies and Sustainability*, *3*(3), 100169. DOI: 10.1016/j.grets.2025.100169

Hina, M., Islam, N., & Luo, X. R. (2025). Towards sustainable consumption decision-making: Examining the interplay of blockchain transparency and information-seeking in reducing product uncertainty. *Decision Support Systems*, *189*, 114370. DOI: 10.1016/j.dss.2024.114370

Javaid, M., Haleem, A., Singh, R. P., Suman, R., & Khan, S. (2022). A review of blockchain technology applications for financial services. *BenchCouncil Transactions on Benchmarks. Standards and Evaluations*, *2*(3), 100073. DOI: 10.1016/j.tbench.2022.100073

Kassa, B. Y., & Worku, E. K. (2025). The impact of artificial intelligence on organizational performance: The mediating role of employee productivity. *Journal of Open Innovation*, *11*(1), 100474. DOI: 10.1016/j.joitmc.2025.100474

Kaur, S., Chaturvedi, S., Sharma, A., & Kar, J. (2021). A research survey on applications of consensus protocols in blockchain. *Security and Communication Networks*, *6693731*. Advance online publication. DOI: 10.1155/2021/6693731

Khan, A., & Ahmad, U. (2024). Multi-criteria group decision-making based on 2-tuple linguistic *q*-rung picture fuzzy sets. *Granular Computing*, *9*, 8. DOI: 10.1007/s41066-023-00441-7

Khan, H. H., Khan, S., & Ghafoor, A. (2023). Fintech adoption, the regulatory environment and bank stability: An empirical investigation from GCC economies. *Borsa Istanbul Review*, *23*(6), 1263–1281. DOI: 10.1016/j.bir.2023.10.010

Kumar, N., Kumar, K., Aeron, A., & Verre, F. (2025). Blockchain technology in supply chain management: Innovations, applications, and challenges. *Telematics and Informatics Reports*, *18*, 100204. DOI: 10.1016/j.teler.2025.100204

Kumar, R., & Khanna, R. (2023). RPA (Robotic Process Automation) in finance & accounting and future scope. In *Proceedings of the 2023 2nd International Conference on Augmented Intelligence and Sustainable Systems (ICAISS)*. IEEE. https://doi.org/DOI: 10.1109/ICAISS58487.2023.10250496

Li, C.-Y., Fang, Y.-H., & Chiang, Y.-H. (2023). Can AI chatbots help retain customers? An integrative perspective using affordance theory and service-domain logic. *Technological Forecasting and Social Change*, *197*, 122921. DOI: 10.1016/j.techfore.2023.122921

Ma, H., & Wei, G. (2025). Exploring the role of AI in shaping social behavior: An intersectional psychological perspective on financial risk assessment through digital platforms. *Environment and Social Psychology*, *10*(1). Advance online publication. DOI: 10.59429/esp.v10i1.3375

Ma, J., & Jia, H. (2022). Application of financial robots based on RPA technology in small and medium-sized enterprises. In *Proceedings of the IEEE International Conference on Knowledge Engineering and Communication Systems (ICKES 2022)*. IEEE. https://doi.org/DOI: 10.1109/ICKECS56523.2022.10060387

Malhotra, D., Saini, P., & Singh, A. K. (2022). How blockchain can automate KYC: Systematic review. *Wireless Personal Communications*, *122*, 1987–2021. DOI: 10.1007/s11277-021-08977-0

Manta, O., Vasile, V., & Rusu, E. (2022). Banking transformation through FinTech and the integration of artificial intelligence in payments. *FinTech*, *4*(2), 279–295. DOI: 10.3390/fintech4020013

Mellouli, S., Janssen, M., & Ojo, A. (2024). Introduction to the issue on artificial intelligence in the public sector: Risks and benefits of AI for governments. *Digital Government: Research and Practice*, *5*(1), 1–6. DOI: 10.1145/3636550

Metawa, N., Afchal, S., & El-Kanj, N. (2025). AI-based model for enhancing credit risk and delinquency management in banks. *Journal of Cybersecurity and Information Management*, *16*(1), 282–300. DOI: 10.54216/JCIM.160120

Najem, R., Bahnasse, A., Fakhouri Amr, M., & Talea, M. (2025). Advanced AI and big data techniques in E-finance: A comprehensive survey. *Discover Artificial Intelligence*, *5*, 102. DOI: 10.1007/s44163-025-00365-y

Odufisan, O. I., Abhulimen, O. V., & Ogunti, E. O. (2025). Harnessing artificial intelligence and machine learning for fraud detection and prevention in Nigeria. *Journal of Economic Criminology*, *7*, 100127. DOI: 10.1016/j.jeconoc.2025.100127

Ostern, N. K., Holotiuk, F., & Moormann, J. (2022). Organizations' approaches to blockchain: A critical realist perspective. *Information & Management*, *59*(7), 103552. DOI: 10.1016/j.im.2021.103552

Papagiannidis, E., Mikaleff, P., & Conboy, K. (2025). Responsible artificial intelligence governance: A review and research framework. *The Journal of Strategic Information Systems*, *34*(2), 101885. DOI: 10.1016/j.jsis.2024.101885

Parambil, M. M. A., Rustamov, J., Ahmed, S. G., Rustamov, Z., Awad, A. I., Zaki, N., & Alnajjar, F. (2024). Integrating AI-based and conventional cybersecurity measures into online higher education settings: Challenges, opportunities, and prospects. *Computers and Education: Artificial Intelligence*, *7*, 100327. DOI: 10.1016/j.caeai.2024.100327

Park, S. (2019). Why information security law has been ineffective in addressing security vulnerabilities: Evidence from California data breach notifications and relevant court and government records. *International Review of Law and Economics*, *58*, 132–145. DOI: 10.1016/j.irle.2019.03.007

Patil, P., & Sangeetha, M. (2022). Blockchain-based decentralized KYC verification framework for banks. *Procedia Computer Science*, *215*, 529–536. DOI: 10.1016/j.procs.2022.12.055

Patrício, L., Varela, L., & Silveira, Z. (2025). Framework for integrating requirements engineering and DevOps practices in robotic process automation with a focus on optimizing human–computer interaction. *Applied Sciences (Basel, Switzerland)*, *15*(7), 3485. DOI: 10.3390/app15073485

Pattnaik, D., Ray, S., & Raman, R. (2024). Applications of artificial intelligence and machine learning in the financial services industry: A bibliometric review. *Heliyon*, *10*(1), e23492. DOI: 10.1016/j.heliyon.2023.e23492 PMID: 38187262

Qatawneh, A. M., & Lutfi, A. (2024). Effect of artificial intelligence (AI) on financial decision-making: Mediating role of financial technologies (Fin-Tech). *HighTech and Innovation Journal*, *5*(3), 759–773. DOI: 10.28991/HIJ-2024-05-03-015

Robertson, J., Botha, E., Oosthuizen, K., & Montecchi, M. (2025). Managing change when integrating artificial intelligence (AI) into the retail value chain: The AI implementation compass. *Journal of Business Research*, *189*, 115198. DOI: 10.1016/j.jbusres.2025.115198

Sadok, H., Sakka, F., & El Maknouzi, M. E. H. (2022). Artificial intelligence and bank credit analysis: A review. *Cogent Economics & Finance*, *10*(1), 2023262. DOI: 10.1080/23322039.2021.2023262

Saleh, A. M. S. (2024). Blockchain for secure and decentralized artificial intelligence in cybersecurity: A comprehensive review. *Blockchain: Research and Applications*, *5*(3), 100193. DOI: 10.1016/j.bcra.2024.100193

Salila, B. M., Fulgen, M., Nambiar, K., Kemanabally, A., Sobhanakumary, L., & Thomas, B. (2025). Analyzing the academic discourse on RPA in banking: A bibliometric approach. *Multidisciplinary Reviews*, *8*(3), e2025064. DOI: 10.31893/multirev.2025064

Shamsuzzoha, A., & Pelkonen, S. (2025). A robotic process automation model for order-handling optimization in supply chain management. *Supply Chain Analytics*, *9*, 100102. DOI: 10.1016/j.sca.2025.100102

Sonik, N., Gupta, D., & Gupta, P. (2024). Robotic process automation in the financial sector. In Gupta, D., Gupta, M., Dhanaraj, R. K., Balusamy, B., & Gupta, P. M. (Eds.), *Computational intelligence for autonomous finance*. Wiley., DOI: 10.1002/9781394233250.ch5

Sutiene, K., Schwendwener, P., Sipos, C., Lorenzo, L., Mirchev, M., Lameski, P., Kabasinskas, A., Tidjani, C., Ozturkkal, B., & Cerveniciene, J. (2024). Enhancing portfolio management using artificial intelligence: Literature review. *Frontiers in Artificial Intelligence*, *7*, 1371502. DOI: 10.3389/frai.2024.1371502 PMID: 38650961

Syed, R., Suriadi, S., Adams, M., Bandara, W., Leemans, S. J. J., Ouyang, C., ter Hofstede, A. H. M., van de Weerd, I., Wynn, M. T., & Reijers, H. A. (2020). Robotic Process Automation: Contemporary themes and challenges. *Computers in Industry*, *115*, 103162. DOI: 10.1016/j.compind.2019.103162

Tang, W., Cao, H., Ye, S., Yang, L., & Chen, F. (2024). Intelligent financial decision support system based on RPA financial robot and artificial intelligence. In *Proceedings of the 2024 International Conference on Power, Electrical Engineering, Electronics and Control (PEEEC)* (Paper No. 00175). IEEE. https://doi.org/DOI: 10.1109/PEEEC63877.2024.00175

Tian, K., Zhu, Z., Mbachu, J., Ghanbaripour, A., & Moorhead, M. (2025). Artificial intelligence in risk management within the realm of construction projects: A bibliometric analysis and systematic literature review. *Journal of Innovation & Knowledge*, *10*(3), 100711. DOI: 10.1016/j.jik.2025.100711

Treacy, S., Adyanthaya, A., Kearney, C., Anand, J., & O'Sullivan, K. (2023). From hype to reality: Navigating the challenges of RPA implementation. In *Proceedings of the European Conference on Innovation and Entrepreneurship (ECIE)*. Cork University Business School, University College, Ireland.

Tripathi, G., Ahad, M. A., & Casalino, G. (2023). A comprehensive review of blockchain technology: Underlying principles and historical background with future challenges. *Decision Analytics Journal*, *9*, 100344. DOI: 10.1016/j.dajour.2023.100344

Vuković, D. B., Dekpo-Adza, S., & Matović, S. (2025). AI integration in financial services: A systematic review of trends and regulatory challenges. *Humanities & Social Sciences Communications*, *12*, 562. DOI: 10.1057/s41599-025-04850-8

Wang, X. (2022). Research on payment settlement mode in cross-border business trade based on blockchain technology. *SAIEE Africa Research Journal*, *113*(3), 129–132. DOI: 10.23919/SAIEE.2022.9853022

Wei, Z., Fang, J., Hong, Z., Zhou, Y., Ma, S., Zhang, J., Liang, C., Zhao, G., & Tang, H. (2024). Privacy protection method for blockchain transactions based on the stealth address and the note mechanism. *Applied Sciences (Basel, Switzerland), 14*(4), 1642. DOI: 10.3390/app14041642

Wieringa, J., Kannan, P. K., Ma, X., Reutterer, T., Risselada, H., & Skiera, B. (2021). Data analytics in a privacy-concerned world. *Journal of Business Research, 122*, 915–925. DOI: 10.1016/j.jbusres.2019.05.005

Xiao, J., Xu, Z., Xiao, A., Wang, X., & Skare, M. (2024). Overcoming barriers and seizing opportunities in the innovative adoption of next-generation digital technologies. *Journal of Innovation & Knowledge, 9*(4), 100622. DOI: 10.1016/j.jik.2024.100622

Zahidi, F., Kaluvilla, B. B., & Mulla, T. (2024). Embracing the new era: Artificial intelligence and its multifaceted impact on the hospitality industry. *Journal of Open Innovation, 10*(4), 100390. DOI: 10.1016/j.joitmc.2024.100390

Zantalis, F., Koulouras, G., & Karabetsos, S. (2023). Blockchain technology: A framework for endless applications. *IEEE Consumer Electronics Magazine, 13*(2), 61–71. DOI: 10.1109/MCE.2023.3248872

Zheng, K., Zheng, L. J., Gauthier, J., Zhou, L., Xu, Y., Behl, A., & Zhang, J. Z. (2022). Blockchain technology for enterprise credit information sharing in supply chain finance. *Journal of Innovation & Knowledge, 7*(4), 100256. DOI: 10.1016/j.jik.2022.100256

Zheng, X., Zhu, Y., & Si, X. (2019). A survey on challenges and progresses in blockchain technologies: A performance and security perspective. *Applied Sciences (Basel, Switzerland), 9*(22), 4731. DOI: 10.3390/app9224731

Zhu, H., Sallnäs Pysander, E.-L., & Söderberg, I.-L. (2023). Not transparent and incomprehensible: A qualitative user study of an AI-empowered financial advisory system. *Data and Information Management, 7*(3), 100041. DOI: 10.1016/j.dim.2023.100041

Zhu, H., Vigren, O., & Söderberg, I.-L. (2024). Implementing artificial intelligence empowered financial advisory services: A literature review and critical research agenda. *Journal of Business Research, 174*, 114494. DOI: 10.1016/j.jbusres.2023.114494

Zirar, A., Ali, S. I., & Islam, N. (2023). Worker and workplace Artificial Intelligence (AI) coexistence: Emerging themes and research agenda. *Technovation, 124*, 102747. DOI: 10.1016/j.technovation.2023.102747

1 The discussion this section is referred to website: https://superiordatascience.com/jp-morgan-coin-a-case-study-of-ai-in-finance/

2 The discussion this section is referred to website: https://www.cnbc.com/2018/04/12/santander-launches-blockchain-based-foreign-exchange-using-ripple-tech.html

3 The discussion this section is referred to website: https://www.ing.com/Newsroom/News/ING-founded-blockchain-trade-finance-platform-launches-in-Singapore.htm

4 The discussion this section is referred to website: https://am.gs.com/en-us/advisors/insights/article/2024/harnessing-the-power-of-ai-to-enhance-investment-decision-making

5 The discussion this section is referred to website: https://www.maybank-am.com.my/w/maybank-asset-management-launches-ai-powered-shariah-compliant-discretionary-portfolio-mandate-with-arabesque-ai-1

Chapter 15

Optimizing Workflow Through Robotic Process Automation:

A Strategic Lever for Digital Business Process Transformation

Mohit Sharma
https://orcid.org/0009-0007-2280-8077
Maharshi Dayanand University, India

Nitika Malik
https://orcid.org/0009-0006-3991-9526
Maharshi Dayanand University, India

ABSTRACT

Robotic Process Automation (RPA) is increasingly becoming a pivotal technology in digital transformation strategies, enabling businesses to automate repetitive, rule-based tasks with minimal human intervention. This chapter explores how RPA serves as a catalyst for optimizing workflows and enhancing the effectiveness of Business Process Management (BPM) in the digital age. By integrating RPA within BPM frameworks, organizations can achieve faster processing times, reduced errors, and improved compliance. The chapter presents a comprehensive discussion on the strategic role of RPA, a synthesis of key implementation models, and case-based insights from various industries. It also critically evaluates the challenges such as change resistance, process fragmentation, and governance risks.

DOI: 10.4018/979-8-3373-3571-1.ch015

INTRODUCTION

In today's era of digital transformation, organizations are constantly exploring innovative approaches to enhance operational efficiency, lower costs, and improve organizational agility. Among the most impactful technologies driving this change is Robotic Process Automation (RPA). RPA refers to the deployment of software robots, or "bots," to perform high-volume, rule-based, and repetitive tasks that would otherwise require human effort (Aguirre & Rodriguez, 2017). By replicating human actions and seamlessly interacting with digital systems, RPA enables enterprises to streamline workflows, improve process accuracy, and increase throughput with minimal disruption to existing IT infrastructures (Willcocks, Lacity, & Craig, 2015). The business environment is experiencing a profound transformation fueled by the convergence of emerging technologies, including artificial intelligence (AI), machine learning (ML), big data analytics, and cloud computing. In this evolving landscape, organizations increasingly prioritize agility, innovation, and the ability to respond quickly to shifting market conditions (Brock & von Wangenheim, 2019). Digital transformation, in this sense, entails the comprehensive integration of digital technologies across all business functions, fundamentally reshaping how organizations operate and deliver value to their customers (Fitzgerald et al., 2014). Within this framework, RPA serves as a pivotal enabler, providing the operational infrastructure necessary to automate complex business processes and delivering significant improvements in speed, accuracy, and scalability.

The Emergence of RPA in Digital Strategy

Traditionally, business process optimization has relied heavily on Business Process Management (BPM), which focuses on analyzing, designing, implementing, and monitoring business processes to improve organizational performance (Dumas et al., 2018). However, conventional BPM approaches often fall short in rapidly changing environments due to their rigidity and dependence on manual interventions. This gap has created an opportunity for RPA to enhance BPM by automating repetitive tasks and enabling continuous improvement without overhauling entire systems (van der Aalst, 2016). RPA serves as a strategic lever for digital business transformation by automating routine and repetitive processes across multiple functions such as finance, human resources, supply chain, and customer service (Syed et al., 2020). Unlike traditional automation tools that require deep integration into core systems, RPA operates at the user interface level, thereby reducing the need for extensive reengineering. This non-invasive characteristic makes RPA an attractive solution for organizations looking to achieve quick wins in their digital transformation journey (Asatiani & Penttinen, 2016). Moreover, RPA aligns well with the principles of lean management by eliminating waste, minimizing errors, and enhancing process transparency. The deployment of bots can significantly reduce the burden on employees, allowing them to focus on higher-value tasks such as problem-solving, innovation, and customer engagement (Willcocks et al., 2017). Consequently, RPA not only improves operational efficiency but also contributes to employee satisfaction and customer experience, both of which are critical components of digital strategy.

Strategic Integration with BPM

The integration of RPA within BPM frameworks creates a symbiotic relationship that enhances overall process effectiveness. While BPM provides a structured methodology for process improvement, RPA offers the execution capability that brings these improvements to life (Lacity & Willcocks, 2018). This fusion enables organizations to move from process modeling and analysis to real-time automation and optimization, thereby accelerating the digital transformation lifecycle. For instance, in financial services, RPA has been effectively used to automate processes such as account reconciliation, loan processing, and fraud detection, leading to significant cost savings and improved compliance (Syed et al., 2020). In the healthcare sector, RPA streamlines administrative functions such as patient scheduling, billing, and insurance claims processing, allowing medical professionals to devote more time to patient care (Brougham & Haar, 2018). Similarly, in the logistics and supply chain industry, RPA automates inventory tracking, order management, and shipment notifications, resulting in better visibility and responsiveness (Hofmann, Samp, & Urbach, 2020). These real-world applications underscore the transformative potential of RPA when strategically integrated with BPM. However, the successful adoption of RPA requires a clear understanding of the process landscape, including the identification of suitable candidates for automation, process standardization, and alignment with business goals (Aguirre & Rodriguez, 2017). Organizations must also invest in building the right governance structures, change management strategies, and talent capabilities to ensure sustainable value from RPA initiatives.

Implementation Models and Success Factors

Several implementation models have emerged to guide organizations in their RPA journeys. The most common approach involves a phased rollout, starting with pilot projects in specific departments, followed by scaling across the enterprise (Willcocks et al., 2015). This model allows organizations to test the technology, measure its impact, and refine their strategies before full-scale deployment. Another model is the center of excellence (CoE) approach, where a dedicated team is responsible for developing RPA capabilities, setting standards, and disseminating best practices across business units (Lacity & Willcocks, 2018). Regardless of the implementation model, certain critical success factors must be considered. These include executive sponsorship, process suitability assessment, stakeholder engagement, and robust change management. Additionally, it is essential to ensure that RPA initiatives are aligned with the broader digital strategy and enterprise architecture. Without this alignment, RPA may become a siloed initiative that fails to deliver long-term value (Syed et al., 2020). Organizations must also anticipate and address potential challenges such as process fragmentation, where disconnected or poorly documented processes hinder automation efforts. Governance risks related to data privacy, compliance, and bot maintenance must be proactively managed through clear policies, monitoring mechanisms, and audit trails (van der Aalst, 2016). Moreover, resistance to change is a common barrier, particularly among employees who fear job displacement. Effective communication, reskilling programs, and involving employees in the automation process can help mitigate these concerns (Brougham & Haar, 2018).

RPA and the Future of Work

As RPA continues to evolve, its integration with cognitive technologies such as natural language processing (NLP), optical character recognition (OCR), and machine learning is giving rise to intelligent automation. This next phase of automation enables bots to handle unstructured data, make decisions, and learn from historical patterns, thereby expanding the scope of automation to more complex processes (Syed et al., 2020). The convergence of RPA and AI is redefining the future of work by creating hybrid workforces where humans and digital workers collaborate to deliver superior outcomes. However, this shift also necessitates a rethinking of organizational roles, skills, and culture. Employees must be equipped with digital literacy, analytical thinking, and problem-solving capabilities to thrive in an environment augmented by automation (Brock & von Wangenheim, 2019). Leadership must foster a culture of innovation, experimentation, and continuous learning to harness the full potential of RPA. From a strategic perspective, RPA is no longer just a tool for cost reduction but a driver of competitive advantage. It enables organizations to be more responsive, customer-centric, and agile in the face of disruption. In an era marked by economic uncertainty, regulatory complexity, and technological disruption, RPA offers a pathway to resilience and scalability (Lacity & Willcocks, 2018). Organizations that proactively embrace RPA as part of their digital transformation strategy are better positioned to navigate change, capitalize on opportunities, and achieve sustainable growth.

In summary, Robotic Process Automation represents a strategic lever for digital business process transformation. It enhances the execution of BPM initiatives by automating routine tasks, improving process efficiency, and enabling rapid innovation. Through effective integration, governance, and change management, RPA can drive significant value across industries and functions. This chapter aims to provide a comprehensive understanding of how RPA can be harnessed to optimize workflows, overcome implementation challenges, and shape the future of work in the digital era. By synthesizing theoretical insights and practical applications, it offers a roadmap for leveraging RPA as a catalyst for organizational transformation.

BACKGROUND OF THE STUDY

The dawn of the digital age has ushered in a paradigm shift in how businesses operate, compelling organizations to seek technologies that enhance efficiency, reduce costs, and deliver superior customer value. One such transformative technology is Robotic Process Automation (RPA), a software-based approach that uses bots to automate high-volume, repeatable tasks traditionally performed by humans (van der Aalst et al., 2016). Unlike traditional automation tools that require significant reengineering of IT systems, RPA integrates seamlessly with existing infrastructure, enabling rapid deployment and minimal disruption. As companies strive to maintain competitiveness and agility in an increasingly digital and globalized marketplace, RPA has emerged as a crucial lever for Business Process Transformation (BPT) and Business Process Management (BPM). The global business environment has witnessed a significant acceleration in digital transformation, especially after the COVID-19 pandemic, c (Brock & Wangenheim, 2019). Organizations have increasingly turned to digital technologies such as Artificial Intelligence (AI), Machine Learning (ML), and RPA to optimize operational workflows, enhance customer experiences, and ensure business continuity. Among these, RPA has garnered attention for its low-code or no-code architecture, cost-effectiveness, and capability to operate round-the-clock with high accuracy (Syed et al., 2020). By mimicking human actions in digital systems, RPA allows organizations to streamline back-office processes, such as invoicing, payroll, data entry, and customer onboarding, thereby creating room for employees to focus on higher-value, strategic activities (Lacity & Willcocks, 2016).

Evolution of Business Process Management and Digital Transformation

To fully appreciate the strategic relevance of RPA, it is vital to contextualize it within the broader framework of Business Process Management (BPM). BPM refers to a systematic approach to improving organizational processes through design, modeling, execution, monitoring, and optimization (Dumas et al., 2018). It has traditionally relied on manual analysis and gradual improvements. However, the digital era has introduced complexities—such as volatile market demands, shorter product life cycles, and the proliferation of data—that necessitate more dynamic and responsive process frameworks. Digital transformation reshapes BPM by incorporating technologies that allow for real-time data analytics, intelligent automation, and continuous process improvement (Jöhnk et al., 2021). Within this context, RPA has become an enabler of digital BPM by offering scalable solutions to automate tasks without the need for deep system integration or costly IT overhauls. It supports BPM in achieving faster throughput, enhancing accuracy, and improving regulatory compliance. RPA's strategic significance lies not only in cost reduction but also in enabling organizations to become more agile and responsive to changing environments. In sectors such as banking, healthcare, retail, and manufacturing, RPA has transformed transactional workflows, reduced process cycle times, and mitigated human error (Moffitt et al., 2018). For instance, banks use RPA to automate loan processing and compliance reporting, while healthcare institutions deploy bots for patient scheduling and medical billing.

Strategic Role of RPA in Business Process Transformation

Business Process Transformation (BPT) involves a fundamental rethinking and redesign of business processes to achieve dramatic improvements in critical performance measures such as cost, quality, service, and speed (Hammer & Champy, 1993). While BPT has historically been driven by management initiatives and enterprise system implementations, the advent of RPA has added a new dimension to this transformation. RPA enables incremental and agile transformation by automating discrete tasks within a process, serving as a bridge between legacy systems and modern digital infrastructures (Aguirre & Rodriguez, 2017). From a strategic standpoint, RPA contributes to operational efficiency, workforce optimization, and enhanced customer experience. Moreover, RPA provides actionable insights through process mining and analytics, thereby enabling continuous improvement and more informed decision-making. For example, integrating RPA with AI can lead to Intelligent Process Automation (IPA), where bots not only execute tasks but also learn from data patterns to make decisions (Davenport & Ronanki, 2018). Such capabilities are central to digital transformation strategies that aim for adaptive, customer-centric, and scalable operations. The integration of RPA within BPM frameworks also promotes organizational alignment, as it aligns operational processes with strategic goals through measurable performance metrics. Companies that have effectively deployed RPA report benefits such as reduced operational costs (up to 30–50%), improved accuracy (with error rates dropping close to zero), and enhanced compliance with regulatory standards (Willcocks et al., 2015). These outcomes significantly contribute to achieving digital transformation goals.

Implementation Models and Case-Based Insights

Despite its benefits, the implementation of RPA is not devoid of challenges. Effective adoption requires a clear roadmap, process suitability analysis, stakeholder alignment, and change management. Common implementation models include the centralized model (where a Center of Excellence oversees RPA deployment across departments), the decentralized model (where departments manage their bots independently), and the hybrid model (a blend of the first two). Each model has its merits and trade-offs, depending on organizational maturity, scale, and digital strategy (Asatiani & Penttinen, 2016). Real-world case studies further elucidate RPA's strategic impact. For instance, a multinational insurance company implemented RPA to automate policy renewals, resulting in a 70% reduction in processing time and an 80% decrease in errors (Lacity & Willcocks, 2016). Similarly, a retail giant used RPA to streamline its supply chain processes, achieving faster delivery cycles and better inventory management. These cases demonstrate how RPA can unlock value across functions such as finance, HR, supply chain, and customer service.

Challenges and Considerations

While RPA offers significant advantages, organizations must navigate several implementation challenges. One of the primary issues is **resistance to change**, especially from employees who fear job displacement. Effective communication and workforce reskilling are essential to mitigate these concerns and foster a culture of innovation (Madakam et al., 2019). Additionally, **process fragmentation**—where workflows span multiple systems and teams—can limit the scalability of RPA unless end-to-end process visibility is ensured.

Another key challenge involves **governance and risk management**. As RPA scales across the organization, maintaining control over bot behavior, data security, and compliance becomes increasingly complex (Syed et al., 2020). Therefore, organizations must implement robust governance frameworks that define bot ownership, access rights, audit trails, and performance metrics. Furthermore, RPA may encounter limitations in **cognitive processing and decision-making**, particularly in unstructured environments. To overcome these limitations, RPA can be integrated with AI technologies such as natural language processing (NLP), optical character recognition (OCR), and machine learning, thereby enhancing its ability to handle more complex tasks (Davenport & Ronanki, 2018).

Bridging Theory and Practice

A significant gap remains between academic literature on automation and its practical applications. While theoretical models often emphasize the technological aspects of RPA, real-world implementations involve organizational dynamics, stakeholder management, and strategic alignment. This chapter aims to **bridge that gap** by combining theoretical frameworks with practical insights, offering a holistic view of how RPA can drive digital business process transformation. The academic discourse on BPM, digital transformation, and RPA is still evolving, with limited consensus on best practices, success metrics, and implementation roadmaps. This study contributes to this discourse by synthesizing interdisciplinary knowledge from information systems, strategic management, and organizational behavior. It also offers a **framework for RPA deployment** that considers organizational readiness, process suitability, change management, and continuous improvement. In conclusion, the background of this study underscores the transformative potential of RPA as a strategic lever in the broader journey of digital business process transformation. As organizations confront unprecedented challenges and opportunities in the digital economy, RPA offers a pragmatic, scalable, and cost-effective tool to optimize workflows, improve efficiency, and enhance competitiveness. However, realizing these benefits requires a thoughtful approach that aligns technology with organizational strategy, process design, and human capital. By addressing both the opportunities and challenges of RPA, this chapter seeks to provide actionable insights for academics, practitioners, and policymakers striving to harness the full potential of automation in the digital era.

RESEARCH METHODOLOGY

This study adopts a qualitative research methodology to explore the nuanced dimensions of consumer trust, data security, and ethical concerns within Internet of Things (IoT) markets. The rationale behind choosing a qualitative approach lies in the complexity and context-specific nature of trust and ethical considerations, which cannot be fully captured through quantitative means alone (Creswell & Poth, 2018). This method allows for an in-depth understanding of consumers' perceptions, behavioral intentions, and the broader socio-technical implications of IoT technologies (Malik et al., 2025a; Malik et al., 2025b; Selvakumar et al., 2025a; Selvakumar et al., 2025b; Selvakumar et al., 2025c). The research design follows a multiple case study approach, focusing on three sectors where IoT adoption is prominent: smart home devices, wearable health technologies, and connected automobiles. These cases were selected due to their high consumer interaction and varying levels of risk and trust implications (Weber, 2010). Each case involves analysis of company practices, consumer reviews, and media reports to identify recurring themes related to trust, security, and ethics. Primary data was collected through semi-structured interviews with 20 IoT users and 10 industry professionals, including data protection officers, product managers, and cybersecurity experts. Interviews were conducted via video conferencing and lasted between 30–45 minutes. This allowed participants to share their experiences, concerns, and insights about data collection practices, consent mechanisms, and perceived ethical dilemmas. All interviews were transcribed and analyzed using thematic analysis (Braun & Clarke, 2006), identifying common patterns and divergent views across stakeholder groups. In addition to interviews, secondary data sources such as white papers, privacy policies, academic journals, and regulatory documents (e.g., GDPR, CCPA) were reviewed to contextualize findings and support triangulation. This mixed-source analysis strengthens the validity and reliability of the results (Yin, 2017). Ethical clearance was obtained prior to data collection, and participants provided informed consent. Anonymity and confidentiality were strictly maintained, and all data were stored securely in compliance with data protection regulations. Overall, this methodology enables a comprehensive exploration of how consumer trust is influenced by security practices and ethical standards in IoT environments, contributing valuable insights for policymakers, designers, and marketers navigating this emerging digital ecosystem.

FINDINGS OF THE STUDY

Robotic Process Automation (RPA) has emerged as a transformative force in the digital age, reshaping workflows, streamlining operations, and driving significant improvements in business process management (BPM). The findings of this study provide an in-depth exploration of the multifaceted impact of RPA across industries, supported by theoretical insights and practical case studies. This section delves into the benefits, strategic roles, implementation frameworks, industry applications, challenges, and future directions of RPA, offering a comprehensive understanding of its role in digital business process transformation.

1. Benefits of RPA in Workflow Optimization

RPA delivers significant benefits that contribute to optimizing workflows, enhancing productivity, and ensuring operational resilience. These include:

1.1 Cost Efficiency

One of the most compelling advantages of RPA is its ability to reduce operational costs. By automating repetitive and labor-intensive tasks, organizations can achieve significant savings. According to Willcocks et al. (2015), organizations implementing RPA reported cost reductions of up to 60%, particularly in high-volume back-office processes such as data entry and invoice processing.

1.2 Enhanced Productivity

RPA bots operate 24/7 without breaks, enabling faster task completion and freeing human workers to focus on strategic and creative tasks. For example, a financial services firm employing RPA to process loan applications reduced processing times by 80%, leading to improved customer satisfaction (Deloitte, 2020).

1.3 Improved Accuracy and Compliance

Automation minimizes human error, ensuring higher accuracy in task execution. In the healthcare sector, for instance, RPA has been used to manage patient records with near-perfect accuracy, reducing the risk of errors in medical documentation (KPMG, 2021). Furthermore, RPA systems can be configured to adhere to regulatory standards, mitigating compliance risks.

1.4 Scalability and Flexibility

RPA systems are highly scalable, allowing businesses to adjust bot capacity based on workload fluctuations. During peak seasons, e-commerce platforms such as Amazon have leveraged RPA to handle increased order volumes seamlessly (McKinsey, 2020).

1.5 Enhanced Customer Experience

By accelerating service delivery and reducing processing times, RPA contributes to improved customer experiences. In the telecom industry, companies using RPA for customer onboarding have significantly reduced onboarding times, leading to higher customer retention rates (PwC, 2019).

2. STRATEGIC ROLE OF RPA IN DIGITAL TRANSFORMATION

RPA is a cornerstone of digital transformation strategies, enabling organizations to:

2.1 Integrate Disparate Systems

RPA acts as a bridge between legacy systems and modern applications, facilitating seamless data integration without extensive reengineering. This capability is particularly beneficial in industries such as banking, where legacy systems remain prevalent (Capgemini, 2021).

2.2 Enable Data-Driven Decision Making

RPA accelerates data collection and processing, providing organizations with real-time insights for informed decision-making. For instance, retail companies utilizing RPA for inventory management gain timely insights into stock levels, enabling more efficient replenishment strategies (Forrester, 2022).

2.3 Drive Innovation

By automating mundane tasks, RPA frees up human resources for innovation and strategic initiatives. In a case study by Gartner (2021), an insurance company used RPA to automate claims processing, allowing its workforce to focus on product development and customer engagement.

3. IMPLEMENTATION FRAMEWORKS FOR RPA

Successful RPA implementation requires a structured approach to maximize its benefits. Key stages include:

3.1 Process Identification and Prioritization

Identifying suitable processes for automation is critical. Ideal candidates are high-volume, rule-based, and repetitive tasks. A manufacturing firm, for example, automated inventory audits, resulting in 70% time savings (EY, 2020).

3.2 Development and Testing

Developing RPA bots involves configuring them to perform designated tasks and conducting rigorous testing to ensure accuracy. In a banking case, RPA bots were tested extensively to handle exceptions in loan approvals, ensuring robustness (Accenture, 2020).

3.3 Deployment and Monitoring

Post-deployment, continuous monitoring and periodic updates are essential to maintain performance. A logistics company used real-time dashboards to monitor RPA bot efficiency, identifying areas for improvement (BCG, 2021).

3.4 Governance and Compliance

Establishing governance frameworks ensures alignment with organizational policies and regulatory requirements. In the pharmaceutical sector, governance frameworks for RPA have been critical in maintaining data integrity and compliance (FDA, 2021).

4. INDUSTRY APPLICATIONS OF RPA

4.1 Healthcare

RPA has revolutionized healthcare operations by automating patient record management, appointment scheduling, and billing. For example, a hospital in the United States reduced patient admission times by 50% using RPA, enhancing patient care (HIMSS, 2020).

4.2 Finance and Accounting

In finance, RPA automates tasks such as invoice processing, reconciliations, and compliance reporting. A multinational bank achieved a 90% reduction in manual errors in its reconciliation processes after implementing RPA (PwC, 2021).

4.3 Manufacturing

Manufacturers use RPA for inventory management, order processing, and quality control. For instance, a global automotive company improved production efficiency by automating parts tracking (Frost & Sullivan, 2022).

4.4 Human Resources

RPA streamlines HR processes such as recruitment, onboarding, and payroll. A tech company reduced its recruitment cycle by 30% after deploying RPA for candidate screening (SHRM, 2021).

4.5 Retail and E-Commerce

Retailers leverage RPA to manage inventory, process returns, and enhance customer engagement. An online retailer automated order processing, achieving a 95% reduction in processing time (Deloitte, 2020).

5. CHALLENGES IN RPA IMPLEMENTATION

Despite its benefits, RPA implementation presents several challenges:

5.1 Change Resistance

Employees often fear job displacement due to automation, leading to resistance. Addressing these concerns through effective communication and training is essential (Kotter, 1996).

5.2 Process Fragmentation

Automating poorly defined or fragmented processes can result in inefficiencies. Organizations must conduct thorough process analyses before automation (Bain & Company, 2020).

5.3 Technical Limitations

RPA bots may struggle with unstructured data or complex decision-making tasks, necessitating the integration of AI. For example, incorporating AI enabled a logistics company to automate exception handling in delivery processes (IBM, 2021).

5.4 Governance Risks

Inadequate governance can lead to compliance violations and operational risks. Establishing robust governance frameworks is critical for mitigating these risks (ISO, 2020).

6. FUTURE DIRECTIONS AND TRENDS

The evolution of RPA is driven by emerging technologies and industry needs:

6.1 Intelligent Automation

Integrating AI with RPA enables intelligent automation, allowing bots to handle complex tasks and make data-driven decisions. For example, AI-enabled RPA in healthcare can analyze patient data to recommend treatment plans (Accenture, 2021).

6.2 Hyperautomation

Hyperautomation involves the use of multiple automation tools, including RPA, AI, and machine learning, to automate entire workflows. This trend is expected to dominate digital transformation strategies in the coming years (Gartner, 2022).

6.3 Focus on Employee Upskilling

Organizations are investing in upskilling employees to work alongside RPA, fostering a collaborative human-automation workforce (World Economic Forum, 2021).

6.4 Sustainability Integration

RPA is increasingly being leveraged to support sustainability initiatives, such as reducing paper-based processes and optimizing resource usage (UNCTAD, 2021)

7. SECTOR-WISE IMPLEMENTATION INSIGHTS OF RPA

The practical value of Robotic Process Automation (RPA) becomes most apparent when examined across different sectors. Various industries have adopted RPA not merely as a cost-saving mechanism but as a broader tool for transforming workflows and digital operations. The following sub-sections provide key findings from sectoral studies and implementations.

7.1 Financial Services Sector: Enhancing Accuracy and Compliance

In the financial services industry, RPA has shown significant improvements in compliance, transaction processing, and customer servicing. For example, a leading European bank implemented RPA to automate Know Your Customer (KYC) and Anti-Money Laundering (AML) checks. This resulted in a reduction in processing time by over 60% and an improvement in data accuracy by 85%, according to a Deloitte case study (Deloitte, 2019). Lacity and Willcocks (2016) further support this by documenting RPA's ability to cut back-office costs by up to **50%**, especially in areas like claims processing, payment reconciliation, and audit reporting. Robotic automation ensures compliance by logging every step, reducing the likelihood of regulatory breaches (Moffitt, Rozario, & Vasarhelyi, 2018).

7.2 Healthcare Sector: Reducing Administrative Overhead

Hospitals and insurance providers are increasingly relying on RPA to streamline administrative and billing operations. One U.S.-based healthcare provider used bots to extract data from incoming medical claims and enter them into Electronic Health Records (EHR) systems, resulting in a 30% improvement in patient onboarding time and increased employee satisfaction due to the reduced manual load (Ernst & Young, 2019). Moreover, RPA has assisted in ensuring compliance with the Health Insurance Portability and Accountability Act (HIPAA) by minimizing human exposure to sensitive patient data, thus reducing privacy violations (Syed et al., 2020).

7.3 Manufacturing: Synchronizing Supply Chains

RPA adoption in the manufacturing industry has targeted areas such as inventory management, invoice processing, and production scheduling. Siemens, for example, implemented an RPA system to automate its procure-to-pay operations. This initiative freed up over 60,000 man-hours annually, translating into cost savings and improved turnaround times (Fersht & Snowdon, 2017). In addition to cost savings, manufacturers benefit from higher production accuracy and improved delivery timelines, enabling greater resilience in supply chain operations (Geyer-Klingeberg et al., 2018).

7.4 Retail and E-Commerce: Enhancing Customer Experience

Retailers such as Walmart and Amazon have used RPA to manage price matching, inventory updates, and order tracking. Automation in these areas ensures real-time updates for customers and significantly reduces cart abandonment rates (Asatiani & Penttinen, 2016). Amazon's implementation of intelligent bots in warehouse management has resulted in a 20% improvement in product picking accuracy, directly contributing to better customer satisfaction metrics (PwC, 2020).

7.5 Public Sector: Improving Citizen Services

Governments across the globe are now using RPA to improve efficiency in departments like taxation, licensing, and citizen grievance handling. The UK's HM Revenue & Customs (HMRC) implemented over 80 bots to process tax records and customer emails, saving nearly £10 million annually and reducing the average response time by 40% (Dery, Sebastian, & van der Meulen, 2017).

This demonstrates how RPA can help public institutions move away from legacy systems, streamline outdated processes, and deliver faster citizen services.

8. COMPARATIVE ANALYSIS: RPA VS. TRADITIONAL AUTOMATION

While RPA is often confused with traditional automation, findings from comparative studies highlight significant differences in scope, scalability, and adaptability. RPA offers the distinct advantage of non-invasive implementation, meaning it can overlay existing IT systems without extensive integration or changes. This reduces both time to market and initial investment, making it ideal for agile business environments (Chugh & Grandhi, 2021).

9. QUANTITATIVE IMPACT: PERFORMANCE METRICS POST-RPA IMPLEMENTATION

Multiple studies have provided empirical evidence of RPA's performance metrics across domains:

- **Reduction in processing time:** 40%–70%
- **Error rate reduction:** Up to 90%
- **Employee productivity improvement:** 35%–50%

- **Cost savings:** 25%–60% (depending on scale)
- **Customer satisfaction uplift:** 20% on average (Zion Market Research, 2022)

For instance, a large insurance firm that deployed 180 bots across its claims processing function observed:

- An annual saving of **$20 million**
- A decrease in average claim resolution time from **16 days to 5 days**
- **Zero SLA violations** over a 12-month post-implementation review (Ernst & Young, 2019)

Such measurable impacts underscore the **tangible business value of RPA** and validate it as a strategic lever for process transformation.

10. HUMAN-BOT COLLABORATION AND WORKFORCE EVOLUTION

One key finding is that RPA doesn't necessarily eliminate jobs—instead, it transforms them. Bots take over routine tasks, allowing human workers to focus on decision-making, analysis, and innovation (Willcocks, Lacity, & Craig, 2017). This "human-bot collaboration" model is particularly evident in financial services. For example, at Xchanging (a UK-based business processing firm), bots handled 120 tasks across 14 functions, while employees were upskilled to manage exceptions and deliver value-added insights (Lacity, Willcocks, & Craig, 2015). The result was not only improved efficiency but higher employee engagement and satisfaction, due to the elimination of monotonous work.

11. REAL-WORLD CASE STUDY SUMMARIES

11.1 Case Study: Deutsche Bank

Deutsche Bank implemented RPA to automate over **1.5 million manual transactions annually**. The initiative resulted in:

- Cost savings of €30 million over 2 years
- Compliance reporting accuracy improved by 92%
- Cycle time for loan processing reduced from 2 days to 2 hours (PwC, 2020)

11.2 Case Study: NHS (UK)

The National Health Service (NHS) used RPA for **staff onboarding, procurement, and appointment scheduling**. Key findings included:

- Reduction in manual effort by 40%
- Decrease in appointment delays by 25%
- Increased efficiency in internal reporting systems (Syed et al., 2020)

11.3 Case Study: Telefónica

Telefónica automated its billing and dispute resolution processes across its Latin America operations. The RPA initiative:

- Handled over 2 million transactions monthly
- Cut down customer service response time by 35%
- Saved over 900,000 hours of manual labor annually (Agostinelli, Marrella, & Mecella, 2020)

12. AI AND COGNITIVE AUTOMATION: CONVERGENCE WITH RPA

One of the most significant advancements in the evolution of Robotic Process Automation (RPA) is its convergence with Artificial Intelligence (AI) and cognitive automation technologies. While traditional RPA excels in rule-based, repetitive tasks, its capabilities are limited when dealing with unstructured data or processes requiring judgment and decision-making. The integration of AI—through natural language processing (NLP), machine learning (ML), and computer vision—extends RPA into what is commonly referred to as Intelligent Automation (IA) (Deloitte, 2019; Lacity & Willcocks, 2016). AI-powered bots can now process invoices by extracting relevant fields using OCR, categorize emails based on sentiment, and even make basic decisions using predictive algorithms. This synergy allows for the automation of more complex, end-to-end workflows that were previously outside the scope of traditional RPA tools (Geyer-Klingeberg et al., 2018). For instance, AI-enhanced RPA can automate customer onboarding by extracting data from forms, cross-referencing databases for validation, and even initiating workflows based on content analysis (PwC, 2020). This transformation is already visible in industries such as banking, healthcare, and insurance. In the insurance sector, for instance, cognitive RPA is being used to assess claims by analyzing claim documents, determining fraud probability, and recommending settlements (Syed et al., 2020). This leads to higher accuracy, faster resolutions, and better compliance, especially in highly regulated industries. However, the convergence also brings technical and ethical concerns. Training AI models requires large volumes of quality data, and there's an inherent risk of bias, lack of transparency, and explainability (Moffitt et al., 2018). Organizations must therefore establish robust governance frameworks that include transparency standards, human oversight, and responsible AI protocols to ensure ethical deployment.

13. GOVERNANCE, CHANGE MANAGEMENT, AND ORGANIZATIONAL READINESS

Implementing RPA is not purely a technical endeavor—it demands a strong governance structure and strategic change management initiatives. One of the recurring themes observed in RPA implementations is that poor governance results in bot sprawl, redundancy, and lack of control, undermining long-term scalability (Lacity et al., 2015). As the number of bots grows, so does the need for centralized orchestration, standardization of deployment practices, and ongoing maintenance. Governance involves defining clear roles and responsibilities, establishing a Center of Excellence (CoE), and implementing lifecycle management practices for bots—from development to retirement. A CoE functions as a hub for best practices, compliance, risk mitigation, and training, ensuring consistency and quality across RPA initiatives (Willcocks et al., 2017). Without a CoE, organizations often struggle with inconsistent ROI and suboptimal performance. Change management is equally critical. Employee resistance, fear of job loss, and lack of awareness about automation's value are common challenges. Studies show that transparent communication, stakeholder engagement, and retraining programs are effective in mitigating resistance (Asatiani & Penttinen, 2016). Organizations that successfully implement RPA prioritize cross-functional collaboration, involving IT, business units, and HR in the change journey. Moreover, organizational readiness determines how well a company can adopt and scale RPA. This includes digital maturity, existing IT infrastructure, leadership support, and a culture of innovation. Firms with agile structures and a data-driven mindset tend to derive greater value from RPA implementations (Dery et al., 2017).

14. CROSS-INDUSTRY CASE ANALYSIS: PATTERNS AND DIFFERENTIATORS

A cross-industry analysis of RPA implementation reveals both universal patterns and sector-specific differentiators:

- **Banking and Financial Services**: This sector is a front-runner in RPA adoption. Use cases include KYC verification, loan processing, account opening, and compliance reporting. These processes are highly rule-based and regulatory-intensive, making them ideal candidates for RPA (Lacity & Willcocks, 2016).
- **Healthcare**: RPA is used for claims management, appointment scheduling, billing, and regulatory compliance. Healthcare institutions particularly benefit from improved data accuracy and patient satisfaction (Ernst & Young, 2019).
- **Retail and E-commerce**: Automation supports inventory management, order processing, and customer service. RPA has enabled 24/7 customer response systems and better personalization via integration with CRM systems (PwC, 2020).
- **Manufacturing**: In manufacturing, RPA complements ERP systems by automating data entry, supply chain processes, and quality checks. Combined with AI, it supports predictive maintenance by analyzing machine logs (Zion Market Research, 2022).
- **Public Sector**: Government agencies use RPA for tax processing, benefits administration, and citizen service requests. Benefits include improved response times and greater transparency (Fersht & Snowdon, 2017).

Across all these industries, key success factors include aligning automation with strategic goals, strong leadership support, and incremental deployment through pilot testing and iterative scaling.

15. FUTURE TRENDS: STRATEGIC ROADMAP FOR INTELLIGENT AUTOMATION

The future of RPA lies in its evolution into a strategic component of the broader intelligent automation ecosystem. Several trends are shaping the next generation of workflow optimization:

a) Hyperautomation

Coined by Gartner, hyperautomation refers to the combination of RPA, AI, analytics, low-code platforms, and other tools to automate increasingly complex tasks (Deloitte, 2019). This allows organizations to automate not just tasks, but entire processes and decision-making chains.

b) Citizen Development

Low-code and no-code RPA platforms are empowering non-technical users—known as "citizen developers"—to build automation solutions. This democratization of RPA reduces dependence on IT and accelerates innovation at the business unit level (PwC, 2020).

c) Automation-as-a-Service

Cloud-based RPA solutions enable small and medium-sized enterprises (SMEs) to adopt automation without large upfront investments. Automation-as-a-Service platforms offer scalability, integration flexibility, and faster deployment, especially for digitally native businesses.

d) Integration with IoT and Blockchain

RPA is also being integrated with IoT devices to automate real-time data-driven decisions, particularly in smart manufacturing and logistics. Additionally, blockchain's auditability can be combined with RPA for secure transaction processing in sectors like banking and logistics (Chugh & Grandhi, 2021).

e) Human-Robot Collaboration

Rather than replacing human labor, the future of RPA lies in augmenting it. Humans will focus on creative, emotional, and strategic tasks while bots handle the repetitive backend operations. This synergy leads to higher employee engagement and better customer experiences.

IMPLICATIONS OF THE STUDY

The findings of this study on Robotic Process Automation (RPA) as a strategic lever for digital business process transformation have profound theoretical, managerial, and policy-related implications. As organizations worldwide embrace digital transformation, the integration of RPA within existing business process management

(BPM) systems emerges not only as a tool for operational efficiency but as a foundational element for future-proofing enterprises against technological disruption.

1. Theoretical Implications

From a theoretical standpoint, this study contributes to the evolving body of literature on digital transformation and BPM by positioning RPA as more than a mere technological enhancement. Instead, it reconceptualizes RPA as a transformative force capable of reengineering value chains, reshaping organizational learning, and shifting the paradigms of work allocation between humans and machines. Traditionally, theories of organizational efficiency and process optimization have focused on lean principles, Six Sigma, and Total Quality Management. However, the emergence of RPA necessitates a rethinking of these frameworks to incorporate intelligent automation as a central variable. The study aligns with socio-technical systems theory by demonstrating how technology interacts with human processes and organizational culture (Bostrom & Heinen, 1977). In doing so, it opens new avenues for research on hybrid human-digital workforces and the co-evolution of technological and organizational capabilities (Janssen et al., 2017). Moreover, the study advances the literature on dynamic capabilities by suggesting that successful RPA implementation enhances an organization's ability to adapt, reconfigure, and innovate business processes in response to a rapidly changing digital environment (Teece, 2010). This reinforces the view that digital tools like RPA are not merely cost-saving solutions but strategic enablers of agility and resilience.

2. Managerial Implications

The practical implications for managers and decision-makers are multifaceted. First, the study underscores the necessity of treating RPA as a strategic investment rather than a back-office tool. While initial RPA deployments often target simple, rule-based tasks such as invoice processing or employee onboarding, the real value is unlocked when organizations scale automation across end-to-end workflows and integrate it with cognitive technologies like artificial intelligence (AI) and machine learning (Lacity & Willcocks, 2016). Secondly, the findings highlight the importance of change management and governance frameworks in ensuring the success of RPA initiatives. Many organizations experience failures not due to technological shortcomings, but due to resistance from employees, fragmented processes, and a lack of strategic vision (Syed et al., 2020). Managers must therefore prioritize communication, upskilling, and stakeholder alignment throughout the RPA lifecycle. A center of excellence (CoE) model may offer a robust mechanism for centralizing governance, disseminating best practices, and ensuring scalability (Deloitte, 2019).

Furthermore, the study offers insights into the role of RPA in enhancing customer experience. By automating repetitive administrative tasks, employees are freed to focus on higher-value, customer-facing activities. This shift aligns with the service-dominant logic of value creation and supports the personalization and responsiveness expected by digitally savvy customers (Huang & Rust, 2018). For industries such as healthcare, banking, and insurance, which are characterized by high volumes of transaction-intensive processes, RPA can be a game-changer. The study's findings suggest that these sectors can leverage RPA to meet compliance standards, reduce operational costs, and minimize errors—key factors in maintaining regulatory trust and operational continuity (Ernst & Young, 2019).

3. Implications for Human Capital and Workforce Strategy

One of the most significant implications of the study pertains to workforce planning and human resource management. As RPA automates routine work, the skill requirements of the workforce are shifting. Roles that were once executional are evolving into supervisory or analytical positions, requiring competencies in digital literacy, critical thinking, and cross-functional collaboration (Madakam et al., 2019, Aradhya et al., 2025; Kaur et al., 2025; Selvakumar at al., 2025). Organizations must prepare for this transition by investing in re-skilling and upskilling programs. The study reveals that employees who are equipped with an understanding of automation technologies are more likely to embrace RPA and find new avenues for growth within the enterprise. Consequently, HR leaders must adopt a proactive stance in workforce transformation by creating learning pathways and career progression models aligned with the digital economy. Moreover, the psychological impact of RPA adoption should not be underestimated. Job displacement fears, identity disruptions, and lack of clarity around career trajectories can lead to disengagement if not properly managed (Kroll & Moynihan, 2018). Therefore, a balanced approach that emphasizes both automation and human empowerment is essential.

4. Policy and Regulatory Implications

The widespread deployment of RPA also carries implications for policy-makers and regulatory bodies. As automation permeates various sectors, concerns around data privacy, ethical AI, and algorithmic accountability become increasingly relevant. The study calls for a coherent regulatory framework that encourages innovation while safeguarding public interest. For instance, the use of RPA in processing sensitive personal data must comply with data protection laws such as the General Data Protection Regulation (GDPR) in the European Union or HIPAA in the United States. Organizations must ensure that bots are designed with built-in auditability, transparency, and security features to maintain regulatory compliance (Moffitt et al., 2018). Furthermore, the study advocates for government policies that support digital skilling and inclusivity. Public-private partnerships can play a pivotal role in ensuring that the benefits of RPA and related technologies are equitably distributed across different regions and socioeconomic groups. This is particularly important in developing economies where automation may disrupt traditional employment models (Muro et al., 2020).

5. Sustainability and Long-Term Business Resilience

Lastly, the findings of this study highlight the long-term sustainability benefits of RPA. By reducing manual interventions, optimizing energy usage, and minimizing human error, RPA contributes to the creation of leaner and more environmentally responsible operations. This aligns with the growing emphasis on Environmental, Social, and Governance (ESG) factors in strategic decision-making. Additionally, RPA enhances organizational resilience by enabling remote process management, business continuity during crises, and agile response mechanisms. For example, during the COVID-19 pandemic, many organizations that had implemented RPA were able to maintain essential services without disruption (PwC, 2020). These experiences underscore the role of automation as a buffer against systemic shocks.

CONCLUSION

The comprehensive exploration of Robotic Process Automation (RPA) within the context of digital business transformation has yielded profound insights into its multifaceted impact on organizational workflows, process efficiency, and strategic outcomes. The findings presented in this study highlight RPA not only as a tactical automation tool but as a strategic lever capable of reshaping business operations, augmenting workforce productivity, and accelerating digital maturity. One of the primary conclusions from the findings is that RPA significantly improves operational efficiency by automating rule-based, repetitive tasks across a variety of business functions. Case-based examples from sectors such as banking, healthcare, and logistics demonstrate consistent reductions in processing time, human error, and operational costs. These efficiency gains, however, are not achieved in isolation. Successful RPA implementations are typically embedded within a broader Business Process Management (BPM) strategy, emphasizing the need for organizations to align automation efforts with clearly defined business objectives and process improvement goals. The findings also reveal that RPA contributes directly to business agility and scalability, particularly in environments characterized by high-volume transactional processes. For example, organizations facing rapid demand fluctuations—such as insurance firms during seasonal claim surges or retail businesses during peak shopping periods—have leveraged RPA bots to dynamically scale operations without proportionally increasing their workforce. This capacity to scale on demand, combined with the low-code nature of modern RPA platforms, empowers organizations to rapidly deploy solutions and adapt to changing business environments. Another key takeaway is the emergence of intelligent automation—a convergence of RPA with cognitive technologies such as Artificial Intelligence (AI), Machine Learning (ML), and Natural Language Processing (NLP). The findings indicate that while RPA alone handles structured tasks, intelligent automation enables businesses to extend automation into semi-structured and unstructured domains, such as customer queries, invoice processing, and fraud detection. Organizations that adopted this hybrid model reported higher levels of digital integration and process optimization, reinforcing the idea that the future of workflow automation lies in intelligent, context-aware systems. Additionally, the findings underscore the importance of process selection and governance in ensuring RPA success. Not all processes are equally suitable for automation; the highest ROI is observed when organizations carefully assess process maturity, rule consistency, volume, and stability before deployment. Similarly, robust governance frameworks—including bot lifecycle management, compliance monitoring, and change control—are critical to maintaining control, avoiding "bot sprawl," and ensuring regulatory alignment. The study also brings to light human-centric challenges in RPA implementation, particularly resistance to change, fear of job displacement, and lack of automation literacy among employees. Several organizations in the findings highlighted that a failure to manage change effectively led to underutilization of RPA or even project failure. Conversely, enterprises that invested in reskilling, stakeholder communication, and collaborative governance were able to foster a culture of digital empowerment. These results align with socio-technical perspectives, emphasizing that technological interventions must be accompanied by cultural and organizational adaptation. From a strategic standpoint, the findings affirm that RPA can be a cornerstone of digital transformation when aligned with enterprise-level goals. This includes supporting customer experience transformation, enhancing back-office productivity, enabling data-driven decision-making, and integrating with legacy and modern systems. In several case examples, RPA acted as a bridge technology, enabling digital capabilities without necessitating full IT overhauls—thereby allowing for phased, cost-effective transformation. Yet, the findings also caution against over-reliance on RPA as a panacea. Automation is most effective when it complements—not replaces—human judgment, innovation, and strategic thinking. Moreover, technical limitations, poor process documentation, and fragmented IT systems can impede RPA performance. These challenges suggest a need for continuous monitoring, iterative scaling, and integration with broader digital strategies. In summation, the findings of this study position RPA as a transformative enabler of operational excellence and strategic agility. It optimizes workflows, enhances decision-making, and equips organizations to navigate complexity with speed and precision. However, realizing its full potential requires more than deploying bots—it necessitates an integrated approach that considers people, processes, and platforms holistically. As organizations continue their journey toward digital maturity, RPA stands as both a foundational tool and a strategic catalyst for sustained innovation and performance.

REFERENCES

Accenture. (2020). RPA deployment strategies in financial institutions. Accenture Industry Report.

Accenture. (2021). *Intelligent automation in healthcare: The next frontier of digital transformation*. Accenture HealthTech Insights.

Agostinelli, S., Marrella, A., & Mecella, M. (2020). Research challenges for intelligent robotic process automation. *AI & Society*, *35*(2), 1–23.

Aguirre, S., & Rodriguez, A. (2017). Automation of business processes using robotic process automation (RPA): A case study. In Applied Computer Sciences in Engineering (pp. 65–71). Springer. https://doi.org/DOI: 10.1007/978-3-319-66963-2_6

Aradhya, S. G. B., Gupta, V. C., Selvakumar, P., Raman, M. S., Singh, J., Manjunath, T. C., & Sharma, M. (2025). Health Crises and Digitalization: Digital Health. In Sustainable Digitalization Strategies in Business and Healthcare (pp. 171-198). IGI Global Scientific Publishing.

Asatiani, A., & Penttinen, E. (2016). Human-robot collaboration in digital transformation. *Journal of Information Technology Teaching Cases*, *6*(2), 89–103.

Asatiani, A., & Penttinen, E. (2016). Turning robotic process automation into commercial success – Case OpusCapita. *Journal of Information Technology Teaching Cases*, *6*(2), 67–74. DOI: 10.1057/jittc.2016.5

Bain & Company. (2020). Process reengineering for successful automation. Bain Insights Report.

Boston Consulting Group (BCG). (2021). *Digital operations and automation monitoring frameworks*. BCG Perspectives.

Bostrom, R. P., & Heinen, J. S. (1977). MIS problems and failures: A socio-technical perspective. *Management Information Systems Quarterly*, *1*(3), 17–32.

Braun, V., & Clarke, V. (2006). Using thematic analysis in psychology. *Qualitative Research in Psychology*, *3*(2), 77–101.

Brock, J. K.-U., & von Wangenheim, F. (2019). Demystifying AI: What digital transformation leaders can teach you about realistic artificial intelligence. *California Management Review*, *61*(4), 110–134. DOI: 10.1177/0008125619862257

Brougham, D., & Haar, J. (2018). Smart technology, artificial intelligence, robotics, and algorithms (STARA): Employees' perceptions of our future workplace. *Journal of Management & Organization*, *24*(2), 239–257. DOI: 10.1017/jmo.2016.55

Capgemini. (2021). RPA and digital transformation in banking. Capgemini Financial Services Analysis.

Chugh, R., & Grandhi, S. (2021). Comparative analysis of traditional and robotic process automation: A systematic review. *International Journal of Business Process Integration and Management*, *10*(3), 205–217.

Creswell, J. W., & Poth, C. N. (2018). *Qualitative inquiry and research design: Choosing among five approaches* (4th ed.). Sage Publications.

Davenport, T. H., & Ronanki, R. (2018). Artificial intelligence for the real world. *Harvard Business Review*, *96*(1), 108–116.

Deloitte. (2019). Intelligent automation in financial services: The evolution of RPA. Deloitte Research Report.

Deloitte. (2019). *RPA in financial services: Redefining the future of compliance and efficiency*. Deloitte Insights.

Deloitte. (2020). *Automation with intelligence: Reimagining business processes*. Deloitte Insights.

Deloitte. (2020). *Intelligent automation in retail and e-commerce*. Deloitte Insights.

Dery, K., Sebastian, I. M., & van der Meulen, N. (2017). RPA in the public sector: Transforming government operations. MIT Center for Information Systems Research Working Paper.

Dumas, M., La Rosa, M., Mendling, J., & Reijers, H. A. (2018). *Fundamentals of business process management* (2nd ed.). Springer., DOI: 10.1007/978-3-662-56509-4

Ernst & Young. (2019). *Intelligent automation in healthcare: Improving operations and compliance*. EY Global Insights.

EY (Ernst & Young). (2020). *Operational excellence through robotic process automation*. EY Global Insights.

Fersht, P., & Snowdon, J. (2017). *RPA in manufacturing: How Siemens and others are driving efficiency through automation*. Horses for Sources Research.

Fitzgerald, M., Kruschwitz, N., Bonnet, D., & Welch, M. (2014). Embracing digital technology: A new strategic imperative. *MIT Sloan Management Review*, *55*(2), 1–12.

Forrester. (2022). The state of intelligent process automation. Forrester Research.

Frost & Sullivan. (2022). Smart manufacturing and RPA adoption trends. Frost & Sullivan Industry Report.

Gartner. (2021). Automation in insurance: RPA and workforce transformation. Gartner Research.

Gartner. (2022). Top strategic technology trends: Hyperautomation and beyond. Gartner Research.

Geyer-Klingeberg, J., Nakladal, J., Baldauf, F., & Veit, F. (2018). Process mining and RPA: An integrated framework for digital transformation. *Process Intelligence Journal*, *6*(3), 34–49.

Hammer, M., & Champy, J. (1993). *Reengineering the corporation: A manifesto for business revolution*. Harper Business.

HIMSS. (2020). Automation in healthcare operations: Case studies and outcomes. Healthcare Information and Management Systems Society Report.

Hofmann, P., Samp, C., & Urbach, N. (2020). Robotic process automation. *Electronic Markets*, *30*(1), 99–106. DOI: 10.1007/s12525-019-00365-8

Huang, M. H., & Rust, R. T. (2018). Artificial intelligence in service. *Journal of Service Research*, *21*(2), 155–172.

IBM. (2021). AI-enhanced RPA in logistics and supply chain management. IBM Business Value Report.

International Organization for Standardization (ISO). (2020). *ISO/IEC 30120:2020 – Governance and compliance in robotic process automation*. ISO Standards.

Janssen, M., van der Voort, H., & Wahyudi, A. (2017). Factors influencing big data decision-making quality. *Journal of Business Research*, *70*, 338–345.

Jöhnk, J., Röglinger, M., Thimmel, M., & Urbach, N. (2021). How to implement agile IT setups: A taxonomy of design options. *Information Systems and e-Business Management*, *19*(2), 473–520. DOI: 10.1007/s10257-020-00474-3

Kaur, H., Sharma, A., & Singh, R. (2025). Upskilling for intelligent automation: The evolving role of HR in Industry 5.0. *Journal of Organizational Change and Innovation*, *29*(2), 134–152.

Kotter, J. P. (1996). *Leading change*. Harvard Business School Press.

KPMG. (2021). Automation in healthcare: Delivering efficiency and safety. KPMG Global Report.

Kroll, A., & Moynihan, D. P. (2018). Automation and accountability in public sector performance: Balancing efficiency and engagement. *Public Administration Review*, *78*(1), 69–79.

Lacity, M., Willcocks, L., & Craig, A. (2015). Governance and change management in RPA implementation. The Outsourcing Unit Working Research Paper Series, 15(08), 1–25.

Lacity, M., Willcocks, L., & Craig, A. (2015). The IT function and robotic process automation. The Outsourcing Unit Working Research Paper Series, 15(05), 1–35.

Lacity, M. C., & Willcocks, L. P. (2016). Robotic process automation: The next transformation lever for shared services. The Outsourcing Unit Working Research Paper Series, London School of Economics and Political Science.

Lacity, M. C., & Willcocks, L. P. (2018). *Robotic process automation and cognitive automation: The next phase*. SB Publishing.

Madakam, S., Holmukhe, R. M., & Jaiswal, D. K. (2019). The future digital work force: Robotic process automation (RPA). *Journal of Information Systems and Technology Management*, *16*, e201916002. DOI: 10.4301/S1807-1775201916002

Malik, N., Nirupama, E., Selvakumar, P., Sankpal, U., Nathsharma, S. K., Sharma, M., & Manjunath, T. C. (2025). 2D Materials for Energy Storage and Conversion. In *Innovations in Next-Generation Energy Storage Solutions* (pp. 1–24). IGI Global Scientific Publishing.

Malik, N., Thakur, D. J., Aradhya, S. G. B., Sudheer, P., Sharma, M., Manjunath, T. C., & Selvakumar, P. (2025). Performance Optimization in Remote Monitoring. In Synergizing Data Envelopment Analysis and Machine Learning for Performance Optimization in Healthcare (pp. 495-516). IGI Global Scientific Publishing.

McKinsey & Company. (2020). *Automation at scale: RPA in e-commerce operations*. McKinsey Digital.

Moffitt, K. C., Rozario, A. M., & Vasarhelyi, M. A. (2018). Robotic process automation for auditing. *Journal of Emerging Technologies in Accounting*, *15*(1), 1–10. DOI: 10.2308/jeta-10589

Moffitt, K. C., Rozario, A. M., & Vasarhelyi, M. A. (2018). The role of AI in audit automation: Risks, ethics, and transparency. *Journal of Emerging Technologies in Accounting*, *15*(2), 1–11.

Muro, M., Liu, S., Whiton, J., & Kulkarni, S. (2020). Digitalization and the American workforce. Brookings Institution Report.

PwC. (2019). Customer experience redefined through automation. PwC Research Report.

PwC. (2020). Automation and the workforce of the future: Insights for retail and e-commerce. PwC Research Report.

PwC. (2020). *Citizen developers and the democratization of RPA: Low-code automation in business transformation*. PwC Research Insights.

PwC. (2020). *Cognitive automation: Driving the next wave of digital transformation*. PwC Global Automation Insights.

PwC. (2021). Automation and compliance in financial services. PwC Global Survey.

Samal, A., Bhargav, J., Selvakumar, P., Sharma, M., & TC, M. (2025). Measuring Brand Performance With AI Tools. In Strategic Brand Management in the Age of AI and Disruption (pp. 279-302). IGI Global Scientific Publishing.

Selvakumar, P., Aradhya, S. G. B., Anute, N., Devi, S., Sharma, M., & Vaissnave, V. (2025). Strategic Collaborations and Partnerships in Entrepreneurship. In Support Networks for Entrepreneurs: Social Capital and Strategic Collaborations (pp. 203-232). IGI Global Scientific Publishing.

Selvakumar, P., Dahiya, R., Manjunath, T. C., & Sharma, M. (2025). Carbon Footprint Analysis of Electronic Products. In Integrated Approaches for Sustainable E-Waste Management (pp. 215-242). IGI Global Scientific Publishing.

Selvakumar, P., Satyanarayana, P., Lahoti, Y., Saipogu, L., Sharma, M., & Manjunath, T. C. (2025). AI Techniques. In AI Methods for Environmental Protection and Resource Conservation (pp. 107-138). IGI Global Scientific Publishing.

Selvakumar, P., Shanthi, M., Pitchiah, R., Sharma, M., Dahake, P. S., & TC, M. (2025). The Role of Artificial Intelligence in Business Model Innovation: Overview of AI Technologies. AI-Driven Business Model Innovation, 219-246.

Selvakumar, P., Sharma, A., Roy, S. S., Sharma, M., & Manjunath, T. C. (2025). Representations in Developing Statistical Thinking in Secondary Education. In Modes of Representation in Developing Statistical Thinking in Education (pp. 59-88). IGI Global Scientific Publishing.

Sharma, M., Kumar, R., Malik, N., & Chaudhry, R. (2026). Digital Maturity and Ecosystems: Catalysts for Achieving the Sustainable Development Goals in the 21st Century. In Impacts of Digital Maturity and Digital Ecosystems on Sustainable Development Goals (pp. 157-188). IGI Global Scientific Publishing.

Society for Human Resource Management (SHRM). (2021). Automation in human resource functions: The RPA revolution. SHRM Research Report.

Syed, R., Suriadi, S., Adams, M., Bandara, W., Leemans, S. J. J., Ouyang, C., ter Hofstede, A. H. M., van de Weerd, I., Wynn, M. T., & Reijers, H. A. (2020). Robotic process automation: Contemporary themes and challenges. *Computers in Industry*, *115*, 103162. DOI: 10.1016/j.compind.2019.103162

Teece, D. J. (2018). Business models and dynamic capabilities. *Long Range Planning*, *51*(1), 40–49.

United Nations Conference on Trade and Development (UNCTAD). (2021). Technology and innovation report 2021: Catching technological waves – Innovation with equity. United Nations.

U.S. Food and Drug Administration (FDA). (2021). *Guidance for RPA use in regulated pharmaceutical environments*. FDA Compliance Document.

van der Aalst, W. M. P. (2020). Robotic process automation and workflow automation: Concepts, technology, and impact. International Conference on Advanced Information Systems Engineering, 19–23. Springer. DOI: 10.1007/978-3-030-49435-3_2

van der Aalst, W. M. P., Bichler, M., & Heinzl, A. (2018). Robotic process automation. *Business & Information Systems Engineering*, *60*(4), 269–272. DOI: 10.1007/s12599-018-0542-4

Weber, R. H. (2010). Internet of Things – New security and privacy challenges. *Computer Law & Security Review*, *26*(1), 23–30.

Willcocks, L., Lacity, M., & Craig, A. (2017). Establishing centers of excellence for automation governance. *MIT Sloan Management Review*, *58*(4), 54–61.

Willcocks, L., Lacity, M., & Craig, A. (2017). Robotic process automation: Strategic transformation lever for organizations. *Journal of Information Technology Management*, *28*(3), 62–78.

Willcocks, L. P., Lacity, M. C., & Craig, A. (2015). The IT function and robotic process automation. The Outsourcing Unit Working Research Paper Series, (15/05), 1–39.

Willcocks, L. P., Lacity, M. C., & Craig, A. (2017). Robotic process automation: Strategic transformation lever for global business services? *Journal of Information Technology Teaching Cases*, *7*(1), 17–28. DOI: 10.1057/s41266-017-0010-4

World Economic Forum. (2021). Upskilling for shared prosperity: Insight report. World Economic Forum Publications.

Yin, R. K. (2017). *Case study research and applications: Design and methods* (6th ed.). Sage Publications.

Zion Market Research. (2022). *Global robotic process automation market report 2022–2030*. Zion Market Research.

KEY TERMS AND DEFINITIONS

Agility: Agility is an organization's ability to quickly adapt to changing market conditions, emerging technologies, or customer demands. By leveraging RPA and digital transformation, businesses can respond faster, launch new services, and reconfigure processes dynamically to maintain competitive advantage.

Artificial Intelligence (AI): AI enables machines to replicate human-like intelligence, such as learning from data, making decisions, and solving problems. In combination with RPA, AI can enhance automation by handling tasks that require judgment, pattern recognition, or decision-making, such as fraud detection or customer support chatbots (Brock & von Wangenheim, 2019).

Big Data Analytics: This refers to analyzing large, complex datasets to identify meaningful patterns and insights. In RPA-enabled processes, big data analytics helps organizations make informed decisions, detect inefficiencies, and optimize automated workflows by understanding process performance at scale.

Digital Transformation: Digital transformation is more than just adopting new technologies—it is a holistic process that redefines how organizations operate, interact with customers, and deliver value. It involves integrating digital tools into all business functions, fostering innovation, enhancing decision-making with data-driven insights, and creating new business models to stay competitive in a rapidly evolving market (Fitzgerald et al., 2014).

Machine Learning (ML): A subset of AI, ML involves training algorithms to learn from historical data and improve over time without explicit programming. ML enables organizations to predict trends, optimize processes, and support intelligent decision-making—for instance, predicting equipment maintenance needs or customer behavior patterns.

Operational Efficiency: This refers to how effectively an organization uses its resources—time, money, and workforce—to achieve its goals. RPA contributes to operational efficiency by eliminating repetitive manual work, reducing errors, and accelerating processes, ultimately lowering operational costs and improving productivity.

Process Accuracy: Process accuracy measures how correctly and consistently business processes are executed. Automation through RPA significantly improves process accuracy by reducing human errors in repetitive tasks, ensuring regulatory compliance, and maintaining consistent outputs, which is critical for high-volume operations such as finance, HR, or supply chain management.

Robotic Process Automation (RPA): RPA is the use of software "bots" to automate repetitive, structured, and rule-based tasks. Examples include data entry, invoice processing, payroll management, and report generation. RPA reduces human error, accelerates task completion, and frees employees to focus on more strategic, creative, and analytical tasks (Aguirre & Rodriguez, 2017).

Rule-Based Tasks: These are tasks that follow a clearly defined set of rules or procedures, making them predictable and suitable for automation. Examples include generating standard reports, processing transactions, or verifying customer data. Rule-based tasks are ideal for RPA because bots can perform them consistently and accurately without supervision.

Software Bots: Software bots are automated programs that mimic human actions within digital environments. They can log into applications, move files, extract and input data, and communicate with other systems. Unlike traditional automation, bots operate without altering existing IT systems, making them non-intrusive and easy to scale (Willcocks, Lacity, & Craig, 2015).

Chapter 16

Robotic Process Automation in Banking: Improving Risk Controls and Fraud Detection

P. Selvakumar
https://orcid.org/0000-0002-3650-4548
Department of Science and Humanities, Nehru Institute of Technology, Coimbatore, India

T. C. Manjunath
https://orcid.org/0000-0003-2545-9160
Rajarajeswari College of Engineering, India

Sumanta Bhattacharya
https://orcid.org/0000-0003-2563-2787
Asian international University, India

Mohit Sharma
https://orcid.org/0009-0007-2280-8077
Maharshi Dayanand University, Rohtak, India

ABSTRACT

The banking industry, thereby enhancing operational efficiency, reducing costs, and minimizing risks. With growing customer expectations, strict regulatory requirements, and the demand for real-time financial services, banks and financial institutions are leveraging RPA to streamline their processes. RPA allows banks to improve their accuracy, speed, and compliance while also freeing human employees to This article explores how RPA is revolutionizing the banking sector, the key benefits it offers, the Unlike traditional automation solutions, RPA does not require complex system

DOI: 10.4018/979-8-3373-3571-1.ch016

integrations as it can interact with existing banking software interfaces just as a human user would. Banks handle millions of transactions and data records daily, requiring meticulous processing and regulatory compliance. Manual processes often lead to errors, inefficiencies, and high operational costs. By implementing RPA, banks can reduce processing time for routine banking operations. Minimize human errors, ensuring greater accuracy and compliance.

INTRODUCTION TO ROBOTIC PROCESS AUTOMATION (RPA) IN BANKING: IMPROVING EFFICIENCY AND REDUCING RISK

The banking industry, thereby enhancing operational efficiency, reducing costs, and minimizing risks.With growing customer expectations, strict regulatory requirements, and the demand for real-time financial services, banks and financial institutions are leveraging RPA to streamline their processes. RPA allows banks to improve their accuracy, speed, and compliance while also freeing human employees to This article explores how RPA is revolutionizing the banking sector, the key benefits it offers, the Unlike traditional automation solutions, RPA does not require complex system integrations—it can interact with existing banking software interfaces just as a human user would.Banks handle millions of transactions and data records daily, requiring meticulous processing and regulatory compliance. Manual processes often lead to errors, inefficiencies, and high operational costs. By implementing RPA, banks can:Reduce processing time for routine banking operations.Minimize human errors, ensuring greater accuracy and compliance.Enhance fraud detection and risk management by monitoring transactions in real time.Improve customer service by automating query resolution and document processing.

How RPA Differs from Traditional AutomationUnlike conventional automation that requires extensive coding and IT support, RPA operates as a non-invasive technology that can be easily deployed on existing banking systems. RPA bots can be trained to perform tasks across multiple applications, allowing banks to automate their operations without overhauling their legacy systems.Customer onboarding in banks involves multiple steps, including identity verification, document validation, and background checks. RPA automates the process by:Extracting data from documents using optical character recognition (OCR).Verifying customer details against regulatory databases.Automating Know Your Customer (KYC) checks to ensure compliance.By using RPA, banks can significantly reduce onboarding time, enhance regulatory compliance, and improve the overall customer experience. With RPA-driven fraud detection, banks can proactively mitigate risks and ensure regulatory compliance. Loan Processing and Credit UnderwritinTraditionally, loan processing involves manual verification of documents, credit history checks, and risk assessments, leading to delays and inefficiencies. RPA speeds up the process by:Extracting applicant data from submitted documents.Cross-checking credit scores and financial statements.Automating loan eligibility calculations and approvals reducing processing time, RPA enables banks to offer faster loan approvals and better customer satisfaction. Regulatory Compliance and ReportingBanks operate under strict regulatory frameworks that require frequent audits, reporting, and compliance checks.(Alexy et al., 2016) RPA simplifies compliance management byAutomating data collection for regulatory reporting.

Ensuring adherence to anti-money laundering (AML) and Basel III requirements. Generating audit trails for better transparency and accountability.RPA reduces the burden on compliance teams while ensuring accurate and timely reporting. Accounts Payable and ReconciliationHandling large volumes of financial transactions manually often leads to errors and delays in payment processing. RPA optimizes accounts payable by:Automating invoice processing and payment approvals.Matching transactions across multiple accounts.Identifying discrepancies in financial records for reconciliation.With RPA, banks achieve faster financial reconciliation and improved cash flow management. Benefits of RPA in Banking Increased Efficiency and Cost SavingsRPA reduces the time required for manual tasks, enabling banks to process transactions fasterand at lower costs. By automating repetitive processes, banks can achieve:60–80% reduction in processing time.Significant cost savings by eliminating human errors and inefficiencies. Enhanced Accuracy and ComplianceRegulatory compliance requires error-free data processing. RPA ensures:100% accuracy in data entry and reporting.Consistent adherence to banking regulations. Improved Customer ExperienceBy automating backend operations, banks can:(Arakelian & Dellaportas, 2012)Process customer requests instantly.Provide real-time assistance through AI-powered chatbots.Enhance service delivery for higher customer satisfaction. Scalability and FlexibilityRPA bots can handle increasing workloads without additional costs, allowing banks to scale operations efficiently. Risk ReductionPA reduces operational risks by:Minimizing manual errors in critical banking processes.Detecting anomalies and preventing fraudulent activities.. Future of RPA in BankingThe future of RPA in banking will be driven by advancements in Artificial Intelligence (AI), Machine Learning (ML), and Cognitive Automation. Emerging trends include:AI-powered RPA for decision-making and predictive analytics.Hyperautomation, combining RPA with AI and data analytics.

THE CURRENT STATE OF BANKING OPERATIONS: CHALLENGES AND OPPORTUNITIES FOR RPA ADOPTION

However, despite its potential, RPA adoption in banking is not without challenges. Issues such as legacy system integration, regulatory complexities, cybersecurity concerns, and workforce adaptation pose significant barriers to seamless implementation. At the same time, banks that successfully adopt RPA can unlock immense opportunities, including enhanced customer service, improved risk management, and streamlined compliance reporting.This article explores the current state of banking operations, identifies the key challenges in traditional banking workflows, and examines the opportunities that RPA presents for modernizing financial institutions. Traditional banking operations involve a significant amount of manual intervention, particularly in areas such as customer onboarding, loan processing, account reconciliation, fraud detection, and regulatory reporting(Baitinger & Papenbrock, 2017). Many of these tasks require employees to process vast amounts of data, leading to bottlenecks, delays, and a high risk of human error.For example:Loan approvals often require extensive document verification, credit checks, and risk assessments—processes that can take weeks to complete.Customer service departments handle thousands of requests daily, leading to long wait times and inconsistent responses.Compliance teams spend countless hours preparing regulatory reports, often struggling with data accuracy and consistency.These inefficiencies result in higher operational costs, slower service delivery, and reduced customer satisfaction.

Deploying RPA requires investments in technology, training, and process reengineering.Many banks struggle to justify these costs without clear short-term returns on investment (ROI).RPA can speed up KYC processes by automating document verification and background checks.Automated reporting ensures that banks stay compliant with regulations and reduce the risk of penalties.RPA can continuously monitor transactions for suspicious activities,(Barucci & Marazzina, 2016) improving fraud detection and prevention. Streamlining Loan Processing and Credit UnderwritingRPA bots can extract applicant data, verify credit history, and calculate loan eligibility within minutes.Banks can process loans faster, leading to higher customer satisfaction and increased revenue.3.4 Improving Customer Service with AI-Powered ChatbotsRPA-driven chatbots can handle customer inquiries instantly, reducing response times.Banks can offer 24/7 customer support, improving service availability and satisfaction. Optimizing Financial Reconciliation and Back-Office OperationAutomated reconciliation of accounts, transactions, and statements eliminates errors and speeds up financial reporting.RPA bots ensure that all financial records are accurate and up to date.

The Role of RPA in Enhancing Risk Management and Fraud Detection in Banking

Banks face an ever-growing number of threats, from cybersecurity risks and money laundering to identity fraud and financial crimes. Traditional risk management frameworks, which rely heavily on manual processes, rule-based systems, and human oversight, struggle to keep pace with the sophistication and volume of modern financial threatsTo address these challenges, Robotic Process Automation (RPA) is revolutionizing risk management and fraud detection in banking. By automating data analysis, transaction monitoring, and regulatory compliance, RPA enables banks to identify and prevent fraudulent activities faster, reduce operational risks, and enhance overall security. management, highlighting how it improves fraud detection, enhances regulatory compliance, minimizes human errors, and optimizes overall security measures.The global financial industry is experiencing an alarming rise in fraud cases, including:Identity theft and account takeoversMoney laundering and financial crimesCredit card fraud and unauthorized transactionsInsider fraud and employee misconduct intervention, making them slow, inefficient, and prone to errors. This delay in fraud detection increases financial losses and exposes banks to legal and reputational risks.(Battiston et al., 2012)The USA PATRIOT Act (for monitoring suspicious transactions).Basel III and Basel IV regulations (for risk management and capital adequacy). Managing compliance manually is resource-intensive and inefficient, making automation essential. handle millions of transactions per day, making it nearly impossible for human employees to manually review each one for suspicious activities or anomalies. Traditional fraud detection systems generate a high number of false positives, requiring extensive human intervention to verify each flagged transaction.Without automation, fraud detection remains slow, reactive, and error-prone, increasing the risk of financial loss. Automated Transaction Monitoring and Fraud DetectionRPA bots can scan thousands of transactions per second to identify unusual patterns, such as:Large, irregular fund transfers.Frequent transactions between high-risk accounts.Unusual login activity or multiple failed authentication attempts.Unlike traditional rule-based fraud detection, AI-powered RPA can adapt to new fraud techniques and evolving financial threats.RPA ensures that fraudulent activities are flagged in real time, reducing financial losses. Know Your Customer (KYC) and Customer Due Diligence (CDD) AutomationRPA automates the KYC process by:Extracting and verifying customer details from ID documents. Checking customer names against global watchlists and sanction lists.Conducting background checks on high-risk individuals or entities.By reducing manual efforts in KYC, banks improve compliance accuracy and onboarding efficiency.

Anti-Money Laundering (AML) Compliance and Suspicious(Battiston et al., 2012) Activity Reporting (SARs)RPA bots track financial transactions to detect potential money laundering activities.When suspicious activity is detected, RPA automatically generates a Suspicious Activity Report (SAR) for regulators.RPA ensures that banks comply with AML laws, avoiding regulatory fines and legal risks. Risk Assessment and Credit ScoringRPA streamlines credit risk management by analyzing customer credit history, income, and financial behavior.Bots can:Calculate credit scores and loan eligibility based on predefined criteria.Flag high-risk applicants for additional review.Reduce bias in credit assessments, improving decision-making transparency. Enhancing Cybersecurity MeasuresRPA strengthens cybersecurity by:etecting unauthorized access attempts.Monitoring network activity for anomalies.Automating password resets and security alerts.By reducing reliance on manual IT security processes, RPA helps banks prevent cyber threats and data breaches.. Benefits of RPA in Banking Fraud Detection and Risk Managemen Faster Fraud Detection and Preventioeal-time monitoring of transactions enables banks to detect fraud before financial losses occurRPA ensures that fraudulent transactions are blocked automatically. Improved Compliance with Regulatory RequirementsRPA reduces compliance risks by automating AML, KYC, and risk reporting processes. Banks avoid fines and penalties by ensuring 100% regulatory compliance. Increased Accuracy and Reduced Human ErrorsManual fraud detection often leads to missed fraudulent activities or false positives.RPA improves accuracy by analyzing data without human bias. Process Automation (RPA) is transforming risk management and fraud detection in banking by automating transaction monitoring, enhancing regulatory compliance, and reducing operational risks. Banks that adopt RPA benefit from faster fraud detection, improved accuracy, cost savings, and enhanced customer trust.As financial threats evolve, banks must embrace(Bernè et al., 2006)

Automating Anti-Money Laundering (AML) and Know-Your-Customer (KYC) Processes with RPA

Bank Secrecy Act (BSA) – Requires banks to report suspicious transactions to FinCEN.AML Directives (AMLD) by the European Union – Mandates risk-based monitoring for financial transactions.The U.S. Patriot Act – (Cerchiello & Giudici, 2015)Enforces customer due diligence (CDD) and suspicious activity reporting (SAR).FATF Recommendations – Global AML compliance framework.Failure to comply with these regulations can result in hefty fines, reputational damage, and legal consequences. Manual AML and KYC Processes are Inefficient(Chen et al., 2018)Customer onboarding takes too long due to mnual identity verification and background checks.High false positive rates in fraud detection require manual intervention, delaying transaction approvals.Human errors and inconsistencies in compliance documentation lead to regulatory risks.Huge operational costs are associated with hiring compliance officers and analysts. Increasing Risk of Money

Laundering and Financial CrimesCriminals use multiple accounts, shell companies, and anonymous transactions to hide illicit activities.Traditional AML systems struggle to detect complex financial crime patterns.Regulatory authorities impose frequent audits and compliance updates, increasing the burden on banks.RPA bots streamline KYC and customer onboarding by:Extracting customer data from application forms, IDs, and supporting documents.Verifying customer identities by cross-checking with public and private databases.Checking names against global sanction lists, politically exposed persons (PEP) databases, and blacklists.Monitoring customer behavior and risk profiles to detect suspicious activities.By automating(Cruz, 2002) KYC, financial institutions reduce onboarding times from weeks to minutes while ensuring 100% compliance with regulatory requirements. Automated AML Transaction Monitoring and Suspicious Activity Reporting (SARs)RPA enhances AML compliance by:Monitoring financial transactions in real-time for suspicious patterns.Applying AI-powered risk-scoring models to flag high-risk transactions. Automatically generating SARs and filing reports with regulatory authorities.

Reducing false positives through intelligent automation, minimizing unnecessary manual reviews.This ensures that money laundering attempts are detected early and reported instantly, preventing financial crime. Automating Risk Assessment and Enhanced Due Diligence (EDDPA bots can assign risk scores to customers and entities based on transaction history, geography, and behavior.High-risk customers undergo additional scrutiny through automated Enhanced Due Diligence (EDD) processes. Continuous risk monitoring ensures ongoing compliance with evolving regulatory requirements. Automating Periodic KYC Updates and RemediationRPA ensures that customer records are regularly updated based on regulatory requirements.ots automatically notify customers about document expirations and missing information.Banks stay compliant with KYC refresh cycles without manual intervention Automating Regulatory Reporting and Audit Trail ManagementRPA compiles and submits compliance reports to financial regulators.Audit trails are automatically generated, ensuring transparency and accountability.educes compliance costs by eliminating manual report preparation.

Using RPA for Fraud Detection and Prevention in Banking: A Case Study

Fraud remains one of the biggest challenges in the banking sector, costing financial institutions billions of dollars annually. Cybercrime, identity theft, money laundering, and insider fraud have become increasingly sophisticated, making it critical for banks to adopt advanced technologies n.

Robotic Process Automation (RPA) is transforming fraud prevention by automating transaction monitoring, risk assessment, anomaly detection, and real-time reporting.

This case study explores how a leading global bank successfully implemented RPA to combat fraud, improve compliance, and enhance customer trust. It highlights the challenges faced, RPA implementation process, key benefits, and lessons learned in using automation to strengthen fraud detection and prevention.Criminals used stolen identities, synthetic accounts, and AI-driven scams to bypass traditional fraud detection systems.(De Filippi & Hassan, 2016)Real-time fraud detection was lacking, leading to financial losses before fraud could be identified.Regulatory bodies such as FATF, FinCEN, and the European Banking Authority imposed strict fraud prevention requirements.Failure to detect fraud early resulted in hefty fines, reputational damage, and increased regulatory scrutiny. Inefficient Manual Investigation Processeraud analysts spent hours manually analyzing transactions, gathering evidence, and filing suspicious activity reports (SARs).The slow fraud response led to financial losses and missed opportunities to prevent repeat fraud. The bank needed an automated, real-time fraud detection and prevention solution that could handle large-scale transactions, improve accuracy, and reduce fraud investigation time.The bank chose an AI-powered RPA platform that could:Monitor transactions in real time and detect anomalies instantly.Integrate with the bank's core systems, AML software, and risk management tools.Reduce false positives by using machine learning models to analyze transaction patterns.

Automate fraud investigations by gathering relevant data and generating reports. Automating (Diebold & Yilmaz, 2014)Transaction Monitoring and Fraud DetectionRPA bots were deployed to:Analyze every transaction in real time, flagging suspicious activities.Cross-check transactions against blacklists, watchlists, and regulatory databases.Identify patterns of fraud, such as unusual fund transfers, rapid withdrawals, or geographic anomalies.Use AI-driven risk scoring to assess transaction legitimacy. Enhancing Identity Verification with RPARPA bots automated Know-Your-Customer (KYC) checks, ensuring that customer identitie were verified before transactions were approved.Bots compared customer details with global fraud databases, biometric authentication records, and government-issued IDs.Suspicious customers were flagged for Enhanced Due Diligence (EDD) before account activation. Automating Suspicious Activity Reporting (SARs)When a fraudulent transaction was detected, RPA bots collected evidence, compiled reports, and submitted SARs to regulatory authorities.raud analysts no longer needed to manually investigate each alert, reducing investigation time by 70%. Cost Savings and Operational EfficiencyAutomating fraud detection reduced compliance costs by 60%, eliminating manual investigations.Fraud teams could focus on high-risk cases, increasing efficiency. Improved Compliance and Regulatory AdherenceThe bank ensured full compliance with AML, KYC, and fraud prevention regulations.RPA bots automated audit trails and regulatory reporting, reducing the risk of non-compliance. (European Commission, 2018) Enhanced Customer Trust and SecurityCustomers felt safer knowing their transactions were monitored and protected.Reduced fraud-related complaints and disputes improved customer satisfaction. Challenges and Lessons Learned from RPA Implementatio Integration with Legacy Systemshe bank faced challenges integrating RPA with old core banking systemshe case study of this multinational bank demonstrates how RPA revolutionized fraud detection and prevention, significantly reducing fraud losses, improving compliance, and enhancing customer security.(Financial Stability Board, 2017a).

RPA IN RISK MANAGEMENT: AUTOMATING RISK ASSESSMENT AND COMPLIANCE PROCESSES

This article explores how RPA is transforming risk management, including its key applications, benefits, challenges, and future directions in automating risk assessment and compliance proceses. The Challenges of Traditional Risk Management and ComplianceRisk management and regulatory compliance involve a complex set of processes, including:Risk assessment and scoring to identify high-risk transactions or customers.Regulatory reporting to comply with international and local financial regulations.Transaction monitoring to detect fraud, money laundering, and suspicious activities. Growing Regulatory ComplexityGovernments and financial watchdogs impose stringent regulations such as GDPR, Basel III, FATCA, and AML/KYC laws.Institutions struggle to keep up with changing compliance requirements, resulting in penalties and reputational risks. Inadequate (Financial Stability Board, 2017b)Fraud Detection CapabilitiesManual risk assessment lacks the ability to detect real-time fraudulent activities in high-volume transactions.False positives in fraud alerts lead to unnecessary investigations and operational inefficiencies.With these challenges in mind, financial institutions need an automated and intelligent solution to enhance their risk management strategies while ensuring full regulatory compliance. The Role of RPA in Risk Management and ComplianceRobotic Process Automation (RPA) is revolutionizing risk assessment and compliance management by automating repetitive, rule-based tasks. By deploying RPA bots, organizations can improve accuracy, efficiency, and compliance adherence while reducing costs and operational risks. Improving Documentation and Audit ManagementRPA can automatically process, validate, and store risk management documents, eliminating manual paperwork.ots generate audit-ready reports, ensuring compliance with regulatory bodies.Automated workflows enhance data accuracy, reducing compliance errors.The adoption of Robotic Process Automation (RPA) in risk management and compliance is transforming the banking and financial sector. By automating risk assessment, fraud detection, compliance reporting, and regulatory monitoring, institutions can enhance efficiency, reduce costs, and improve accuracy As financial institutions face increasingly complex risks and regulatory pressures, leveraging AI-driven RPA will be key to maintaining robust risk management frameworks, preventing financial crimes, and ensuring long-term regulatory compliance.

The Benefits of RPA Adoption in Banking: Improved Efficiency, Reduced Risk, and Enhanced Customer Experience

Through the deployment of software robots or "bots," banks can streamline operations, enhance compliance, and improve service delivery—all while maintaining a sharp focus on customer experience.In the context of banking, the benefits of RPA are profound. This article delves into the key advantages of RPA adoption in banking, including improved efficiency, reduced risk, and enhanced customer experience, showcasing how automation is reshaping the way banks deliver services, handle back-office functions, and interact with customers. routine transactions, and customer interactions. Many of these tasks are highly repetitive and prone to human error, making them ideal candidates for automation. By adopting RPA, banks can significantly enhance operational efficiency across various areas. document validation are among the most common activities that RPA handles in banks. These tasks can take a significant amount of time when performed manually, but with RPA, they can be executed faster and more accurately. For example, in loan processing, RPA bots can automate document verification, data extraction, and entry into core banking systems.(Foley et al., 2018) This automation allows for faster approvals, reducing the time required to close a loan application from days to mere hours. mundane processes such as reconciliation, reporting, and compliance checks.

This not only speeds up workflows but also frees up human resources to focus on higher-value tasks, such as strategy development or customer service improvements. For instance, automating reconciliation processes between various accounts and systems ensures that discrepancies are caught early and corrected before they can lead to more significant issues season, automation can handle increased workloads without the need for additional human staff. This enables banks to respond quickly to market conditions, changes in customer demand, and new regulatory requirements.RPA helps financial institutions meet regulatory compliance standards more effectively. For instance, when it comes to KYC and AML processes, RPA can automatically collect customer data, compare it with external databases for verification, and flag suspicious activities(Giudici, 2003). By automating these processes, banks can ensure compliance with anti-money laundering laws, prevent financial crimes, and adhere to evolving regulatory frameworks. Moreover, RPA enables timely reporting, ensuring that banks can submit necessary documentation and filings on schedule, minimizing the risk of penalties due to non-compliance. example, bots can track unusual patterns in account activity, such as large, frequent transfers or international money transfers to high-risk jurisdictions. When suspicious activity is detected, the bot can trigger alerts to human auditors or compliance teams for further investigation, mitigating risks associated with fraud and ensuring that suspicious transactions do not go unnoticed. banking sector. In an industry where customer expectations are continually rising, banks must find ways to deliver faster, more personalized services while maintaining the highest levels of security processing times for customer requests. For example, when a customer applies for a loan, RPA can streamline the entire process, from application to approval, by automatically verifying documentation and assessing eligibility based on pre-defined criteria. This leads to faster decision-making, allowing customers to receive loan approval or denial more quickly than through manual processes. Additionally, RPA can support real-time communication with customers, providing them with instant responses to inquiries about their accounts or services customer's history and preferences. For instance, if a customer regularly deposits funds into a savings account, the bot may suggest new investment opportunities or savings plans that align with their financial goals.RPA also contributes to the growth of self-service banking. By using automated systems, customers can independently perform routine tasks, such as checking balances, updating contact information, or transferring funds between accounts, without needing to interact with bank staff. This empowers customers, offering them more control over their banking activities and improving their overall satisfaction.

Overcoming the Challenges of RPA Adoption in Banking: Addressing Data Quality Issues, Integration Challenges, and Change Management

These challenges include data quality issues, integration complexities, and the need for effective change management. Overcoming these barriers is crucial for banks to harness the full power of RPA and realize its transformative potential(Giudici, 2015). This section delves into the primary obstacles banks face during RPA adoption and outlines strategies for addressing data quality issues, overcoming integration challenges, and navigating the complexities of organizational change. banks face when implementing RPA is ensuring data quality. RPA is designed to work seamlessly with data-driven processes, and the accuracy of the data it processes directly impacts the performance of the automation. However, many banking systems still operate with legacy infrastructure or outdated data management practices, resulting in data that may be incomplete, inconsistent, or inaccurate. For RPA to function optimally, banks must ensure that the data it relies on is clean, accurate, and structured properly RPA bots may perform tasks incorrectly, leading to potential errors in financial reporting, transaction processing, and customer communications. For example, if customer records are missing important information, such as a valid address or updated contact details, RPA bots may fail to complete a transaction or initiate the correct action, resulting in delays, financial discrepancies, and customer dissatisfaction. Improving Datand consistency. Banks can then implement data cleansing techniques such as removing duplicateBanks should also consider data governance frameworks that include clear policies and procedures for data management across the organization. This will ensure that data is continuously monitored, cleaned, and updated in real-time, making it suitable for RPA-driven processes.Another significant challenge in the adoption of RPA in banking is integration with existing systems. Many financial institutions still rely on a combination of legacy systems and modern applications, creating a fragmented IT ecosystem. Integrating RPA tools with these disparate systems can be complex and time-consuming, particularly if the legacy systems were not designed with automation in mind.Legacy System IntegrationIn banks, legacy systems such as core banking platforms, transaction processing systems, and customer relationship management (CRM) tools may not be compatible with newer automation technologies.

This mismatch can make it difficult for RPA bots to access the data and functionality needed to automate key processes. For example, RPA bots may struggle to extract data from legacy databases or work with outdated application interfaces that do not support automation. smaller, more manageable areas before expanding to the entire organization. By prioritizing processes that involve less integration with legacy systems, such as account opening, payment processing, and report generation, banks can avoid the complexity of integrating RPA into legacy systems from the outset. allows modern RPA bots to interact with legacy systems by translating requests and responses between the two technologies. Additionally, APIs can enable real-time data exchange between systems, allowing RPA tools to access and automate tasks across multiple platforms Effective change management is essential to overcoming resistance to RPA adoption. Banks must create a culture of transparency and communication, where employees are informed about the benefits of automation and how it will improve their roles. Rather than focusing solely on job displacement, banks should emphasize how RPA can improve the quality of work by reducing monotonous and repetitive tasks, thereby allowing employees to focus on more strategic and value-added activities.(Giudici & Abu-Hashish, 2018)It is also important for banks to engage employees early in the RPA implementation process. Training programs and workshops should be organized to help employees understand how RPA works, how it will impact their daily operations, and what role they will play in the automation process. By providing employees with the necessary skills and knowledge, banks can foster a more positive outlook on RPA and reduce anxiety about job displacement.By involving employees in this process, banks can ensure that RPA remains aligned with business needs and that employees feel supported throughout the transition.

A Step-by-Step Guide to Implementing RPA in Banking: From Planning to Deployment

When selecting processes for RPA, banks should consider the following criteria:Repetitiveness: Processes that are performed frequently and consistently, such as account reconciliations, data entry, and report generation.Rule-Based Tasks: Processes that follow clear, predefined rules without much human judgment, such as loan application verification or compliance checks.Data-Driven: Processes that rely on structured data inputs, such as customer account management or transaction processing.Volume: High-volume processes that require significant manual effort and time, such as credit card fraud detection or KYC (Know Your Customer) verification.By focusing on processes with these characteristics, banks can ensure that RPA delivers quick wins and tangible benefits early in the implementation process(Jabłecki et al., 2015). Prioritizing Automation OpportunitieOnce the processes have been identified, banks should prioritize them based on factors such as impact, ease of implementation, and potential return on investment (ROI). High-impact processes that offer significant time savings, cost reduction, or risk mitigation should be automated first. These processes will also serve as proof of concept for future RPA deployments, demonstrating the benefits of automation to the wider organization.Banks need to evaluate their existing IT infrastructure to ensure that RPA can be seamlessly integrated into their systems. This includes reviewing their software platforms, databases, legacy systems, and cloud environments to determine whether RPA tools can be easily deployed and scaled. Additionally, banks should assess the data quality and structure to ensure that RPA bots will be able to access and process the necessary information without significant barriers.part of the business case is defining clear business objectives for RPA implementation. These objectives should be measurable, such as reducing transaction processing time by 30%, increasing operational efficiency by 20%, or improving customer satisfaction by 15%. Having specific goals will help measure the success of the RPA deployment and demonstrate the ROI to stakeholders.(Kersting, 2018) and ongoing maintenance. The expected savings from increased efficiency, reduced error rates, and lower labor costs should be factored in to calculate the potential ROI Scalability: The solution should be able to scale as the organization grows, with the ability to automate additional processes as neededIn addition to the software itself, banks must choose the right vendor to implement and support the RPA solution. The vendor should have experience in the banking sector, with a proven track record of successful RPA implementations(Hochreiter, 2015).

Vendors should also offer customized support to address the specific needs of financial institutions, such as regulatory compliance and data security RPA workflows should be designed with a clear understanding of the processes they are automating. Banks should work closely with subject matter experts (SMEs) from business units to ensure that the bots are designed to follow established procedures and rules. Workflow design involves specifying the steps the bot will take, the data it will access, and how it will interact with various systems.A pilot deployment is an essential part of the RPA testing process. This involves deploying the bots in a small, controlled environment to monitor their performance and identify any issues before full-scale deployment. During this phase, banks should gather feedback from users, monitor key performance indicators (KPIs), and make any necessary adjustments to improve bot performance.Banks should start by deploying RPA bots in the business units that were initially identified for automation. The bots should be integrated into the daily workflows of these departments, ensuring that they can handle tasks autonomously while interacting with other systems and processes. Full deployment should be carried out in phases to ensure a smooth transition and avoid disrupting business operations.ensuring that the bots continue to operate smoothly after deployment. RPA is not a one-time implementation but rather an ongoing process that requires continuous support and maintenance.Banks must also provide ongoing training and support for employees who interact with RPA bots. As business processes evolve, employees must be educated on how to work with new bots and how to troubleshoot any issues that arise. Change management initiatives should continue to support the workforce, ensuring that employees are comfortable with the automation tools.

Best Practices for RPA Implementation in Banking: Ensuring Success and Minimizing Risk

A successful RPA implementation in banking hinges on selecting the right processes for automation. The most suitable tasks for RPA are typically rule-based, repetitive, and time-consuming processes that involve structured data. For instance, routine tasks like transaction reconciliation, account updates, loan processing, and customer verification are ideal candidates for RPA. It is essential to start small with well-defined, low-risk processes, and gradually expand automation to more complex workflows as the organization becomes more comfortable with the technology.Best Practice:Use data-driven insights and collaboration with process owners to identify tasks that are repetitive, time-intensive, and prone to errors.Avoid automating overly complex or unstructured processes in the early stages of RPA adoption.Conduct a thorough process audit to assess automation feasibility and understand the potential impact on the workforce.. Involve Key Stakeholders Early and OftenEngaging key stakeholders early in the RPA implementation process is critical to ensuring the technology addresses the right needs and delivers value across the organization. This includes collaborating with IT, compliance teams, business unit leaders, and end-users who will interact with the automated processes. Their involvement helps to ensure that the RPA solution is tailored to the bank's specific needs, supports regulatory compliance, and integrates seamlessly with existing systems.(Makrichoriti & Moratis, 2016) Focus on Data Quality and IntegratioData is at the heart of RPA, and without high-quality, structured data, the automation process will not be effective. Financial institutions must ensure that their data is clean, accurate, and well-organized before automating any processes. Additionally, integrating RPA with existing core banking systems and other IT infrastructure is crucial to the success of automation. Data should be readily accessible and consistent across platforms to ensure RPA bots can perform tasks without interruptions.(Kabašinskas et al., 2017) Best Practice:Ensure RPA tools can integrate seamlessly with legacy systems, cloud platforms, and other applications used within the bank.Regularly update and clean databases to prevent errors in automated processes.

Build a Robust Change Management PlanRPA adoption in banking often requires significant changes in workflows, employee roles, and organizational structures. To minimize resistance and ensure a smooth transition, it is essential to have a robust change management plan in place. This plan should address how RPA will impact employees, outline new responsibilities, and provide training and upskilling opportunities for staff. Proper communication and transparency about the benefits of RPA will help ease concerns and create a positive atmosphere for change.A successful RPA implementation in banking does not happen overnight. It is essential to start with small-scale, low-risk processes to demonstrate the effectiveness of RPA before expanding to more complex workflows. This approach allows for testing, refining, and optimizing the bots, ensuring that the technology delivers the expected results and minimizing the risk of failure.Best Practice:Begin with pilot projects or proof of concept (PoC) to validate the effectiveness of RPA in real-world scenarios.Gradually scale up automation efforts, expanding to more complex processes as the organization becomes more confident in its RPA capabilities.Continuously track performance metrics, gather feedback, and make adjustments to improve RPA workflows. Implement Robust Security and Compliance ProtocolsBanking is a highly regulated industry, and RPA solutions must be designed and implemented in compliance with industry standards, data protection laws, and regulatory requirements. Financial institutions must implement robust security protocols to ensure that automated processes are secure and do not expose the bank to cybersecurity risks. Additionally, RPA should support compliance with regulations such as Anti-Money Laundering (AML), Know Your Customer (KYC), and data privacy laws like GDPR.Once RPA is implemented, ongoing monitoring and optimization arecrucial for maintaining its effectiveness. Financial institutions should track key performance indicators (KPIs) such as process efficiency, error rates, cost savings, and customer satisfaction to assess the impact of RPA. Regular performance reviews help identify bottlenecks, improve automation accuracy, and optimize the bots to handle new tasks or changing business needs.Best Practice:Set up a centralized RPA monitoring system to track the performance of bots in real time.Conduct regular audits to ensure that RPA processes are meeting the defined objectives and delivering value.

Case Study: Implementing RPA in a Large Bank to Automate AML and KYC Processes

To address these challenges, a large international bank sought to implement Robotic Process Automation (RPA) to automate its AML and KYC workflows. This case study explores the bank's journey through the RPA implementation, the challenges faced, the benefits realized, and the lessons learned in transforming the bank's compliance processes.Background: The Need for AutomationThe bank, with a global presence, handles millions of transactions daily, each of which must be subjected to rigorous AML checks and KYC verifications. Historically, these compliance tasks were carried out manually, involving extensivewas also experiencing significant operational costs associated with compliance activities, which were escalating as regulatory scrutiny intensified.In addition to these operational inefficiencies, the increasing regulatory pressure for faster reporting, improved audit trails, and enhanced customer verification processes highlighted the need for a more streamlined and scalable solution. The bank recognized that the adoption of RPA could be the key to reducing the time and costs involved in compliance while ensuring that it remained compliant with ever-evolving regulatory standards.Implementing RPA for AML and KYC AutomationThe decision to automate AML and KYC processes using RPA was driven by the need for greater efficiency, accuracy, and consistency across the bank's compliance functions. The bank's RPA implementation involved several phases: (Mantegna, 2009)identification of processes for automation, tool selection, integration with existing systems, and scaling the solution across the organization.The first step was to identify which specific AML and KYC tasks were suitable for automation. These included routine activities such as customer identity verification, data extraction from documents, screening customers against sanctions lists, updating customer profiles, and monitoring suspicious activities. Tasks that were repetitive, involved structured data, and required minimal human judgment were prioritized for automation.Best Practices for Process Selection:Start with processes that are rule-based and well-defined to ensure quick wins.Conduct a thorough assessment of the manual tasks to gauge their impact and potential for automation.Avoid automating tasks that require complex decision-making or subjective interpretation.After identifying the key processes for automation, the bank evaluated various RPA tools and selected one that was compatible with its existing IT infrastructure and regulatory frameworks. The chosen RPA tool offered advanced capabilities such as AI-based document recognition, machine learning for transaction monitoring, and real-time data processing, making it an ideal fit for automating the bank's AML and KYC workflows.The integration of RPA with the bank's core banking systems, customer databases, and regulatory reporting platforms was a critical aspect of the project. Ensuring seamless data flow between different systems was essential for ensuring that the RPA bots could perform tasks such as screening client profiles, extracting relevant data from customer documents, and flagging suspicious transactions without causing delays or data integrity issuesThe bank began by running a pilot phase, deploying RPA bots on a limited set of tasks, such as customer screening and document verification.

The goal was to assess the effectiveness(Mahadev Madgule et al., 2023) of the bots, validate the accuracy of the automation, and identify any technical or operational issues that could arise during full-scale deployment. During this phase, the bots were trained to handle various types of customer documentation, such as passports, utility bills, and bank statements, while ensuring compliance with regional and international KYC regulations.he testing phase also involved rigorous validation against the bank's compliance requirements to ensure that the automated processes met the necessary regulatory standards. With positive results from the pilot phase, the bank moved forward with scaling the RPA solution across its various branches and operations.Best Practices for Testing and Deployment:Start with a pilot program to test the bots on a small scale before full deployment.Ensure that bots are thoroughly tested against regulatory standards and real-world scenarios. Use feedback from pilot users to optimize bots and identify areas for improvement.. Scaling and OptimizationOnce the RPA solution was fully deployed for AML and KYC automation, the bank scaled the solution to handle larger volumes of transactions and customer profiles. The bots were able to process thousands of customer records per day, screening and verifying data more efficiently than human agents. Additionally, RPA bots could work 24/7 without interruptions, significantly reducing the processing time for KYC verification and AML checks. that could be reallocated to higher-value activities, such as customer service and relationship management. Regulatory Compliance: Ensuring compliance with the regulatory framework was an ongoing concern. The bank had to continuously update its RPA bots to comply with new and changing AML and KYC regulations. This required regular audits and updates to the automation processes to ensure that the bots remained compliant.. Change Management(Nakamoto, 2009)

Example of RPA in Risk Management: Automating Risk Assessment and Compliance Processes in a Small Bank

In the banking sector, risk management is a fundamental aspect of ensuring financial stability and regulatory compliance. For small banks, the complexity of managing various risks— This case study explores how a small community bank implemented RPA to automate its risk assessment and compliance processes, significantly improving efficiency, reducing operational risks, and ensuring regulatory compliance.Background:(Papenbrock & Schwendner, 2015) The Need for Automation in Risk Management The small bank in question, serving a regional market, faced several challenges related to its risk management processes. The manual nature of its risk assessment procedures, coupled with an increasing volume of regulatory requirements, made it difficult for the bank to maintain efficiency and consistency in its operations. Risk management tasks such as assessing loan applications, evaluating customer creditworthiness, ensuring compliance with Anti-Money Laundering (AML) regulations, and conducting periodic risk assessments were time-consuming and often prone to human error. These inefficiencies not only slowed down operations but also exposed the bank to the risk of non-compliance and financial penalties.Implementing RPA for Risk Management and Compliance Automation The first step in the bank's RPA implementation was to identify the specific risk management and compliance processes that were suitable for automation. The bank collaborated with its risk management and compliance teams to assess which tasks were most repetitive and rule-based, making them ideal candidates for automation. Some of the tasks identified for automation included Credit Risk Assessment: The bank manually assessed the creditworthiness of loanapplicants by reviewing credit reports, financial statements, and historical transaction data. This process was time-intensive and susceptible to human error.Customer Due Diligence (CDD) and KYC Compliance: The bank conducted Know Your Customer (KYC) checks as part of its AML compliance procedures. This included verifying customer identities, cross-referencing data with global watchlists, and ensuring that customers were not involved in financial crimes.Regulatory Reporting: The bank was required to generate various reports for regulators, including transaction monitoring reports, suspicious activity reports (SARs), and AML compliance documentation. These reports needed to be generated quickly and accurately to avoid regulatory penalties.Risk Monitoring and Alerts: The bank had a manual system for monitoring risk-related activities, such as unusual transactions, large withdrawals, and non-compliance with lending guidelines. This process involved sifting through vast amounts of transactional data, which was inefficient and prone to oversight.Once the processes were identified, the bank selected an RPA solution that could integrate seamlessly with its existing core banking systems, databases, and risk management platforms. The RPA tool was designed to handle structured and semi-structured data, which made it well-suited for the bank's requirements.

Additionally, the tool had built-in capabilities for integrating with external data sources, such as credit bureaus, government databases, and regulatory reporting platforms.Key Stages of RPA Implementation The implementation of RPA at the small bank was carried out in several stages, ensuring that the automation process was gradual, scalable, and efficient. Process Mapping and Analysis The first stage of the RPA implementation involved process mapping and analysis. The bank's risk management and compliance teams worked closely with RPA experts to map out each of the identified processes, documenting every step, decision point, and data flow. This step was crucial for understanding the intricacies of each task and determining how best to automate it. Additionally,(Pontiveros et al., 2018) the bank's IT team collaborated with the RPA developers to ensure that the automation could integrate smoothly with the bank's core systems.est Practices:Conduct a detailed process audit to understand where inefficiencies and bottlenecks occur in the existing workflows. Involve key stakeholders in process mapping, such as risk managers, compliance officers, and IT specialists, to ensure the automation meets the bank's needs.Identify tasks that are rule-based and repetitive to ensure they are the best candidates for automation.. Pilot Automation ImplementationThe next stage was to implement RPA on a small scale. The bank decided to start by automating its credit risk assessment process, which involved evaluating loan applications. The bank integrated(Post & Potì, 2016) the RPA tool with its core banking system and external credit bureaus to automate the collection of credit reports and financial data. The RPA bots were programmed to evaluate the creditworthiness of applicants by applying predefined rules based on the bank's credit policy.During the pilot phase, the bank closely monitored the performance of the RPA bots, collecting feedback from users, and making adjustments as needed. This phase allowed the bank to test the automation in real-world conditions and identify any potential issues before scaling it up across the entire organization.(Ramakrishnan et al., 2022)Best Practices: Start with a pilot project that automates a single, well-defined process to test the effectiveness of the RPA solution.Gather feedback from stakeholders involved in the pilot and refine the bots based on their inpu Continuously monitor bot performance to ensure accuracy and compliance with regulatory requirements. Full-Scale ImplementationAfter successfully testing the RPA solution in the pilot phase, the bank moved forward with full-scale implementation. This involved automating additional processes such as customer due diligence (CDD), KYC compliance, and regulatory reporting. The RPA bots were programmed to verify customer identities, cross-reference data with global watchlists, and generate required compliance reports. By automating these processes, the bank was able to significantly reduce the manual effort involved in compliance, improve accuracy, and reduce the time spent on risk assessments and reporting.The implementation of RPA for risk management and compliance processes yielded significant benefits for the small bank. the time spent on manual tasks(Sreenivasulu et al., 2022). The bank had to ensure that the RPA bots adhered to changing regulatory standards, which required frequent updates to the automation scripts.hange Management: Employees were initially concerned about job displacement. However, the bank managed this change effectively by focusing on how RPA could improve their work and reduce the burden of repetitive tasks.The small bank's experience with RPA in risk management and compliance processes demonstrates the power of automation in transforming the way financial institutions handle risk and adhere to regulations. By automating credit risk assessments, KYC verifications, and regulatory reporting,

FUTURE DIRECTIONS IN RPA FOR BANKING: EMERGING TRENDS AND OPPORTUNITIES

In this section, we explore the future directions of RPA in banking, focusing on emerging trends, opportunities, and the transformative impact they will have on the industry. where processes become more dynamic, adaptive, and capable of handling unstructured data. This shift allows banks to automate tasks that require cognitive decision-making, pattern recognition, and predictive analytics. customer communications, and scanned documents, which are often difficult to handle with traditional RPA. By using AI algorithms to extract relevant data from these unstructured sources, RPA bots can complete tasks like contract management, customer sentiment analysis, and fraud detection more efficiently. Additionally, AI and ML can enable RPA bots to learn from past actions and adapt to new scenarios, improving their performance over time without the need for constant reprogramming. support: AI chatbots, when integrated with RPA, can handle complex customer inquiries, process transactions, and resolve issues faster, improving customer satisfaction., RPA has been primarily used for front-office tasks, such as customer service, compliance reporting, and transaction processing. However, the future of RPA in banking will see its expansion into more complex back-office functions, such as credit risk assessments, financial forecasting, and portfolio management. These areas, traditionally reliant on human expertise and judgment, will increasingly benefit from RPA's ability to manage vast volumes of data and perform complex calculations with greater speed and accuracy. For instance, RPA can be integrated with advanced analytics tools to automate the process of credit risk analysis by aggregating data from multiple sources, assessing borrower creditworthiness, and providing real-time risk scores. Similarly, RPA can streamline financial forecasting by automatically pulling financial data, analyzing historical trends, and generating accurate predictions for future financial performance. Credit risk automation: Automating the risk assessment process by analyzing multiple data points, including credit history, financial reports, and market conditions, allows for faster and more accurate credit decisions.Portfolio management: By automating portfolio monitoring and adjusting investment strategies based on real-time market conditions, RPA can help wealth managers provide personalized services to clients more efficiently.Financial reporting: RPA can automate the creation and submission of regulatory reports, ensuring that they are accurate, timely, and compliant with evolving regulations.. Process Mining and Intelligent Process (Sugumaran et al., 2024)DiscoveryThe next frontier for RPA in banking involves integrating process mining and intelligent process discovery techniques to identify and optimize processes for automation. Process mining uses data from enterprise systems to map out business processes, uncover inefficiencies, and identify bottlenecks.

Recommendations for Banks: Implementing RPA to Enhance Risk Management and Fraud Detection

To fully leverage RPA in risk management and fraud detection, banks must implement a strategic approach that combines the right technologies with effective governance and a focus on continuous improvement. The following recommendations outline best practices for banks looking to integrate RPA into their risk management and fraud detection frameworks. For RPA to be most effective, it must be applied to repetitive, rule-based tasks that are prone to human error or inefficiencies. Banks should begin by mapping out their current risk management and fraud detection processes to uncover areas where RPA can add the most value. KYC (Know Your Customer) checks: Automating KYC processes, such as verifying customer identities and cross-referencing data with global sanctions lists, helps banks mitigate the risk of financial crimes, such as money laundering, and ensures compliance with regulatory requirements. can continuously scan customer transactions against predefined criteria, such as large withdrawals, rapid transfers, or frequent small payments, which are indicative of money laundering activities.Credit risk assessments: Automating credit risk assessments ensures faster and more accurate evaluations of loan applications. By pulling in data from multiple sources, such as credit reports and financial statements, RPA bots can create real-time credit scores and flag high-risk customers for further review.Fraud detection systems: Fraud detection processes, such as monitoring card transactions for anomalies or screening digital payments for irregularities, can benefit from RPA's ability to analyze vast amounts of data rapidly. RPA bots can flag suspicious activities, initiate investigations, and trigger alerts to human analysts.Recommendation:Process Mapping: Work closely with risk managers and compliance officers to map out all relevant risk management processes. Identify tasks that are highly repetitive, data-intensive, and governed by clear rules, making them ideal candidates for RPA automation. Collaborative Analysis: Engage cross-functional teams, including(Talonen et al., 2016) IT, compliance, and operations, to identify pain points and gaps in existing processes where RPA can make the most impact.In fraud detection, for instance, machine learning algorithms can be trained on historical transaction data to identify patterns and predict fraudulent behavior. RPA bots can then use these predictions to automatically flag suspicious transactions, escalating only high-probability fraud cases for human review. Similarly, in risk management, AI-powered analytics can assist RPA bots in assessing the likelihood of risk based on complex datasets, such as market conditions, customer behavior, and external risk factors.One of the most significant advantages of RPA is its ability to streamline compliance tasks, such as reporting and documentation,(Tasca & Hayes, 2016) which are crucial for maintaining regulatory adherence in the banking sector.

The Role of Human-AI Collaboration in RPA for Banking: Balancing Efficiency and Effectiveness with Human Judgment

AI-powered RPA bots is essential for ensuring that banks not only achieve operational excellence but also maintain a high standard of ethical decision-making, customer service, and strategic insight. The role of human-AI collaboration in RPA for banking involves a delicate balance between leveraging the efficiency of automation and maintaining the effectiveness of human oversight, creativity, and critical thinking. Leverage RPA to automate rule-based tasks, while using AI to enhance the bot's cognitive abilities, such as detecting fraud or assessing credit risk, thereby creating a more efficient and intelligent automation environmentContinuous Feedback Loop: Establish mechanisms for human employees to provide feedback on AI outputs, ensuring that the system adapts and improves over time, learning from human oversight and decision-making.. Addressing Complex Decisions with Human JudgmentWhile AI and RPA excel in automating routine tasks and processing large volumes of data, there are still many banking tasks that require human judgment, intuition, and experience. AI may struggle with understanding context, recognizing subtle nuances, or making ethical decisions in complex situations, areas where human expertise and empathy are irreplaceable.Human judgment is also crucial when RPA bots are tasked with tasks that involve ethical considerations, such as ensuring fairness in lending decisions or adhering to complex regulatory guidelines. The Role of Humans in RPA Monitoring and OptimizatioWhile RPA can automate tasks, it still requires ongoing human oversight to ensure its efficiency and accuracy. RPA bots are highly effective at handling well-defined tasks, but they need to be constantly monitored and adjusted to handle exceptions, adapt to changing conditions, and ensure that they remain aligned with organizational goals. By encouraging collaboration between humans and AI, banks can foster a work environment where both parties complement each other, leading to enhanced performance and job satisfaction.(Vijayakumar et al., 2024).

CONCLUSION

The transformative power of Robotic Process Automation (RPA) in the banking industry cannot be overstated. As financial institutions face mounting pressure to improve operational efficiency, mitigate risk, and provide exceptional customer service, RPA has emerged as a pivotal tool that addresses these challenges while driving significant advancements across various banking functions. From automating routine tasks to enhancing fraud detection, RPA's role in streamlining operations is reshaping the future of banking. This conclusion summarizes the key findings on the impact of RPA in banking, reflecting its capacity to deliver operational excellence, mitigate risks, and elevate the customer experience.operational efficiency that RPA enables also extends to back-office functions, where tasks such as compliance reporting, regulatory filing, and audit preparation can be automated, ensuring a quicker turnaround and higher accuracy in meeting regulatory standards. This not only leads to significant reductions in operational costs but also ensures that banks are operating at peak efficiency, ultimately benefiting their bottom line.Moreover, RPA can scale effortlessly to meet increased demand, particularly during periods of high transaction volumes, such as financial year-ends or tax season. The scalability of RPA ensures that banks can maintain high service levels without having to expand their workforce or invest in additional infrastructure. Through automation, banks can handle large volumes of work with consistent quality and speed, positioning themselves for greater profitability and resilience.nhancing Risk Management and Fraud DetectionRisk management is a core function for any banking institution, and RPA has proven invaluable in enhancing both risk assessment and fraud detection processes. RPA, in conjunction with RPA systems can promptly alert human employees, enabling quicker responses to mitigate risks. For example, automated bots can identify transactions that deviate from standard behavior patterns, such as large transfers or high-risk geographical locations, allowing banks to prevent fraud before it occurs. These bots not only increase the speed of fraud detection but also reduce the chances of human oversight, which is often a source of vulnerability. meet these demands by automating tasks that directly impact the customer journey. Automated systems, for instance, can handle routine inquiries, process simple transactions, and resolve basic issues quickly, providing customers with immediate responses and reducing wait times.The automation of customer service tasks using RPA ensures that human agents are available for more complex or specialized inquiries, such as resolving complaints, handling high-value transactions, or offering personalized financial advice. This dual approach — automation for routine tasks and human intervention for complex issues — creates a more efficient and satisfying customer experience.Additionally, RPA enables faster processing of loan applications, credit assessments, and to RPA's ability to automate data gathering, document verification, and risk assessment.Another area where RPA improves customer experience is through personalization. Automated bots can analyze vast amounts of data about customer behavior, transaction history, and preferences, providing insights that allow banks to offer tailored services. Personalized financial advice, targeted promotions, and customized offers can all be delivered through RPA systems, making the customer experience more relevant and engaging.To address these challenges, banks should prioritize the creation of a comprehensive RPA strategy that includes stakeholder engagement, staff training, and the selection of the right technology partners. Starting with smaller, low-risk processes and gradually expanding RPA implementation will allow banks to test the waters before scaling up. Furthermore, investing in the training and development of existing staff to handle more complex tasks will help mitigate concerns about job displacement. personalized customer interactions, and intelligent decision-making. For example, AI-powered RPA bots will be able to make real-time decisions, such as identifying high-value customers or recommending financial products based on a customer's behavior and preferences.

REFERENCES

Alexy, M., Georgantzis, N., Káčer, M., & Péliová, J. (2016). Risk attitude elicitation methods: Do they tell similar stories? *Ekonomicky Casopis*, *64*, 847–877.

Arakelian, V., & Dellaportas, P. (2012). Contagion determination via copula and volatility threshold models. *Quantitative Finance*, *12*, 295–310. DOI: 10.1080/14697680903410023

Baitinger, E., & Papenbrock, J. (2017). Interconnectedness risk and active portfolio management. *The Journal of Investment Strategies*, *6*, 63–90. DOI: 10.2139/ssrn.290983

Barucci, E., & Marazzina, D. (2016). Asset management, high water mark and flow of funds. *Operations Research Letters*, *44*, 607–611. DOI: 10.1016/j.orl.2016.07.002

Battiston, S., Delli Gatti, D., Gallegati, M., Greenwald, B., & Stiglitz, J. E. (2012). Liaisons dangereuses: Increasing connectivity, risk sharing, and systemic risk. *Journal of Economic Dynamics & Control*, *36*, 1121–1141. DOI: 10.1016/j.jedc.2012.04.001

Bernè, F., Ciprian, M., Fanni, N., Marassi, D., & Pediroda, V. (2006). Multi criteria credit rating (MCCR): A credit rating assignment process for Italian enterprises according to BASEL II. *J. Finan. Dec. Making*, *2*, 1–26.

Cerchiello, P., & Giudici, P. (2015). How to measure the quality of financial tweets. *Quality & Quantity*, *50*, 1–19. DOI: 10.1007/s11135-015-0229-6

Chen, C. Y. H., Härdle, W. K., Ai, J. H., & Wang, W. (2018) Pricing cryptocurrency options: the case of CRIX and Bitcoin. SFB DP. doi: DOI: 10.2139/ssrn.3159130

Cruz, M. G. (2002). *Modelling, Measuring and Hedging Operational Risk*. Wiley Finance.

De Filippi, P., & Hassan, S. (2016). Blockchain Technology as a Regulatory Technology: From Code is Law to Law is Code. Chicago, IL. *First Monday*. Advance online publication. DOI: 10.5210/fm.v21i12.7113

Diebold, F., & Yilmaz, K. (2014). On the network topology of variance decompositions: Measuring the connectedness of financial firms. *Journal of Economics*, *182*, 119–134. DOI: 10.1016/j.jeconom.2014.04.012

European Commission. (2018). *Fintech Action Plan: For a More Competitive and Innovative European Financial Sector*. European Commission.

Financial Stability Board. (2017a). *Artificial Intelligence and Machine Learning in Financial Services: Market Developments and Financial Stability Implications*. Financial Stability Board.

Financial Stability Board. (2017b). *Financial Stability Implications for Fintech: Supervisory and Regulatory Issues that Merit Authorities' Attention*. Financial Stability, Board.

Foley, S., Karlsen, J. R., & Putnins, T. J. (2018). Sex, Drugs and Bitcoin: How Much Illegal Activity is Financed Through Cryptocurrencies. London: Working Paper.

Giudici, P. (2003). *Applied Data Mining*. Wiley.

Giudici, P. (2015). Scorecard models for operational management. *Int. J. Data Sci.*, *1*, 96–101. DOI: 10.1504/IJDS.2015.069055

Giudici, P., & -Hashish, A. (2018). What determines bitcoin exchange prices: A network VAR approach. [Epub ahead of print]. *Finance Research Letters*. Advance online publication. DOI: 10.1016/j.frl.2018.05.013

Hochreiter, R. (2015). Computing trading strategies based on financial sentiment data using evolutionary optimization. *Adv. Intell. Syst. Comp.*, *378*, 181–191. DOI: 10.1007/978-3-319-19824-8_15

Jabłecki, J., Kokoszczynski, R., Sakowski, P., Slepaczuk, R., & Wójcik, P. (2015). *Volatility as an Asset Class, Obvious Benefits and Hidden Risks*. Peter Lang.

Kabašinskas, A., Šutiene, K., Kopa, M., & Valakevičius, E. (2017). The risk–return profile of Lithuanian private pension funds. Econ. Res. *Ekonomska Istrazivanja*, *30*, 1611–1630. DOI: 10.1080/1331677X.2017.1383169

Kersting, K. (2018). Machine learning and artificial intelligence: Two fellow travelers on the quest for intelligent behavior in machines. *Frontiers in Big Data*, *1*, 6. DOI: 10.3389/fdata.2018.00006 PMID: 33693322

Madgule, M., Vinayaka, N., Sonkhaskar, Y. M., Patel, D. R., Karthikeyan, R., & Selvakumar, P. (2023). Mechanical properties and microstructure of activated TIG welded similar joints of Inconel alloys by desirability approaches. *Materials Today: Proceedings*, *77*, 528–533.

Makrichoriti, P., & Moratis, G. (2016). BitCoin's Roller Coaster: Systemic Risk and Market Sentiment. Athens: Working Paper.

Mantegna (2009). The European Physical Journal B: Condensed Matter and Complex Systems, Vol. 11. Berlin, 193–197.

Nakamoto, S. (2009). Bitcoin: A Peer-to-Peer Electronic Cash System. White Paper.

Papenbrock, J., & Schwendner, P. (2015). Handling Risk On/Risk Off dynamics with correlation regimes and correlation networks. *Financial Markets and Portfolio Management*, *29*, 125–147. DOI: 10.1007/s11408-015-0248-2

Pontiveros, B. B. F., Norvill, R., and State, R. (2018). Monitoring the Transaction Selection Policy of Bitcoin Mining Pools. To appear in NOMS 2018 - IEEE/IFIP Man2Block

Post, T., & Potì, V. (2016). Portfolio analysis using stochastic dominance, relative entropy, empirical likelihood. *Management Science*, *63*, 153–165. DOI: 10.1287/mnsc.2015.2325

Ramakrishnan, T., Mohan Gift, M. D., Chitradevi, S., Jegan, R., & Subha Hency Jose, P. Nagaraja, Rajneesh Sharma H.N., Selvakumar P., Sintayehu Mekuria Hailegiorgis, (2022). "Study of Numerous Resins Used in Polymer Matrix Composite Materials", Advances in Materials Science and Engineering, vol. 2022, Article ID 1088926, 8 pages, 2022. https://doi.org/DOI: 10.1155/2022/1088926

Sreenivasulu, A., & Manimozhi, T. Satheesh kumar P., Selvakumar P. and Singh P.,(2022). "The Model of Smart Sensing Device For Sensitive Nanoclusters Modification in Sensing Properties," 2022 2nd International Conference on Advance Computing and Innovative Technologies in Engineering (ICACITE), Greater Noida, India, 2022, pp. 1043-1047, doi: DOI: 10.1109/ICACITE53722.2022.9823846

Sugumaran, K. M. D., Pandiyarajan V., Y. B S, Selvakumar P. and M. S, (2024). "Analysis of Electromyography Signals for Control of Mechanical Prosthesis using Machine Learning Techniques," 2024 2nd International Conference on Intelligent Data Communication Technologies and Internet of Things (IDCIoT), Bengaluru, India, 2024, pp. 1059-1067, doi: DOI: 10.1109/IDCIoT59759.2024.10467707

Talonen, A., Kulmala, J., & Ruuskanen, O. P. (2016). "Co-operative platforms: harnessing the full potential of crowdfunding," in European Conference on Innovation and Entrepreneurship (Jyväskylä: Academic Conferences International Limited), 810.

Tasca, P., & Hayes, P. (2016). *"Blockchain and crypto currencies," in The Fintech Book*. Wiley., Available online at https://thefintechbook.com/

Vijayakumar, G., Rajkumar, M., Chandar, N. R., Selvakumar, P., & Duraisamy, R. (2024). Environmentally friendly TDS removal from waste water by electrochemical ion exchange batch-type recirculation (EIR) technique. *Environmental Science. Water Research & Technology*, *10*(4), 826–835.

About the Contributors

Nik Herda Binti Nik Abdullah holds a PhD in Accountancy from Universiti Teknologi MARA and an MBA in Finance from Universiti Malaya, Malaysia. She is a Senior Lecturer and Programme Director for Accounting programmes at the School of Accounting and Finance, Taylor's University. Her research interests encompass management accounting, digitalization, fintech, blockchain, Big Data, and sustainability. Dr. Nik Herda has published widely in Scopus-indexed journals and received multiple accolades, including Best Paper and Best Presenter awards at international conferences. She contributes to academic governance through her roles in the University Human Ethics Scientific Review Panel. At the community level, her involvement as a trainer in SME entrepreneurship programmes reflects her commitment to translating research into capacity building. Collectively, these contributions illustrate how her research not only advances academic knowledge but also creates meaningful value for policymakers, industry stakeholders, students, and the wider community.

Kelvin Lee is a Senior Lecturer and Programme Director at Taylor's University, Malaysia. He earned a First-Class Degree Honours in Bachelor of Finance (Hons) in 2014, followed by a Master of Science in Finance, supported by the prestigious Zamalah and MyBrain Scholarships in 2016. Dr. Kelvin also completed his Ph.D. in Finance from 2017 to 2021 and was honored with the Best Postgraduate Award at the end of his Ph.D. program. In 2023, Dr. Kelvin showcased his commitment to financial education by conducting webinars and workshops for over 500 SMEs across Malaysia. These sessions focused on enhancing SME's financial literacy and contributing to their long-term success and sustainability. Beyond teaching, Dr. Kelvin is actively engaged in research and publication. He consistently contributes to scholarly literature with papers published in reputable Scopus and WOS indexed journals, showcasing his dedication to advancing knowledge in the field of finance. In 2023, Dr Kelvin also achieved a major milestone by securing a national research grant for his project titled " Revolutionizing Climate Finance for Malaysian SMEs in the Food and Beverage Manufacturing Sector". This project reflects Dr. Kelvin's commitment to addressing crucial issues at the crossroads of finance and climate. His success in obtaining the grant underscores his expertise and contributions to cutting-edge research in the field. Dr. Kelvin also plays a significant role in promoting responsible management education and sustainability initiatives at Taylor's University. He serves as a committee member for TBS UNPRME (Taylor's Business School - Principles for Responsible Management Education), which focuses on reporting annual Sharing Information on Progress (SIP) and advancing sustainability management activities across the university. Additionally, Dr. Kelvin contributes his expertise as part of the Taylor's University Environmental, Social, and Governance (TU-ESG) committee, where he actively engages in fostering a culture of sustainability within Taylor's University. Through his involvement in these committees, Dr. Kelvin demonstrates his commitment to integrating sustainability principles into teaching and learning, research, and community engagements, contributing to the university's efforts towards responsible management practices and environmental stewardship. Recently, Dr. Kelvin has also been invited to deliver a keynote speech at three conferences, including: (i) the International Conference on Increased Use of Artificial Intelligence (AI) & Its Impact (21 March 2024); and (iii) the 2024 Annual International Conference on "Harmony in Innovation" (3 April 2024 to 5 April 2024).

Irfah Najihah Binti Basir Malan is a Senior Lecturer at Universiti Teknologi MARA Cawangan Melaka Kampus Bandaraya Melaka.

Sumanta Bhattacharya from Salkia, Howrah, West Bengal is an Expert commissioner member IUCN Environmental, economics and social policy, Individual Policy Analyst (ADB) . He is among the top 2% world scientist (AD Index), Special civilian member Ministry of Defence, (USI) Member at Ministry of Culture (AS) . Member of Ministry of Defence { [Defence and strategic analysis] IDSA }, Member of World Economic Forum, littérateur at National Economic Research Bureau, Member of Ministry of Personnel, Pension and Public grievance (IIPA) He completed his B.Tech, M.Tech in Textile Technology along with that he has MA in Development studies, M.Sc in Environmental Science, LLB, MA in security and defence law from NALSAR, MA in Economics, MBA, Post Graduate Diploma in Environment and Sustainable Development, DIA&D, DS&DS, MPI(Oxford University) . Being a research scholar he has 290 research papers published in International and Scopus index journal, Wiley, Springer, Hindawi in various sectors like law, economic, Development studies, Textiles, Climate Change, SDGs, Public policy, Nanotechnology, Biotechnology, cancer, environment studies,

100 Book chapters published, Edited 2 books, wrote 4 book, 95 patents International and National, 4 copyrights, attended 150 International and National conference and presented papers and won 29 awards for his excellent in education, policy making and innovation which includes Best Water Management Solution Award 2022, Young Researcher Award 2020, Bravo International Book of Records, Mahatma Gandhi Ratna Award 2020, Sustainable Technology Award 2022, Global Change Maker Award 2020, Best Water Management Solution Award at the Rural & Urban Development Summit & Awards 2022, New Delhi, The Progress Global Awards 2023 for Excellence in Research, GIS -Person of the year – Water Conservation, Green India award, 2023, Global Legend Award 2023 by GSIL Educational Society . He is also a member of various International and National organization which includes IEI, IICHE, Asiatic Society, IIPA, IDSA, YOUNGO, IUCN, USI and many more .

Sweta Goel is associated with Jaypee business school, JIIT, Noida as an Assistant Professor (Sr. Grade) in finance and accounting. She has around 15 years of teaching and research experience. She has contributed research papers in various national and international journals of repute. She has been actively engaged in guiding the PhD research scholars, reviewing the research papers in reputed indexed journals and evaluating the PhD thesis.

Hamsatulazura Hamzah is a lecturer at the Faculty of Management, Multimedia University, Cyberjaya, Selangor. She holds a Masters in Accounting (Hons) from Curtin University, Australia and a Bachelor of Science (Hons) in Accounting from Cardiff University, United Kingdom. She has extensive experience in teaching and research in the field of financial accounting and management accounting. Her main research areas are corporate governance, management accounting systems and management control systems where some research results have been published in several selected journals and conference proceedings. She is also involved in research related to entrepreneurship and financial literacy. She is also an 'HRDF Certified Trainer' and has a 'Professional Certificate for Entrepreneurship Educators'. Often invited as a speaker and trainer to share entrepreneurship and management knowledge by non-governmental organizations and government agencies such as Kementerian Pembangunan Wanita dan Keluarga, UKM and INTI-UC. She was also involved with the 'Start-up' Scheme at the university level and had won an award as a result of the scheme. Actively involved and had organized various social responsibility activities, and holds several important positions in several non-governmental organizations such as the Chairman of the Pertubuhan Ikatan Wanita Prihatin (PIWP) and the Treasurer of the Persatuan Penduduk Bandar Baru Bangi (BANGIKU). Ever held positions as the Treasurer of the Main Committee of the

Islamic Women's Council of YADIM (MAYA) and YAPEIM's Special Procurement Committee.

Amirah Fatini Johari is a postgraduate student at Taylor's University

Manjit Kour is working as Professor at University Business School at Chandigarh University. She holds a PhD in Business Management from IK Gujral Punjab Technical University, India. She has a good research record with several research papers published in reputed journals. She is an editorial board member and reviewer of many reputed journals including Scopus and ABDC-indexed journals. Her areas of interest include sustainability, green practices, Fintech, online marketing trends and business ethics.

T.C. Manjunath was born on 6th February 1967. He did his B.E. from RV College of Engg. in 1989 (B'lore Univ), M.E. in Robotics & Automation from Govt.'s LD College of Engg., (Gujarat Univ., Ahmedabad) in 1992 with FCD-First Rank & PhD in Systems & Control Engg. from IIT Bombay in the year 2007, got best research scholar award for the entire IITB. He is currently working as the Dean Research (R & D) and as Professor in Dept. of Computer Science & Engg. with specialization in IoT, Cyber Security, Blockchain Technology in Rajarajeswari College of Engineering, Bengaluru, Karnataka. He has got a teaching (academic), research & administrative experience of more than 35+ years (10 years as Principal & 8 years as HoD, 1 year as Dean Research & rest as faculty) and has published more than 1500+ research papers in various national & international journals & conferences, with 175+ patents, 100+ book-chapters, 25 awards, 100s of Keynote talks and also published 25 text books in various field of science, engineering & technology, completed projects worth Rs. 9 Crores till date & guided 10 PhD scholars in VTU. He is a member of 30 Professional Society Chapters and is a fellow of IE, IETE, AMIEE, Sr. Member of IEEE and a chartered engineer.

Azliza Azrah Mohd Zakaria is a senior lecturer at INTI International University, Malaysia, with nearly two decades of experience in the education sector, specializing in accounting and management disciplines. Her teaching expertise spans financial accounting, management accounting, performance management, integrated case study and public sector accounting. She is deeply involved in curriculum design, assessment moderation, and academic development for undergraduate accounting programs aligned with the Malaysian Institute of Accountants (MIA) and higher education standards. Her current research focuses on the intersection of performance measurement, Islamic social finance, sustainability, and digital transformation—particularly the integration of Robotic Process Automation (RPA) and Artificial Intelligence (AI) into performance management frameworks such as the Balanced

Scorecard. Her published works reflect these interests. For example: She co-authored "Strategic management accounting practices in business: A systematic review of the literature and future research directions" (Nik Abdullah, Shamala Krishnan, Azliza Azrah Mohd Zakaria, Grace Morris, 2022), a study that reviews the adoption, benefits, and gaps in strategic management accounting (SMA) practices in business contexts. She is also the lead author of "Venture Philanthropy – Waqf Practices and Its Implementation: Scenario in Malaysia", which conceptualizes how the venture philanthropy model can interface with Islamic philanthropic instruments like waqf to create sustainable social impact. Beyond academia, Azliza actively engages in community and educational outreach, organizing mental health awareness initiatives, student development programs, and sustainability-driven community service projects. Her work bridges theory and practice, reflecting a strong commitment to nurturing ethical, future-ready accounting professionals and advancing research that aligns with Malaysia's sustainability and Islamic finance agenda.

Mohammad Asif Nawaz holds a Ph.D. in Computing and Information Systems from the University of North Carolina at Charlotte, USA, as well as an MBA and BSBA from Oklahoma City University, USA. Additionally, he has completed the Political/Economic Tradecraft Course from the U.S. Foreign Service Institute and the Instructor Course and Diagnostic Assessment Certifications from the U.S. Defense Language Institute. He has also received a Hospitality Management Program Certification from Wyndham University and is a Certified SAFe® 6 Agilist. His diverse roles have included Chief Operating Officer, Director of Operations, General Manager, Deputy Director, Economic Specialist, Professional Associate, Academic Specialist, Assistant Professor, and Math Teacher. He has worked with esteemed institutions such as the U.S. Department of Defense, the U.S. Department of State, and the U.S. corporate sector. Dr. Nawaz is a recipient of the Commander's Award for Civilian Service and the Achievement Medal for Civilian Service. He contributed to this research paper while serving as the Chief Operating Officer at Vibrant Hospitality."

P. Selvakumar is currently an Associate Professor in the Department of Science and Humanities at Nehru Institute of Technology, Coimbatore, Tamil Nadu, India. Prior to joining Nehru Institute of Technology, he held positions as an Assistant Professor in the Department of Science and Humanities various engineering colleges in Coimbatore, Tamil Nadu, India, spanning from 2011 to 2023. Dr.P. Selvakumar's academic and research interests lie primarily in the field of phytochemistry. He has made significant contributions to his field, with an impressive record that includes 30 journal publications, 12 patents, and 20 conference papers. Additionally, he has organized 5 seminars and conferences, further demonstrating his commitment to academic and scientific exchange. Throughout his career, Dr. P.Selvakumar has been recognized for his dedication and excellence in teaching, having been honored with several awards, including the Best Teacher Awards. He holds a prominent position in the academic community, contributing actively to research.

Cheng Rui, Lecturer, Department of Financial Management at Nanjing Technology University Pujiang Institute. She obtained a master's degree in professional accounting from Monash University in Australia and a bachelor's degree in management from Anhui Normal University in China.

Soliha Sanusi is a Senior Lecturer at the Faculty of Economics and Management, Universiti Kebangsaan Malaysia. She obtained her Phd from the Universiti Teknologi MARA Malaysia in Financial Criminology. Her research interests are in the areas of taxation, financial criminology, small medium enterprises, cash waqf and public sector accounting.

Mushtaq Ahmad Shah is a finance expert with a PhD in Infrastructure Finance. He teaches banking, finance and economics at Lovely Professional University in India. Dr. Shah has over 8 years of experience teaching and researching at various institutions. He's written articles in academic journals and presented his work at conferences on banking, partnerships between public and private sectors, and behavioral finance.

Mohit Sharma is a research scholar at the Institute of Management Studies and Research, Maharshi Dayanand University, Rohtak, Haryana, India. His research focuses on the dynamic relationship between brand image and purchase intentions, particularly in the context of brand extensions within the FMCG sector. He has co-authored several scholarly works and continues to contribute to the academic community. Mohit can be reached at mohitattri.rs.imsar@mdurohtak.ac.in

Rahul Sharma is an Associate Professor and Discipline Leader in Accounting and Finance at Horizon University College, Ajman, UAE. He has over 22 years of experience in academia. He is an Associate Financial Accountant (AFA) with the UK's Institute of Financial Accountants and a Member (MIPA) of the Institute of Public Accountants, Australia. Dr. Sharma is an active researcher and Ph.D. supervisor, guided many doctoral candidates toward their successful completion. His community engagement includes delivering financial education sessions to underprivileged groups, participating in media panel discussions, and conducting training for industry practitioners.

Mujahid Tabassum has completed his PhD at USIM, Malaysia. He also obtained an MSc Computer System Engineering (Specialization in Computer Network) degree from Halmstad University, Sweden, and a B.Sc. in Computer Science from Wollongong, Australia. He is a qualified "Chartered Engineer (CEng)" registered with the UK's Engineering Council. He has 16 years of working experience and has worked at various international universities. His research interests include Computer Networks, AI, Wireless Sensor Networks, IoT, Security, and Applications. He has published several conference papers, journals, and book chapters.

Xiaoli Yu is an Associate Professor of Accounting at Yiwu Industrial & Commercial College, China. Her research centers on robotic process automation (RPA), artificial intelligence, and digitalization in management accounting, with publications in Accounting Horizons. She holds a PhD in Accounting from Taylor's University and has consulted for Chinese SMEs, implementing digitalization strategies that enhanced efficiency. Dr. Yu's work explores automation's role in strategic decision-making, contributing to cost savings and process optimization in management accounting.

Index

A

B

C

D

E

F

H

R

S

T

W